PHYSICS

SECOND EDITION

PHYSICS

SECOND EDITION

AKRILL • BENNET • MILLAR

Hodder & Stoughton

A MEMBER OF THE HODDER HEADLINE GROUP

British Library Cataloguing in Publication Data

Akrill, T. B.
  Physics. – 2Rev.ed
  I. Title
  530

ISBN 0 340 54242 X
(ISBN 0 7131 0297 7 1st edition)

First published 1994
Impression number 10   9   8   7   6   5   4   3   2   1
Year   1998   1997   1996   1995   1994

Typeset by Wearset, Boldon, Tyne and Wear.
Printed in Great Britain for Hodder & Stoughton Educational, a division of Hodder Headline Plc, Mill Road, Dunton Green, Sevenoaks, Kent TN13 2YA by Thomson Litho Limited.

Preface

This is a one-volume textbook which covers the material needed for Advanced Level courses in Physics. The original version of this book has been in print since 1979. Since then there have been many changes in the Advanced Level syllabuses, and this new edition has been almost entirely rewritten so that it is suitable for the new A and AS courses beginning in September 1994. We assume that you will, like most students, have studied science at GCSE.

Although much of the writing in the book is new, we have kept to our original principle – that you cannot study Physics without trying to understand (not just learn) it and that our job as authors is to try to explain it. Between us we have a lot of experience in teaching A Level Physics, and in setting and marking examination papers. We hope that the expertise which has gone into producing this new edition is going to help you understand Physics.

In addition to this book there is also a book of questions which will provide the practice you need to check that you have understood what you have studied. It is called *Practice in Physics* and the questions in it are arranged so that they match the sections in this book. For example, in *Practice in Physics* you will find a section '2.4 Newton's Second Law' which provides about 30 questions covering the section '2.4 Newton's Second Law' in this book. Answers are provided for very nearly all the questions.

Tim Akrill
George Bennet
Chris Millar

Contents

Acknowledgements

We are grateful to the following companies, institutions and individuals who have given permission to reproduce photographs in this book. Every effort has been made to trace and acknowledge ownership of copyright. The publishers will be glad to make suitable arrangements with any copyright holders whom it has not been possible to contact.

Action-Plus (8, 17, 20, 22, 28, 84, 139, 141 left, 171); Anderson & Neddermyer (1936) *Phys. Rev.* **50** (361); Arrow Dynamic Inc. (169); Associated Press (357); Barrie Schwartz (356 right); Bill Rose (169); Bromhead (Bristol) Ltd. (226); BT Pictures; a BT photograph (331, 412); BT Telefocus pictures (178 left); Builders Group (74 bottom); C. T. Hutchinson (36, 213); Central Electricity Generating Board (274); CERN (356 left); *Chamber Photographs*, Genter, Maier-Leibnitz & Bothe, (Pergamon Press), 1954 (348); Colin Taylor Productions (46 lower, 14 middle and right, 144, 237, 276, 295, 335, 336, 338, 341); *College Physics 4th Edition*, Sears, Zemansky & Young (Addison-Wesley) 1960 (319); D. A. Hodges 'Microelectronic Memories' in *Microelectronics*, Scientific American Inc. (382 right); Dartford River Crossing (61); Ealing Beck Ltd. (305); Electricity Association Services (167); European Space Agency, Paris (281); G. G. Scott *Journal of Applied Physics* **28** 1957 (231); G. R. Graham, Cambridgeshire College of Arts and Technology, *Physics Education* **7** no 6, July 1972 (317, 320); Giselle Harvey (279); H. Hashimoto, Science Museum, London (49); Hulton-Deutsch Collection (284); J. Allen Cash (12, 32, 74 left, 80, 83 right, 152 left, 292, 341); J. Cobby & C. Mijovic, North London Collegiate School (351); J. S. Hey *The Radio Universe Pergamon*, 1971 (299); Jim Jardine (45); Jodrell Bank (319); K. W. C. Watson (40 both, 190, 196, 199, 218, 212, 216, 220, 257, 283, 303, 311 both, 313, 314, 316, 326, 327, 339, 379, 383, 387, 391, 398); Kodansha Ltd. (272); Leybold Heraeus Ltd. (227); Mansell Collection (361); Marie Curie Cancer Care (343); Metropolitan Police (302); National Medical Slide Bank (149 lower); National Power (234, 255); P. M. S. Blackett (1923) *Proc. Roy. Soc.* A **103** (349); P. M. S. Blackett (1925) *Proc. Roy. Soc.* A **107** (353); Paul Brierley (86); Philips (374); *Physics, Concepts and Models*, Wenham, Dorling, Snell & Taylor (Addison-Wesley); L. Meitner, provided by B. Taylor (43); Prof. A. P. French (297) and in *Mechanics Vibrations and Waves*, Akrill & Millar John Murray (285); Prof. H. E. Edgerton (39); Professor Fowler from Akrill & Millar: *Mechanics, Vibrations and Waves* John Murray (47); *PSSC Physics 4th Edition* © D. C. Heath & Co 1976 (293); R. S. Longhurst *Geometric and Physical Optics*, Longman, 1973 (315); *Radioactivity* J. L. Lewis, E. J. Wenham (Longman Physics Topics) (349); Rex Features Ltd. (289); Robert Harding Picture Library (29, 127, 176); Science Photo Library (34, 142 left, 178, 186, 202, 287, 324, 377, 382 left, 402); *Scientific American* September 1967, courtesy of G. C. Smith (68); *Seeing Beyond the Visible* ed. Hewish, (Hodder & Stoughton) (57); Sir James Menter (69); Sir Lawrence Bragg & J. F. Nye (1974) *Proc. Roy. Soc.* A **190** (56, 57, 69 left); *Special Relativity* A. P. French, (Nelson), reproduced from the film 'The Ultimate Speed' (46 upper); *The Project Physics Course* Holt Rinehart & Winston 1970 (380); Vauxhall Motors Ltd. (30); *Waves*, D. C. Chaundy (Longman) (294); ZEFA Picture Library (9); Zeneca Agrochemicals/CTC Publicity (204).

Cover photograph by kind permission of Dr Jeremy Burgess/Science Photo Library.

Introduction

Physical quantities

A physical quantity is represented by a number and a unit: e.g. 12 N, or 5.3 m. In the **Système Internationale** (SI) there are seven **base quantities**. One of these (luminous intensity) we shall not use at all: the other six are shown in Table i, together with their symbols and the **base units** and the symbols for the units.

TABLE i   Base quantities in the SI

Physical quantity	Symbol	SI unit	Symbol
length	l	metre	m
mass	m	kilogram	kg
time	t	second	s
current	I	ampere	A
temperature	T	kelvin	K
amount of substance	n	mole	mol

The SI units for the base quantities are defined with reference to particular reproducible physical situations: e.g. 'the second is the time taken for there to be 9 192 631 770 periods of the radiation corresponding to the transition between two particular levels within the ground state of the caesium atom', but you do not need to remember facts like these (though the definitions of the ampere, the kelvin and the mole will be referred to later in the book).

Derived quantities and units

There are other quantities which may be *derived* from the base quantities: for example, **average speed**, which is defined by the equation

$$\text{average speed} = \frac{\text{distance travelled}}{\text{time taken}}$$

Each of these derived quantities has its own derived unit: the **defining equation** always enables you to see what this unit is. For example, if you know that the distance travelled was 240 m, and the time taken was 48 s, then the average speed v is given by

$$v = \frac{240\,\text{m}}{48\,\text{s}} = 5.0\,\text{m/s or } 5.0\,\text{m s}^{-1}$$

So if we use the SI units for length and time, we see that the SI unit for speed must be the m/s, which may be written m s^{-1}. You probably already know that some of these derived units may be quite a mouthful: e.g. the unit of force is the kg m s^{-2}. In such cases we give the unit a special name: the kg m s^{-2} is called the **newton** (N). But it is important to remember that these 'short' names can in each case be replaced by a combination of the base units.

Homogeneity of equations

Any equation connecting physical quantities must be **homogeneous** when you look at the units of the quantities. For example, if F, m and v have their usual meanings of force, mass and velocity, the equation

$$F = mv$$

cannot be correct. This can be tested by looking at the units: the units of the left-hand side are kg m s^{-2}, and the units of the right-hand side are the units of mass × velocity, i.e. $(kg)(m\,s^{-1})$, i.e. $kg\,m\,s^{-1}$. Since the units of the two sides of the equation are not the same, the equation cannot be correct, since it is stating that an amount of one physical quantity is equal to an amount of a different physical quantity. A convenient way of saying 'the units of' is to use square brackets [. . .]: e.g. [velocity] means 'the units of velocity'.

You can see that you need to know how to express the 'short' form of a unit in terms of the base units if you are going to be able to check the homogeneity of equations in this way. You can usually find out what this relationship is by looking at an equation which contains the quantity that has the short form as its unit. For example, if you wanted to know what the newton was short for, you could remember the equation $F = ma$:

$$[\text{left-hand side}] = N$$
$$[\text{right-hand side}] = (kg)(m\,s^{-2}) = kg\,m\,s^{-2}$$

so the newton must be short for $kg\,m\,s^{-2}$.

The fact that the units are homogeneous does not mean that the equation is correct. You will recognise the equation 'kinetic energy = $\frac{1}{2}mv^2$':

$$[\text{right-hand side}] = (kg)(m\,s^{-1})^2 = kg\,m^2\,s^{-2}$$

and that will be so whether or not the '$\frac{1}{2}$' is present in the equation. If the units do not match, the equation *cannot* be true: if they do match, the equation *may* still be false.

EXAMPLE

Is it possible that the equation $c = \sqrt{(g\lambda/2\pi)}$ is correct, where c is the speed of deep-water waves, g is the gravitational field strength, and λ is the wavelength of the waves?

[gravitational field strength]
$\qquad$ = [gravitational acceleration] = $m\,s^{-2}$;

$[\lambda] = m$, and the number 2π has no unit, so

$$[\sqrt{(g\lambda/2\pi)}] = \{(m\,s^{-2})(m)\}^{1/2} = m\,s^{-1}$$
But $\qquad\qquad [c] = m\,s^{-1}$

so the equation *may* be correct.

This method does not tell us whether the unitless factor '2π' is correct. Nor does it tell us whether the speed depends on these quantities in some other way, or on some other quantities.

Dimensions

Before the SI was introduced there were other systems of units in use. Using units to check on equations was not then possible. Instead the **dimensions** of a quantity were used. There is nothing that using dimensions can do that cannot be done using units, but you may still come across the use of dimensions in some textbooks or examination papers, so we shall explain what dimensions are.

The dimensions of the base quantities mass, length and time are M, L and T. We can write these as dimensional equations

$$[\text{mass}] = M, \ [\text{length}] = L, \ [\text{time}] = T$$

where the square brackets are now used as shorthand for 'the dimensions of'. Note that capital letters must be used for the dimensions M, L and T. For checking equations dimensions are used just as units are: e.g. $[v] = LT^{-1}$ and $[a] = LT^{-2}$.

Units in equations

In this book when we substitute a quantity into an equation we put in the unit as well as the number. For example, in calculating the acceleration a when a car reaches a velocity of $20\,m\,s^{-1}$ after $5.0\,s$ we write

$$a = \frac{\text{change of velocity}}{\text{time elapsed}}$$

$$= \frac{20\,m\,s^{-1}}{5\,s} = 4.0\,m\,s^{-2}.$$

You can see that we have not only divided 20 by 5.0 to obtain 4.0, but we have also divided $m\,s^{-1}$ by s to obtain $m\,s^{-2}$. This is a trivial example, but it does serve as a check that we are doing the right thing: if we had obtained $m\,s^{-3}$, or $m\,s^{-1}$ we would have known that we must have had the wrong equation, or were using it wrongly.

Prefixes

The SI allows certain standard **prefixes** to be placed in front of the symbols for units. These are shown in Table ii.

TABLE ii Standard prefixes in the SI

Multiple	Prefix	Symbol
10^{-18}	atto	a
10^{-15}	femto	f
10^{-12}	pico	p
10^{-9}	nano	n
10^{-6}	micro	μ
10^{-3}	milli	m
10^{3}	kilo	k
10^{6}	mega	M
10^{9}	giga	G
10^{12}	tera	T

There are also prefixes centi (10^{-2}) and deci (10^{-1}) but we do not usually need to use these, unless we are dealing with areas and volumes, when the cm^2 and cm^3 are useful units.

Significant figures

Your calculator often gives you a large number of figures as the result of a calculation: e.g. the average speed of a car which travels 233 m in 15.9 s may be given by a calculator as 14.65408805 m s^{-1}. But scientists use the number of figures as an indication of how reliable the result of the calculation is, and if you use a large number of significant figures in an answer you may give the wrong impression. So you should 'round off' a number on your calculator display to take account of the uncertainty in the information on which the calculation was based. A good working rule is

the number of significant figures in the answer should be the same as the least number of significant figures in any of the data

where we start counting significant figures from the left, after any initial zeros. So in this example, the speed should be given as 14.7 m s^{-1}. A well-set question should provide data all of which have the same number of significant figures (e.g. here, both the distance and the time are given to three significant figures). If the distance travelled was 233 m, in a time of 16 s, the calculator would have given us 14.56250000 but we should give the average speed as 15 m s^{-1}, i.e. to two significant figures, since the time was given to only that number of significant figures. Do not confuse significant figures with **decimal places**, which have no relevance to scientific calculations. Table iii shows some raw data given to 3 and 2 significant figures:

TABLE iii Data given to 3 and 2 significant figures

Raw number	To 3 s.f.	To 2 s.f.
12.642	12.6	13
14.031	14.0	14
382.32	382	3.8×10^2 (*not* 380)

Note that '14' is not the same as '14.0'. Note also that writing '380' is ambiguous, since it is not clear whether the '0' is significant. You can write '380' if you add 'to two significant figures' but it is easier to write '3.8×10^2' which is unambiguous.

Standard form

Writing 3.8×10^2 is an example of using a number in **standard form**. Numbers in standard form always have one significant figure before the decimal point: e.g. 3400000 is written 3.4×10^6 and 0.0000045 is written 4.5×10^{-6}. Numbers written in standard form are easier to write and recognise, and you should have a calculator which can accept numbers entered in standard form, and can give you answers in standard form. This is sometimes called **scientific notation**.

It is of course acceptable to use the SI prefixes instead of giving a number in standard form: simply use your common sense to judge which of the two methods presents the information best. For most people 43 km probably means more than 4.3×10^4 m. You will also have to interpret what the prefixes mean, especially when entering data in a calculator: for example a question may give a measurement as 0.5 mm. You will need to enter this as 0.5×10^{-3} m: *do not* think of it as 0.0005 m, since this wastes time and it is easy to make mistakes.

Small changes

A useful way of indicating the change in a quantity is to use the symbol Δ (the Greek capital delta). Placed in front of the symbol for a quantity it means 'a small increase in'. For example, if a spring is stretched so that its length l increases from 100 mm to 112 mm, then we could write

$$\Delta l = 12 \text{ mm}$$

In the same way Δv is the increase in the velocity of a body, ΔI is the increase in electric current. In a particular case Δl, or Δv or ΔI might be negative: e.g. $\Delta v = -15$ m s^{-1} would mean that the velocity had decreased by 15 m s^{-1}.

Labels for graphs and tables

When we state the value of a quantity we give a number and a unit: e.g. we write

$$d = 2.34\,\text{m}$$

This is the same as $d/\text{m} = 2.34$, because '/' means 'divide' so if we head a table, or label a graph, we need not put the units *in* the table or *on* the axis if we use a label like this.

Another example: if $d = 2.34 \times 10^{-3}\,\text{m}$ we could write $d/10^{-3}\text{m} = 2.34$, and the label would then be $d/10^{-3}\,\text{m}$.

Drawing graphs

At the end of some practical work the results may need to be presented as a graph. The purpose of doing this may be merely descriptive, since most people can more quickly understand information given in a picture than in words. But it may be that the graph is an intermediate step in the calculation process. Here is an example. Suppose you had been measuring the speed of sound in air, by measuring the time t taken by the sound in travelling different distances d: your measurements might be shown in the table shown in figure 1 (notice how the labels at the head of the table save us from having to enter the units again).

The quantity in the first column is the **independent** variable (the one you chose the values of); the second column shows the values of the **dependent** variable. On the graph the independent variable is always plotted on the horizontal (or x-) axis: the dependent variable is plotted on the vertical (or y-) axis.

You should normally include the origin on your graph. Your graph may give a false impression if you do not. What is more important is that the origin may be one of the points on your graph: e.g. if you were measuring values of current I through a resistor when a p.d. V was applied to it, the point $V = 0$, $I = 0$ is one of the points, even if you did not measure it. Or sometimes, as in this example, including the origin reveals that there is a systematic error in the measurements. You would expect that when $d = 0$, then $t = 0$, since the sound would take zero time to travel zero distance, but the graph does *not* pass through the origin: it appears that for some reason all the measurements are overestimated by 15 mm, a fact which you would not have noticed if the origin had not been there.

You should choose scales which are easy to use: the main markings on the graph paper should represent simple numbers like 2, 5 or 10, or multiples such as 20, 50, 100, and never 3, 7 or other awkward numbers which it is hard to sub-divide.

When you have plotted the points you should draw attention to them: preferably by circling them with a small, neat circle. Alternatively you could mark a cross on top of the point, but the actual point is then not as obvious.

When you draw the line which represents how the two quantities vary, remember that the line is almost certainly going to be a straight line or a curve which has no rapid changes of curvature. It may be impossible to draw such a line or curve through all, or even, most of the points, but this does not matter. You must use your judgement to draw the **best-fit** line or curve, i.e. the line or curve which best fits the variation, assuming that there are random errors, caused by the measuring techniques. When you do this, do not assume that the first or last plotted point must lie on the line: these points do not have any special status. At this point you may decide that the measurements for one or more of your plotted points must be wrong,

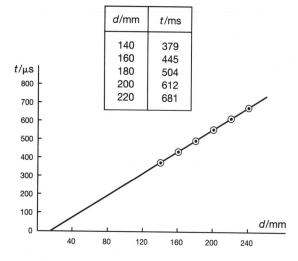

d/mm	t/ms
140	379
160	445
180	504
200	612
220	681

FIGURE 1

4

and you want to go back to the apparatus to repeat the measurements. Or you may find there is a region of the graph where you need more information to enable you to decide how to draw the graph: you should take more measurements to fill in the gaps.

You should draw the line in pencil so that you can erase it if you then think you should have drawn it in a different position. If you think the best-fit line is straight, you should use a ruler to draw it. As you will often need to calculate the **gradient** or **slope** of the line, then the longer the line the better it is for this purpose. In symbols, and using the delta notation

$$\text{gradient} = \frac{\Delta y}{\Delta x}$$

and Δy and Δx should both be made as large as possible to reduce uncertainty in their measurement.

Calculating from graphs

One way in which numerical information can be derived from a graph of, say, y plotted against x, is to find a value of y for a value of x which was not measured in the experiment. The graph is then acting as a sort of continuous table of values in which the value of a quantity which corresponds to any value of the other quantity may be looked up. When we deduce a value *within* the range of values which we have measured the process is called **interpolation**. **Extrapolation**, which is the deducing of values *outside* the range of measured values, is more risky: the stretching of a spring is a situation in which extrapolation *cannot* be used, since we do not know at what point the stretch ceases to be proportional to the stretching force.

A common task is to find the value of one quantity when the other is zero. These values can be read off the graph axes: in our example, we say the **intercept** on the x-axis is $10\,\text{mm}$.

Another important way in which a straight-line graph may be used to derive information is by measuring its gradient. The gradient gives the average value of the ratio of the two plotted quantities: in our example the gradient is the average value of the ratio time/distance and, making measurements from *the whole of* the line, its value is

$$\frac{741\,\mu s}{(260-15)\,\text{mm}} = \frac{741 \times 10^{-6}\,\text{s}}{245 \times 10^{-3}\,\text{m}} = 3.02 \times 10^{-3}\,\text{s}\,\text{m}^{-1}$$

The gradient often has some physical significance: here the speed of the sound, which is the ratio distance/time, is the reciprocal of the gradient, so

$$\text{speed of sound} = \frac{1}{3.02 \times 10^{-3}\,\text{s}\,\text{m}^{-1}}$$
$$= 331\,\text{m}\,\text{s}^{-1}$$

The gradient is always a physical quantity, and will therefore have units, which can be deduced, as was done here, by dividing the units of the quantity on the y-axis by the units of the quantity on the x-axis.

Proportionality

When the relationship between two quantities is such that the graph of one plotted against the other is a straight line passing through the origin, the quantities are said to be **directly proportional**, as shown in the first graph in figure 2. When the graph is a straight line which does not pass through the origin, the quantities are *not* proportional: there is a **linear relationship** between them, as in the second graph in figure 2.

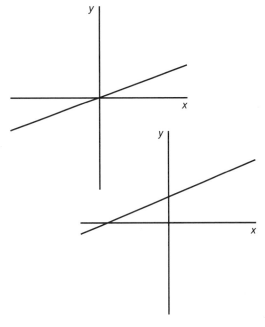

FIGURE 2

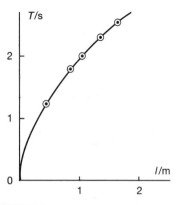 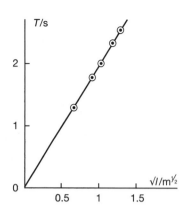 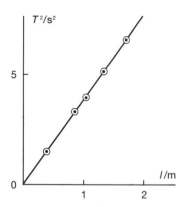

FIGURE 3

If there is a non-linear relationship between two quantities it is still possible to plot a straight line graph if the relationship is known or suspected. For example, the period T of a pendulum is given by $2\pi\sqrt{(l/g)}$, so a graph of T against l will not give a straight line, but a graph of T against $\sqrt{l}$ will, and so will a graph of T^2 against l (as shown in figure 3). You should check that the gradients of these two graphs will give the values of $2\pi/(\sqrt{g})$ and $4\pi^2/g$ respectively.

Rates of change

When a graph is not straight its gradient is not constant, so it cannot be calculated by finding the value of $\Delta y/\Delta x$, since the value of this will depend on which part of the graph is chosen. In this case the gradient at a particular point on a curve is the gradient of *the tangent to the graph* at that point. We now write the gradient as

$$\frac{dy}{dx}$$

where this is a *single symbol* meaning 'the **rate of change** of y with respect to x'. This is dealt with in a branch of mathematics called **differential calculus** but all you need do is to understand what the symbol means. You will nearly always be using rates of change with respect to *time*: e.g. the rate of change of velocity v with respect to time t (which is the acceleration), or the rate of flow of electric charge Q with time t: the symbols for these would be

$$\frac{dv}{dt} \quad \text{and} \quad \frac{dQ}{dt}$$

Uncertainties

Manufacturers of measuring instruments (rulers, thermometers, ammeters, etc.) do not sub-divide their scales more than is justified by the **precision** of the instrument. So, as a general rule, a measurement is subject to an **uncertainty** of the smallest division of the instrument.

It is therefore important to make a measurement as large as possible. That is why we time many swings of a pendulum, or measure the thickness of many sheets of paper. It is also why we use different shunts and mutipliers on electrical meters. If the quantities to be measured are small, then we really need a more *sensitive* measuring device. For example, if a temperature rise is only about 3°C we need a thermometer calibrated in 0.1°C, and if a time interval is about only 1 ms we need a timer calibrated in internals of $10\,\mu s$ or $1\,\mu s$. If we cannot make the measurement larger, or use a more sensitive instrument (and even if we can) there will be an uncertainty in the measurement which should be stated, along with the result: e.g. 'current = 4.6 ± 0.1 A' or 'current = 4.6 A with an uncertainty of 2.2%'.

Where the result is derived from a calculation there is a simple rule for calculating the approximate *maximum* percentage uncertainty in the result:

if the measured quantities are multiplied together, or divided by each other, the approximate maximum percentage uncertainty in the result is the sum of the percentage uncertainties in the individual measurements.

For example, if

$$x = \frac{ab^2}{c}$$

the maximum percentage uncertainty in x
    = the percentage uncertainty in a
      + 2(percentage uncertainty in b)
      + percentage uncertainty in c.

So if these are 0.5%, 2% and 1% respectively, the maximum percentage uncertainty in x is approximately 5.5%. This technique deals with **random** uncertainties arising from measurement: there may also be **zero errors**, or **systematic errors** caused by the method of doing the experiment. It gives the *maximum*, not the *probable*, uncertainty. Also, it is only approximate: it is not reliable for calculating percentage uncertainties of 10% or more. In any case it is more important to go back to the experiment and try to think of ways of reducing the uncertainties than to calculate precisely the size of a very large uncertainty!

The micrometer screw gauge

A useful instrument for measuring small widths (e.g. the diameter of a wire) is the **micrometer screw gauge**, which is shown in figure 4.

Assume it starts with the jaws closed. You would then unscrew the sleeve of the instrument. As you do so, each complete rotation of the sleeve uncovers one half-millimetre of a scale marked in millimetres and half-millimetres. The sleeve, where it uncovers the scale, is marked in 50 divisions round its circumference, so you can tell when you have rotated the sleeve one-fiftieth of a revolution, which amounts to one-fiftieth of a half-millimetre, which is one-hundredth of a millimetre, i.e. 0.01 mm. Some examples of how to read the screw gauge are shown in the figure. It is a sensitive device, but nevertheless the uncertainty in the measurement may still be large if the length being measured is small. For example, there will be an uncertainty of 0.01 mm in a measurement of 0.20 mm, i.e. a percentage uncertainty of 5%.

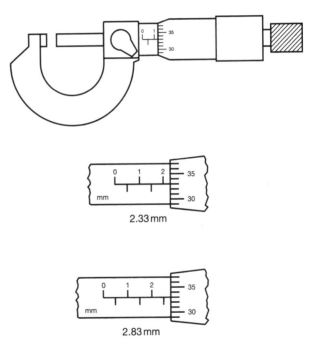

2.33 mm

2.83 mm

FIGURE 4

1 Describing Motion

Travelling fast without help from a machine is common to many of our modern pastimes. A sky-diver drifts down to Earth at about 125 m.p.h., nearly 200 km h^{-1} or 50 m s^{-1}. The skier in the photograph, seen breaking the world record for the flying kilometre at Les Arcs in 1988, was timed at 16.09 s, a speed of 233.7 km h^{-1}; this is faster than the sky-diver! Often it is not simply how fast you travel but the rate at which your speed increases or decreases which is of significance. An astronaut might experience an acceleration of 120 m s^{-2} soon after lift off, which is equivalent to an increase of 270 m.p.h. per second!

Physicists, more perhaps than other scientists and certainly more than most people, are aware of how big things are. They are skilful in making estimates and thinking quantitatively. Enrico Fermi, who led a team of physicists in building the world's first nuclear reactor in Chicago in the 1940s, was well known for asking his students questions such as 'Estimate how many male hairdressers there are in Chicago who wear spectacles', in order to give them practice at estimating. Throughout this book you are going to be developing your own ability to answer, and to ask, such questions.

1.1 Measuring speed

To find the speed at which an object is moving you need to know how long it takes to move a known distance. Its **average speed** is then defined by the equation

$$\text{average speed} = \frac{\text{distance moved}}{\text{time taken}} = \frac{s}{t}$$

If we measure a small distance then the distance over which the speed is found is written as Δs. Similarly the time taken is written as Δt, so that

$$v_{\text{av}} = \frac{\Delta s}{\Delta t}$$

The unit of speed is the metre per second (m s^{-1}). Speeds are often quoted in other units, e.g. miles per hour, but for all calculations in physics we must first convert to the basic SI unit. It is useful to know that 50 m.p.h. $\approx$ 80 km h^{-1}, and is nearly 25 m s^{-1}. (You should read Δs in the above equation as delta s, meaning a small change in s – it does *not* mean delta times s. Similarly Δt is delta t, the corresponding small change in t.)

A *word of caution:* when you use clocks or rulers, or other measuring instruments you have to assume that they are **accurately calibrated**. You may make mistakes when you use them but the marks on the ruler and the movement of the clock are assumed to be exactly right.

Measuring speed during a fixed time interval

One convenient and readily available fixed time interval is the fiftieth of a second or 0.02 s provided by the 50 Hz a.c. mains. A ticker-timer uses this to provide dots on a moving strip of paper and this can provide you with a lot of information about a moving object, e.g. a foot – figure 1.1. To find the average speed of the foot between any two adjacent dots we need to measure the distance Δs between the dots. If the maximum value of Δs in this experiment is about 50 mm, then the uncertainty in this may be ± 1 mm or $\pm 2\%$. Taken together

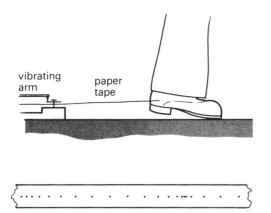

vibrating arm

paper tape

FIGURE 1.1

about $10\,\mu\text{s}$ per flash and can operate at frequencies between 1 and $250\,\text{s}^{-1}$. Providing the objects move close to a metre rule or other calibrated scale, measurements from the photograph give the distance moved between flashes. Assuming the stroboscope to be correctly calibrated, the precision with which v can be found depends on the uncertainty in Δs, which will vary with the size and quality of the print. As with the tape we can not expect an uncertainty better than a few per cent. For a discussion of uncertainties see page 6 in the Introduction.

with any doubt about the 0.020s we are unlikely to be able to measure the maximum speed of the foot during a step as precisely as this, and anyway the issue is confused by the fact that the foot moves up and down a little. When a **ticker-tape** technique of this sort is used to measure the average speed of an object which is moving with a uniform speed then the uncertainties can be reduced by analysing the tape over longer time intervals Δt, e.g. measuring Δs over a five-space length of tape. The technique is simple and the pull of the tape is usually so slight as not to affect the motion being studied, but it can only be used for motion in a straight line.

EXAMPLE

By taking measurements for 5-space lengths from the tape in figure 1.1 calculate the speed of the foot when it is (a) fastest and (b) slowest. Estimate the percentage errors in your results.

(a) Fastest five dots cover 24 mm

$$\therefore \text{speed} = \frac{24\,\text{mm}}{5 \times 0.02\,\text{s}} = 240\frac{\text{mm}}{\text{s}}$$

Uncertainty in distance measured is 0.5 mm, i.e. $(0.5 \div 24) \times 100\% = 2\%$. There is no uncertainty in the time interval. So final error = 2%.

(b) Slowest five dots cover 6 mm

$$\therefore \text{speed} = \frac{6\,\text{mm}}{5 \times 0.02\,\text{s}} = 60\frac{\text{mm}}{\text{s}}$$

Uncertainty in distance measured is 0.5 mm, i.e. $(0.5 \div 6) \times 100\% = 8\%$.

Beware when counting dots or stroboscopic images: start 0, 1, 2 . . .

FIGURE 1.2

For objects moving in two dimensions a flashing **stroboscope** lamp can be used – figure 1.2. The lamp, a small xenon discharge tube, is 'on' for

Measuring speed over a fixed distance

The 1000 m covered by the skier in the introduction can be measured very precisely as it is a large distance. Even an error of 1 m would be only 0.1%. The timing, which is done electronically will have an uncertainty of only 0.01 s, again better than 0.1%, so we have a very *precisely* measured speed. (Avoid using the word *accurate*.)

In the laboratory things are much less precise as the times and distances measured are much smaller. The air-track glider in figure 1.3 breaks the light beam for only a fraction of a second. The small time interval, Δt, could be recorded on a VELA or a computer. These **datalogging** instruments convert the voltages from the sensor, here the photodiode, from analogue to digital form before storing them. They can process and display the data and can perform calculations to give you values of, for example, velocity directly.

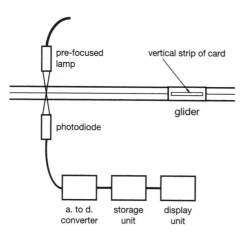

FIGURE 1.3

If the beam is only broken for 0.30 s then the uncertainty might be ±0.01 s or ±3%. Measuring the length of the card involves a smaller uncertainty, perhaps 1 mm in 200 mm or 0.5%, so in this experiment the timing introduces the greatest errors.

1.2 Velocity, a vector quantity

It matters whether you walk *towards* or *away* from a cliff edge at a given speed. We say that the speed of the motion is the same in each case but that the velocity is different. To define velocity we must include the direction of the motion: we say that velocity is a **vector** quantity. All vector quantities obey a special rule for addition and subtraction.

Force and magnetic field strength are other examples of vector quantities. Physical quantities which obey the normal rules of arithmetic are called scalars. Time and energy are examples of **scalar** quantities.

Adding velocities

Figure 1.4 illustrates how a horizontal velocity of $3.0\,\text{m s}^{-1}$ and a vertical velocity of $2.0\,\text{m s}^{-1}$ are added. You can find the size and direction of the resultant velocity either:

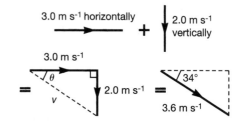

FIGURE 1.4

- ◆ by scale drawing, in which the velocities to be added are represented by lines proportional to their lengths; or

- ◆ by calculation, which is easy here, since the velocities to be added are perpendicular to one another.

$$v^2 = (3.0\,\text{m s}^{-1})^2 + (2.0\,\text{m s}^{-1})^2$$

hence $v = 3.6\,\text{m s}^{-1}$

and $\tan\theta = \dfrac{2.0}{3.0} \Rightarrow \theta = 34°$

This vector $v = 3.6\,\text{m s}^{-1}$ at an angle of 34° below the horizontal is known as the **resultant** of the two original vectors. Notice that the vectors in this example have been 'slid' so that they can be added head-to-tail. We would get the same resultant velocity had we added them in the other order.

Subtracting velocities is an extension of the addition process. The velocity being subtracted is first reversed in direction ($-20\,\text{m s}^{-1}$ due north is equivalent to $+20\,\text{m s}^{-1}$ due south) and then added as above. Subtraction is illustrated in the example that follows. Once the problem is properly understood all we need is a sketched triangle showing the relative directions of the two vectors to be added and the resultant vector.

EXAMPLE

A car is initially travelling at a speed of $20\,\mathrm{m\,s^{-1}}$ along a road in a direction of 20° S of E. A few minutes later it is still travelling at the same speed but is now moving in a direction of 20° N of E. What is the change of velocity of the car?

The change of velocity means the final velocity minus the initial velocity. The figure shows the solution. The change of velocity is due north and is of size Δv, where

$$\Delta v = 2(20\,\mathrm{m\,s^{-1}})\cos 70°$$

$$= 14\,\mathrm{m\,s^{-1}}$$

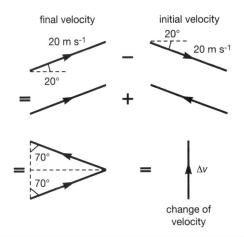

If you are not studying mathematics, do not be frightened by the calculations. A good scale diagram is often the best solution and will give answers to a precision of a few per cent if drawn carefully. Try this example by drawing for yourself.

Resolving velocities

Just as two velocities can be added to give a single resultant velocity, so it is sometimes convenient to break one velocity up into two parts. Where these parts are perpendicular the process is called **resolving** and the two velocities are called the **two resolved parts** or perpendicular components of the original velocity.

The process of resolving is useful with all vector quantities and is often the key to seeing how a problem about velocity or force can be solved. Thus a pull of 80 newtons (80 N) acting at 30° above the horizontal can be resolved into:

- a horizontal pull of $(80\,\mathrm{N})\cos 30° = 69\,\mathrm{N}$
- a vertical pull of $(80\,\mathrm{N})\sin 30° = 40\,\mathrm{N}$

i.e. our 80 N pull, perhaps on a dog lead, is producing a horizontal pull of 69 N but is at the same time producing an upward pull of 40 N.

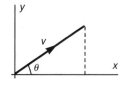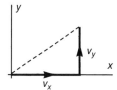

FIGURE 1.5

In general, if we want to resolve v in figure 1.5 along the x- and y-axes, then the resolved parts v_x and v_y are given by

$$v_x = v\cos\theta$$

$$v_y = v\sin\theta$$

or, if v_x and v_y are known, then

$$v^2 = v_x^{\,2} + v_y^{\,2}$$

and

$$\tan\theta = \frac{v_y}{v_x}$$

We must stress that the methods of calculating *all* vectors are the same as those used here for velocity vectors or force vectors. When you meet any new quantity you must ask yourself whether it is a vector or scalar.

Displacement

The average velocity v of a particle is defined by an equation just like that for average speed:

$$v_{\mathrm{av}} = \frac{\Delta s}{\Delta t} \text{ or simply } v_{\mathrm{av}} = \frac{s}{t}$$

where Δs is the displacement of the particle in the time interval Δt. Displacement, a vector quantity, measures the separation of two points A and B in size and direction. A person might undergo a displacement of 3.3 m vertically upwards in going to bed. If the journey took him 30 s then his average velocity would be $3.3\,\text{m}/30\,\text{s} = 0.11\,\text{m s}^{-1}$ vertically upwards. But in going upstairs he may have walked a distance of 18 m, so that his average speed is $18\,\text{m}/30\,\text{s} = 0.60\,\text{m s}^{-1}$. In travelling round and round a circular track a runner can cover a great distance but whenever he is at the starting point his displacement is zero. The importance of displacement is that, knowing where you start, it tells you where you get to (but it doesn't tell you about the path you may have followed).

The **instantaneous velocity** of a body is the value obtained when the time interval Δt is made very small; we then write

$$v = \frac{ds}{dt}$$

i.e. velocity is the *rate of change of* displacement. For motion in a straight line it is equal to the gradient of the displacement–time graph. See the Introduction (page 6) for a discussion on the gradient of a graph.

It is perhaps unfortunate that we use the symbol s for both distance and displacement and the same symbol v for both speed and velocity. But we do!

Mathematics

Physics does involve a lot of calculations (it is a quantitative subject) but you should not be concerned if you are not now studying mathematics.

You *do* need to get $\cos\theta$ or $\sin\theta$ right when you resolve a velocity or a force but for much of physics it is more important to *understand* an idea, like rate of change, than to be able to write what are called calculus statements and to use advanced mathematics. What mathematicians call differentiation is really only finding the gradient of a graph, while integration is finding the area under a graph. Computers can achieve both tasks at the push of a key, once the data has been stored in their memories. The expression for the instantaneous velocity, dv/dt, you should read as 'dee-vee by dee-tee'.

1.3 Acceleration

An aeroplane which is taking off (figure 1.6) increases its speed by about 5 m.p.h. every second. It accelerates. When it lands it slows down by about 8 m.p.h. every second. We also describe this as an acceleration though this time the acceleration is negative.

FIGURE 1.6

5 m.p.h. is equal to $2.2\,\text{m s}^{-1}$, so the acceleration during take off is $2.2\,\text{m s}^{-1}$ per s, or $2.2\,\text{m s}^{-2}$. In general the **average acceleration** a_{av} of an object is defined by the equation

$$a_{av} = \frac{\Delta v}{\Delta t}$$

where Δv is the change of velocity of the object in the time interval Δt. More simply

$$a_{av} = \frac{v - u}{t}$$

where u and v are the initial and final velocities. As it is a change of velocity in the equation, then a change of direction at a constant speed also involves acceleration.

Usually acceleration results from a change in speed: think of a sprinter in a 60 m dash who accelerates from rest to a speed of $8.4\,\text{m s}^{-1}$ in 2.4 s. She moves in a straight line and her acceleration is

$$a_{av} = \frac{(8.4 - 0)\,\text{m s}^{-1}}{2.4\,\text{s}}$$

$$= 3.5\,\text{m s}^{-2} \text{ along the track.}$$

But acceleration is a vector quantity and its direction is that of Δv, the change of velocity. Referring to the example on page 11, if the car's change of velocity occurred in 4 minutes, then the average acceleration would be

$$a_{av} = \frac{14 \, \text{m s}^{-1}}{240 \, \text{s}}$$

$$= 0.058 \, \text{m s}^{-2} \text{ due north.}$$

In this example the speed of the car was unchanged: the acceleration was wholly the result of changes in its direction of motion.

Obviously motion in a circle at a constant speed will be an important example of this second kind of acceleration. In this chapter we will stick to the simpler kind where the acceleration involves only a change of speed.

The instantaneous acceleration of a body is the value obtained when the time interval Δt is made very small; we then write

$$a = \frac{dv}{dt}$$

i.e. the acceleration is the rate of change of velocity and is equal to the gradient of the velocity-time graph.

EXAMPLE

A piece of ticker-tape attached to a foot is analysed to give the following information:

dot	0	5	10	15	20	25
	30	35	40	45		
distance moved /mm	0	7	80	254	488	802
	1133	1347	1380	1388		

Estimate the maximum acceleration of the foot.

Each 5-space length of tape represents a time interval of $5 \times 0.02 \, \text{s} = 0.10 \, \text{s}$ so that the average speeds v_{av} over each 5-space length of tape are:

| v_{av}/mm s^{-1} | 70 | 730 | 1740 | 2340 |
| | 3140 | 3310 | 2140 | 330 | 80 |

The greatest acceleration occurs when the speed changes most rapidly: this is between speeds of $2140 \, \text{mm s}^{-1}$ and $330 \, \text{mm s}^{-1}$, giving an average

---

acceleration there of

$$a = \frac{(0.33 - 2.14) \, \text{m s}^{-1}}{0.10 \, \text{s}}$$

$$= -17.1 \, \text{m s}^{-2}$$

The maximum acceleration of the foot thus occurs as it slows down prior to touching the ground and is about $17 \, \text{m s}^{-2}$ in the opposite direction to the movement of the foot.

Motion in a straight line: graphs

In the above example the ticker-tape data enables us to visualise the motion of a walking foot to some extent. But it is much easier to visualise the motion if the information is presented in graphical form. Three types of graph are useful and it is important to be able to deduce what the others are like when given only one of them.

Displacement–time graphs

The gradient of an s-t graph at any point is by definition the velocity of the body at that point. Where the gradient is constant, for example in figure 1.7, then so is the velocity; if the gradient is zero (when there is a horizontal line on the graph), the velocity is zero, i.e. the body is stationary. If the graph has a negative gradient, this means that the motion is backwards towards the place from which the displacement is measured.

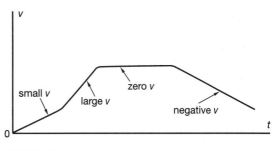

FIGURE 1.7

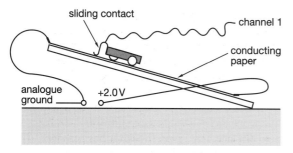

FIGURE 1.8

To produce *s-t* graphs experimentally we can use a sliding electrical contact like that shown attached to a trolley in figure 1.8. The position of the trolley as it moves down the runway is logged by a computer. The runway is lined with conducting paper (or has a wire stretched along its length) and a voltage is set up between the top and bottom. The voltage between the sliding contact and the top of the track is fed to the input channel of the analogue to digital converter, the top being connected to analogue ground, 0 V. The data can be presented directly as a graph, figure 1.9, and the computer can now be programmed to process the *s-t* data and produce a *v-t* graph – a task which would be very laborious if you had to draw tangents to the *s-t* graph at a series of points.

you have an **ultrasonic motion sensor** the business of producing an *s-t* and then a *v-t* graph for a moving object becomes even simpler. You aim the ultrasonic beam along the line of motion of the trolley, and the computer program does the rest.

Velocity–time graphs

The gradient of a *v-t* graph at any point is by definition the acceleration of the body at that point. Where the gradient is zero there is zero acceleration, and where negative the body is slowing down (or perhaps speeding up in the negative direction).

The area between the graph-line and the time-axis measures the change of displacement of the body. Consider figure 1.10; during the time interval $t = 0$ to $t = 1.5$ s the graph shows the body's velocity was constant at $20\,\text{m s}^{-1}$. The displacement during this time interval is therefore $(20\,\text{m s}^{-1})(1.5\,\text{s}) = 30\,\text{m}$. The large shaded area has an area of 12 large squares. As each large square represents $(5\,\text{m s}^{-1})(0.5\,\text{s}) = 2.5\,\text{m}$, we see that the displacement can be found from the area, i.e. $12 \times 2.5\,\text{m} = 30\,\text{m}$.

When the velocity changes *continuously* the area under the graph-line is perhaps the only way of calculating the displacement. For example between t_1 and t_2 in figure 1.10 the area under the graph-line is approximately equal to the sum of the areas of the rectangles ($\Delta s = v\,\Delta t$) and so represents the displacement.

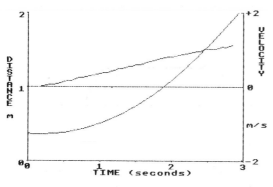

FIGURE 1.9

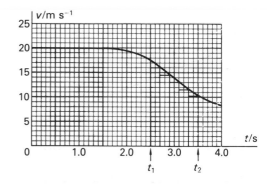

FIGURE 1.10

In this case the velocity increases steadily so the *v-t* graph (figure 1.9) is a straight line through the origin. Can you predict what a graph of acceleration against time will be like in this case? If

By making the reactangles narrower and narrower their total area more and more closely approximates to the area under the graph-line, and in practice it is easier just to count the squares under the graph-line. The area from t_1 to t_2 is 4 large squares plus about 37 small squares, i.e. 137 small squares. One small square represents 0.10 m, so that the displacement between 2.5 s and 3.5 s is 13.7 m. This v-t graph has positive values for v throughout. When v is negative the body is going backwards so the displacement is negative, e.g. the body could be returning to its starting point.

We have already seen that you can produce v-t graphs by using a computer. Another method is to make a ticker-tape chart. This involves cutting up a piece of tape at 0.1 s intervals and sticking them side-by-side. As the length of each piece of tape is proportional to the speed during that 0.1 s interval, then the top of the strips form a graph of velocity against time.

Figure 1.11 shows two sets of graphs for motion in a straight line. In (a), looking at the bottom

graph first: the acceleration is constant and negative; it is the graph for a freely falling body, measuring positive upwards. The velocity gets steadily less but is instantaneously zero when the stone is at the top of its flight at $t = t_1$. The area under the v-t graph-line from $t = 0$ to $t = t_2$ is zero (area below time-axis is negative); at that moment the stone once again passes the thrower and has zero displacement, even though it has perhaps travelled a considerable distance up and down.

In (b), looking at the top graph first: the displacement increases throughout the journey. It increases steadily between $t = t_3$ and $t = t_4$; and the velocity is constant between these times. The area under the v-t graph-line is equal to the train's net displacement (here equal to the distance covered).

Motion graphs for a stone thrown vertically upwards by a child standing on a wall

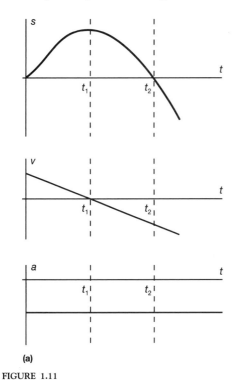

Motion graphs for an underground train moving from one station to the next

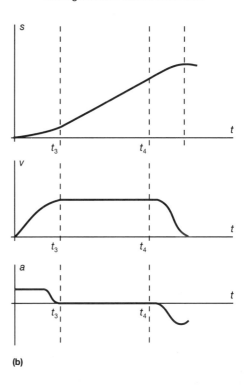

(a)

(b)

FIGURE 1.11

It takes a lot of practice to be able to predict the shape of these graphs for a particular motion. You should start with the simplest of the three and then deduce the other two. For example think of a snooker ball bouncing from side to side of the table: here the easiest graph to start with is the v-t graph. Can you sketch it and then add the s-t graph and the a-t graph? (The v-t graph will consist of a series of horizontal lines first above and then below the time axis.)

Motion in a straight line: equations

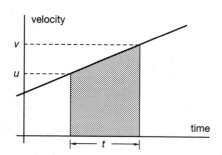

FIGURE 1.12

Figure 1.12 shows a v-t graph for a body which is moving in a straight line with *constant acceleration*. Using the notation from the figure we can deduce that during the time interval t:

$$a = \frac{v - u}{t}$$

$$\Rightarrow \quad v = u + at \quad (1)$$

$$v_{av} = \frac{s}{t}, \therefore s = v_{av}t$$

$$\Rightarrow \quad s = \left(\frac{u + v}{2}\right)t \quad (2)$$

Equation (1) is derived from the definition of acceleration and equation (2) from that of average velocity. Almost all problems in straight-line motion can be solved using only these two equations.

In the examples that follow, notice how the units balance, so that each stage makes 'grammatical' sense.

EXAMPLE

A train slows down at a constant rate from $50\,\mathrm{m\,s^{-1}}$ to $10\,\mathrm{m\,s^{-1}}$ in $80\,\mathrm{s}$. Find its acceleration and the distance travelled in this time. Take the positive direction to be in the direction of the train's motion.

$$a = \frac{v - u}{t} = \frac{(10 - 50)\,\mathrm{m\,s^{-1}}}{80\,\mathrm{s}}$$

$$= \frac{-40\,\mathrm{m\,s^{-1}}}{80\,\mathrm{s}} = -0.50\,\mathrm{m\,s^{-2}}$$

$$s = \frac{(10 + 50)\,\mathrm{m\,s^{-1}}}{2} \times 80\,\mathrm{s}$$

$$= 30 \times 80\,\mathrm{m} = 2400\,\mathrm{m}$$

EXAMPLE

A pram rolls down a ramp which is $12\,\mathrm{m}$ long. If it started from rest and took $5.0\,\mathrm{s}$ to reach the bottom, what is its acceleration?

It looks as if equations (1) and (2) are not useful here, but don't search for another one until you are sure.

$$\text{Pram's average velocity} = \frac{12\,\mathrm{m}}{5.0\,\mathrm{s}}$$

$$= 2.4\,\mathrm{m\,s^{-1}}$$

$$\therefore \text{ as } u = 0, \text{ final velocity} = 2v_{av}$$

$$= 4.8\,\mathrm{m\,s^{-1}}$$

$$\text{and } a = \frac{v - u}{t} = \frac{4.8\,\mathrm{m\,s^{-1}}}{5.0\,\mathrm{s}}$$

$$= 0.96\,\mathrm{m\,s^{-2}}$$

By rearranging equations (1) and (2) you can show that

$$v^2 = u^2 + 2as$$

and

$$s = ut + \tfrac{1}{2}at^2$$

For motion in a straight line with constant acceleration and starting from rest, this last equation becomes

$$s = \tfrac{1}{2}at^2 \qquad (3)$$

which is very useful, especially for bodies dropped or released from rest. Equation (3) could be used to solve the problem in the example about the pram.

1.4 Free fall

Any object thrown upwards or downwards or simply released close to the Earth's surface will, in the absence of air resistance, move with a constant downward acceleration of about $10\,\mathrm{m\,s}^{-2}$, which is called g. It is useful and important to know g precisely at a given place. Stroboscopic photographs or ticker-tapes for bodies moving freely in a vertical line can provide the information from which we can measure g. Whether the object moves up or down we say it is in a state of free fall.

By dropping a card of known height past an illuminated photodiode, g can be measured using modern computer-based techniques. Another method, which allows systematic errors to be eliminated, involves an arrangement for measuring the time taken for a steel ball to fall a measured distance from rest – figure 1.13. For a fall of about 1 m, $t \approx 500\,\mathrm{ms}$ and can be read from the clock to $\pm 1\,\mathrm{ms}$. Several values of t should be taken for the same h, which can itself be measured with a ruler to $\pm 2\,\mathrm{mm}$. As the initial velocity of the ball is zero, we have

$$h = \tfrac{1}{2}gt^2$$

from which g can be found.

The uncertainties in the experiment include the action of the two-way switch and the electromagnet-release and trap-door-opening action. By plotting a graph of $\sqrt{h}$ against t for different values of h and measuring the gradient of this graph a value of g, free from these systematic errors, can be achieved. The gradient, which is equal to $\sqrt{g}/2$, and should give g with an uncertainty of less than $\pm 1\%$.

EXAMPLE

If the diver jumped up 0.40 m from the 10-metre board calculate (a) the speed at which she reached the water and (b) for how long she was in the air. Take $g = 9.8\,\mathrm{m\,s}^{-2}$.

(a) She hit the water after falling from rest at the top of the dive for 10.4 m.

Using $s = \tfrac{1}{2}at^2$, $10.4\,\mathrm{m} = \tfrac{1}{2}(9.8\,\mathrm{m\,s}^{-2})t^2$

$\Rightarrow \qquad\qquad t = 1.46\,\mathrm{s}$

So her entry speed is given by

$$v = at = (9.8\,\mathrm{m\,s}^{-2})(1.46\,\mathrm{s}) = 14.3\,\mathrm{m\,s}^{-1}$$

(b) The time to fall from the top of the dive was 1.46 s. The time to rise the 0.40 m is given by

$$0.40\,\mathrm{m} = \tfrac{1}{2}(9.8\,\mathrm{m\,s}^{-2})t^2$$

$$t = 0.29\,\mathrm{s}$$

So the total time in the air is 1.75 s.

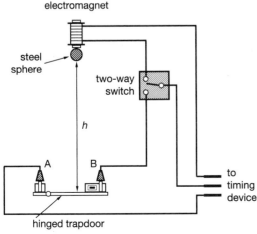

electromagnet

steel sphere

two-way switch

h

A   B

to timing device

hinged trapdoor

FIGURE 1.13

Projectiles

When an object falls freely its vertical motion is *independent* of its horizontal motion. The diver in the above example probably didn't move in a vertical line – she must have moved forward a bit. But the calculation of her vertical velocity when she hit the water and how long the dive took are still valid.

The diver using a 1-metre spring board in figure 1.14 moves forward horizontally at a constant speed. (We ought to say his *centre of gravity* (the blob) moves forward with constant speed.) He moves vertically with a fixed acceleration of $9.8 \, \mathrm{m \, s^{-2}}$ downwards. The path he follows is called a **parabola**. Any diver or jumper follows a parabolic-shaped path as do cricket balls

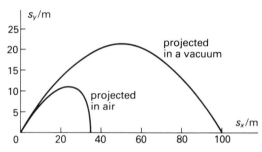

FIGURE 1.15

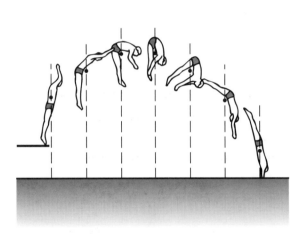

FIGURE 1.14

or netballs, provided they are not affected noticeably by air resistance. Figure 1.15 shows the parabola a tennis ball would follow in a vacuum and the real path it follows in air when struck at $30 \, \mathrm{m \, s^{-1}}$ at just above 45° to the horizontal.

Exercises on each section of this chapter may be found in the companion textbook, *Practice in Physics*.

SUMMARY

At the end of this chapter you should be able to:

◆ describe how speeds can be measured by timing over a known distance or by finding the distance travelled in a given time interval.

◆ use the equations:

$$\text{average velocity} = \frac{\Delta s}{\Delta t}$$

$$\text{average acceleration} = \frac{\Delta v}{\Delta t}$$

◆ understand that velocity and acceleration are vector quantities and obey special rules for addition.

◆ remember that vectors can be resolved into two perpendicular components, $v_x = v\cos\theta$ and $v_y = v\sin\theta$, where θ is the angle between v and the x-axis.

◆ understand that the gradient of a displacement–time graph is the velocity and that of a velocity–time graph is the acceleration.

◆ understand that the area between a velocity–time graph and the time axis represents the displacement.

◆ draw graphs for the displacement, velocity or acceleration of a body against time for straight-line motion when given only one of them.

◆ use the following equations for uniform acceleration:

$$v = u + at$$

$$s = \left(\frac{v+u}{2}\right)t$$

$$s = \tfrac{1}{2}at^2 \quad \text{when} \quad u = 0$$

◆ remember that the free fall acceleration at the Earth's surface is $9.8\,\text{m s}^{-2}$.

◆ describe how to measure the free fall acceleration in the laboratory.

◆ understand that when an object is falling freely its vertical motion is independent of its horizontal motion.

2 Momentum and Force

The sprinter in the photograph will 'explode' when she hears the gun. Her brain will send messages to her limbs telling them to drive hard against the blocks so that the blocks will suddenly push her forward. To get the size of this push as large as possible and to ensure that it is in exactly the right direction will have been the focus of hours of training. A split-second delay off the blocks makes all the difference at the end of a short sprint.

As the sprinter accelerates forwards, so something must accelerate backwards. Providing the blocks are firmly attached to the track it is the whole Earth which 'recoils' and, of course, that is not noticeable. You may have wondered what would happen if every one of the Earth's ten billion inhabitants all tried a sprint start – to the West – at exactly the same instant. Would we feel the Earth move? No, for it is far too massive compared even with all those people; though if you thought that it might, you would have been right in principle.

2.1 Mass and momentum

When you push a supermarket trolley to link with a stationary one, what happens depends not only on the velocity of the moving one at impact but also whether your trolley is empty or loaded. We say that a loaded trolley has a greater mass or **inertia** than the empty trolley.

The product of the mass and the velocity of a body is called the **linear momentum** p or simply the momentum of the body.

$$p = mv$$

For an empty trolley, of mass 8.0 kg moving at $2.0\,\text{m s}^{-1}$, for example,

$$mv = 8.0\,\text{kg} \times 2.0\,\text{m s}^{-1} = 16\,\text{kg m s}^{-1}$$

The unit is awkward but is not given its own name. The reason for studying momentum is that, if you find the momentum of the linked trolleys straight after this collision you will find it is still $16\,\text{kg m s}^{-1}$. This is why momentum is important – it is *conserved*, i.e. we can make predictions about what happens when things collide because the momentum before is equal to the momentum after the collision. This **principle of conservation of momentum** applies to all collisions or explosions; to *all* interactions.

Experiments with air-track gliders

The linear air track allows us to study collisions in one dimension in a situation where there are no other horizontal influences on the interacting objects. The gliders sit astride a track which consists of a tube into which air is pumped. The air continuously emerges through rows of tiny holes and supports the glider on a cushion of air. The gliders are not therefore in contact with the track and move freely along it. Their speeds can be measured using the photodiodes and the datalogging equipment described in the previous chapter.

In such experiments the necessary speeds can be measured to about ±2% and the conservation principle demonstrated. The gliders and photocells could be replaced by dynamics trolleys and ticker-tape. Trolleys often have built-in springs to produce the recoil and you can use pins and corks to make two trolleys stick together.

Of course in experiments of this kind you do not *prove* that the principle of conservation of momentum is true but that it seems to work well enough for simple collisions and explosions in our laboratories.

Experiment 1 – collision

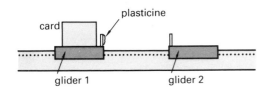

card
plasticine
glider 1
glider 2

FIGURE 2.1

You can set glider 1 in motion towards glider 2, which is stationary and has no card attached, and arrange for them to stick together. Plasticine will be adequate for this. You can show that the momentum before the collision (all in glider 1) is equal to the momentum after the collision (in the coupled gliders).

EXAMPLE

A boy of mass 45 kg jumps into a stationary boat of mass 215 kg. If the boy and boat move off at $0.50\,\mathrm{m\,s^{-1}}$, calculate the horizontal velocity of the boy as he jumped.

momentum before = momentum after

$$(45\,\mathrm{kg})\,v = (45\,\mathrm{kg} + 215\,\mathrm{kg})\left(0.50\frac{\mathrm{m}}{\mathrm{s}}\right)$$

$$\Rightarrow v = \frac{260}{45} \times 0.50\frac{\mathrm{m}}{\mathrm{s}} = 2.9\frac{\mathrm{m}}{\mathrm{s}}$$

Experiment 2 – explosion or recoil

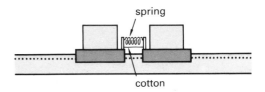

spring
cotton

FIGURE 2.2

You can arrange for both the gliders to be stationary and then have them fly apart, each recoiling from the other. A light spring attached to one glider can provide the necessary energy, release being achieved by burning through the cotton connecting thread. You can show that the gliders move apart with equal sized momentums so that the total momentum is still zero after the explosion (as it was before the string was burned). Momentum is a vector quantity so one glider has positive momentum and the other negative momentum, the total being zero. In calculations you must be careful: choose a positive direction and give each momentum a plus or a minus sign.

EXAMPLE

A pair of skaters, a man of mass 65 kg and a woman of mass 50 kg, are together moving in a straight line across the ice at a speed of $6.0\,\mathrm{m\,s^{-1}}$. They push each other apart along their line of motion so that after they are separate the man is moving in the same direction at $4.0\,\mathrm{m\,s^{-1}}$. What is the woman skater's velocity?

Suppose the woman's velocity is v.

momentum before = momentum after

$$(65 + 50)\,\mathrm{kg} \times 6.0\frac{\mathrm{m}}{\mathrm{s}} = (65\,\mathrm{kg})\left(4.0\frac{\mathrm{m}}{\mathrm{s}}\right) + (50\,\mathrm{kg})\,v$$

$$\therefore 50v = (115 \times 6.0 - 65 \times 4.0)\frac{\mathrm{m}}{\mathrm{s}}$$

$$\Rightarrow \qquad v = 8.6\,\mathrm{m\,s^{-1}}$$

(ii) The change of momentum of the box is

$$p_2 = m_2 v_2 = 12\,\text{kg}(1.4 - 0)\,\text{m}\,\text{s}^{-1}$$
$$= 16.8\,\text{kg}\,\text{m}\,\text{s}^{-1} \text{ to the right}$$

while that of the ball is

$$p_1 = m_1 v_1 = 0.8\,\text{kg}(-9 - 11)\,\text{m}\,\text{s}^{-1}$$
$$= -16.8\,\text{kg}\,\text{m}\,\text{s}^{-1} \text{ to the right}$$
$$\text{i.e. } 16.8\,\text{kg}\,\text{m}\,\text{s}^{-1} \text{ to the left}$$

Notice that the change of momentum of the ball is the same size as the change of momentum of the box.

EXAMPLE

A ball of mass 0.80 kg is kicked at a speed of 12 m s^{-1} to the right and hits a wooden box of mass 12 kg which is at rest on a frictionless floor. The box moves off with an initial velocity of 1.4 m s^{-1} to the right after the collision. What is (i) the new velocity of the ball and (ii) the change of momentum of (a) the ball, (b) the box, as a result of the collision?

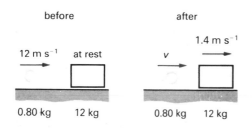

(i) Referring to the diagram and applying the principle of conservation of momentum,

$$12 \times 0.80\,\text{kg}\,\text{m}\,\text{s}^{-1} = (0.80\,\text{kg} \times v)$$
$$+ (12 \times 1.4\,\text{kg}\,\text{m}\,\text{s}^{-1})$$
$$\therefore 0.80\,v = (9.6 - 16.8)\,\text{m}\,\text{s}^{-1}$$
$$= -7.2\,\text{m}\,\text{s}^{-1}$$

which gives $\qquad v = -9.0\,\text{m}\,\text{s}^{-1}$

The minus sign means that the ball has rebounded from the box and is now travelling to the left.

In solving problems using the principle of conservation of momentum you must be sure that in the direction in which momentum conservation is to be applied either there are no external forces acting or the time for the interaction is very small. It is also useful to sketch the situation both before and after the interaction, adding all the relevant data to the diagram. But the vital idea is to appreciate that the principle is a universal one. It applies to gas molecules colliding and to exploding galaxies.

Experiment 3

When the colliding objects don't stick together but bounce off each other with some loss of energy the principle still applies. As momentum is conserved then the change of momentum of one of them is the same as the change of momentum of the other but in the opposite direction.

2.2 Forces

We all know what forces are. They are **pushes** and **pulls**. The SI unit of force is the **newton** (N), named after Sir Isaac Newton (1642–1727). For instance, the pull of the Earth on a mass of 60 kg is about 600 N and typical pushes and pulls made by the hand are between 0.1 N (pushing the key of an electronic calculator) and 100 N (pulling open a very large door).

All the apparently different types of force can be traced to only three basic interactions: gravitational, electromagnetic (both electric and magnetic forces) and nuclear. Of these **nuclear** forces exist only between particles in the nucleus and become negligible when these particles are more than about 10^{-15} m apart. Of the other two, **gravitational** forces are much weaker than **electromagnetic** forces. To quantify this consider two protons: the electric force which each exerts on the other is about 10^{36} times greater than the gravitational forces which they exert on each other. Thus although all bodies do attract each other gravitationally, these forces are only noticeable if one of the bodies is very massive, e.g. like the Earth or the Moon.

Everyday contact forces are electromagnetic in origin, i.e. they result from the repulsion or attraction between the outer electrons of all atoms. When we stamp on the ground or pull open a door it is the electromagnetic interaction (push or pull) between groups of atoms which we feel.

Representing forces

It is *always* possible to describe a force by the phrase

the push (or pull) of A on B

where A is a body exerting a force on body B. For example, we talk of the 'push of the pen on the paper' or 'the pull of the lamp on its cable'. The consistent use of this sort of phrase prevents statements like 'the force of friction' or 'the pull of gravity' and makes us identify B (the body on which the force acts) and A (the body which exerts a force). To identify B is often the key to starting a problem and to thinking clearly about it. For this reason, if we are drawing a sketch to show the forces acting on body B, it is best not to include any other bodies in the diagram. Such a diagram is called a **free-body force diagram**. Figure 2.3 shows an example of a simple situation (a) involving forces. Suppose we wish to consider the forces acting on the box; we draw a free-body force diagram of the box (b) showing the push P of the ground, the pull W of the Earth, and the pull T of the rope – these all being pushes or pulls acting *on the box*. The point of drawing such a diagram is

that you can then easily see *all* the forces acting on the body.

If possible the forces in a free-body force diagram should be drawn so that their lengths are proportional to the size of the forces. There are other bodies in figure 2.3 in which we may be interested. You should try to draw free-body force diagrams for the pulley block or for the man.

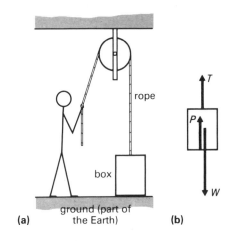

FIGURE 2.3

Forces obey the rules of vector calculations. In particular they can be resolved.

Newton's first law

This is usually stated as follows:

a force is required to change the velocity of a body, that is, to alter its speed or the direction of its motion.

It is sometimes called the law of inertia and was almost fully formulated by Galileo Galilei (1564–1642). A body is said to be **in equilibrium** if it is either at rest (relative to the Earth's surface) or moving with a constant speed along a straight line. The box in figure 2.3 is in equilibrium and so we can deduce that $T + P = W$. If a car is moving along a straight road at $25 \, \text{m s}^{-1}$ there may be an air resistance force (the push of the air on the car) of 2000 N. As the car is in equilibrium there must be a forward driving force (the forward push of the road on the wheels of the car) which is also 2000 N. Notice that it is *not* the engine which pushes the car forward, though the energy conversions necessary for the car to move take place there.

Some common forces

Figure 2.3 showed three forces acting on a box. Forces of each type are commonly met in problems in mechanics and they are generally referred to as tension T, weight W, and contact force (or reaction) P.

Tension   Any flexible rope (or chain or string) being used to pull a body is said to be in a state of tension. The molecules of the rope are very slightly further apart than they would be if it was slack. The tension in the rope at a point X is the size of the force which one part of the rope cut at point X exerts on the other part. For chains or thick ropes the tension varies from place to place, but this is often ignored and the size of the tension is taken to be constant. For light threads and strings it is a reasonable assumption. Where the rope ends, the pull of the rope on the object to which it is attached is equal in size to the tension in the rope at the point of connection. When a rope passes round a free-running pulley or a smooth post the tension is usually taken to be the same on both sides of the pulley or post. This again is only approximately true in practice. Pulleys can thus change the direction of a force without altering its size and it is for this reason that they are useful. In figure 2.3(b), if the pull of the man on the rope is 70 N downwards, then the tension in the rope is 70 N and the pull of the rope on the box is 70 N upwards.

Weight   The Earth exerts an attractive force, a gravitational pull, on every particle. This pull is directed toward the centre of the Earth. For a body which is not a point object, the resultant of all the tiny forces is a *single* pull called its weight. Whichever way the body is turned its weight is found always to act through a particular point: this is called the **centre of gravity** of the body. The weight of a body varies slightly from place to place on the Earth's surface but the variations are always less than 0.5%, so you should usually treat the weight of a body on the Earth's surface as constant.

Contact force   Any two bodies pressed together by some external agency repel one another. The repulsion is the result of the compression of the surface layers of material, i.e. the surface molecules are very slightly closer together than when they are not pressed together. The contact force is the resultant of all the tiny electrostatic pushes and can be represented by a single force which is perpendicular to the surfaces in contact. If a body is being pushed against a surface in such a way as to make it slide over the surface the contact force will no longer be perpendicular to the surface. It is, however, convenient to talk of the two resolved parts of the contact force in this case: the *perpendicular* push of the surface on the body and the *tangential* push of the surface on the body. The perpendicular push we will call the contact force and the tangential push we will call the frictional force.

Equilibrium

If a body *is* in equilibrium then the vector sum of the forces acting on it is zero. The sum of the resolved parts of the forces acting on it in any convenient direction must therefore also be zero.

In order *to show that* a body is in equilibrium it is necessary to show that the sum of the resolved parts of the forces acting on it in two separate directions is zero. It is in situations like this that a free-body force diagram is particularly helpful.

EXAMPLE

A kite of weight 2.0 N is held stationary by the push of the wind and the pull of the string on it. The tension in the string, which makes an angle of 35° to the vertical at the place where it meets the kite, is 9.0 N. What is the size and direction of the push of the air on the kite?

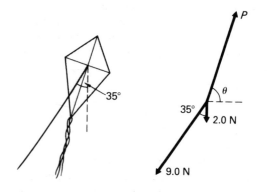

EXAMPLE

After the sketch of the kite is a free-body force diagram of it represented as a particle. The unknown push of the air on the kite is shown as P at an angle θ above the horizontal. Resolving horizontally

$$P\cos\theta - (9.0\,\text{N})\sin 35° = 0$$

so that $\qquad P\cos\theta = 5.17\,\text{N}$

Resolving vertically

$$P\sin\theta - (9.0\,\text{N})\cos 35° - 2.0\,\text{N} = 0$$

so that $\qquad P\sin\theta = (7.37 + 2.0)\,\text{N} = 9.37\,\text{N}$

Thus $\qquad \dfrac{P\sin\theta}{P\cos\theta} = \dfrac{9.37}{5.17} = 1.81$

i.e. $\qquad\qquad \tan\theta = 1.81$

$\Rightarrow \qquad\qquad\qquad \theta = 61°$

Substituting for θ gives

$$P\cos 61° = 5.17\,\text{N}$$

so $\qquad\qquad\qquad P = 10.7\,\text{N}$

You could find the sum of the 9.0N and 2.0N forces by scale drawing (the addition of vectors). Their resultant would be equal in size to P and parallel to it, but opposite in direction to it so both P and θ could be deduced from the drawing.

2.3 Moments and equilibrium

The effect of a force is often to cause rotation. The pull of a hand on a doorknob produces rotation of the door about the vertical door hinges and the push of a foot on the starting pedal of a motor cycle produces rotation of the engine about the horizontal engine axle. To study the *turning effect* of a force in the laboratory it is only necessary to support a rod or ruler on a pivot and to exert known forces on it – a see-saw type of experiment. The smaller push of a girl on the left of a real see-saw, e.g. 300N, the further she must be from the centre in order to balance the greater push of her brother on the right of the see-saw, perhaps 700N. We say that the moment of the 300N force *about the axis of rotation* increases as its distance from the axis decreases (figure 2.4).

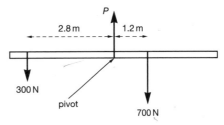

FIGURE 2.4

We define the **moment** M **of a force** about a chosen axis by the equation

$$M = rF$$

where F is the force and r is the perpendicular distance from the chosen axis to the line of action of the force. The unit of M is the N m. So in the see-saw above, figure 2.4, we might have:

- moment of push of girl on see-saw about the pivot = $300\,\text{N} \times 2.8\,\text{m} = 840\,\text{N m}$ *anticlockwise*, i.e. twisting the see-saw clockwise as we see it about the pivot.

- moment of push of her brother on see-saw about the pivot = $700\,\text{N} \times 1.2\,\text{m} = 840\,\text{N m}$ *clockwise*.

- moment of push of pivot on see-saw = 0, because the distance from the pivot is zero.

These calculations illustrate how carefully we must describe the moment of force. A value, e.g. 840N m, tells us little unless the pivot is referred to and there is a statement of clockwise ⌢ or anticlockwise ⌢. If we call ⌢ positive and ⌢ negative then we notice that the moments on the see-saw add up to zero.

The principle of moments

For a body which is in equilibrium the sum of the moments, about any axis, of the external forces acting on the body is zero.

$$\text{sum of M} \curvearrowright = \text{sum of M} \curvearrowleft$$

To make use of this principle in solving problems the crucial issue, after drawing a free-body force diagram of the body which is being considered, is to decide on the axis or pivot you are going to use. Any axis will do but it usually helps to choose one which reduces the moment of an unknown force to zero.

EXAMPLE

A gate has a top hinge H in working order but the bottom one is missing, so that the gate post may be assumed to push horizontally against the bottom of the gate. The gate is of symmetrical design and has a weight of 2000 N. Find the forces exerted by the gatepost on the gate.

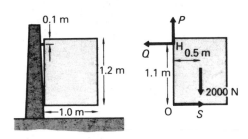

When you do not know the direction of a force – here the pull of the hinge on the gate – it is best to draw two resolved parts in convenient directions.

The moment of P about H = zero
The moment of Q about H = zero
The moment of S about H = (1.1 m) S, anticlockwise
The moment of W about H = (0.5 m)(2000 N), clockwise

The sum of the moments about H must be zero. Taking clockwise moments to be positive

$$1000\,\text{N m} - (1.1\,\text{m})S = 0$$
$$\Rightarrow S = 910\,\text{N}$$

The sum of the forces in any direction must be zero so,

vertically	$P - 2000\,\text{N} = 0$
horizontally	$Q - 910\,\text{N} = 0$
so that	$P = 2000\,\text{N}$
and	$Q = 910\,\text{N}$

The size of the resultant pull R of the hinge on the gate is given by

$$R = \sqrt{(2000^2 + 910^2)}\,\text{N}$$
$$= 2200\,\text{N}$$

and the angle R makes with the horizontal is θ where

$$\tan \theta = \frac{2000\,\text{N}}{910\,\text{N}}$$

$$\Rightarrow \qquad \theta = 66°$$

Centre of gravity

The previous example assumes that we know where to put the 2000 N force, the pull of the Earth on the gate. The Earth exerts a gravitational pull on every molecule of the gate, but we can replace all these by a single resultant force which acts through a point G called the centre of gravity of the gate. Its position is that point about which the sum of the moments of the pull of the Earth on the molecules of the body is zero however the body is placed in the Earth's gravitational field.

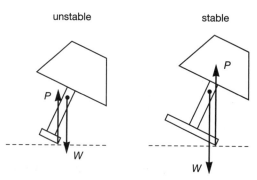

FIGURE 2.5

For flexible bodies G can change its position depending on how, for example, a diver is tucking or stretching. The motion of G for a freely falling body such as a diver – see page 17 – or high-jumper follows a parabola, the same path as a ball.

Bodies such as double-decker buses or table lamps are more **stable** when their centres of gravity are low. On a free-body force diagram the weight force arrow will lie within the base even when the bus or the lamp is tilted – figure 2.5.

Structures

In analysing the equilibrium of structures such as skeletons or bridges it is often the internal forces, the pushes and pulls of one part of the structure on another, that concern us. This being the case we must take particular care in choosing the object or bodies for which we draw free-body force diagrams. Consider a bent arm which is holding a 1 kg mass as in figure 2.6.

This is a complicated structure and we need to simplify it. By representing the lower arm and the forces on it as shown we can take moments about the pivot and, given the weight of the lower arm and various distances, can find F, the pull of the biceps on the lower arm. It will be a large force, perhaps 200 N and *much* larger if the arm is supporting a 10 kg mass.

2.4 Newton's second law

A force changes the velocity of a body. Newton's second law uses this idea to *measure* forces by stating that for a body of fixed mass:

the mass of a body multiplied by its acceleration is equal to the resultant force acting on it.

$$ma = F_{res}$$

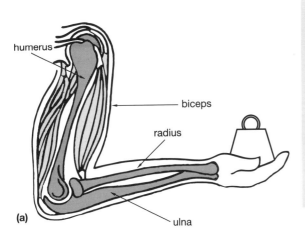

humerus

biceps

radius

ulna

(a)

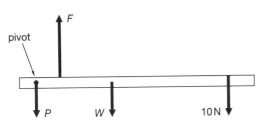

F

pivot

P W 10 N

(b)   simplified free-body force diagram of the lower arm

FIGURE 2.6

EXAMPLE

A javelin thrower accelerates a javelin of mass 0.80 kg from $5.5\,\mathrm{m\,s^{-1}}$ to $31.5\,\mathrm{m\,s^{-1}}$ in 0.30 s. What is the average pull he exerts on the javelin?

$$\text{acceleration } a = \frac{(31.5 - 5.5)\,\mathrm{m\,s^{-1}}}{0.30\,\mathrm{s}}$$

$$= 87\,\mathrm{m\,s^{-2}}$$

$$\therefore \quad \text{pull } F = (0.80\,\mathrm{kg})(87\,\mathrm{m\,s^{-2}})$$

$$= 70\,\mathrm{N}$$

From this simple example we see that a **newton**, the unit of force, is simply a name for a kilogram metre per second squared (quite a mouthful), i.e.

$$1\,\mathrm{N} = 1\,\mathrm{kg\,m\,s^{-2}}$$

We should also note that there are other forces acting on the javelin, but they are small. The pull we have calculated is the resultant of all the forces and this resultant or **unbalanced** force acts along the line of the javelin, the line in which the acceleration took place.

A freely falling body, e.g. a high diver, of mass 56 kg, accelerates downward at $9.8\,\mathrm{m\,s^{-2}}$. We can therefore calculate the resultant force on her as 550 N. As the only force acting is the pull of the Earth then we have calculated her weight. In general, a body of mass m, has a weight W where

$$mg = W$$

because all bodies accelerate at g when in free fall. This can be used to calculate weight from mass or mass from weight whether the body is accelerating or is at rest. One way of remembering this is to realise that $9.8\,\mathrm{m\,s^{-2}}$ can also be written as $9.8\,\mathrm{N\,kg^{-1}}$. This fact enables you to calibrate a spring balance using some known masses and a knowledge of the local value of g. In reverse it enables you to find the mass of something on a top pan balance which measures its weight.

Applying Newton's second law

In solving more difficult problems using $ma = F_{res}$ it is best to start with a as it is the direction of the acceleration which is usually known. The forces can then be resolved in this chosen direction and

the resolved parts added to give the resultant force producing the acceleration, i.e.

ma = sum of forces resolved parallel to a

In applying the law you should:

◆ choose a body (as a first step this act of choosing cannot be emphasised too strongly).

◆ draw a free-body force diagram of the chosen body, marking all the external forces on it. Ensure that each is a genuine force on the body B by expressing it as the pull (or push) of something on B.

◆ mark the acceleration alongside the body.

◆ in numerical problems ensure that all forces are expressed in newtons.

◆ apply Newton's second law in the form given above.

EXAMPLE

A man stands in a lift which is moving upwards and slowing down at $0.80\,\mathrm{m\,s^{-2}}$. If his mass is 70 kg, what is the push of the floor on his feet? Take $g = 10\,\mathrm{m\,s^{-2}}$.

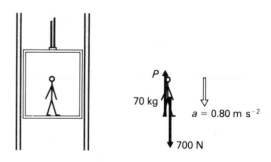

The diagram shows the situation and a free-body force diagram of the man. Using $ma = F_{res}$

$$(70\,\mathrm{kg})(0.80\,\mathrm{m\,s^{-2}}) = 700\,\mathrm{N} - P$$
$$\therefore P = (700 - 56)\,\mathrm{N} = 644\,\mathrm{N}$$

It does not matter that the lift is moving upwards. It is the direction of the acceleration only which matters; if it was moving downwards the force would still have been found to be 644 N upwards.

Impulse

Newton's second law states that

$$ma = F$$

As acceleration is $\Delta v/\Delta t$ then

$$m\frac{\Delta v}{\Delta t} = F$$

or
$$m\,\Delta v = F\Delta t$$

which is often called the **impulse–momentum equation**, the product $F\,\Delta t$ being called the impulse of the force. If you can calculate the impulse you have found the change of momentum, and vice versa.

If a graph of F (up) against t (along) is drawn then the impulse of a force during a given time interval is represented by the area under the graph line.

The impulse–momentum equation is useful in problems involving steady forces acting for known lengths of time, for instance in calculating the effect of firing a rocket. Suppose a rocket motor is switched on for 3.0 s and during that time produces a steady thrust of 12 000 N as in figure 2.7: then the impulse is 36 000 N s.

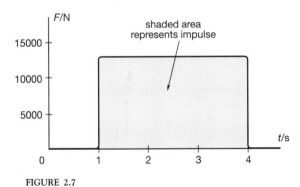

FIGURE 2.7

The N s (*not* N/s), is identical with our earlier unit of momentum, the $kg\,m\,s^{-1}$. If the rocket has a mass of 2000 kg (assumed constant), then its change of velocity is given by

$$\Delta v = \frac{36\,000\,\text{N s}}{2000\,\text{kg}} = 18\,\text{m s}^{-1}$$

This might be a course correction on a Voyager mission; obviously the direction of F and of the change of velocity would be critical. Where F and Δt are not known at every instant (e.g. when a racket hits a tennis ball – figure 2.8), then the impulse of the push of the racket on the ball can still be found by calculating the change of momentum of the ball. If an electric circuit can be arranged to give a measure of the time interval Δt during which the racket is in contact with the ball, then the average value of F can be found. Sudden large forces which last for small time intervals are sometimes referred to as **impulsive forces**.

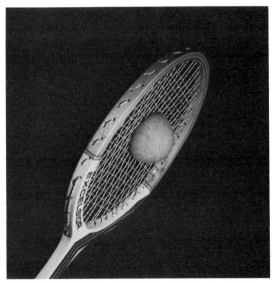

FIGURE 2.8

This impulse–momentum way of using Newton's second law is often the best method of giving a physical explanation in answer to questions such as 'Why are eggs packed in crushable egg-boxes?' or 'How should you try and catch a cricket ball?' In each case we are dealing with an object, a dropped egg or a moving ball, which has to have its momentum reduced to zero. $m\,\Delta v$ is predetermined for us and so, therefore, is the impulse $F\,\Delta t$ that is needed. Clearly the larger Δt can be made, the smaller F, the force which is used to slow down the egg or the ball, will be. Thus the egg is put in a box which will crush slightly on hitting the ground; the cricketer allows his hands to 'give' with the ball as he catches it.

Car passenger safety

Packaging eggs to prevent breakage during handling is illustrative of a much more significant problem: the packaging of human beings in motor cars. If the car is to crash, let us say head-on into a large concrete block, then there are two main factors which affect the safety of the driver. Firstly the front of the car will crush, acting like the egg box (but nevertheless producing very high accelerations of about $-400\,\mathrm{m\,s^{-2}}$ for impact velocities of about $20\,\mathrm{m\,s^{-1}}$). Figure 2.9 shows a crumpling test with a seat-belt holding a dummy figure.

FIGURE 2.9

Secondly, the passenger is not snugly fitted into a box with moulded supporting materials around him. He is sitting about 0.5 m away from a solid windscreen, dashboard and steering wheel.

The second factor is illustrated by graphs A and B of figure 2.10 which show the variation of the force F acting on the driver against time t (note the scales). For curve A the driver is wearing a seat-belt and F is the pull of the belt on his chest. For curve B he is not wearing a seat-belt and F is the push of the windscreen, etc. on his chest. The impulse must be the same in each case as $F\Delta t = m\Delta v$, regardless of how he is brought to rest. The area between graph-line A and the time-axis must therefore be the same as the area between graph-line B and the time-axis.

In the design of seat belts the objective is to tie the passenger to the car body and to hold him there with a restraining force something like the dashed graph-line C in figure 2.10. To achieve this the crumpling characteristics of the front of the car

can be altered, the space between the seated driver and the windscreen should be as large as possible and drivers and passengers must be encouraged to wear their seat-belts.

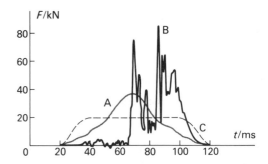

FIGURE 2.10

EXAMPLE

Estimate the impulse on the driver described by curve A in figure 2.10. If the driver was of mass 80 kg, what was the speed of the car before it hit the concrete block?

Curve A is roughly triangular with a base of 100 ms and a height of 35 kN. The average force is thus 17.5 kN.

$$\text{impulse} = (17.5 \times 10^3\,\mathrm{N})(100 \times 10^{-3}\,\mathrm{s})$$
$$= 1750\,\mathrm{N\,s}$$

∴ change of momentum = $1750\,\mathrm{kg\,m\,s^{-1}}$ and hence change of speed

$$\Delta v = \frac{1750\,\mathrm{kg\,m\,s^{-1}}}{80\,\mathrm{kg}} = 22\,\mathrm{m\,s^{-1}}$$

As the final velocity is zero the initial velocity was $22\,\mathrm{m\,s^{-1}}$.

In using impulse = FΔt remember it is the average force which is needed for F.

2.5 Animal and vehicle propulsion

No animal or vehicle can pull *itself* along: an *external* force is always needed. If you doubt this perhaps you could try to lift yourself up to the ceiling. No, you cannot have a rope for then it would be the pull of the rope which was lifting you; nor a stepladder, for then it would be the push of the steps, and so on. The external forces result from the animal or vehicle itself exerting a force on the ground, water or air with which it is in contact.

The ground, water or air in their turn exert forces on the animal or vehicle in a way described by **Newton's third law**.

Newton's third law

Newton was the first to appreciate that whenever two bodies interact, *two* forces are involved. One force acts on each body and there is a simple relationship between them at every instant of the interaction which can be expressed as follows:

if body A exerts a force F on body B, then body B exerts a force −F on body A, that is, a force which has the same size but is in the opposite direction.

This law tells us that a single force is an impossibility. Forces always occur in pairs, one force acting on each of the two interacting bodies. They will never therefore *both* appear on the same free-body force diagram. And the law holds *at every instant* of the interaction. The principle of conservation of momentum results directly from this idea.

If the Earth pulls you down with a force of 600 N, then you pull the Earth up with a force of 600 N. If, later, the Earth's surface pushes your feet up with a force of 650 N, then, at that instant, your feet are pushing the Earth down with a force of 650 N. Notice that the resultant force on *you* is now 50 N upward; you must be accelerating upward. Perhaps you are in the process of jumping. If you jump off the ground both the push of the Earth on you *and* the push of you on the Earth are, of course, then zero.

Frictional forces

Frictional forces occur in a wide variety of situations but always act so as to oppose or prevent the onset of relative motion between the two solid surfaces which are in contact. Though this seems to imply that friction will always be a nuisance, quite the reverse is true. Our ability to move about on the Earth's surface is almost entirely the result of frictional forces. Let us consider the forces

acting on a motor car which is driving at a steady speed along a straight road.

The car shown in figure 2.11 has a rear-wheel drive. So the wheels are pushing backward on the road with a frictional force F. By Newton's third law there is a forward push F on the car. The frictional force which opposed the motion of the wheel becomes the car's external driving force. F is called a *static* friction force.

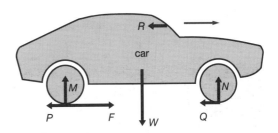

FIGURE 2.11

The size F depends on the nature of the road and wheel surfaces. A lack of tread on the tyres and ice or an oily film on the road will reduce the maximum possible value of F before slipping occurs. It also depends on the size of the perpendicular contact push M of the ground on the rear wheels. The larger M is, the larger the size of F before slipping occurs. In an ordinary car $M + N = W$, the pull of the Earth on the car, but racing cars are fitted with aerofoils which produce a greater downward push on the car.

Rolling friction

In figure 2.11 two other frictional forces are shown, P and Q. These are forces of rolling friction. They are also described as the frictional push of the ground on the wheels but must be distinguished from F as they cannot be utilised as driving forces. Rolling friction results from the deformation of the tyre and the ground and has the effect of making the wheels always seem to be rolling slightly uphill. The other force opposing motion is an air resistance force R, the push of the air on the car.

If the car is to accelerate then the engine must attempt to rotate the rear wheels faster. *F*, the forward push of the ground on the car, thus increases. To brake the car the discs or drums attempt to prevent the wheels rotating. Because of the inertia of the car, the direction of *F* is reversed so that the braking force is the (external) backward push on the ground on the car.

The above discussion is based on a car. But the operation of all vehicles can be understood similarly. Electric or diesel trains, motorcycles and bicycles all use the same principle, the main difference being the nature of the energy source.

EXAMPLE

A bicycle and its rider have a combined mass of 90 kg. Assuming that the maximum frictional force between the rear wheel and the ground is 80% of the perpendicular contact force at that wheel, estimate the maximum possible acceleration of the bicycle.

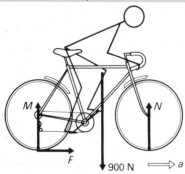

Suppose the weight of the bicycle and rider is evenly distributed, then, taking $g = 10\,\text{N}\,\text{kg}^{-1}$ gives the size of the contact push of the ground on each wheel to be $N = M = 450\,\text{N}$. The maximum frictional push of the ground on the wheel is thus

$$F = 0.8 \times 450\,\text{N} = 360\,\text{N}$$

and the acceleration (assuming rolling friction and air resistance to be negligible) is given by Newton's second law

$$(90\,\text{kg})\,a = 360\,\text{N} \quad \Rightarrow \quad a = 4\,\text{m}\,\text{s}^{-2}$$

If the rider wants a higher acceleration he does so by lifting the front wheel thus increasing the size of M and thus F.

In fact all animals, including man, achieve a forward motion by pusing the ground backwards and relying on Newton's third law of motion. In all such propulsion momentum is conserved if we take into account not only the momentum of the vehicle or animal but also the momentum of the ground (which is part of the Earth). As the mass of the Earth is so huge we do not notice it recoiling when we accelerate forward.

2.6 Jets and rockets

So far all the bodies to which we have applied Newton's laws of motion have had a constant mass. Newton's second law $ma = F_{\text{res}}$ can be generalised to be

$$\frac{\text{d}p}{\text{d}t} = F_{\text{res}}$$

The rate of change of momentum of a body is equal to the resultant force acting on it and takes place in the directon of the force.

Using this statement of the law we can calculate the thrust needed to produce or to stop jets of air or water and to understand how rockets and jet engines work.

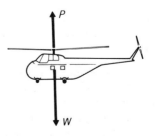

FIGURE 2.12

Consider first a stationary hovering helicopter of mass 800 kg – figure 2.12. By Newton's first law we see that $P - W = 0$, i.e. P and W are each 8000 N in size, taking g as $10\,\mathrm{m\,s^{-2}}$. P is the upward push of the air on the helicopter. By Newton's third law the push of the helicopter on the air is 8000 N downwards at such a rate as to achieve, by Newton's second law, a rate of change of momentum of the air of 8000 N, i.e.

$$F = \frac{dp}{dt} = \frac{d(mv)_{\mathrm{air}}}{dt} = 8000\,\mathrm{N}$$

If the air is moving at constant speed v, then

$$v\left(\frac{dm}{dt}\right)_{\mathrm{air}} = 8000\,\mathrm{N}$$

Suppose the air is projected at a speed of $25\,\mathrm{m\,s^{-1}}$; then we have

$$\left(25\frac{\mathrm{m}}{\mathrm{s}}\right)\left(\frac{dm}{dt}\right)_{\mathrm{air}} = 8000\,\mathrm{N}$$

$$\Rightarrow \quad \frac{dm}{dt} = \frac{8000\,\mathrm{kg\,m\,s^{-2}}}{25\,\mathrm{m\,s^{-1}}} = 320\,\mathrm{kg\,s^{-1}}$$

As air has a density of about $1\,\mathrm{kg\,m^{-3}}$, the helicopter blades are gathering and projecting about $320\,\mathrm{m^3}$ of air each second.

Jet engines

In principle a jet engine is very simple:

♦ air is taken in at the front; this air is compressed.

♦ fuel (usually paraffin vapour) is mixed with the compressed air and ignited. In the resulting explosion the air expands.

♦ the exhaust gases rush out of the back, some of their momentum being used to turn a turbine which drives the compressor in a turbo-jet.

The turbo-jet is the basic propulsion unit for modern jet aeroplanes. The jet engine, figure 2.13 uses the oxygen in its air intake as well as the fuel it carries to cause the explosive expansion which ensures that the exhaust gases rush out at a much higher speed than the air enters.

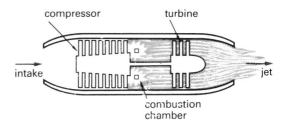

FIGURE 2.13

EXAMPLE

A horizontal jet of water emerges from a nozzle of cross-sectional area A at a speed v. It strikes a vertical wall and does not rebound. If the density of water is ρ, the push of the water on the wall is $v^2 \rho A$.

Show that this expression has the unit of force and derive it from Newton's laws.

Units of $v^2 \rho A$ are $(\mathrm{m\,s^{-1}})^2(\mathrm{kg\,m^{-3}})(\mathrm{m^2})$ which gives $\mathrm{kg\,m\,s^{-2}}$, i.e. newton.

In a time Δt a stream of water of length $v\Delta t$ strikes the wall. Its volume is $Av\Delta t$ and hence its mass is $\rho Av\Delta t$.

Momentum of water hitting the wall in time Δt

$$= (\rho A v\Delta t)v = \rho A v^2\Delta t$$

and the rate of change of momentum $= \rho A v^2$ which, by Newton's second law, is the force.

In all calculations of this kind it is vital to discuss what happens during a definite time period Δt or 1.0 s.

Rocket propulsion

A jet engine collects air but a rocket carries all its propellant materials with it. Figure 2.14 is a schematic diagram of a liquid-fuel rocket motor. The mass of the rocket is m and will decrease as the exhaust gases, ejected at v_0 relative to the rocket, escape at a rate dm/dt. The push F of the rocket on the gases is given by

$$F = \frac{dp}{dt} = \frac{d(mv)}{dt} = v_0\frac{dm}{dt}$$

and by Newton's third law, the *thrust* (the push of the exhaust gases on the rocket) is thus of size $v_0 \, dm/dt$.

At the launch of a space shuttle the gases could emerge from the combustion chamber at over $10^4 \, \mathrm{kg \, s^{-1}}$. If they emerge at $5000 \, \mathrm{m \, s^{-1}}$ then the thrust is $(10^4 \, \mathrm{kg \, s^{-1}})(5000 \, \mathrm{m \, s^{-1}}) = 50 \, \mathrm{MN}$. This must be greater than the initial weight of the rocket, or else it cannot accelerate from the ground. But as it uses fuel the mass of the rocket decreases and so the unbalanced force increases. The acceleration of a rocket therefore increases as it moves upwards.

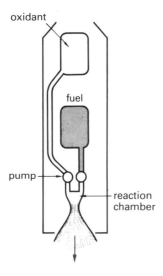

FIGURE 2.14

Exercises on each section of this chapter may be found in the companion textbook, ***Practice in Physics***.

SUMMARY

At the end of this chapter you should be able to:

◆ remember that linear momentum, defined as mass times velocity, is a vector quantity.

◆ understand that linear momentum is conserved in collisions and explosions provided no external forces act during the interaction.

◆ understand the meaning of the words weight and tension and of frictional and contact forces.

◆ describe experiments to demonstrate the validity of the principles of conservation of momentum.

◆ draw sketches when applying the principle of conservation of momentum which show what is happening immediately before and after the interaction.

◆ understand that all forces are pushes or pulls and use the phrase 'the push or the pull of A on B' when describing any particular force.

◆ remember that forces outside the nucleus are either gravitational or electromagnetic in origin and that the electromagnetic forces between surfaces have a perpendicular component called a contact force and a tangential component called a frictional force.

◆ understand that when a body is in equilibrium the sum of the forces acting on it, resolved in any direction, is zero.

◆ use the equation: moment of a force about an axis = the force times the perpendicular distance from the axis to the line of action of the force.

◆ use the principle of moments: for a body which is in equilibrium the sum of the moments of the forces acting on it, about any axis, is zero.

◆ describe experiments to demonstrate the principle of moments.

◆ draw free-body force diagrams for bodies in equilibrium when solving problems involving the principle of moments.

◆ understand that the pull of the Earth on a body, its weight, acts through a point called the centre of gravity of the body.

◆ use the equation for accelerated motion:

$$ma = \text{sum of forces resolved parallel to } a$$

and understand that a newton is equivalent to a $\mathrm{kg\,m\,s^{-2}}$.

◆ draw a free-body force diagram of the accelerated body when solving problems involving Newton's second law.

◆ describe experiments to demonstrate the validity of Newton's second law of motion.

◆ use the impulse–momentum equation $F\,\Delta t = \Delta(mv)$.

◆ understand Newton's first law of motion: the velocity of a body remains constant when the resultant force on it remains zero.

◆ use Newton's second law in the form:

$$\frac{\text{rate of change of momentum of a body}}{} = \frac{\text{resultant force acting on the body}}{}$$

◆ understand Newton's third law: that if a body A exerts a force F on body B then body B exerts a force $-F$ on body A and that this is true at every instant of any interaction.

◆ understand that all animals and vehicles accelerate forwards, by pushing the ground, some water or some air backwards.

3 Energy and its Conservation

We read a great deal about an energy crisis yet we know that energy is conserved. We hear of clean energy, of alternative energy, of low-grade energy. We talk about the efficiency of light bulbs, of rechargeable cells, of diesel and petrol engines. We eat low-calorie foods, pay for therms of natural gas, read about the horsepower of cars. Energy is clearly of major political, social and general public interest; but what precisely is it?

In physics you learn how to make calculations involving energy transfer and once you can calculate you have a good chance of understanding just what it is you are discussing. This is an important generalisation: before scientists could measure energy transfer they were not at all sure what energy was. To test this think of, for example, telepathy and ask yourself what you would need to be told before you could understand it or begin to believe in it.

3.1 Fuels and energy

Throughout history we have burned fuels to cook, to keep warm and more recently to drive steam engines and turbines. It is chemical energy that is stored in coal, oils and natural gas. These form a group of primary energy sources called the fossil fuels. Many attempts have been made to estimate the available store of **fossil fuels** in the Earth's surface. Table 3.1 gives some approximate statistics. The first two columns of numbers show how fossil fuel **reserves** are believed to be distributed between the solid, liquid and gaseous states. The figures for readily and less readily recoverable supplies are, of course, estimates.

The most significant thing to be learned from statistics such as these, however, is that at present rates we are going to run out of oil and gas in the not very distant future. To appreciate this we need to look at quantities of energy. The numbers given in the table are percentages of total amounts of energy as follows:

Recoverable supplies	3000×10^{20} J
Supplies recoverable at no more than twice the present cost of extraction	400×10^{20} J
Fuel used worldwide in 1932	0.7×10^{20} J
Fuel used worldwide in 1992	3.5×10^{20} J

Thus, for example, the total energy value of all recoverable supplies of natural gas is 5% of 3000×10^{20} J $= 150 \times 10^{20}$ J, and is only 10% of 400×10^{20} J $= 40 \times 10^{20}$ J if we extract all those supplies which are readily accessible. The implications of these numbers are immense. Consider petroleum liquids:

Recoverable reserves of oil

$$= \frac{13}{100}(400 \times 10^{20} \text{J}) \approx 50 \times 10^{20} \text{J}$$

World use of oil during 1992

$$= \frac{43}{100}(3.5 \times 10^{20} \text{J}) \approx 1.5 \times 10^{20} \text{J}$$

TABLE 3.1 Fossil fuel reserves (first two columns) and world fuel uses. The figures in each column are percentages for that column. The other sources were mainly wood in 1932 and hydroelectric or nuclear in 1992.

	Readily recoverable supplies (%)	Recoverable at twice present cost (%)	Fuel used worldwide in 1932 (%)	Fuel used worldwide in 1992 (%)
coal and lignite	88	77	72	27
petroleum liquids	7	13	14	43
natural gas	5	10	6	22
other sources	–	–	8	9

Therefore at the 1992 rate of use, the readily available reserves of oil will last us for a further

$$\frac{50 \times 10^{20}\,\text{J}}{1.5 \times 10^{20}\,\text{J year}^{-1}} < 33 \text{ years}$$

and some of those years are already behind us!

This is not as ridiculous an estimate as it may seem for although we may ultimately extract five times as much petroleum liquid as in the calculation (including such sources as tar-sands and shales), the world consumption of oil was rising in 1992 at about 5% per year. Taking everything into account it is estimated that by the year 2040 90% of the available petroleum liquids will have been extracted and used. The total quantities of fossil fuels which exist in the Earth are referred to as fossil fuel **resources**.

Alternative energy sources

The primary energy sources discussed so far have all originated in sunlight which has been stored in fossil fuels. Some other possible sources are:

- direct solar energy
- wind energy
- wave energy
- hydroelectricity
- tidal energy
- geothermal energy

These are often called alternative energy sources. They avoid the atmospheric pollution produced by the burning of fossil fuels and the problems of waste disposal posed by the use of nuclear fuels.

Hydroelectricity, wind and **wave** energy derive from sunlight stored over short periods of time; **tidal** energy is transformed from gravitational energy and **geothermal** energy is transformed from nuclear energy in the Earth's interior and from the movement of continental plates. If the best use were made of all these alternative energy sources then it is conceivable that a large proportion of the world's supply of energy could derive from them, particularly if cheap and efficient methods of transforming **solar** energy directly into electrical energy are developed. In the UK central government is concentrating mainly on wind energy as the best alternative energy source.

Nuclear energy sources

In the twenty-first century nuclear energy will become a major energy source. Nuclear energy is transformed when there are changes in the nuclear composition of matter, just as chemical energy is transformed when matter undergoes chemical changes, such as oxidation. Nuclear energy may be released in two ways:

- by the breakdown of the nuclei of heavy elements, called fission; and
- by the building up of heavier nuclei from light ones, called fusion.

Fission The fission of uranium-235 is used in nuclear power stations. The kinetic energy of the fission products is transferred to internal energy and then used to generate steam and so drive turbines. A power output of 2000 megawatts for a day, that is a total of about 10^{14} J, can be obtained from just over 10 kilograms of uranium. Mined uranium is, however, a limited resource and there are also problems in storing the waste from nuclear fission power stations so that their development is under question in some countries.

Fusion The energy transformed in the fusion of light nuclei, e.g. two deuterium ($^2_1$H) nuclei, has so far not been achieved in a controlled way but it is hoped that the nuclear power stations of the mid-twenty-first century will contain fusion reactors. The only way in which we have so far made use of fusion is in the hydrogen bomb, and you should realise that the Sun is a huge fusion bomb. It is estimated that if we could build a fusion reactor the energy released by using only 1% of the deuterium in the world's oceans would amount to about 500 000 times the energy of the world's initial supply of fossil fuels!

On page 47 the principle of conservation of mass–energy is stated as $\Delta E = c^2 \Delta m$. In chemical reactions Δm is a very, very small fraction of m, the mass of coal etc. which burns; but in nuclear reactions Δm is a much larger fraction (10^{-4} to 10^{-5}) of the mass of the fissionable material. All energy conversions, ΔE, could be expressed as mass conversions using this famous equation. Our Sun is a vast fusion reactor which is converting some four million tonnes of matter into radiant energy *per second*!

3.2 The principle of conservation of energy

Energy can be neither created nor destroyed; it is only ever converted from one form to another.

This widely known statement hides a lot of physics. A man on a bicycle free-wheeling along a level road has kinetic energy. If he brakes suddenly and comes to rest mechanical energy is *not* conserved but the brake blocks and wheel rims become hot. We say that the brake blocks and wheel rims are gaining **internal energy**, or heat energy as it is sometimes called, but this use of the word heat is best avoided. The system, the man on the bicycle in this case, does not lose energy if we take mechanical *and* internal energy into account.

Forms of energy

The forms of energy which are met in this book are:

- mechanical energy: kinetic energy (k.e.); graviational potential energy (g.p.e.); elastic potential energy (e.p.e.); sound energy.

- energy in matter: chemical energy; nuclear energy (this is often wrongly called atomic energy); internal energy.

- electromagnetic energy: electrical energy; magnetic energy; radiant energy.

When mechanical energy is transformed to internal energy the process can be seen in terms of the nature of the contact between the two surfaces or between a surface and a fluid, e.g. air. Before the surfaces come into contact all the molecules of the moving body are travelling in the same direction. As the surfaces slide over one another this energy becomes random energy associated with individual molecules or groups of molecules. The relative motion of the bodies is reduced and the surfaces become slightly warmer.

TABLE 3.2 Conversion efficiency of various devices

	Energy from	Energy to	Efficiency % approx
night storage heater	electrical	internal	100
dry cell or battery	chemical	electrical	90
small electric motor	electrical	mechanical	65
steam turbine	internal	mechanical	45
liquid fuel rocket	chemical	kinetic	45
internal combustion engine	chemical	mechanical	25
fluorescent lamp	electrical	radiant	25
solar cell	radiant	electrical	10
thermocouple	internal	electrical	8

Conversion efficiency

Though it is possible to convert all forms of energy to internal energy *completely*, the extent to which the reverse can be achieved, e.g. in power stations or petrol engines, is limited (Table 3.2).

Devices which transform energy from one form to another are called **energy transducers**. An efficiency of 80% means that 80% of the energy input has been usefully converted to the desired form, the other 20% being converted to internal. When it becomes internal energy it is usually of little further use.

$$\text{Efficiency} = \frac{\text{useful energy output}}{\text{total energy input}}$$

It is often given the symbol η. Efficiency has no units – it is a fraction, always <1, or is expressed as a percentage.

$$\eta = \frac{W_{out}}{W_{in}}$$

EXAMPLE

Describe the energy changes which occur when a man sprints uphill from rest.

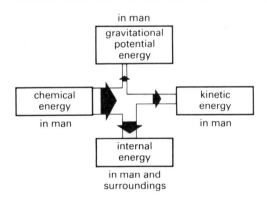

An energy flow diagram, sometimes called a Sankey diagram, is often the best way to answer this sort of question.

Consider another example: a pole vaulter runs along the runway (he then has k.e.), plants his pole and swings into it bending the pole (he and the pole then have k.e. + e.p.e.). The pole straightens as he rises (he then has g.p.e. + k.e.), he just clears the bar (he then has mainly g.p.e.) and falls towards the pit (he then has k.e.) – the stroboscopic photograph of figure 3.1 shows just what is happening.

FIGURE 3.1

It thus seems possible to see the vault as an example of the conservation of mechanical energy. After landing the vaulter and the pit gain some internal energy. But where does his initial kinetic energy as he runs down the runway come from? And does he not use his muscles during the vault itself? In asking these questions we have taken it for granted that we do not expect energy to disappear. On the contrary we expect that **the total energy of a closed system is constant**. For the pole vault, the closed system is the vaulter, his pole and the ground (which is part of the Earth). His initial k.e. came from chemical changes in his body which enabled his muscles to do internal work. We say that he transforms chemical energy to kinetic energy. At all stages of the vault his body transforms chemical energy to internal energy so that he becomes warmer as a result of the effort made in the vault.

3.3 Work and power

What changes the kinetic energy of a body? We find that it changes when an unbalanced force F acts on it and when there is a displacement s in the direction of the force. For instance when F and s are in the same direction the body speeds up; it gains kinetic energy: but when F and s are perpendicular the speed and therefore the kinetic energy of the body are unchanged. To find out what energy 'is' we need first to define **work**.

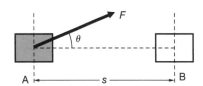

FIGURE 3.2

Consider a body which undergoes a displacement s while a constant force F acts on it – figure 3.2. The **work** W_{AB} *done by the force* is defined by the equation

$$W_{AB} = Fs\cos\theta$$

where θ is the angle between F and s. We can think of $Fs\cos\theta$ either as the force times the resolved part of the displacement in the direction of F or as the displacement times the resolved part of the force in the direction of s. W_{AB} is a scalar quantity. The unit of work is the N m which is called a **joule** (J) after a British scientist, J.P. Joule (1818–98).

If you push a car to try and start it, the push is in the direction of motion, so a push of 200 N for 6 m does 1200 J of work and the kinetic energy of the car increases by 1200 J. A book sliding across a desk, however, slows down. If the frictional push on the book is 2.0 N and it slides 0.8 m the work done on it is −1.6 J, and it loses this amount of kinetic energy.

Calculating work

Here are two slightly more complicated situations which illustrate how to calculate the work done by commonly met forces.

FIGURE 3.3

Firstly think of a girl on a swing – figure 3.3. The push of the seat on the girl on the swing is *always* perpendicular to the direction of motion ($\cos\theta = 0$). The work done by P is thus zero, or, as we sometimes say, P *does no work* on the girl. If the girl swings so as to have moved a vertical distance of 0.8 m, then if $W = 250$ N the work done by W is simply

$$250\,\text{N} \times 0.8\,\text{m} = 200\,\text{J}$$

In this case where the girl follows a curved path, but where the force considered is constant, it is easier to find the resolved part of the displacement.

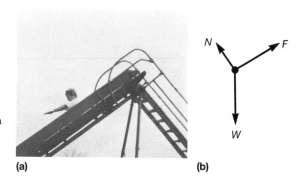

FIGURE 3.4

Secondly, consider a child on a playground slide – figure 3.4. The work done by the contact force N on the child is zero as N is perpendicular to the displacement. The work done by W can easily be calculated: e.g. if $W = 350$ N and it is a 45° slope, then in sliding 10 m the work done is

$$(350\,\text{N} \times 10\,\text{m})\cos 45° = 2500\,\text{J}$$

It does not matter whether you think of the force or the displacement as being resolved in this case. The work done by the frictional push of the slope as the child slides is *negative*, for the angle between F and the displacement is 180°, and $\cos 180°$ $= -1$. Thus if in this case $F = 64\,\text{N}$, the work done by F on the slide is

$$-64\,\text{N} \times 10\,\text{m} = -640\,\text{J}$$

so the total work done is $2500\,\text{J} - 640\,\text{J}$ $= 1860\,\text{J}$, and this is the increase in the child's k.e.

EXAMPLE

A skier is being pulled up a 20° slope by a drag lift which is inclined at an angle of 15° to the slope. The forces acting on the skier are:

the pull P of the drag lift, 320 N
the contact push N of the snow, 620 N
the frictional push F of the snow, 55 N
the pull W of the Earth, 850 N

Calculate the work done by each force as he is dragged 20 m up the slope.

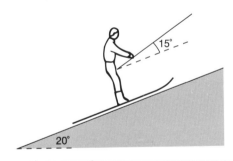

Work done by $P = (320\,\text{N} \times 20\,\text{m}) \cos 15°$
$= 6200\,\text{J}$

Work done by N = zero $(\cos 90° = 0)$

Work done by $F = -55\,\text{N} \times 20\,\text{m}$
$= -1100\,\text{J}\ (\cos 180° = -1)$

Work done by $W = (750\,\text{N} \times 20\,\text{m}) \cos 110°$
$= -5110\,\text{J}$

In this example the total work done on the skier is zero. You can check, by adding the forces vectorially, that the sum of the forces is zero, i.e. he is in equilibrium: he moves at a constant speed and so has a constant kinetic energy.

Work done by a variable force

With a variable force we can calculate the value of $Fs \cos \theta$ for a *small* displacement and then add all the values over the path we wish to consider.

This can be done graphically, so the area between the graph-line and the x-axis of a graph of F against x represents the work done. Consider the case of a spring for which the extending force F is proportional to the extension x,

$$F \propto x \quad \text{or} \quad F = kx$$

where k is a constant called the **stiffness** of the spring. Let us consider the work done on the spring by the extending force.

The graph of figure 3.5 shows how the force varies with x. For an extension from 0 to x_1, we have

W_{OA} = heavily shaded area
$\quad$ = (average force)(displacement OA)
$\quad = \frac{1}{2}F_1 x_1$

Similarly
W_{CB} = lightly shaded area
$\quad$ = (average force)(displacement BC)
$\quad = \frac{1}{2}(F_2 + F_3)(x_3 - x_2)$

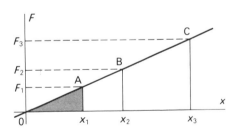

FIGURE 3.5

In both cases the work done is given by the area under the graph of force against extension. This is generally true when both the force and the displacement are along a line.

Power

How quickly a force does work is often more significant than the total amount of work done. The rate at which an agent does work is called the **power** of the agent.

$$P = \frac{W}{t}$$

P is a scalar quantity with unit $\mathrm{J\,s^{-1}}$ which is called a **watt** (W) after the British engineer James Watt (1736–1819) who first developed the steam engine. kW, MW and GW are common multiples and the kWh ($= 3.6 \times 10^6\,\mathrm{J}$) is often used in commercial contexts as a convenient unit of energy. When work W is done by a force F, $W = Fs$, and P can be expressed as

$$P = F\frac{\Delta s}{\Delta t}, \text{ which is}$$

$$= Fv$$

when F and v are in the same direction.

EXAMPLE

A sack of mass 45 kg is lifted 6.0 m in 10 s at a steady speed. What is the average power needed?

Either: work done $= (450\,\mathrm{N})(6.0\,\mathrm{m}) = 270\,\mathrm{J}$

$$\text{power} = \frac{W}{t} = \frac{2700\,\mathrm{J}}{10\,\mathrm{s}} = 270\,\mathrm{W}$$

Or $$\text{speed} = \frac{6.0\,\mathrm{m}}{10\,\mathrm{s}} = 0.60\,\mathrm{m\,s^{-1}}$$

$\Rightarrow$ $$\text{power} = Fv = (450\,\mathrm{N})(0.60\,\mathrm{m\,s^{-1}})$$
$$= 270\,\mathrm{W}$$

3.4 Kinetic energy

Whenever a body has the **ability to do work** we say that the body possesses **energy**. The sort of test we need is to ask the question, 'Could I, in principle, by any series of devices (machines, pulleys, etc.) arrange for the body to push a piston or pull a rope?' A moving truck could readily be made to do either and we say that it has energy of motion – kinetic energy. Water in a stream could do either, perhaps by letting it first turn a paddle wheel.

We define the kinetic energy E_k of a moving body by the equation

$$E_k = \tfrac{1}{2}mv^2$$

Kinetic energy is a scalar quantity and is measured in joules (J).

When work is done on a moving body it gains kinetic energy. The **work–energy equation** states that

$$\genfrac{}{}{0pt}{}{\text{work done}}{\text{on a body}} = \genfrac{}{}{0pt}{}{\text{change in its}}{\text{kinetic energy}}$$
$$W_{AB} = \tfrac{1}{2}mv^2 - \tfrac{1}{2}mu^2$$
$$Fs = \Delta(\tfrac{1}{2}mv^2)$$

and this can be represented by an energy–flow diagram as in figure 3.6.

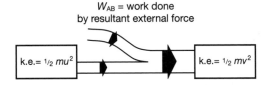

W_{AB} = work done by resultant external force

k.e.= $\frac{1}{2}\,mu^2$ k.e.= $\frac{1}{2}\,mv^2$

FIGURE 3.6

The work–energy equation can be derived as follows:

Suppose that a constant force F accelerates a body of mass m from a speed u. In so doing the body is displaced a distance s.

The work done on the body $= Fs = mas$ (as $F = ma$). From the equations on page 16 we have

$$v^2 = u^2 + 2as$$
so that
$$as = \tfrac{1}{2}v^2 - \tfrac{1}{2}u^2$$
$$mas = \tfrac{1}{2}mv^2 - \tfrac{1}{2}mu^2$$
and
$$W_{AB} = \tfrac{1}{2}mv^2 - \tfrac{1}{2}mu^2$$

Just as experiments cannot 'prove' the principle of conservation of momentum, you cannot do experiments to prove that k.e. $= \tfrac{1}{2}mv^2$. But you can take measurements, e.g. on the speed of a ball fired from a catapult, to show that the expression 'works'.

EXAMPLE

Show that $\tfrac{1}{2}mv^2$ has the unit of energy and estimate the kinetic energy of (a) a sprinter and (b) a high-velocity bullet.

Units of $\frac{1}{2}mv^2$ are $(kg)\left(\dfrac{m}{s}\right)^2 = kg\,m^2\,s^{-2}$

which is $(kg\,m\,s^{-2})(m) = N\,m$ or J

(a) Mass of sprinter, $80\,kg$; maximum speed of sprinter, $10\,m\,s^{-1}$.

$$\text{k.e.} = \tfrac{1}{2}(80\,kg)\left(10\dfrac{m}{s}\right)^2 = 4000\,J = 4\,kJ$$

(b) Mass of bullet, $90\,g = 0.09\,kg$; speed of bullet, $400\,m\,s^{-1}$

$$\text{k.e.} = \tfrac{1}{2}(0.09\,kg)\left(400\dfrac{m}{s}\right)^2 = 7200\,J \approx 7\,kJ$$

In estimates, give the final answers only to one significant figure.

EXAMPLE

The diagram shows a rough slope up which a body of mass $2.0\,kg$ is moving with a speed of $12\,m\,s^{-1}$. The frictional push of the slope on the block is $2.0\,N$. Find the speed of the block when it has travelled $10\,m$ along the slope and has thus risen $4.0\,m$ vertically.

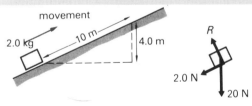

Taking $g = 10\,m\,s^{-2}$, the pull of the Earth on the block is $20\,N$. The work done on the block:
(i) by the pull of the Earth is

$$-(20\,N)(4.0\,m) = -80\,J$$

(ii) by the perpendicular contact push R of the slope is zero
(iii) by the frictional push of the slope is

$$-(2.0\,N)(10\,m) = -20\,J$$

Using
$$W = \tfrac{1}{2}mv^2 - \tfrac{1}{2}mu^2$$
$$-100\,J = \tfrac{1}{2}(2.0\,kg)v^2 - \tfrac{1}{2}(2.0\,kg)(12\,m\,s^{-1})^2$$
$$\Rightarrow \quad 44\,J = \tfrac{1}{2}(2.0\,kg)v^2$$
so that $\quad v = 6.6\,m\,s^{-1}$

3.5 Potential energy

Gravitational potential energy

When a body of mass m moves in the Earth's gravitational field the pull of the Earth, $W = mg$, does work on it. For a vertical displacement h between two levels A and B, the work done by mg is

$$W_{AB} = mgh$$

If g is constant we can write

$$W_{AB} = \text{change in } (mgh)$$
$$W_{AB} = \Delta(mgh)$$

The change in (mgh) is called the change in the body's gravitational potential energy (usually abbreviated g.p.e.).

If the body moves down, the body loses g.p.e.; if it moves up, it gains g.p.e.

EXAMPLE

The diagram is a stroboscopic photograph of a swinging pendulum. The flash rate was $100\,Hz$ and the pendulum $0.325\,m$ long (the photo is printed to one-fifth scale). If the mass of the pendulum bob was $0.10\,kg$, show that the gain of k.e. is equal to the loss of g.p.e.

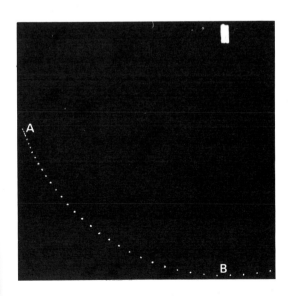

Consider the portion of the path A to B. Take $g = 9.8\,\text{N}\,\text{kg}^{-1}$.

(i) g.p.e.: A is 41 mm on the photo higher than B, so with ×5 scale

$$h = 5 \times 41\,\text{mm}$$
$$= 0.205\,\text{m}$$
$$\Delta(mgh) = (0.10\,\text{kg})(9.8\,\text{N}\,\text{kg}^{-1})(-0.205\,\text{m})$$
$$= -0.210\,\text{J}$$

No other forces do work on the pendulum bob, since the tension in the string is always perpendicular to the motion of the bob.

(ii) k.e.: the speed at A is zero; at B, the speed over 3 flashes, which cover 12 mm on the photo, is:

$$v = \frac{5 \times 12.0\,\text{mm}}{3 \times 0.01\,\text{s}} = 2.00\,\text{m}\,\text{s}^{-1}$$

$$\Delta(\tfrac{1}{2}mv^2) = \tfrac{1}{2}(0.10\,\text{kg})(2.00\,\text{m}\,\text{s}^{-1})^2$$
$$= 0.200\,\text{J}$$

So 0.201 J of gravitational potential energy are lost while 0.200 J of kinetic energy are gained which, to better than 1%, confirms our expectations.

Elastic potential energy

Just as a body can have potential energy by being high up so it can have potential energy when it is on a compression spring or at the end of a stretched rubber cord.

If the spring obeys Hooke's law, $F = kx$, then the work done in stretching the spring from zero extension is given by $W = F_{av}\,x = \tfrac{1}{2}(kx)x$ so $W = \tfrac{1}{2}kx^2$. Thus the work done by the spring on the body when the spring alters its extension is

$$W = \text{change in } (\tfrac{1}{2}kx^2)$$
$$W = \Delta(\tfrac{1}{2}kx^2)$$

The change in $(\tfrac{1}{2}kx^2)$ is called the change in the body's elastic potential energy (usually abbreviated e.p.e.).

EXAMPLE

A very bouncy ball is dropped onto a fixed flat surface. It bounces and is caught. Describe the energy changes of the ball.

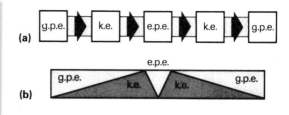

The best descriptions will involve some sort of sketch showing the energy changes in diagrammatic or graphical form. Two possible answers are given.

In (a) in the above example energy transfers are shown as completed steps while (b) shows the relative proportions of the three forms of energy as the bouncing proceeds. In (b) the horizontal axis roughly indicates the time scale. In both it is assumed that mechanical energy is conserved, though in practice this is an approximation as the ball is likely to warm up slightly as it bounces.

3.6 Collisions

An interaction in which mechanical energy is conserved is called a **perfectly elastic** collision or, more simply, an elastic collision. The elastic forces do not convert any mechanical energy to internal energy during the process of deformation and recovery; and the total kinetic energy of the system before the collision is equal to the total kinetic energy of the system after the collision. For such collisions in one dimension the principles of (i) conservation of linear momentum and (ii) conservation of mechanical energy are sufficient to predict the velocities of the bodies after the interaction if their masses and their velocities before the interaction are known. (i) will give a relation between v_1 and v_2, the final velocities of the colliding bodies, and (ii) will give a relationship between v_1^2 and v_2^2.

The case where a body of mass m_1 moving with velocity u which directly strikes a stationary body of mass m_2 is of special interest. If the collision is perfectly elastic we can show that the velocities of the bodies after the collision, v_1 and v_2, depend upon the ratio of their masses. In particular

m_1/m_2	v_1	v_2
$\gg 1$	$\approx u$	$\approx 2u$
1	0	u
$\ll 1$	$\approx -u$	≈ 0

Let us consider the three cases separately.

♦ For $m_1 \gg m_2$, the maximum velocity v_2 of the body which is struck is only twice that of the striking body. Thus in kicking a football or hitting a golf ball the ball cannot have a speed greater than twice that of the foot or club-head respectively.

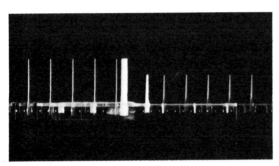

FIGURE 3.7

♦ When the masses are equal the moving body stops and *all* its momentum and kinetic energy are transferred to the body which was struck. Two air-track gliders of equal mass and fitted with repelling magnetic buffers illustrate this, as the stroboscopic photograph in figure 3.7 shows; and so do two snooker balls colliding head-on, providing the moving ball is skidding and not rolling. The energy transfer in this case is the maximum possible.

♦ If the struck body is very massive, i.e. $m_1 \ll m_2$ (e.g. the Earth against which you bounce a very elastic ball), the recoil of the massive body is undetectable and the velocity of the moving body is reversed. The moving body loses very little of its initial kinetic energy as a result of the collision. When an electron hits an atom of helium, the ratio $m_1/m_2 \approx 10^{-4}$, so that an electron could bump elastically into hundreds of helium atoms and keep nearly all its kinetic energy.

EXAMPLE

In a nuclear reactor, high-speed neutrons are slowed down as they bump into the carbon atoms of the moderator. If the masses of a neutron and a carbon atom are m and $12m$ respectively, and assuming that the collisions are perfectly elastic, calculate the speed of a neutron after one collision. Take the initial speed of the neutron to be u and the carbon atom to be at rest.

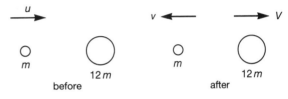

before   after

Using the principle of conservation of momentum:

$$mu = 12mV - mv$$

If the collision is perfectly elastic, then k.e. is conserved:

$$\tfrac{1}{2}mu^2 = \tfrac{1}{2}(12m)V^2 + \tfrac{1}{2}mv^2$$

Beware of + and − signs: momentum is a vector, k.e. a scalar. The physics is now done – the rest is algebra.

Removing m and $\tfrac{1}{2}m$ from the two expressions gives

$$u = 12V - v$$
$$u^2 = 12V^2 - v^2$$

Substituting $v = 12V - u$

$$u^2 = 12V^2 + (12V - u)^2$$
$$\Rightarrow \qquad 0 = 156V^2 - 24uV$$

so that $\qquad V = \dfrac{2u}{13}$ and hence $v = \dfrac{11u}{13}$

i.e. the speed of the neutron is only 85% of what it was.

Many interactions are not elastic. Energy is converted to internal energy during the collision. Most energy is lost when the two colliding objects stick together. Take, for example, a boy jumping into a boat: his k.e. as he jumps in is $\frac{1}{2}(45\,\text{kg})(2.9\,\text{m s}^{-1})^2 = 190\,\text{J}$ and the boat is at rest. After he lands in the boat they move off together with k.e. of $\frac{1}{2}(260\,\text{kg})(0.50\,\text{m s}^{-1})^2 = 3\,\text{J}$. So $157\,\text{J}$ of k.e. has become internal energy.

Or, sometimes, energy is injected during a 'collision' – it is then an explosion. A similar analysis for the skaters on the same page will show that there is *more* k.e. after they separate than there was before. This comes from the chemical energy the man uses to push his partner away from him.

Collisions in two dimensions

Snooker players – figure 3.8 – know from experience how to judge the angles at which colliding spheres of equal mass will separate. If they are not spinning the angle between the paths after the collision will be 90°.

Collisions between nuclei enable physicists to learn a great deal about the properties of the building blocks of matter. This is looked at further in Chapter 25.

FIGURE 3.8

3.7 High-energy physics

When particles are accelerated to high speeds their behaviour is not that predicted by the rules of mechanics developed so far in this book. Figure 3.9 records a pulse of electrons at the beginning and end of an 8.4 m flight path in a vacuum. The time base of the oscilloscope is set at $1.0 \times 10^{-8}\,\text{s cm}^{-1}$, so that the electrons took about $3.3 \times 10^{-8}\,\text{s}$ to travel 8.4 m, which gives a speed of just under $2.6 \times 10^{-8}\,\text{m s}^{-1}$. The mass of an electron is $9.1 \times 10^{-31}\,\text{kg}$ so that, using our rules, the kinetic energy of each electron is

$$\frac{1}{2}mv^2 = \frac{1}{2}(9.1 \times 10^{-31}\,\text{kg})(2.6 \times 10^8\,\text{m s}^{-1})^2$$
$$\approx 3 \times 10^{-14}\,\text{J}$$

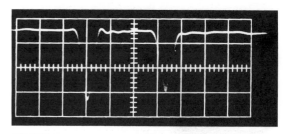

FIGURE 3.9

But the experimenter (the photograph is taken from a film made by W. Bertozzi at M.I.T.) transferred $8 \times 10^{-14}\,\text{J}$ to each electron in accelerating it from rest before measuring its speed as above. The energy was certainly all transferred to the electron and yet $3 \times 10^{-14} \neq \text{J } 8 \times 10^{-14}\,\text{J}$ by any stretch of the imagination. Our rule for calculating kinetic energy *does not work* at very high speeds. Further tests with the same apparatus show that no matter how much energy you give to an electron you cannot accelerate it beyond a speed of $3 \times 10^8\,\text{m s}^{-1}$ (the speed of light in a vacuum).

Special relativity

The rules of high speed or **relativistic mechanics** were developed by Albert Einstein (1879–1955) in the early years of the twentieth century. In them, the kinetic energy E_k of a particle of rest mass m is given by

$$E_k = (m - m_0)c^2$$

where its mass m varies with its speed v relative to the observer in the following way

$$m = \frac{m_0}{\sqrt{1 - v^2/c^2}}$$

c being the speed of light and m_0 the value of m at $v = 0$.

At low speeds, $v \ll c$, these equations lead to $E_k = \frac{1}{2}mv^2$ as in Newtonian mechanics. In fact we need only really use relativistic mechanics when $v > 10^8\,\mathrm{m\,s^{-1}}$. As the highest speed at which anyone has so far travelled is $\approx 10^4\,\mathrm{m\,s^{-1}}$ you can see that the Newtonian rules continue to be adequate for ordinary purposes.

Changes in kinetic energy ΔE_k, can be seen to be given by $\Delta E_k = c^2 \Delta m$, and this is a special case of the more general principle of conservation of mass–energy.

$$\Delta E = c^2 \Delta m$$

Suppose, for example, a nucleus of rest mass $2.0 \times 10^{-25}\,\mathrm{kg}$ emits a γ-ray of energy $2.7 \times 10^{-14}\,\mathrm{J}$. This photon energy is equivalent to a mass of

$$\frac{2.7 \times 10^{-14}\,\mathrm{J}}{c^2} = \frac{2.7 \times 10^{-14}\,\mathrm{J}}{(3.0 \times 10^8\,\mathrm{m\,s^{-1}})^2} = 3.0 \times 10^{-31}\,\mathrm{kg}$$

or about a third of the mass of an electron. As the photon is moving at $3.0 \times 10^8\,\mathrm{m\,s^{-1}}$ and thus has a momentum mc ($=9.0 \times 10^{-23}\,\mathrm{N\,s}$), the nucleus must recoil; so experiments which detect such recoil are confirming the equivalence of mass and energy.

The quantities of energy handled in everyday life are far too small to cause detectable changes of mass. Thus 1 kg water needs $4.2 \times 10^5\,\mathrm{J}$ energy to raise its temperature from 0°C to 100°C. $\Delta E = c^2 \Delta m$ predicts that its mass will increase by less than $10^{-11}\,\mathrm{kg}$ or 0.01 µg!

A further consequence of the equivalence of mass and energy is that matter, that is particles with measurable rest masses, *can be produced from energy*. Figure 3.10 shows a high-energy cosmic ray particle which has hit a nucleus in the Earth's atmosphere. The shower of particles produced has a total rest mass which is many times the rest mass of the incident particle.

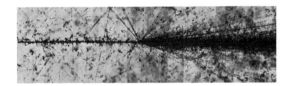

FIGURE 3.10

Exercises on each section of this chapter may be found in the companion textbook, **Practice in Physics**.

SUMMARY

At the end of this chapter you should be able to:

♦ explain that energy sources include fossil fuels and nuclear fuels which are gradually used up and that there are alternative energy sources which are continuously renewable.

♦ understand that although energy is always conserved, the effect of transferring energy is nearly always to produce some which is effectively wasted.

♦ draw energy flow diagrams (Sankey diagrams) to illustrate energy transfer processes.

- remember that the efficiency of energy (or power) conversion is defined as the useful energy (or power) output divided by the total energy (or power) input and is usually expressed as a percentage.

- remember that energy and power are both scalar quantities.

- use the equation: work done by a force = the force times the distance moved in the direction of the force.

- understand that the work done by a force can, in certain circumstances, be zero or have a negative value.

- understand how to calculate the work done by a variable force.

- use the equations for power:

$$P = \frac{W}{t} \quad \text{and} \quad P = Fv$$

- describe experiments to demonstrate the validity of the work–energy equation.

- use the work–energy equation $Fs = \Delta(\frac{1}{2}mv^2)$ where the kinetic energy of a body is calculated as $\frac{1}{2}mv^2$.

- understand that changes in gravitational potential energy close to the Earth's surface are calculated as mgh where h is the vertical displacement.

- describe experiments to measure kinetic energy and elastic potential energy.

- use the principles of conservation of momentum and of mechanical energy to analyse collisions in one dimension.

- use the principle of conservation of mass–energy:

$$\Delta E = c^2 \Delta m$$

where c is the speed of light in a vacuum.

4 Structure of Matter

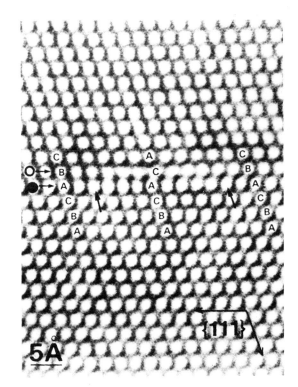

The photograph is an image of the surface of a crystal of gold. You can see the almost perfectly regular arrangement of the atoms, which makes it (like most metals) a crystalline material. This regular arrangement is possible because the atoms are all alike. One defect has crept in: you can see it near the middle of the photograph, where one row of atoms comes to an end: this sort of defect is called a dislocation, and makes the metal much less able to resist forces. The scale superimposed on the photograph shows a distance of 50 nm, i.e. 50×10^{-9} m or 0.00005 mm, so this cannot be a photograph, since the wavelength of light is much larger than the diameter of the atoms. Instead the image was produced using a field ion microscope which records the positions of the atoms in a totally different way. Substances like gold, which contain only one kind of atom, are called elements.

4.1 Atoms and molecules

Atoms have a structure; they can be broken down into smaller parts still. The most obvious of these smaller parts are the **electrons**, **protons** and **neutrons**. Figure 4.1 shows how these parts join together to form typical atoms: the neutrons and protons are in a *very* small nucleus, with the electrons in **orbitals** at different distances from the nucleus.

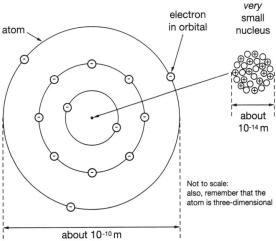

FIGURE 4.1

Atoms are electrically neutral because there are exactly as many electrons as there are protons and each has the *same amount* of electric charge, but the charges are negative and positive respectively. Sometimes an atom gains or loses an electron which makes it a negative or a positive **ion**.

All atoms of a particular element behave **chemically** in the same way, because all have the same number of electrons and protons, and it is the number of electrons which decides what an atom is going to react with. For example, all atoms of iron have 26 electrons and 26 protons. An atom with a different number is not an atom of iron. But not all atoms of iron have the same number of neutrons: although it is usual for them to have 30 neutrons, there are iron atoms with anything from 28 to 32 neutrons. These different versions of iron atoms are called **isotopes** of iron: literally this means they have the same place in the periodic table. They are *physically* different (e.g. the atoms do not have the same mass) but *chemically* they are identical.

You must be careful not to think that atoms are like the diagrams shown in figure 4.1. The diagrams just represent some features of atoms in a way which helps us form a picture of them: they are **models**. It is impossible for us to know, for example, just where the electrons are. A slightly better way of drawing an atom is shown in figure 4.2 but as you learn more and more you must be willing to change your picture of what atoms are like, just as you have already deepened your understanding of some of the ideas which you met earlier in science.

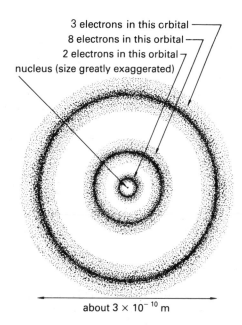

3 electrons in this orbital
8 electrons in this orbital
2 electrons in this orbital
nucleus (size greatly exaggerated)

about 3×10^{-10} m

FIGURE 4.2

The sizes of atoms

How do we know how large atoms are? Nowadays X-rays are used to probe inside crystals: the directions in which the X-rays emerge can be used to tell us how far apart the atoms are. In an earlier kind of experiment, a simpler version of which you can do in your own laboratory, a drop of a fatty acid (an oil) is placed on a water surface.

Each oil molecule consists of a long chain of carbon atoms, but one end has a group of atoms which is attracted to water: the atoms at the other end are not, so the atom stands on end, rather like a fibre in a carpet. The oil spreads to form a roughly circular patch on the surface. Drops of the same size always produce a patch of the same size and so we assume that the oil always spreads until it can spread no further, and that the patch is then one molecule thick.

EXAMPLE

In a particular experiment $1.00 \, cm^3$ of palmitic acid $C_{16}H_{32}O_2$ was added to $499 \, cm^3$ of benzene.

The point of doing this is to produce a dilute solution which is more bulky and therefore easier to measure. The benzene dissolves in the water, so there is no effect on the size of the patch.

One drop of the diluted oil was placed on the water surface, and the diameter of the circular patch was measured to be 222 mm. 100 such drops were found to have a volume of $4.72 \, cm^3$. Estimate the thickness of the layer.

Volume of one drop

$$= \frac{4.72 \times 10^{-6} \, m^3}{100} = 4.72 \times 10^{-8} \, m^3$$

Volume of palmitic acid in one drop

$$= \frac{4.72 \times 10^{-8} \, m^3}{500} = 9.44 \times 10^{-11} \, m^3$$

Area of the patch $= \pi(0.111 \, m)^2 = 3.87 \times 10^{-2} \, m^2$

$$\frac{\text{Thickness of}}{\text{layer}} = \frac{\text{volume of drop}}{\text{area of patch}}$$

$$= \frac{9.44 \times 10^{-11} \, m^3}{3.87 \times 10^{-2} \, m^2} = 2.44 \times 10^{-9} \, m$$

The molecular formula shows that there are 16 carbon atoms, so diameter d of carbon atom

$$= \frac{2.44 \times 10^{-9} \, m}{16} = 1.6 \times 10^{-10} \, m.$$

The masses of atoms

There are several ways of measuring the masses of atoms. Usually the atoms are first ionised. Then they can be accelerated using electric forces. In a **time-of-flight mass spectrometer** a group of ions is accelerated by a p.d. of about 1 kV and travels down an evacuated tube which is about 0.5 m long. At the moment when the group of ions starts a spot starts to move across an oscilloscope screen. The arrival of the ions at the end of the tube is indicated by a 'blip' on the oscilloscope trace, as shown in figure 4.3.

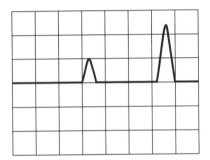

FIGURE 4.3

In this experiment chlorine was used: the more massive chlorine ions arrived later. The time of flight can be deduced from the oscilloscope trace, and hence the speed (= distance/time) calculated. The k.e. of all the ions is the same, and is equal to the accelerating p.d. multiplied by the electronic charge, so the masses of the atoms can be calculated.

Atomic mass units

The masses of atoms are so small that it is not always convenient to give them in kilograms. Instead we often give their masses in **atomic mass units**, for which the symbol is u. By definition, one atomic mass unit is one-twelfth of the mass of the commonest type of carbon atom, which is 19.92×10^{-27} kg. So

$$1 u = \frac{19.92 \times 10^{-27}}{12} kg = 1.66 \times 10^{-27} kg$$

Although this type of carbon atom will have a mass of exactly 12 u, other atoms will not have masses which are whole numbers of u; e.g. the mass

of the commonest isotope of iron is 55.935 u. We say that the **atomic mass** of this isotope is 55.935 u and its **relative atomic mass** (r.a.m.) is 55.935. The r.a.m. has no unit. It is a ratio. It is how many times greater the atomic mass is than one atomic mass unit.

The commonest isotope of hydrogen consists of just one proton and one electron. The r.a.m. of this isotope of hydrogen is 1.0078, i.e. very nearly 1. As the mass of the electron is *much* less than the mass of the proton, the masses of the hydrogen atom and the proton are very nearly the same, so we can say that the masses of the hydrogen atom and the proton are both nearly 1u. The mass of the neutron is also nearly the same as the mass of the proton, so

mass of hydrogen atom ≈ mass of proton
≈ mass of neutron ≈ 1u = 1.66×10^{-27} kg.

Molecules

You know that atoms join together to form **molecules**. For example, atoms of the elements carbon and oxygen can join together to form carbon monoxide (CO) or carbon dioxide (CO_2). A molecule of carbon monoxide is the smallest possible part of this substance. You can also use the word molecule to describe the combination of atoms of the same element: for example, hydrogen gas usually occurs as molecules each of which consists of two atoms of hydrogen, written H_2. The relative molecular mass (r.m.m.) of a molecule is the number of times larger its mass is than u, so, for example, the r.m.m. of methane (CH_4) is $12 + (4 \times 1.0078) = 16.0352$.

The mole

You sometimes use words to describe numbers of items: for example, you talk about a dozen eggs, or two dozen eggs. The word we use to describe large numbers of atoms and molecules is the **mole**, which is 6.02×10^{23}, so if we talked about 3 moles of hydrogen atoms, we would mean $3 \times 6.02 \times 10^{23}$ hydrogen atoms (i.e. 1.81×10^{24} hydrogen atoms). Of course we must say what we are counting: we cannot just say '3 moles of hydrogen'. That might mean '3 moles of hydrogen atoms' or '3 moles of hydrogen molecules', and these are obviously two different things. The mole is said to measure the **amount of substance**. Amount of substance is one

TABLE 4.1 Some calculations on amount of substance

Quantity	Element	R.A.M.	Atomic mass/g	Number of atoms
1 mole	carbon	12	12	6.02×10^{23}
1 mole	oxygen	15.994	15.994	6.02×10^{23}
1 mole	iron	55.847	55.847	6.02×10^{23}
1 mole	uranium	238.030	238.030	6.02×10^{23}
5 moles	carbon	12	60	30.1×10^{23}
5 moles	oxygen	15.994	79.970	30.1×10^{23}

of the base quantities in the SI, so the mole (abbreviation **mol**) is one of the base units.

The number 6.02×10^{23} is called the **Avogadro constant** and is given the symbol L or N_A. It is usually written $L = 6.02 \times 10^{23} \, \text{mol}^{-1}$ to remind us that it is the number of objects per mole. The number is chosen because it is the number of atoms in 12 g of carbon-12 (i.e. The commonest isotope of carbon) but it follows that it is also the number of atoms in the atomic mass of any element. Table 4.1 gives a few simple calculations.

EXAMPLE

How many moles of iron atoms are there in 5.00 kg of iron? The r.a.m. of iron is 55.8.

The fact that the r.a.m. of iron is 55.8 tells us that 55.8 g of iron contain 1 mole of atoms so 5.0 kg ($= 5000$ g) contains

$$\frac{5000 \, \text{g}}{55.8 \, \text{g}} = 89.6 \, \text{moles}.$$

Notice that it is 55.8 grams of iron that contain one mole of atoms. A common mistake is to treat it as if it was 55.8 kilograms.

EXAMPLE

Find the diameter d of a copper atom, if $L = 6.02 \times 10^{23} \, \text{mol}^{-1}$, the r.a.m. of copper is 63.5 and the density of copper is $8.93 \times 10^{-3} \, \text{kg m}^3$.

You cannot give a precise answer to this question without knowing how the atoms are packed together, but you can give a rough idea by thinking of the atoms as packed together in a simple 'cubical' arrangement, like lumps of sugar in a box. Then the volume of one of these 'cubical' atoms is d^3.

One mole of copper atoms has a mass of 63.5 g, so 6.02×10^{23} copper atoms have a mass of $63.5 \times 10^{-3} \, \text{kg}$

and each atom has a mass of $\dfrac{63.5 \times 10^{-3} \, \text{kg}}{6.02 \times 10^{23}}$

$$= 1.05 \times 10^{-25} \, \text{kg}$$

$$\text{Volume of atom} = \frac{\text{mass}}{\text{density}} = \frac{1.05 \times 10^{-25} \, \text{kg}}{8.93 \times 10^3 \, \text{kg m}^{-3}}$$

$$= 1.18 \times 10^{-29} \, \text{m}^3.$$

So $\quad\quad\quad\quad d^3 = 1.18 \times 10^{-29} \, \text{m}^3$

and $\quad\quad\quad\quad d = 2.3 \times 10^{-10} \, \text{m}.$

Because the atoms are packed together more closely than we have imagined, the actual diameter is larger than this estimate: it is $2.6 \times 10^{-10} \, \text{m}$.

4.2 Atomic forces

Atoms attract each other

In a tug-of-war contest the rope does not usually break. At any cross-section (figure 4.4) the atoms are being pulled by the neighbouring atoms in the other piece of rope, and the total of all these atomic pulls is equal to the pull of the tug-of-war

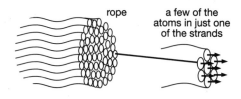

rope   a few of the atoms in just one of the strands

FIGURE 4.4

team on the rope. How do atoms attract each other? There are different methods, all involving the outermost electrons. Here are some examples.

◆ A *sodium atom* has one electron in its outermost orbital. If it comes near a fluorine atom, which has 7 electrons in its outermost orbital, the sodium atom gives the electron to the fluorine atom. Now the sodium atom has become a positive sodium ion, and the fluorine atom has become a negative fluorine ion, and they attract each other. This is called **ionic** bonding.

◆ A *hydrogen atom* has one electron in its outermost orbital; it needs another to make it complete. If another hydrogen atom is brought near they share their electrons: the electrons act like a kind of glue holding the positive hydrogen ions together. This is called **covalent** bonding.

◆ *Metals* are elements which have one or more of their outermost electrons bound only loosely to the atoms, so the electrons are free to move amongst the positive ions, and hold them together. It is like covalent bonding, but now the electrons hold together many ions, not just two. This is called **metallic** bonding.

◆ *Neon atoms* have outermost orbitals which are complete and the atoms therefore cannot be held together by any of the methods already mentioned. But from time to time the electrons are not symmetrically arranged, and then the atom is said to be **polarised**, and can

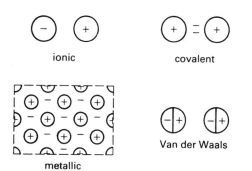

FIGURE 4.5

exert electric forces on neighbouring atoms and hold them together. This is called **Van der Waals** bonding.

These four possibilities are summed up in figure 4.5. It must also be possible for atoms to repel each other. You know this is true because you have to squeeze very hard to make solids and liquids take up less room.

How forces vary with distance

The cause of forces between molecules is the same as the cause of forces between atoms and from now on we shall talk about inter**molecular** forces. We need to think how these forces must vary with the distance apart of the molecules. What do you know from your everyday experience?

◆ Molecules in a solid must, on average, be *in equilibrium*, so the sum of all the attracting forces must be equal to the sum of all the repelling forces. If in equilibrium they are a distance r_0 apart, the resultant force on them then must be zero, and you can plot this point on the graph in figure 4.6(a), on which a resultant repelling force is positive, and a resultant attracting force is negative.

◆ If you pull on the ends of a rod to make it longer, the molecules – which are now further apart – are still in equilibrium, so must be pulling on each other with larger attracting forces. So you know that when the molecules are further apart, the intermolecular attracting forces increase. You can also push inwards on a rod and make it shorter: the greater the push, the greater the compression, so just as with attracting forces, we know that the intermolecular repelling forces must increase when the molecules are closer together, so you can draw a short length of line on the graph in figure 4.6(b). must slope as shown, because when the distance of the molecules is less than forces are repelling; and when gr r_0, the forces are attracting.

◆ The amount of extension solid is proportional to

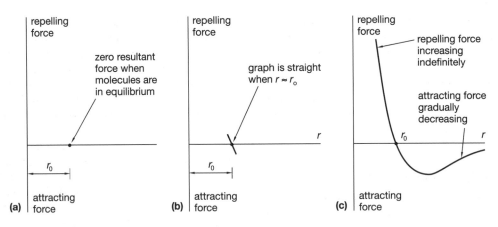

FIGURE 4.6

the force is not too great (you know this as Hooke's law). So you should draw the short length of line *straight* provided r does not differ much from r_0.

◆ You also know that you cannot break a solid rod by pushing inwards on it more and more, so the *repelling* forces must increase indefinitely as the distance apart of the molecules decreases. But if you pull on the ends of the rod with larger and larger forces, the rod breaks, so there *is* a limit to the size of the *attracting* forces. You can pull on the molecules with a larger force than they can pull on each other. So you can complete the graph by drawing the curved parts shown in figure 4.6(c).

Potential energy

Imagine that one of the molecules is fixed in position, and another one is placed a distance r away. Suppose that in this position the second one has *zero* potential energy. That is a choice you are free to make, just as you may find it helpful to be able to say that a lump of matter has zero gravitational potential energy when it is on the floor of your laboratory, even if it is on the first floor. It is only *differences* in potential energy that ever matter.

If r is greater than r_0, the molecules attract each other, and if left to themselves the second one will accelerate towards the first, gaining kinetic energy and losing electric potential energy, as shown in figure 4.7(a), where the potential energy is

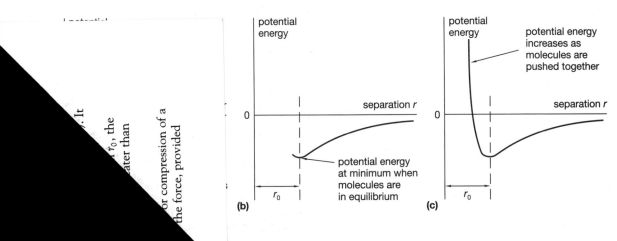

becoming more and more negative. Its speed will stop increasing when $r = r_0$, because then the molecules are in equilibrium (figure 4.7(b)), and from then on it will be slowing down, losing kinetic energy and gaining potential energy. The potential energy will therefore be a minimum when $r = r_0$, and when r is less than r_0 the potential energy will increase indefinitely as shown in figure 4.7(c).

Using the potential energy curve

We worked out the shape of the potential energy curve by thinking about just two molecules, but the same sort of argument could be used for one molecule surrounded by many more, so we can use the curve to help us understand better what is happening to molecules in a solid.

You know that the molecules of a solid are vibrating: they are oscillating about their equilibrium positions, rather like marbles rolling backwards and forwards in the bottom of a bowl, with their kinetic energy and (electric) potential energy continually being converted from one kind to the other, as shown in figure 4.8.

You can see that if a molecule is given more kinetic energy, it can move further from its equilibrium position. There must come a point where it has so much k.e. that it can escape from the other molecules altogether. This is what happens when you heat a solid and it melts and evaporates. Figure 4.9 shows the extreme positions A and B for molecules in a solid at four different temperatures: T_1 is the lowest temperature, T_4 is the highest. You can see that the average position M moves outwards as the temperature increases because the graph is not symmetrical, so the average separation of the molecules increases. This is why solids expand when they are heated.

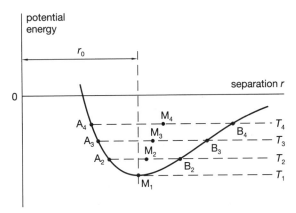

FIGURE 4.9

4.3 The solid phase

Crystals

The molecules in a solid are often arranged in a regular manner: much more often than is sometimes realised. Such solids are called **crystals**, and we describe them as **crystalline**. There may be a million or more molecules occurring in regular rows and the crystal is said to have **long-range order**. Figure 4.10 is a photograph showing how a crystalline substance, mica, can easily be split. The smoothness of the face is not the result of careful sawing and polishing: the surface and edges were produced by **cleaving** a larger (and probably rather

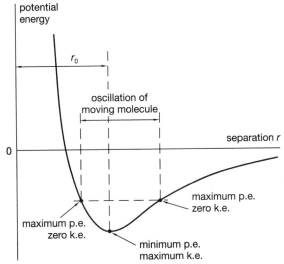

FIGURE 4.8

irregular) crystal. Figure 4.10 also illustrates the cleaving of a crystal: the edge of a razor blade is pressed against the crystal where the cleavage is expected to form, and only a small force is then needed to separate one part of the crystal from another. These newly formed surfaces occur naturally, and it is these *naturally-occurring plane faces* which are the distinguishing feature of crystals. Crystals seem to be rare, and large ones will be found only in geological museums (where they are worth seeing) but small ones are very common. A close look at some grains of salt, or sugar, shows that these substances, like all salts, and metals and some non-metals, are crystalline. Indeed there are only a few types of substance (the glasses and the polymers) which are not crystalline. The smallness of most crystals usually arises because in a solidifying liquid the growth of particular crystals is hindered by the closeness of other growing crystals, or the edge of the vessels. But crystals of side about 10 mm can be grown by anyone with a little care and patience.

glass dish, and a piece of glass tubing with its end drawn into a jet is held at a constant depth below the surface of the liquid and connected to a constant-pressure source (e.g. a bench gas supply). Figure 4.11 shows the apparatus and the result (which is best displayed by placing the dish on an overhead projector). The bubbles collect on the surface in a pattern in which they fit together as closely as possible. The bubbles make a good model of molecular behaviour since in their equilibrium positions they, like the molecules, are under the influence of two opposing forces – the *attracting* surface tension forces and the *repelling* air pressure forces.

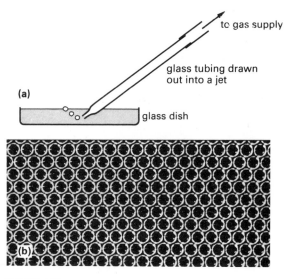

FIGURE 4.11

FIGURE 4.10

Another characteristic of crystalline solids is their *sharp melting-points*. Indeed, the melting-points are used as fixed points in defining temperature scales and are quoted to five significant figures.

Molecular patterns in a crystal

Crystallography, the study of the different arrangements of atoms in crystals, is a science of its own, and here we shall give only some indication of how different crystalline shapes arise, using models to help us. One such model is the **bubble raft**. A detergent solution is placed in a shallow

Figure 4.12(a) shows a very simple (and therefore probable) way for some bubbles to come together to form a group with straight boundaries and you can see that the angles between the boundary edges are either 60° or 120°. Obviously the bubbles will serve as a good model only for single atoms, and we can imagine that the angles would not be as simple in a crystal which is formed from molecules each containing several atoms. Even the bubbles can come together to produce boundaries which make angles other than 60° and 120°, as you can see in figure 4.12(b). The problems of the different shapes which can be formed using the same building blocks are not easy

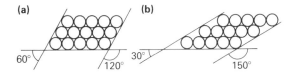

FIGURE 4.12

FIGURE 4.14

to unravel, and this is not the place to do so. Looking at the bubble raft you can at least more easily believe that molecules will fit themselves together in a particular way to give smooth surfaces and characteristic angles between those surfaces.

A three-dimensional model can help us to see the different arrangements when molecules come together and to visualise the many kinds of planes of atoms within a crystal. This spacing of layers is of great importance in X-ray crystallography, and the use of 3D models is a great help to anyone who is studying the way in which the X-rays are reflected.

Polycrystalline materials

We have already mentioned that when a liquid freezes the crystals which form are likely to be small because crystal growth has started simultaneously in many different places. A metal or a salt in the solid phase will therefore consist of many regions in which the orderliness is perfect, but there is no relation between the orderliness of one region and the orderliness of the next. These regions are called **grains**. The bubble raft model illustrates this very well, and in figure 4.13 you can

clearly see the grain boundaries. Occasionally the grain boundaries can easily be seen: for example, on the surface of a galvanised iron bucket or a worn brass door handle or, as in the photograph of figure 4.14, on the surface of an aluminium sheet. The existence and size of the grains has an important bearing on the mechanical properties of materials.

Glasses

There are some substances which, unlike crystals, pass from the liquid to the solid phase and vice versa almost imperceptibly: these glasses

- do not have a sharp melting-point

- do not undergo a sudden change in density

- do not release energy when they melt

These differences show that although glasses are often shiny, they are quite different from crystals. Their molecules are not arranged in any sort of pattern, and they cannot be cleaved: they simply shatter if you try to cleave them.

Polymers

Very many organic materials, and an increasing number of man-made materials, are formed from molecules which consist of very long chains of atoms. One of the simplest of these polymers (from the Greek words for many-parts) is poly(ethene) or polythene. As its name suggests it is formed by

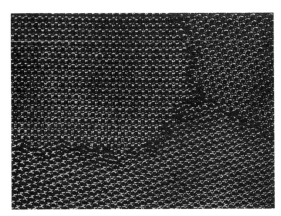

FIGURE 4.13

linking together many molecules of ethene (C_2H_4) to form a long chain of perhaps 20 000 carbon atoms to each of which are attached two hydrogen atoms. Its length might therefore be of the order of 10^{-6} m, and its relative molecular mass about 300 000. Figure 4.15(a) shows a photograph of a model of part of this molecule. However, the molecule is not rigid, and in practice the chain will be twisted (figure 4.15(b)), doubled back on itself, and in something of a tangle. Indeed it is the tangling of the chains which gives a polymeric material much of its strength.

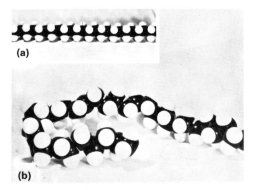

(a)

(b)

FIGURE 4.15

Polymers include many natural materials such as hair, wool, silk, cotton and cellulose (the main ingredient of wood); but since about 1920, when polymer science began to expand, many new polymeric materials have been created, some of which serve as alternatives to natural polymers and others provide us with totally new materials.

Every home, school, office and factory contains many examples of these new materials: e.g. Perspex, polythene, PVC, polyurethane foam, nylon, polystyrene, Melamine. They are called **plastics** because they are moulded plastically into the required shapes. It is hard to imagine everyday life without them. Their properties differ so widely that we cannot make any deductions about polymers from looking at them as a group: what we can do is to explain their properties in terms of their molecular structure.

Thermoplastics

Perspex is a typical **thermoplastic** material. Such materials exist in a glassy state at low temperatures, soften and become rubbery at a higher temperature (the glass transition temperature), and at still higher temperatures soften and melt. In the softened state they can be moulded. This happens because at low temperatures the long-chain molecules are trapped in fixed positions by the weak Van der Waals forces, but when energy is supplied to the substance the molecules gain enough energy to be able to move.

Rubber is a naturally-occurring material which can be thought of as a thermoplastic material. It seems to be different from Perspex only because room temperature is above the glass-transition temperature for rubber ($-75°C$) and below the glass-transition temperature ($100°C$) for Perspex.

Thermosets

Some other polymers are hard and rigid: when heated they do not soften, but eventually decompose. To create these materials, chemicals are mixed to produce the polymers and then the polymer is either heated strongly, or a catalyst is added, so that chemical (covalent) bonds or cross-links are formed between the polymeric chains (compare this with the weak Van der Waals forces in thermoplastics). This chemical action is not reversible: the material cannot be softened by re-heating. The amount of energy needed to break a covalent bond cannot be obtained by heating.

A good example of this is the **vulcanisation** of rubber. Untreated (natural) rubber is a flexible thermoplastic material which would not be a useful engineering material: the very weak forces between the atoms in different chains would mean that there was very little resistance to deformation, and it would not return to its original shape when the force was removed. So sulphur atoms are added to the rubber and the covalent bonds between sulphur atoms in neighbouring chains makes the rubber much more rigid, and suitable, for example, for vehicle tyres. Varying the proportion of sulphur to rubber is what gives tyres the different properties which are needed for different conditions in motor racing. In this way a thermoplastic material can be converted into a thermoset. Adding even more sulphur produces the material called ebonite: a hard, shiny electrical insulator which has no 'rubbery' properties.

4.4 The liquid and gaseous phases

The liquid phase

A substance in the liquid phase is usually only slightly less dense than it is in the solid phase (though water, being *denser*, is a notable exception). A typical metal, for example, is only 3 per cent less dense as a liquid. A liquid is not rigid, but it is hard to compress. This all suggests that the molecules are nearly as close together in a liquid as they are in the solid phase, but that they do not just vibrate about a fixed position. They must be free to move about but presumably they still vibrate, as they are still close enough to their neighbours to be within range of the attracting and repelling forces.

There will not be the long-range order which exists in crystals, where there might be rows of about a thousand regularly arranged molecules (in each direction), but there may well be **short-range order** extending over a few near neighbours. The study of liquids has not advanced nearly as far as that of solids and gases. Much of the work now being done consists of making models of liquids in order to investigate how far from a particular molecule other molecules might be expected to be.

We can think of a molecule in a liquid as being like a man in a dense football crowd. Individual members of the crowd are never still. They are always moving backwards and forwards, but never moving very far, unless they make a determined effort to change neighbours or change places.

The gaseous phase

A substance in the gaseous phase is much less dense than it is in either the liquid or solid phases – perhaps by a factor of 1000. It is also highly compressible, and fills the whole of any closed space in which it is put. We can also observe:

- **Brownian motion**, first noticed in 1827 by Robert Brown, though his first observations were on small particles suspended in a *liquid*. This can be seen through a microscope when ash particles are blown into an enclosed space. The ash particles have a random, jerky motion: see figure 4.16, where the positions

of one ash particle, at intervals of one second, are shown. The particle does not move straight along the lines. Between the positions shown it will travel a random path similar to that shown in the diagram. The particles move as if they were being pushed about by something we cannot see, which we believe to be air molecules. The molecules are too small to see, even under a microscope, so presumably they must be moving fast (and randomly) to be able to produce that effect on the much more massive ash particles.

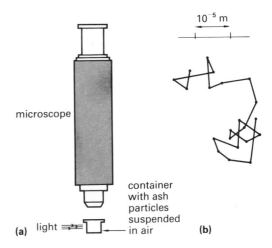

FIGURE 4.16

- *The diffusion of one gas into another.* Diffusion does not mean the bodily mixing of gases, but the intermingling of the molecules of a gas, one molecule at a time. It can be studied by releasing liquid bromine at the bottom of a vertical glass tube containing air. The liquid evaporates, producing the brown bromine gas, which is seen to make its way slowly up the tube. We know that the bromine molecules must actually be moving at high speeds because when released into a similar evacuated tube, they reach the top in an immeasurably short time. So when diffusing they must be taking a very irregular path, colliding with many air molecules and each other, since their progress in any direction is so slow.

The **mean free path** (the average distance travelled between collisions) of the bromine molecules is of the order of 10^{-7} m and since they are typically moving at about $200\,\mathrm{m\,s}^{-1}$ they must be making of the order of 10^9 collisions each second. It can also be worked out that the molecules must be about 10^{-9} m apart, i.e. about 10 times the diameter of a molecule, as shown in figure 4.17. So each molecule is, on average, surrounded by a volume of about 1000 times its own volume (remember that figure 4.17 is only a two-dimensional view of a three-dimensional situation). A substance in gaseous form can therefore be expected to occupy a *much* larger volume than it does in solid or liquid form.

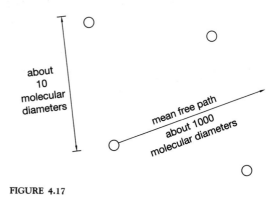

FIGURE 4.17

Exercises on each section of this chapter may be found in the companion textbook, *Practice in Physics*.

SUMMARY

At the end of this chapter you should be able to:

◆ remember that an atom consists of electrons, protons and neutrons.

◆ remember that the protons and neutrons are contained in the nucleus, which has a diameter which is about 1/10000th of the diameter of the atom.

◆ understand what is meant by an isotope.

◆ remember that the diameter of an atom is about 10^{-10} m.

◆ understand what is meant by an atomic mass unit.

◆ understand that the mass of a mole of carbon atoms is exactly 12 g.

◆ understand that the mass in grams of one mole of atoms of any other element is equal to the element's atomic mass.

◆ convert a number of moles into a number of particles, using the Avogadro constant, and vice versa.

◆ draw a graph to show how the resultant force on a molecule varies with distance.

◆ draw a graph to show how the potential energy of two molecules varies with their separation.

◆ understand what is meant by saying that a material is crystalline.

◆ describe the nature of glasses and polymers and understand in what ways they differ from crystalline materials.

◆ explain how Brownian motion, and diffusion, provide evidence for the idea that the molecules of gases are moving fast and randomly.

5 Performance of Materials

This photograph shows the bridge across the River Thames at Dartford (opened in October 1991). You can see that the deck of the bridge is supported by cables running from the towers at each end of the bridge. This is called a cable-stayed bridge: its construction is cheaper than a conventional suspension bridge, which has large cables draped from two towers at each end of the bridge. Here the cables are up to 164 mm in diameter, and consist of many steel wires which are each a few mm in diameter, spirally-wound like the strands in a rope. Until recently it was thought that it would not be possible to build cable-stayed bridges with spans of more than 350 m, but the main span of the Dartford bridge is 450 m long. This type of construction is being used for the second River Severn crossing, which will have a main span of 456 m. A bridge of 856 m span is planned across the Seine in France.

The engineers who build bridges like this need to know how the steel will behave when it is stretched: will it be strong enough? How much will the bridge sag under its own weight and when there is traffic on it? How well will it resist wind forces? This chapter deals with the strength, stiffness, elasticity and toughness of materials.

5.1 Materials in tension

In the last chapter we dealt with the *structure* of materials; in this one we shall deal with their *performance*. A solid can be formed into many different shapes, and have forces exerted on it in many different ways: we shall consider only materials in the form of wires or rods or strips which are pulled lengthways, like the cables in a cable-stayed bridge or a suspension bridge.

Tension

A specimen which is stretched is said to be in a state of tension, and the forces are called tensile forces. Figure 5.1 shows a horizontal wire running over two pulleys: equal weights W hang from the ends of the wire. We assume that the pulleys are frictionless. Imagine the wire cut at any point such as A. The pull of either piece of wire on the other must be the same size (by Newton's third law) and equal to W, so throughout the wire each part pulls on the other with a force W. We say the tension in the wire is W: the tension is the force with which each part of the wire pulls the other (note that the

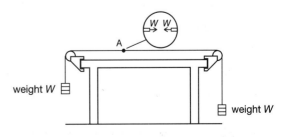

FIGURE 5.1

tension is not $2W$). Even if the cross-sectional area of the wire was different in different places, the tension would still be the same throughout the wire.

Strength and stress

The most obvious question to ask about any wire is 'how strong is it?' That is, what is the largest force with which it can be pulled without breaking it? The force will not depend on the length of the wire, but it will depend on the cross-sectional area: twice as much force will be needed to break a wire which has twice the cross-sectional area. So if we want to compare the strengths of two *materials* we need to consider the force *per unit area*. We define the **tensile stress** σ in a specimen by the equation

$$\sigma = \frac{F}{A}$$

where F is the tensile force and A the cross-sectional area. You can see from the equation that the unit of stress is the $N\,m^{-2}$, which for convenience is called the **pascal** (abbreviation Pa).

When materials are stretched their cross-sectional area decreases: eventually quite considerably. It is not usually easy to measure the new cross-sectional area, so that the **nominal** or **engineering** stress is used. This is defined by the equation

$$\text{nominal stress} = \frac{\text{tensile force}}{\text{original cross-sectional area}}$$

The **ultimate tensile stress** σ_{max} of a material (i.e. the maximum stress before the material breaks) is defined in a similar way, i.e.

$$\text{ultimate tensile stress} = \frac{\text{maximum tensile force}}{\text{original cross-sectional area}}$$

At the point where a wire thins the actual stress will be greater than the nominal stress.

EXAMPLE

A crane uses a cable consisting of 20 strands of steel wire to support a load. If the ultimate tensile stress of the steel is 300 MPa, what will the diameter of each strand need to be if the cable is to support a load of 25 tonnes, and if, for safety reasons, the tensile stress in the cable is not to exceed one-quarter of the ultimate tensile stress? Take $g = 9.8\,N\,kg^{-1}$.

$$\sigma = \tfrac{1}{4}(300\,MPa) = 75 \times 10^6\,N\,m^{-2} \text{ and}$$

$$F = \text{the weight of 25 tonnes}$$
$$= (25 \times 10^3\,kg)(9.8\,N\,kg^{-1})$$
$$= 2.5 \times 10^5\,N.$$

So if A is the cross-sectional area of the cable,

$$A = \frac{F}{\sigma} = \frac{2.5 \times 10^5\,N}{75 \times 10^6\,N\,m^{-2}} = 3.3 \times 10^{-3}\,m^2$$

and the area of one strand $= \tfrac{1}{20}(3.3 \times 10^{-3}\,m^2)$
$$= 1.67 \times 10^{-4}\,m^2.$$

If d is the diameter, $1.67 \times 10^{-4}\,m^2 = \tfrac{1}{4}\pi d^2$, so

$$d = 1.5 \times 10^{-2}\,m = 15\,mm$$

Strain

All wires and rods stretch when they are pulled. The extension depends on the material and the force, but it also depends on the original length of the specimen, so to make fair comparisons we calculate the fractional extension or **tensile strain** ϵ which occurs. This is defined by the equation

$$\text{tensile strain } \epsilon = \frac{\Delta l}{l}$$

where Δl is the extension which occurs from an original length l. Because it is a fractional extension, tensile strain has no unit.

Hooke's law and the Young modulus

For many materials the strain is proportional to the stress applied, at least for small values of stress: this is known as **Hooke's law**. A graph of stress against strain is a straight line passing through the origin. The point at which the material ceases to obey Hooke's law is known as the **limit of proportionality**. For materials which obey Hooke's law we define the **Young modulus** E by the equation

$$E = \frac{\text{tensile stress}}{\text{tensile strain}}$$

In symbols, $E = \dfrac{\sigma}{\epsilon} = \dfrac{F}{A} \div \dfrac{\Delta l}{l} = \dfrac{Fl}{A\Delta l}$.

You should be able to see that its unit must be the same as the unit of tensile stress, i.e. the $N\,m^{-2}$ or pascal. The greater the value of E, the less the material stretches when a given stress is applied, i.e. the *stiffer* the material is, and E is sometimes called the **tensile stiffness** of the material. It is often important that a material should be stiff. You would not feel very comfortable walking or driving over a bridge which sagged under your weight, even if it did support you!

EXAMPLE

If, in the previous example, the Young modulus of steel is 200 GPa, what will the strain be under these conditions? If the length of the cable is 45 m, how much will it stretch?

$$\epsilon = \frac{\sigma}{E} = \frac{75 \times 10^6\,N\,m^{-2}}{200 \times 10^9\,N\,m^{-2}} = 3.8 \times 10^{-4}.$$

The extension $\Delta l = \epsilon l = (3.8 \times 10^{-4})(45\,m)$

$$= 1.7 \times 10^{-2}\,m = 17\,mm.$$

Distinction between strength and stiffness

The strength and stiffness of a material are independent, and different, properties. Figure 5.2 shows stress–strain graphs for two materials A and B. Each material breaks at the final point on its graph. For the same strain, A needs more force, so A is stiffer than B, but it takes more force to break B, so B is stronger than A.

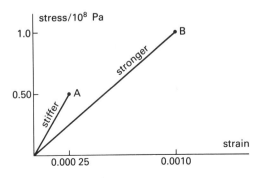

FIGURE 5.2

EXAMPLE

An aluminium wire of length 1.0 m has a cross-sectional area of $0.10\,mm^2$ and a chromium wire of length 2.0 m has a cross-sectional area of $0.050\,mm^2$. The Young moduli of aluminium and chromium are 70 GPa and 280 GPa respectively. (a) When the same mass is hung from each wire, which will have the larger extension? (b) Each metal has the same ultimate tensile stress: if increasing loads are applied, which will break first?

(a) The chromium wire has twice the length and half the cross-sectional area. These factors together would make its extension four times greater. But the Young modulus of chromium is four times greater, so that it is four times stiffer, and therefore the extension is just the same as for the aluminium.

(b) Now the length and Young modulus are irrelevant, and the two ultimate tensile stresses are the same. All that matters is the cross-sectional area. The aluminium wire has the greater cross-sectional area so it can withstand the greater force.

Elasticity

Most materials used in engineering need to be strong and stiff: they also need to be **elastic**, i.e. they should return to their original shape when the stress is removed. If you again imagine yourself driving over a bridge, you would expect it not to break (i.e. to be strong), not to sag very much (i.e. to be stiff) and to be the same shape after you had crossed it. Most materials are elastic at least for small stresses: we say that the **elastic limit** is the point on the stress–strain graph where elastic behaviour ceases. For greater stresses the material behaves **plastically** and retains at least some of the strain when the load is removed: it has become a different shape. The elastic limit is often close to (but need not coincide with) the limit of proportionality.

Beams and cantilevers

Any structural member, whether horizontal or vertical, is liable to have forces exerted on it which will bend it and therefore stretch and compress it. Figure 5.3 shows a beam and a cantilever and the regions where these structures are in tension and compression. Wood and steel are strong in tension and can be used for beams. Concrete and stone are not strong in tension, and so concrete needs to be reinforced with steel; and the roofs of Greek temples needed to be supported by many columns, close together, because the span of a stone beam cannot be very great.

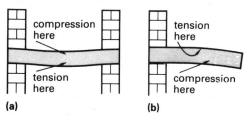

(a)   **(b)**

FIGURE 5.3

Stress–strain graphs

We shall now look at the performance of four different specimens chosen to represent most of the commonest possibilities: a copper wire, a glass rod, a rubber cord and a polythene strip. Figure 5.4 shows graphs of stress against strain for the four specimens. X marks the point where the specimen breaks. There are two graphs for the copper wire since the first of the two graphs cannot show enough detail when the strain is small. Before considering the graphs individually, there are some general points you should note.

◆ These are *typical* results: the behaviour of a specimen depends on its purity, its previous treatment, and many other factors.

◆ The scales on the axes cover very different ranges.

◆ We cannot usually tell from the graph whether the specimen is behaving elastically: both copper and polythene (of the materials illustrated) behave plastically if the stress is great enough.

◆ The slope of the graph gives the stiffness, or Young modulus, of the material. We can see that different materials have very different,

and not always constant, stiffnesses. Table 5.1 gives the value of the Young modulus E and also values of the ultimate tensile stress σ_{max} for the specimens discussed.

TABLE 5.1   Young modulus (E) and ultimate tensile stress σ_{max} for the specimens discussed

Specimen	E/MPa	σ_{max}/MPa
glass (rod)	75 000	75
copper (wire)	130 000 where constant	200
rubber (cord)	≈0.7 average value	4
polythene (strip)	200 where constant	20

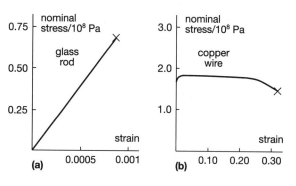

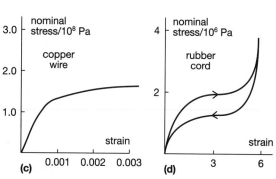

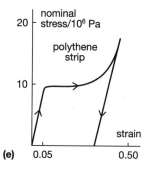

FIGURE 5.4

(a) The glass rod

You can see from figure 5.4 and the table that glass is about half as strong as our representative metal (copper in wire form) and about half as stiff: it also behaves elastically. It would therefore seem to be a useful structural material, but the abrupt end to the graph shows that it is brittle, and is therefore not tough. Toughness is yet another property of materials, and is distinct from strength and stiffness. We shall discuss this further in the next example. Some metals (like cast iron) behave like glass in this brittle way.

(b, c) The copper wire

The graph for the copper wire is more complicated:

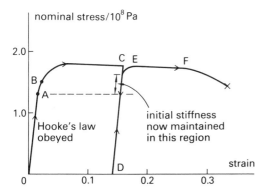

FIGURE 5.5

see figure 5.5, which shows the graph in more detail. For small stresses the copper wire obeys Hooke's law: in figure 5.5 A marks the limit of proportionality. Up to and slightly beyond this point the wire behaves elastically, until the elastic limit B is reached. The stiffness then decreases considerably. In fact because the cross-sectional area is decreasing, less force is now needed to produce further extension. This is why the graph (which is a graph of *nominal* stress) from now on has a negative, and not a positive, slope. The *actual* stress is still increasing. If the stress is removed at a point such as C the wire does *not* return to its original length, but the graph follows a line parallel to AO, and retains a permanent strain OD. If a stress is then applied to the wire again, the graph retraces the line from D towards C until it curves round to rejoin the curve which would have passed direct from C to E if the original loading had been maintained.

You can see that when reloaded the wire obeys Hooke's law again up to greater stresses than before, and its stiffness remains at its original high value up to greater stresses. The wire is said to have been **work-hardened**, and this obviously increases the usefulness of a material.

The change in direction of the curve at F is the result of a constriction or **neck** occurring at one point in the wire. At this particularly narrow part the actual stress is very high, and here the wire will eventually break. The plastic behaviour of the copper makes it a tough material: copper objects are not *designed* so that plastic flow is likely to occur, but the ability to behave plastically is useful in an emergency. A copper saucepan, if dropped, or a steel motor car body, if it collides with a wall, will be only dented: a glass bowl, if dropped, will break. We say also that copper is **ductile**: it can be drawn, rolled and hammered into shape. It is typical of many metals, and it is particularly the toughness of metals (not their strength or stiffness, which other materials also possess) which makes them so useful.

(d) The rubber cord

Rubber is unusual in that it may undergo *very* large strains and still behave elastically, although as the graph shows, the stress–strain curve for unloading is not the same as for loading. This effect is called **hysteresis**. Nevertheless the rubber returns to its original length when the stress is finally removed.

Since the strain is so large, the cross-sectional area decreases considerably, and plotting the nominal stress gives a particularly false impression for rubber. Its actual stress (force/actual cross-sectional area) is much larger than the nominal stress. In fact the actual stiffness of rubber is almost constant at the start, and stiffer than the graph seems to show (but still *much* less stiff than other materials). But the actual stiffness does increase considerably towards the end of the graph.

(e) The polythene strip

The strip could have been cut from the low-density polythene which is used to make transparent plastic bags. You can see that its stiffness is constant at first: in this region it also behaves elastically. Then there is a sudden increase in strain for virtually no increase in stress: what

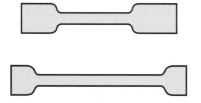

FIGURE 5.6

happens is illustrated in figure 5.6. The part which has narrowed becomes no narrower; but it does become longer, and all the extension occurs in this section. When the whole strip has narrowed the material becomes about as stiff as it was originally. If at this stage the stress is removed, the strip is left with a large permanent strain.

Measuring the Young modulus

You can measure the Young modulus E for a material by taking a wire made from it, stretching it with measured forces, and recording the extensions. There is apparatus specially designed for this purpose but it is possible (and perhaps more instructive) to measure E with simpler apparatus. Suppose you want to measure E for copper: your laboratory will have reels of copper wire of different diameters, so which should you choose? With a thin wire you will be able to get large stresses without using large forces, but the wire should not be too thin or else you will not be able to measure its diameter precisely enough. A wire of diameter 0.50 mm would be a good compromise.

How long should the wire be? The longer the wire, the greater the extension and the easier it will be to measure the extension precisely, so we should have a wire which is as long as is conveniently possible – perhaps about 3 m in an ordinary laboratory.

It will be easier to measure the extension of the wire if it is horizontal, so we could clamp one end of the wire, very securely, between wooden blocks at one end of a bench, and run the other end over a freely running pulley fixed to the other end of the bench (or another bench) and hang loads on that end, as shown in figure 5.7. Safety spectacles must be worn, because a stretched wire has energy stored in it, and if the wire breaks the broken ends will be moving dangerously fast. A simple marker (made from a sticky label) can be fixed near the end of

the wire next to the pulley. If a ruler is placed on the bench just underneath the wire, the extension can be measured as the marker moves over the ruler.

The results shown in Table 5.2 were recorded in an actual experiment in which the copper wire was loaded and unloaded for three cycles, increasing the maximum load each time. You can see that for the loads used in the first cycle the wire was behaving **elastically**, i.e. when the loads were removed, the wire went back to its original length. Also the extension was proportional to the force.

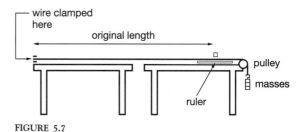

FIGURE 5.7

TABLE 5.2 Results for loading and unloading of copper wire

First cycle		Second cycle		Third cycle	
load/g	l/mm	load/g	l/mm	load/g	l/mm
0	0	0	0	0	4.3
300	1.0	300	0.9	300	5.1
600	1.9	600	1.9	600	6.0
900	3.0	900	3.0	900	6.9
1200	4.0	1200	4.0	1200	7.9
900	3.0	1500	5.7	1500	9.0
600	1.0	1800	7.9	1800	10.0
300	0.9	2100	10.9	2100	11.4
0	0	1800	10.3	2400	15.9
		1500	9.2	2500	18.8
		1200	8.5		
		900	7.3		
		600	6.5		
		300	5.3		
		0	4.3		

From the second cycle you can see that when large loads are used the wire no longer behaves elastically. When the load is reduced to zero you can see that there has been a **plastic** increase in length of 4.3 mm. For the larger loads the extension is no longer proportional to the force.

The third cycle shows clearly the effect of **work-hardening** caused by the loading in the second cycle. For loads from 1200 g to 1800 g the *additional* extensions are much less than they were in the second cycle. However, eventually the additional extensions begin to increase again. Eventually, even with only an extra 100 g added, a large extension occurs, and the wire begins to increase in length even if no more load is added. The measurement of 18.8 mm was made as soon as the load was added. 30 minutes later the extension was 20.3 mm and still increasing: at this stage the wire is thinning at some point and the stress is increasing. The wire eventually broke, at an extension of 21.8 mm, without any more load being added.

From the measurements for the first cycle we can calculate the Young modulus for copper. A graph of load against extension gives a straight line passing through the origin: for 1200 g the extension was 4.0 mm ($=4.0 \times 10^{-3}$ m). The weight of 1200 g is $1.2 \text{ kg} \times 9.8 \text{ N kg}^{-1} = 11.8$ N. The length of the wire was 4.35 m, and four measurements of the diameter, made with a micrometer screw gauge at different points, were all the same: 0.46 mm. So

$$E = \frac{\sigma}{\epsilon} = \frac{Fl}{A\Delta l} = \frac{11.8\,\text{N} \times 4.35\,\text{m}}{\pi(0.23 \times 10^{-3}\,\text{m})^2(4.0 \times 10^{-3}\,\text{m})}$$
$$= 7.7 \times 10^{10}\,\text{N m}^{-2} = 77\,\text{GPa}.$$

How reliable is this result? It depends on the reliability of the individual measurements. There is very little uncertainty in the measurements of the mass or the length: probably less than 0.1%. But you could probably measure the extension to the nearest 0.1 mm, at best: this gives a percentage uncertainty of (0.1 mm)/(4.0 mm) × 100% = 2.5%. You could probably measure the diameter of the wire to the nearest 0.01 mm, and this gives a percentage uncertainty of (0.01 mm)/ (0.46 mm) × 100% = 2.2%, so the *maximum* uncertainty (obtained by adding the percentage uncertainties) in the result is 4.7%. 4.7% of 77 GPa is 3.6 GPa, so we could give the value of E as

(77 ± 4) GPa. The probable uncertainty would be a bit less than this.

5.2 A molecular view

Why do these materials behave as they do? Our model of the molecular structure of matter should be able to explain. It will be justified as a good model to the extent to which it can. Here is a summary of the behaviour of solid materials, together with the explanation which the model provides.

◆ *They obey Hooke's law* (i.e. have a constant stiffness) up to a point, when failure of some kind usually occurs. The material breaks, or deforms considerably.

Explanation: What happens on a large scale, to the material, happens at the molecular level, too. When we increase the length of a copper wire, we are increasing the distance between the molecules, and the strain is the same, whether we measure it by the fraction $\Delta l/l_0$ or the fraction $\Delta r/r_0$ where l_0 is the original length of the wire and r_0 the original separation of the molecules. At first sight the graph of figure 4.7(a) (page 54) suggests that strain is not proportional to stress – until we remember that we expect Hooke's law to be obeyed for very small strains only: up to 0.001 for most materials. A closer look at the graph shows that the graph is very nearly straight for strains of this order. So our model explains why Hooke's law will be obeyed, as nearly as shall be measurable, for strains up to 0.001. Incidentally, the fact that the force– separation graph is a straight line through the point $(r_0, 0)$ into the compression region explains why materials have a constant stiffness in compression, and the same stiffness as they have in tension.

◆ *Rubber behaves unusually* because its stiffness increases considerably as the strain increases.

Explanation: Rubber is a polymer and we know that the long-chain molecules are not usually arranged in an orderly way. They double back on themselves, and are tangled. Figure 5.8 gives some

FIGURE 5.8

idea of how three such molecules might be arranged. Stretching a rubber cord merely straightens out the long chains, and it needs little force to do this (in the absence of cross-linking, which makes some other polymers stiff): also, a considerable extension can be produced just by straightening out the chains. However, when the chains have been straightened, the length of the rubber can be increased only by increasing the separation of the molecules, and it is as hard to do this with rubber as it is for any substance. We could expect the rubber cord then to be flexible up to much larger strains than other materials, but to be stiff eventually.

◆ *Materials are elastic for small stresses.*

Explanation: When external forces are used to keep material stretched, its molecules are kept in equilibrium at a greater separation then usual. The forces of attraction between the molecules are then greater than usual. When the external forces are removed, the molecular forces pull the molecules back to their normal positions again. This elasticity can occur even when the material does not have a constant stiffness (e.g. with the rubber cord). Some specimens (e.g. the glass rod and the rubber cord) are elastic up to the point where they break: we shall discuss the plasticity of copper wire when we discuss its failure mechanism.

5.3 Failure mechanisms

There is nothing in our model which explains why, with a relatively small stress, the glass rod breaks and the copper wire flows plastically. Why do these two materials behave differently?

Plastic flow

Part of the explanation is that in a crystalline or polycrystalline material it is easier to slide whole planes of molecules over each other than it is to pull molecules directly apart. Figure 5.9 shows what would happen if two layers of molecules slid over each other. The upper layer has slid over the lower layer and has settled down again in a similar position.

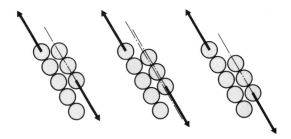

FIGURE 5.9

FIGURE 5.10

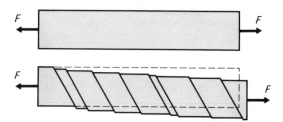

FIGURE 5.11

Crystalline materials do have layers of molecules which can slide over each other in this way. Figure 5.10 is a photograph of the surface of a piece of aluminium in which this has happened. It shows the rows of parallel steps formed within the grains of the aluminium after the layers have slipped over each other. Figure 5.11 explains why the slipping results in considerable elongation. The idea that layers of molecules slid over each other goes some way to explaining why crystalline materials are not as strong as we could expect them to be if they could be broken only by the molecules being pulled directly apart.

Dislocations

But in practice we find that crystalline materials flow under stresses which are perhaps 10 or 100 times smaller than the stresses which we can calculate should be needed. In 1934 G.I. Taylor put forward the idea of **dislocations**. There are different types of dislocation: the simplest to describe and explain is shown in the photograph at the beginning of Chapter 4, where one row of gold atoms ends abruptly in the middle of one grain of a crystal. Figure 5.12 shows a photograph of a bubble raft model. You will see the dislocation most easily if you raise the page, keeping it horizontal, to eye level and look in the direction of the arrow. The presence of a dislocation allows plastic deformation to occur relatively easily, since only one row of molecules need move at a time, and so plastic flow will occur when the applied forces are relatively small.

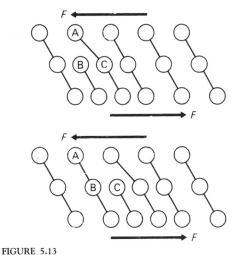

FIGURE 5.12

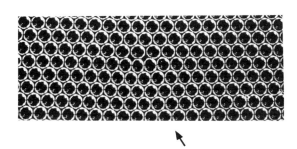

FIGURE 5.13

Figure 5.13 explains how a dislocation moves. Under the action of the pair of forces *F*, the row of

molecules A is pushed to the left, and a bond is formed between them and row B instead of between them and row C. The dislocation has now moved one molecular diameter to the right, and this will continue to happen until it reaches the edge of the grain, or another dislocation.

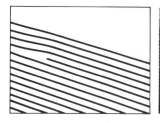 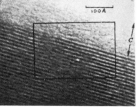

FIGURE 5.14

Figure 5.14 shows a photograph, taken with an electron microscope, of a dislocation. It was possible to do this only because of the large spacing of the molecules in this crystal. An analogy of the dislocation mechanism may be seen in the pulling of a large carpet across a floor. It is hard to pull the whole carpet all at once. A technique which requires smaller forces is to make a ruck at one edge, and to move the ruck across the floor a little at a time.

Tangled dislocations

Dislocations in the arrangement of the molecules within the grain of a polycrystalline material are inevitable. In practice the arrangement of molecules, even within a grain, will not be perfect. Molecules will not all come together in the right place at the right time. If the dislocations are inevitable and we want to make the material stronger, we make the best of the situation: if we *increase* the number of dislocations (which we can do by stressing the material) the materials become stronger. This happens because if there are more dislocations, the increased number of dislocations does not enable the layers of molecules to slip very far before their progress is hindered by other layers sliding in other directions within the same grain. We say that the dislocations become *tangled*. This is what happens when materials like copper are **work-hardened**. The number of dislocations is increased and it therefore becomes a stronger and stiffer material. Another way in which the movement of dislocations can be hindered is by introducing atoms of a different element. These will generally be of a different size, and this interruption in the regularity of the layers prevents the dislocations from moving far. Work-hardening, and the deliberate introduction of foreign atoms (as when carbon atoms are added to iron to make steel), are two examples of ways in which we now *design* materials to do a particular job. Engineers used to have to make do with whatever materials occurred naturally, or whose composition and structure occurred by chance.

Fracture

The failure of the glass rod was quite different from the yielding of the copper wire. A.A. Griffith suggested in 1920 that the cause of all brittle failures is a **crack** within or on the surface of a material. The crack may be very narrow: indeed it is more dangerous if it is. The theory of this is too advanced for this book, but if you have watched a glass cutter at work you will know that he runs his cutter very lightly over the glass to score it, and the glass then breaks easily when tapped or flexed. A deeper, blunter scratch makes the glass harder to break. The danger of a crack can be seen from figure 5.15. The circular shaded area in (a) is where the stress will be large, and (b) shows why: the force in bond A will be large because this bond has to balance the forces exerted on molecules X and Y from above and below. In (c) **photoelastic stress analysis** shows, by the closeness of the spacing of the interference fringes, the high stress at the tip of a cut in the material.

For this reason a crack is sometimes called a **stress-multiplier**. The molecular bonds at the tip of the crack are more likely to break than any others, and thus the crack spreads through the material, the local stress becoming even greater as the top of the crack moves on. Newly drawn fibres of brittle materials are strong because cracks are initially absent; but handling them, and chemical action, will cause cracks to form before long and even these fibres will become susceptible to brittle fractures.

It is clearly not possible for glassy, non-crystalline, materials to yield, since this kind of flow mechanism occurs only when there is a long-range order of planes of molecules which can slip over each other (and more easily still where there are dislocations). But why do crystalline materials seem not to be susceptible to failure through cracks? It might be thought that if a crack existed on the surface of a copper wire it, like glass, would also break in a brittle manner. When a crack does occur in a ductile material the atoms at the tip of the crack start to slide past each other, rather than pulling apart, so that plastic flow occurs and the stress concentration is relieved and the crack does not extend further. This argument is summed up in table 5.3.

TABLE 5.3   Comparison of glasses and crystalline materials

	Glasses	Crystalline materials
brittle fracture?	yes	no, because flow at the tip of the crack relieves the stress concentration
ductile yielding?	no, because there are no planes of molecules which slip over each other	yes

There is no rigid dividing line between materials which are ductile and those which are brittle. Crystalline materials possess different amounts of ductility, and any of them may behave in a brittle manner under different conditions (e.g. at a lower temperature).

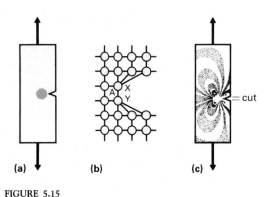

(a)   **(b)**   **(c)**   = cut

FIGURE 5.15

5.4 Energy stored in stretched materials

Comparison between E and k

Robert Hooke's investigations were actually into the stretching of springs, not wires. In its construction a spring is more complicated than a wire, and the forces which make it longer are what are called shearing forces (like the force exerted by a guillotine slicing through a pad of paper) not tensile forces. But the behaviour of springs is as simple as the behaviour of wires and rods, if the forces are not too large. As already stated in Chapter 3

the extension is proportional to the force

if the forces are not too large. We write

$$F = kx$$

where F is the force, x the extension and k the **stiffness** of the spring. The larger k is, the stiffer the spring, i.e. the less it stretches for a particular force. How does k compare with E, the Young modulus, which measures the stiffness of the material of a wire or rod? E refers to a material, and to find the extension Δl we need to know the length and cross-sectional area A of the wire. k refers to a particular spring. The material the spring is made from, the thickness of the wire, the size of the coils, etc., are all factors which affect the value of k. You will understand this if you compare the two equations

$$F = kx \quad \text{and} \quad F = \frac{EA}{l}\Delta l$$

You can see that if F is measured in N, and x in m, the unit of k must be the $N\,m^{-1}$. (In mathematics textbooks the equation $F = kx$ is sometimes written as $F = (\lambda/l)\Delta l$, where $\lambda = EA$; its unit is the newton.)

Suppose we take a particular spring of stiffness k and cut it in half: one of the half-size springs will have stiffness $2k$, since the same force will stretch it only half as much. Similarly, two springs of stiffness k, joined end-to-end, will have stiffness $\frac{1}{2}k$.

Energy stored in a spring

It was also shown in Chapter 3 that when a spring is stretched the elastic potential energy stored in it is given by the area beneath the graph of force against extension. (Notice, incidentally, that force–extension graphs are always drawn with extension on the x-axis, even though the force is probably the independent variable: if this is not done, the area beneath the graph does *not* give the energy stored.)

For a spring with no initial extension

$$\text{elastic p.e.} = \tfrac{1}{2}Fx, \text{ and } F = kx, \text{ so}$$
$$\text{elastic p.e.} = \tfrac{1}{2}kx^2$$

For a wire with no initial extension

elastic p.e. $= \tfrac{1}{2}F(\Delta l)$ and $F = \sigma A$ and $\Delta l = \epsilon l$,

so   elastic p.e. $= \tfrac{1}{2}\sigma A\epsilon l = \tfrac{1}{2}\sigma\epsilon V$

i.e.   elastic p.e. $= \tfrac{1}{2}(\text{stress})(\text{strain})(\text{volume})$

or,   elastic p.e. per unit volume $= \tfrac{1}{2}(\text{stress})(\text{strain})$.

EXAMPLE

How much energy is stored in a steel wire of length 2.0 m and diameter 0.80 mm when it is pulled with a force of 50 N? (E for steel $= 200$ GPa: assume that the wire obeys Hooke's law.)

$$\Delta l = \frac{Fl}{EA} = \frac{(50\,\text{N})(2.0\,\text{m})}{(200\times10^9\,\text{N\,m}^{-2})(\pi)(0.40\times10^{-3}\,\text{m})^2}$$

$$= 9.9\times10^{-4}\,\text{m}$$

$$\text{Elastic p.e.} = \tfrac{1}{2}F(\Delta l)$$
$$= \tfrac{1}{2}(50\,\text{N})(9.9\times10^{-4}\,\text{m})$$
$$= 2.5\times10^{-2}\,\text{J} = 25\,\text{mJ}.$$

This does not seem much, but it is stored in a small volume. That much energy, stored in a small mass, makes it move fast. The density of steel is 7700 kg m^{-3}, so

$$\begin{array}{l}\text{mass} \\ \text{of wire}\end{array} = \rho V$$

$$= (7700\,\text{kg\,m}^{-3})(2.0\,\text{m})(\pi)(0.40\times10^{-3}\,\text{m})^2$$
$$= 7.7\times10^{-3}\,\text{kg}$$

If the wire breaks, and if we guess that the mass of the moving ends is about one-tenth of the total mass, then using elastic p.e. = k.e. with k.e. = $\frac{1}{2}mv^2$ where v is the speed of the ends, we have

$$2.5 \times 10^{-2}\,J = \frac{1}{2}(7.7 \times 10^{-3}\,kg)v^2$$

$$\Rightarrow v = 8\,m\,s^{-1}$$

which would be dangerous if one of the ends hit your eye.

Even if the extension of the wire is not proportional to the force, the energy stored in a stretched spring or wire or rod or any other specimen will be equal to the area beneath the force–extension graph. The calculation of the area will obviously be more difficult, but it can be done.

EXAMPLE

The graph shows how the extension varies with the force when a rubber band is stretched. Estimate the work done in stretching it.

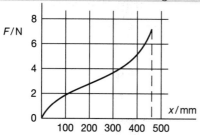

The area of each square represents work done equal to 2 N = 100 mm = 0.2 J.

The areas of each whole or part square beneath the curve are shown in table 5.4.

TABLE 5.4

				0.1
			0.1	0.5
	0.2	0.6	1.0	0.6
0.6	1.0	1.0	1.0	0.6

Total number of squares = 7.3, so work done = 7.3 × 0.2 J = 1.46 J.

This technique is quick. You can estimate fractions of a square accurately without taking very much time over it.

Energy stored in different materials

Looking at the area under the graph can provide useful information about a material even if you do not actually do a calculation. Look at the three graphs in figure 5.16. The first is for cast iron (a brittle material), the second for aluminium (a ductile material) and the third for rubber. For the first two, the graphs are drawn up to the breaking point of the specimen. You can see that the area beneath the first graph is much smaller than the area under the second graph, so if the scales used are the same, you can say that much more work needs to be done to break the aluminium than to break the cast iron. (The energy is used to slide layers of atoms over each other.) We say that aluminium is *tougher* than cast iron. How much work needs to be done to break a material is a measure of its **toughness**.

For rubber, the area under the upper line tells us how much work we must do to stretch it, and the area under the lower line tells us how much work

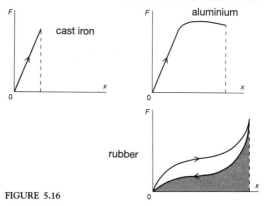

FIGURE 5.16

the rubber can do as it contracts again. We get less back than we put in. The difference is converted wastefully into internal energy in the rubber, and the rubber becomes warm. The loop is called a **hysteresis loop**; its area measures the energy converted into internal energy. You can experience this for yourself by stretching and releasing a rubber band a few times and then putting it to your cheek or lips (since they are sensitive detectors of temperature change). Vehicle tyres are continually being compressed and relaxed as they roll, so it is important that the rubber is designed so that the hysteresis loop is as small as possible. This will limit the proportion of chemical energy which is wasted and therefore save on fuel.

Exercises on each section of this chapter may be found in the companion textbook, *Practice in Physics*.

SUMMARY

At the end of this chapter you should be able to:

◆ understand what is meant by the tension in a stretched wire or rod and that it is the same throughout.

◆ use the equation $\sigma = F/A$, which defines tensile stress σ.

◆ use the equation $\epsilon = \Delta l/l$, which defines tensile strain ϵ.

◆ understand what is meant by saying that a wire or spring obeys Hooke's law.

◆ understand what is meant by the limit of proportionality, and the elastic limit of a material.

◆ use the equation $E = \sigma/\epsilon$, which defines the Young modulus E.

◆ understand what is meant by stiffness, and how it differs from strength.

◆ understand what is meant by elastic and plastic behaviour.

◆ explain where tension and compression occur in a beam or cantilever.

◆ draw stress–strain graphs for typical ductile and brittle metals, rubber and other polymers.

◆ describe how to measure the Young modulus for a material in the form of a wire.

◆ use molecular ideas to explain why wires obey Hooke's law.

◆ use molecular ideas to explain why rubber's stiffness increases as the strain increases.

◆ use molecular ideas to explain why materials are elastic for small stresses.

◆ use molecular ideas to explain how dislocations weaken a crystalline material.

◆ use molecular ideas to explain how a crack causes brittle failure in non-crystalline materials.

◆ use the equation $F = kx$, which defines the stiffness k of a spring.

◆ calculate the stiffness of two identical springs of stiffness k when arranged in series or in parallel.

◆ use the equation $W = \frac{1}{2}kx^2$ to calculate the elastic potential energy W stored in a spring.

◆ use the equation elastic p.e. per unit volume $= \frac{1}{2}$(stress)(strain) to calculate the elastic p.e. stored in a wire.

◆ estimate the energy stored in a spring or wire when the force–extension graph is not straight.

◆ understand that the toughness of a material is related to the work that must be done to break it, and that toughness is often a desirable property.

6 Fluid Behaviour

Parachuting has come a long way from being just a very useful way of escaping from a doomed aircraft! Parachuting and sky-diving are becoming increasingly popular sports, and hang-gliding is being overtaken in popularity by paragliding. This paraglider, moving about in a 'sea' of air, is manipulating the controls so that she can move horizontally or vertically: she can move upwards even if there is no upward air current. The upward push which balances her weight when she is moving horizontally is provided by the movement of air over the surfaces of the parachute, which is exactly the same as, though on a much smaller scale than, the movement of air over an aircraft's wings. One of the topics dealt with in this chapter is the explanation of how aerofoils keep aircraft in the air and how hydrofoils keep the hulls of boats out of the water, how sailing 'into the wind' is possible, and how putting spin on golf balls and tennis balls enables them to swerve through the air.

6.1 Pressure

We have all found a piece of meat hard to cut with a knife – and then found that we were using the wrong edge of the blade! The 'right' edge is the thinner one, because it has the smaller area of contact, and, for a particular force, exerts a larger **pressure**. Many objects are designed to create large pressures: pins, chisels, saw blades, etc. Figure 6.1 shows a drawing-pin being pushed into a board: the push of the pin on the board is the same size as the push of the pin on the thumb, but the pressure is large at the tip of the pin because it is sharp there, so the pin pierces the board rather than the thumb.

FIGURE 6.1

Other objects are designed to create low pressures: caterpillar tracks on tanks and bulldozers and wide tyres on earth-moving machines (figure 6.2) and farm tractors. On all of these there are also ridges to create regions of higher pressure so that there is also a good grip to provide forward motion.

FIGURE 6.2

Definition of pressure

When we place a block on a table, as in figure 6.3, the block pushes down on the table and the table pushes up on the block. These two contact forces, are, of course, the same size because of Newton's third law. If the block is placed on another of its faces, of different area, these forces will stay the same size as before, but the area of contact will be different and so the force on each unit area of surface will be changed: the *pressure* will be different. We define the pressure p by the equation

$$p = \frac{\text{contact force}}{\text{area of contact}} \text{ or in symbols, } p = \frac{F}{A}$$

You can see that the unit of pressure must be the $N\,m^{-2}$. It is the same as the unit of stress, since this is also a force per unit area. Again we call the $N\,m^{-2}$ the pascal (Pa).

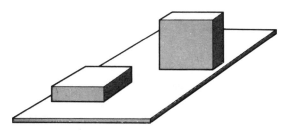

FIGURE 6.3

EXAMPLE

A girl of mass 52 kg stands on the ground. If the area of contact of her feet with the ground is $110\,cm^2$ what pressure does she exert? If she then stands 'on tip toe' so that the area of contact is reduced to $45\,cm^2$, what does the pressure become? Take $g = 9.8\,N\,kg^{-1}$

Area of contact $= 110\,cm^2 = 110 \times 10^{-4}\,m^2$

$$p = \frac{F}{A} = \frac{(52\,kg)(9.8\,N\,kg^{-1})}{110 \times 10^{-4}\,m^2}$$

$$= 4.6 \times 10^4\,N\,m^{-2} = 46\,kPa.$$

The area of contact is now reduced in the ratio 45/110, so the new pressure is increased in the ratio 110/45:

$$\text{it is } 46\,kPa \times \frac{110}{45} = 113\,kPa$$

Notice that it was much quicker to think what effect the change of area would have on the pressure than to start the question again from the beginning. This is a very useful time-saving technique for doing calculations.

Pressure caused by liquids and gases

Pressures are exerted on surfaces not only by solids but also by liquids and gases. Gas pressure is caused by the bombardment of a surface by the molecules of the gas. A model can be used to demonstrate this. It consists of a vertical transparent cylinder (figure 6.4) which contains a disc with a mass of about 1 gram. A motor drives a crank which pushes a piston against the flexible rubber sheet which forms the bottom of the cylinder. There are about 50 ball-bearings in the cylinder: these represent the gas molecules. When the piston pushes against the bottom of the cylinder the ball-bearings are knocked upwards and strike the underside of the disc.

Each time a ball-bearing hits the disc it pushes upwards on it (the ball-bearing has its momentum changed, so there must be a force on it and also on

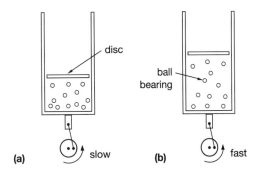

FIGURE 6.4

the disc). The size of the force depends on the speed of the ball-bearings. The ball-bearings have different, unpredictable, speeds and the impacts do not occur at regular intervals, so the total upward force on the disc is not constant; but even with just a small number of ball-bearings in the cylinder the impacts keep the disc steady to within a few millimetres of its average position. If there were more, smaller, ball-bearings, the force would be even steadier. Figure 6.5 shows a graph of how the force might vary with time for the two situations shown in figure 6.4. In (b) the impulses are larger but occur less often, but the average force exerted by the ball-bearings is the same in (a) and (b), as it must be to balance the weight of the disc on each occasion.

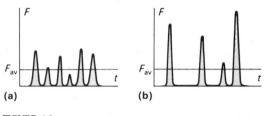

(a)   **(b)**

FIGURE 6.5

This is a model of the way in which individual molecules exert a force and the way in which a whole collection of molecules, which form a gas, exerts a pressure. In the volume which the ball-bearings occupy in the model there might be 10^{23} gas molecules, so you can understand why gases exert pressures in which you can detect no fluctuation even though the pressure is the result of impacts of varying size and which occur randomly.

Atmospheric pressure

We live at the bottom of a sea of air: the molecules are kept near the surface of the Earth by the gravitational forces pulling on them (very few are ever moving fast enough to escape). The pressure decreases with height above the surface because at greater heights the weight of air above is less. There is less to support: figure 6.6 illustrates this.

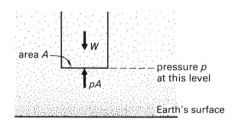

FIGURE 6.6

The pressure does not decrease *linearly* with height for two reasons. Air is compressible, and so the

density decreases with height; and, secondly, the temperature varies. The variation of temperature (which occurs in no regular way) makes it impossible to state laws about the variation of pressure with height (though it may be helpful to know that at a height of 20 km the pressure has fallen to about one-tenth of the pressure at the Earth's surface). In any case, for small differences in height the change of pressure is small compared with atmospheric pressure. For example, in moving from the floor to the ceiling of an average room the decrease in pressure would be about 40 Pa, which is about 0.04% of atmospheric pressure. So in laboratory experiments we assume that the atmospheric pressure, or the pressure in a gas, is the same throughout the apparatus, despite any differences in height.

A **barometer** is used to measure atmospheric pressure. An **aneroid** barometer consists of a sealed flexible metal box which contains gas at low pressure. When the atmospheric pressure changes the height of the box changes and a system of levers and cog-wheels makes a pointer move over a scale graduated in units of pressure. This sort of barometer has been widely used as an **altimeter** in aircraft to measure height above the ground, but of course the altimeter must be recalibrated before each flight to allow for daily changes in atmospheric pressure.

Weather

Atmospheric pressure also varies with time at a particular place on the Earth, and this is the basic cause of changes in the weather. The Sun heats the Earth, but not evenly, because of variable cloud cover and the fact that radiation falls more or less obliquely on different parts of the Earth at different times. Also, the radiation is absorbed differently by land and ocean, and even differently by different kinds of land surface. The situation is also complicated by the presence of water vapour mixed with the air, but for our purposes it will be enough to realise that convection currents of warm air will rise in some places, and currents of cold air will fall in others – thus causing differences in atmospheric pressure which will make large volumes of air travel across the surface of the Earth, carrying with it more or less water vapour, and at high or low speeds. A knowledge of atmospheric pressure at different places is therefore of great importance since it can be used to predict the forthcoming weather.

A **cyclone** or **depression** is a region where the atmospheric pressure is less than that in the surrounding region, and an **anti-cyclone** is a region where it is higher. It might be expected that air would flow directly into a depression and directly away from an anti-cyclone, but the rotation of the Earth modifies the flow so that in the northern hemisphere the air flows anti-clockwise round a depression and into it; and clockwise round an anti-cyclone, and away from it. In the southern hemisphere, the rotations are opposite.

We have all seen weather maps like that in figure 6.7 in newspapers and on television. They often have lines drawn on them which look like the contour lines on land maps. These lines join places which have the same atmospheric pressure, and are called **isobars**. The numbers on them give the atmospheric pressure in **millibars** (mb). One **bar** (1000 mb) is defined to be exactly equal to 1.0×10^5 Pa. Atmospheric pressure is rarely less than 950 mb or more than 1050 mb, i.e. it is unlikely to vary more than ±5% from the pressure of 1000 mb. You may also meet an **atmosphere** used as a unit of pressure (though it is being used less frequently nowadays). This is equal to the pressure created by a column of mercury 760 mm high, and is equal to 1013.25 mb, i.e. about 101 kPa. This pressure is called **standard atmospheric pressure**.

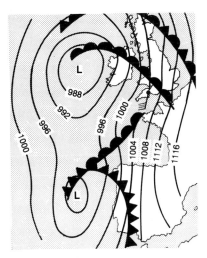

FIGURE 6.7

Pressure at a point in a fluid

Consider a small surface of area A in a fluid. The pressure at that point in the fluid will exert a force F on the surface. We define the **pressure at a point in a fluid** by the equation

$$p = \frac{F}{A}$$

The area A does not need to be horizontal. It could be vertical, or inclined at any angle to the horizontal: the size F of the force is the same in all cases. We say that 'pressure has no direction'. It exerts a force at right angles to *any* surface on which it acts. For example, the air underneath a table pushes *upwards* on the underneath side of the table.

EXAMPLE

Calculate the force exerted by the air in a room on a window which measures 0.80 m by 1.2 m on a day when the atmospheric pressure is 103 kPa.

$F = pA = (103 \times 10^3 \, \mathrm{N\,m^{-2}})(0.80 \, \mathrm{m} \times 1.2 \, \mathrm{m})$
$= 99$ kN horizontally outwards on the window.

Here atmospheric pressure pushes horizontally on a surface, since the surface is vertical. A force of 99 kN is equal to the weight of a mass of 9.9 tonnes, but the window does not collapse because the air outside is pushing inwards with a force of just the same size.

Pressure change with depth

How does the pressure change as we move deeper in a liquid? In figure 6.8 we look at the forces on a cylinder of liquid of density ρ, which is part of the liquid in a tank. The cylinder has cross-sectional area A and height Δh. The weight W of the cylinder is given by

W = (volume)(density)(gravitational field strength)

i.e. $$W = A(\Delta h)\rho g$$

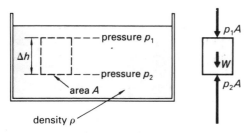

FIGURE 6.8

Suppose the pressures at the top and bottom of the cylinder are p_1 and p_2 respectively. These pressures produce a downward force p_1A on the top of the cylinder and an upward force p_2A on the bottom, and these forces, together with the weight of the cylinder, keep it in equilibrium. Considering the free body diagram for the cylinder we have

$$p_1A + W = p_2A$$
$$(p_2 - p_1)A = (\Delta h) A\rho g$$
$$\Delta p = \rho g(\Delta h)$$

where Δp is the difference in pressure between the two levels. So the increase in pressure $= \rho g \times$ the increase in depth. The area of the cylinder, or of the vessel, is irrelevant: this makes this result very useful. In using it we assume that the density of the liquid is uniform (i.e. its temperature is uniform, and the pressure does not become so great that the liquid at the bottom becomes compressed).

You have already seen that with gases, when Δh is small, Δp is negligible. When Δh is large, their density is not usually constant, so that this result cannot be used with gases.

EXAMPLE

A capsule, containing air at a pressure of 150 kPa is sealed by a valve which can withstand a pressure difference of 650 kPa. How deep in sea water of density $1.02 \times 10^3 \, \text{kg m}^{-3}$ can it be lowered if no water is to enter? (Pressure at surface of sea = 101 kPa; $g = 9.8 \, \text{N kg}^{-1}$.)

Maximum pressure outside capsule
= 150 kPa + 650 kPa = 800 kPa.
Atmospheric pressure = 101 kPa, so additional pressure due to sea-water must be $\leqslant$ 699 kPa.

$$\Delta p = \rho g(\Delta h)$$
$$699 \times 10^3 = (1.02 \times 10^3 \, \text{kg m}^{-3})(9.8 \, \text{N kg}^{-1})(\Delta h)$$
$$\Rightarrow \quad \Delta h = 70 \, \text{m}$$

The pressure at all points at the same horizontal level in a liquid or gas (in equilibrium) is the same. Look at figure 6.9 which shows a horizontal tube connecting two points X and Y: if the pressure at X were greater than the pressure at Y there would be a resultant force in the direction XY on the liquid in the tube, and this is clearly impossible since the liquid is in equilibrium. The fact that the pressures at X and Y are equal means that they must be equally far below the surface of the liquid, which means that the surface of a liquid *in equilibrium* must be horizontal.

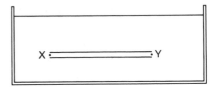

FIGURE 6.9

We can use the result $\Delta p = \rho g(\Delta h)$ to calculate Δp between two points in the same liquid which are not in the same vertical line: Δh is simply the difference in height between the two points.

EXAMPLE

One arm of a U-tube containing oil of density 800 kg m^{-3} is connected to a gas supply, and the other is left open to the atmosphere. The difference in level of the oil in the two sides of the tube is measured to be 200 mm. By how much does the pressure of the gas exceed atmospheric pressure?

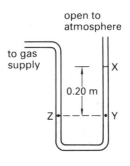

Pressure of gas
= pressure in liquid at Z
= pressure in liquid at Y
= pressure in liquid at X
    + pressure caused by 200 mm of oil
= atmospheric pressure
    + $(800 \, \text{kg m}^{-3})(9.8 \, \text{N kg}^{-1})(0.200 \, \text{m})$
= atmospheric pressure + 1.6 kPa

So gas pressure is 1.6 kPa greater than atmospheric pressure.

Note that (i) you do not need to know the cross-sectional areas of the tubes. There would be the same difference in level even if the arms had had different cross-sectional areas. (ii) A reason for using oil rather than water (density 1000 kg m⁻³) might have been that its smaller density gives a larger difference in levels, which makes the device more sensitive. The difference in levels, using water, would have been 0.16 m.

Manometers and barometers

The device described in the last example is called a manometer and conveniently measures the difference in pressure between two regions. For example, one simple piece of apparatus for verifying Boyle's law is shown in figure 6.10. The volume of the gas is measured directly in the graduated tube, and its pressure is calculated by adding Δp ($= \rho g \Delta h$, where ρ is the density of mercury, and Δh is the difference in the mercury levels) to the measured value of atmospheric pressure. When we want to use this kind of device to measure atmospheric pressure we need to have zero pressure (and therefore no gas) in one of the arms of the manometer. This kind of **barometer** always uses mercury as the liquid because its high density makes the differences in level as small as can be. Even so the difference in level (typically about 0.75 m) is large enough to make a mercury barometer cumbersome, which is why the aneroid barometer is often used.

Two versions of a mercury barometer are shown in figure 6.11. You should realise that they are theoretically identical. The second one is more convenient, as it needs just a straight tube and bowl to hold the mercury. Inevitably the space above the mercury in the closed end is not a true vacuum, because the mercury evaporates. However, at room temperature the pressure exerted by this vapour is about 0.2 Pa, so that the error is very small: it is equivalent to about 1/1000 mm of mercury. The **Fortin** barometer is a mercury barometer with additional refinements: a vernier device to make the barometer more sensitive, and a means of ensuring that the level of the mercury in the bowl is constant.

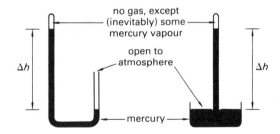

FIGURE 6.11

Safety precautions

A disadvantage of using mercury is that it is a cumulative poison, which can be absorbed by the human body both by inhaling the vapour, and also if the liquid comes into contact with the skin. It should therefore never be touched or handled. Mercury apparatus should always be placed in a tray, so that any liquid spilt can be contained in a small area and afterwards collected. Apparatus containing mercury should be fitted with rubber bungs and kept in a separate well-ventilated cupboard.

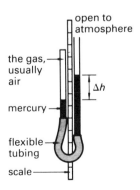

FIGURE 6.10

Archimedes's principle

If you look back at figure 6.8 you will remember that the resultant of the pressure forces is the same size as the weight of the liquid. Now imagine a cylinder of any substance taking the place of the cylinder of liquid: the pressure forces will be just the same as they were (they depend only on the depth in the liquid, and the cross-sectional area of the cylinder) so the pressure forces will cause a resultant upward force on the cylinder which is the same size as the weight of the liquid which would have been there. The shape does not have to be cylindrical: the argument works for any shape. The argument also works for gases as well as for liquids. This is called Archimedes's principle:

> the upthrust on a body immersed in a fluid is the same size as the weight of the fluid displaced by the body

Of course the upthrust will probably not be equal to the weight of the body which has been put in the fluid: often the weight of the body is greater than the upthrust. If these are the only two forces acting the body accelerates downwards (e.g. a stone in air or water).

If the upthrust is greater than the weight, the body accelerates upwards until one of three things happens, as shown in figure 6.12. In (a) a body which has been pushed below the surface of some liquid accelerates upwards ($U_1 > W$) until it floats. Then it is in equilibrium; the upthrust U_2 on it will be less than it was before and now equal to its weight ($U_2 = W$). In (b) the same body is in a closed tank. Again it accelerates upwards ($U > W$) until the top of the tank pushes down on it and it is in equilibrium ($R + W = U$). In this case equilibrium could be achieved by any downward force R (e.g. the push of someone's hand). In (c) an airship has $U_1 > W$, so it accelerates upward until it reaches a region where the air is less dense and the upthrust (on the same volume) U_2 is less. In equilibrium $U_2 = W$.

You may have wondered how it is possible to say that the mass of an oil tanker is 250 000 tonnes. How could it be measured? In equilibrium the weight of the tanker is the same size as the upthrust on it, so to find its weight we need only find the upthrust on it. This can be done by finding the volume of water it displaces (by

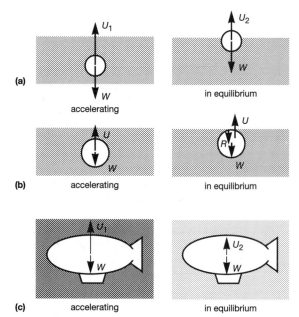

(a) accelerating / in equilibrium

(b) accelerating / in equilibrium

(c) accelerating / in equilibrium

FIGURE 6.12

estimation from the drawings of the ship): then upthrust = volume × density × gravitational field strength. In fact the weight of a ship is usually called its **displacement**.

The more a ship is loaded, the greater the displacement of water needs to be, i.e. the deeper it sinks. Figure 6.13 shows the Plimsoll line markings on a ship. These are the levels above which the water must not rise, for safety reasons. For example, TF and F stand for tropical fresh water and normal fresh water: tropical fresh water is warmer and less dense so the ship has to sink further to displace the same weight of water.

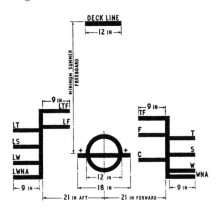

FIGURE 6.13

EXAMPLE

An airship has a volume of $45000\,\text{m}^3$ and contains helium of density $0.18\,\text{kg}\,\text{m}^{-3}$. If the weight of its fabric and gondola (the structure which contains passengers and cargo) is $150\,\text{kN}$, and it floats in air of density $1.3\,\text{kg}\,\text{m}^{-3}$, what is the weight of the load of passengers and cargo which it can carry? Take $g = 9.8\,\text{N}\,\text{kg}^{-1}$.

The weight of the helium is
$(45000\,\text{m}^3)(0.18\,\text{kg}\,\text{m}^{-3})(9.8\,\text{N}\,\text{kg}^{-1}) = 79.4\,\text{kN}$,
so the total unladen weight W is

$$W = 150\,\text{kN} + 79.4\,\text{kN} = 229\,\text{kN}$$

The upthrust U on the airship is equal to the weight of air displaced, which is (volume)(density of air) g, so

$$U = 45000\,\text{m}^3 \times 1.3\,\text{kg}\,\text{m}^{-3} \times 9.8\,\text{N}\,\text{kg}^{-1} = 573\,\text{N}$$

If L is the load it can carry, $L + W = U$ in equilibrium, so

$$L = U - W = (573 - 229)\,\text{kN} = 344\,\text{kN}.$$

Atmospheric pressure forces

The size of the atmospheric pressure is so great that it exerts very large forces on anything which the air is in contact with: on a table top of area $2\,\text{m}^2$, for example, the force is about $20\,\text{kN}$, which is the weight of a mass of 20 tonnes. Why does the table not collapse? Figure 6.14 shows a table resting on a horizontal floor, and a free body diagram for the table (although the *horizontal* pressure forces have been omitted, since these balance). What other forces are there?

There are two forces which we would normally consider: the pull W of the Earth, and the push R of the ground. There are also three pressure forces:

F_1 is the push of the air on the top of the table.
F_2 is the push of the air on the underside of the table.

There is also air in the tiny gap between the legs and the floor, so

F_3 is the push of the air on the underside of the legs of the table.

Are these three forces too small to matter? Is that why we usually ignore them? No! If the table has an area of about $2\,\text{m}^2$, W and P will each be about $500\,\text{N}$, whereas F_1, for example, will be about $200\,000\,\text{N}$. F_1, F_2 and F_3 are usually ignored because F_1 is very nearly equal to $F_2 + F_3$ so the three forces have a resultant which is very nearly zero.

The pressures p_2 and p_3 (which cause F_2 and F_3) are slightly greater than p_1 (which causes F_1) so $F_2 + F_3$ is slightly greater than F_1: in this case the difference would be about $1\,\text{N}$ and this small difference can usually be ignored, compared with the weight of the body and the other forces acting on it. So we usually ignore the upthrust of the air on a body unless we need to be very precise, or the volume of the body is very large (as it is for a balloon or an airship).

6.2 Flow

We shall now consider fluids which are moving. We shall assume that the flow is **streamline** or **laminar**, i.e. that layers of fluid slide over each other, with their velocity constant at a particular point in the fluid, i.e. the velocity of the fluid does not change at that point as time goes on.

Viscous flow

You should understand that all fluids are **viscous**: that is, that as these layers of fluid slide over each other, or over a surface (like the bank of a stream or the inside of a pipe), they exert forces on each other. Figure 6.15(a) shows liquid flowing through a pipe, with arrows showing the sizes of the velocities at different places. This makes the point that the liquid at the centre of the pipe is flowing

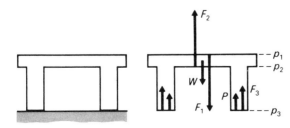

FIGURE 6.14

faster than the liquid at the edge. In fact the liquid immediately next to the pipe must be stationary. You can see this for yourself in any stream or river: the water at the edge is hardly moving even if there is rapid flow at the middle of the stream. Figure 6.15(b) shows the direction of the viscous forces on the layer of liquid shaded in figure 6.15(a). The faster liquid on top is pulling the layer forward, and the slower liquid below is dragging the layer back.

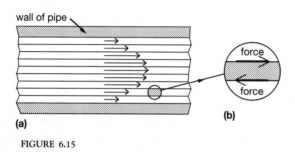

wall of pipe

(a)

(b)

FIGURE 6.15

You can see that viscous forces resist relative motion of the fluid, but there is a difference between them and frictional forces (exerted by *solid* objects on each other): frictional forces may exist even when there is no relative movement. Viscous forces exist only when there is relative movement, and this makes them useful for **damping**, i.e. for providing forces which convert unwanted kinetic energy into internal energy (e.g. in getting rid of the unwanted oscillations in the springs of a car being driven over a rough road). Some fluids (e.g. treacle, and automobile engine oil) are obviously viscous, but all fluids are to some extent: this is why the coffee in a cup does not spin round endlessly when you stop stirring.

Flow in pipes

Because a fluid is viscous it will not flow through a pipe without being pushed, because the walls of the pipe exert resisting forces on it. Pushing forces can be provided by making the pressure in the fluid greater at one end of the pipe. The rate of flow of fluid V/t (i.e. volume per unit time) is proportional to the pressure difference Δp. What else will affect the rate of flow? The rate of flow will be greater if the pipe is wide (has large cross-sectional area A)

and short (has small length l). These facts enable us to say that

$$\frac{V}{t} \propto \frac{A\Delta p}{l} \quad \text{and we write} \quad \frac{V}{t} = \frac{1}{\eta}\frac{A\Delta p}{l}$$

where η is the **viscosity** of the fluid. The equation holds only for streamline flow (i.e. not if the flow is turbulent). In practice this means that the rate of flow must be very small.

Bernoulli's equation

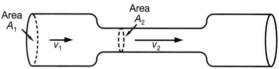

Area A_1

Area A_2

v_1

v_2

FIGURE 6.16

Figure 6.16 shows a pipe which becomes narrower at one point and then widens out again. If liquid is flowing through the pipe (and fills it) the liquid must flow faster where the pipe is narrower if there is to be the same rate of flow of mass through the pipe. If the two cross-sectional areas are A_1 and A_2, and the two speeds v_1 and v_2 then

$$A_1 v_1 = A_2 v_2$$

This is sometimes called the **equation of continuity**.

Since $A_2 < A_1$, $v_2 > v_1$, so the liquid accelerates as it enters the narrow pipe. We must therefore expect there to be an unbalanced force on the liquid, accelerating it to the right. This is provided by a difference of pressure in the two parts of the tube. If the pressure is measured in the wider part of the tube it is found to be higher than it is in the narrow part of the tube. If we analysed this situation in detail we could arrive at **Bernoulli's equation** which states that for streamline non-viscous flow

$$p + \tfrac{1}{2}\rho v^2 = \text{constant}$$

at all points on the same horizontal level where p is the pressure, ρ is the density of the liquid and v its speed. (For a gas flowing through the pipe the equation must be modified to take account of density changes in the gas as it is compressed, so it is not as simple as this.)

Applications of the Bernoulli effect

Bernoulli's equation shows us that the pressure in fast-moving fluid is lower than the pressure in the same fluid moving slowly (and as a special case of this, the pressure in a fluid moving at any speed is less than the pressure in the same fluid when it is stationary). This behaviour is of great importance. Without it aeroplanes could not fly.

When an aeroplane is flying at a steady speed at constant height its weight is balanced by the lift force. The lift force is provided partly by the under-surface of the wing striking the air as it moves through it: the air is pushed down, and the wing pushed up: the extent of this effect depends on the **angle of attack** of the wing surface, i.e. on the angle to the horizontal of the underside of the wing. But the lift force is also partly provided by the shape of the wing section. Aeroplane wings are **aerofoils**, i.e. their surfaces are shaped in such a way that air flows faster over the upper surface than over the lower surface. Where the air flows faster the pressure is less: you can prove this to yourself by blowing over a sheet of paper, as in figure 6.17.

The size of the lift force on an aerofoil is proportional to the area A of the wing, the density ρ of the fluid and to (approximately the square of) the speed v of the aerofoil through the fluid: i.e. $F \propto \rho A v^2$.

Some boats are supported by **hydrofoils**, which are aerofoil-shaped surfaces which are below the water surface. When the boat is at rest, it is supported by the upthrust on the hull. When it is moving, the lift on the hydrofoils pushes the hull out of the water, as shown in figure 6.18. Then there is much less resistance to motion. The speed of a boat is of course much less than the speed of an aircraft, and the hydrofoils have a much smaller area than the wings of an aircraft, but the density of water is about 800 times greater than the density of air, which means that the lift force is in practice large enough to support the weight of the boat.

FIGURE 6.18

The same effect makes it possible to sail a boat, or windsurf, 'into' the wind. Figure 6.19 shows the wind blowing across the front of the sail which takes up the curved shape of an aerofoil. Notice that the wind is blowing mainly in the opposite direction to that in which the boat or board is moving. There is a 'lift' force L as shown in the diagram (and also a drag force D). These can be resolved into a force P in the forward direction, and a force Q at right angles to that direction. Q is balanced by a push, R, from the water, on the hull and keel. So the boat or board is propelled forwards. The flexible surfaces of hang-gliders and sport parachutes also become aerofoil shaped when in flight, and so their forward motion provides lift.

FIGURE 6.17

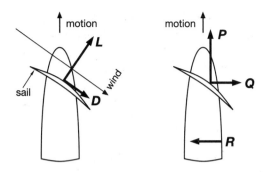

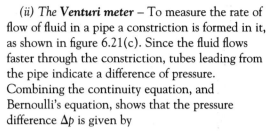

FIGURE 6.19

Racing cars have aerofoil surfaces at the front and rear, but they are fixed upside down (compared with those on an aircraft), as shown in figure 6.20, since their purpose is to push the car on to the road and increase the grip of the tyres on the road. A damaged aerofoil therefore affects a car's handling considerably.

(ii) The Venturi meter – To measure the rate of flow of fluid in a pipe a constriction is formed in it, as shown in figure 6.21(c). Since the fluid flows faster through the constriction, tubes leading from the pipe indicate a difference of pressure. Combining the continuity equation, and Bernoulli's equation, shows that the pressure difference Δp is given by

$$\Delta p = \tfrac{1}{2}\rho v^2 (r^2 - 1)$$

where r is the ratio of the larger cross-sectional area to the smaller cross-sectional area, ρ is the density of the fluid and v is the speed of the fluid in the unconstricted pipe. So the speed of the liquid is proportional to the square root of the measured pressure difference.

FIGURE 6.20

Other practical applications

(i) The carburettor and Bunsen burner – The air in a carburettor, and the gas in a Bunsen burner, are made to move through a constriction, as shown in figure 6.21(a, b), so that they move faster: there the pressure is less than atmospheric. In a carburettor the atmospheric pressure can then push petrol vapour into the air stream; in a Bunsen burner it pushes air into the gas stream. (The float in the carburettor is part of the mechanism for keeping the level of petrol constant.)

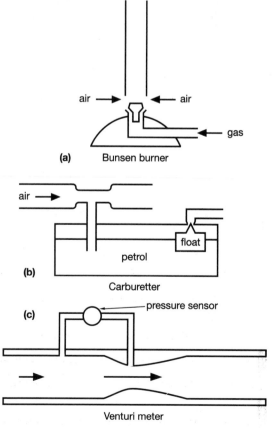

FIGURE 6.21

(iii) A ***spinning tennis ball*** moves in a curved path through air. Its rough surface drags round the air near it, and produces a difference in speed of the air on opposite sides of the ball. Figure 6.22(a) shows the actual situation and figure 6.22(b) shows the situation as 'seen' by the ball: air comes towards it, and the net effect of the air being dragged in the direction in which the ball rotates is to make the air move slowly past the top of the ball and more quickly past the bottom: hence a downward force. So a tennis player who puts top spin on a ball as he or she hits it fast just over the net can make the ball swerve downwards into the other court more steeply than normal so that the ball does not go out of court beyond the baseline. Golf balls rise up in the air as they do because the angle at which the club hits them makes them

spin. This is the opposite kind of spin to that described for the tennis ball. The dimples on the spinning ball drag the air round with the ball, and this makes the ball swerve upwards.

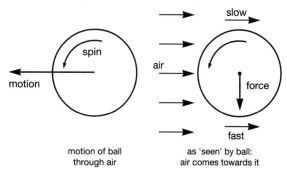

motion of ball
through air

as 'seen' by ball:
air comes towards it

FIGURE 6.22

Exercises on each section of this chapter may be found in the companion textbook, ***Practice in Physics***.

SUMMARY

At the end of this chapter you should be able to:

- use the equation $p = F/A$, which defines pressure.

- understand that gas pressure is caused by fast-moving gas molecules hitting a surface and that the pressure is steady because there is a very large number of individual randomly-occurring impacts.

- understand that atmospheric pressure is caused by the layer of air which is attracted to the Earth by gravitational forces.

- understand that changes in weather are caused by unequal atmospheric pressures in different parts of the world.

- understand that the pressure in a fluid exerts a force at right angles to any surface with which the fluid is in contact.

- use the equation $\Delta p = \rho g (\Delta h)$.

- calculate pressure differences in manometers and barometers.

- understand how Archimedes's principle can be used to calculate the upthrust of a body immersed in a fluid.

- understand what it means to say that a fluid is viscous.

- understand the difference between frictional and viscous forces.

- understand that in a pipe fluid flows faster where the pipe is narrower, and that the pressure there is less.

- understand how the Bernoulli effect is used in situations where air flows over a curved surface (aerofoils, hydrofoils, sails, windsurfing, paragliding).

7 Electric Currents, Energy and Power

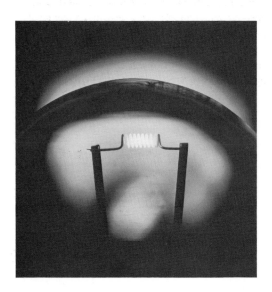

When we turn on a light switch we take it for granted that the filament will heat up and give us light. But why does the filament become that hot but no hotter? Why is tungsten used for the wire? Why is the filament coiled?

We want the light to be as 'white' as possible – to imitate the surface of the Sun which is at a temperature of about 6000 K. All metals melt before that temperature is reached so for everyday use we have to be content with a temperature of about 2600 K. We use tungsten because it has a high melting point (about 3700 K). What decides the temperature that the filament reaches? There is no thermostat! The temperature it reaches is the temperature at which its rate of loss of energy (mainly by radiation) is exactly the same as the rate of supply of energy. But its rate of loss of energy also depends on the surface area of the wire, so we need to use thin wire so that the temperature has to rise to a high value before these two rates are the same. To avoid making the wire excessively thin, and therefore too fragile, the wire is coiled, so that effectively the exposed surface area is reduced.

7.1 Electric current

There are many substances in which there are charged particles which can wander freely through the substance. In such a substance electric charge can be transferred from one point to another by a general drift of the charged particles within it. Such a movement of charge is called **an electric current**. Materials through which an electric current can pass are called **conductors**. Materials which do not contain charged particles which are free to move are called **insulators**.

A continuous flow of electric charge can happen only in a closed path or **circuit**. The circuit must also contain some device such as a cell or dynamo which maintains the flow. The device does this by the internal transference of electrons from one terminal to the other. In a cell chemical energy is used to do this. One terminal gains a surplus of electrons and so carries a negative charge: the other has a deficit of electrons and therefore a positive charge. When a conducting path is provided between the terminals the charge flows and tends to reduce both the surplus and the deficit, but the cell or dynamo then acts so as to maintain their initial values. So charge circulates continuously, as shown in figure 7.1, where a circuit diagram using conventional symbols is shown alongside a picture of the apparatus.

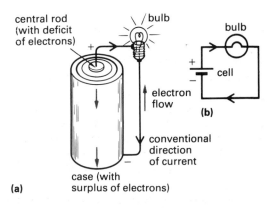

FIGURE 7.1

The direction of the current

The same electric current may be carried by different means even in the same circuit: at one point it may be caused by a flow of positively-charged particles in one direction, and at another point by negatively-charged particles moving in the opposite direction. We have to make a choice about which way we say the current flows, and we choose to say that the direction is that in which *positive* charge would flow. In figure 7.1(b) the arrow marked *on* the wire shows the direction of the current. In conducting wires the moving charges have a negative sign, so they flow in the opposite direction to the current.

Effects of an electric current

When there is an electric current in a circuit, two physical effects are observed.

- A *heating effect* – The kinetic energy of the moving charged particles is converted to internal energy when they collide with the stationary particles within the material, and the temperature of the material rises.

- A *magnetic effect* – In most positions close to a current-carrying wire a compass needle will be affected: this is evidence that moving charged particles create a magnetic field.

Ammeters

You will already have used ammeters to measure electric current. They may be **analogue** instruments, which show the reading by means of a pointer moving across a scale. One kind of analogue meter consists of a small coil pivoted between the poles of a magnet; the current to be measured passes through this coil. The reading of the instrument depends on the twisting effect caused by the magnetic forces on the coil. Or they may be **digital** meters, in which an analogue-to-digital converter chip, together with other electronic logic chips, provide a digital reading on a display panel.

Measurements taken with ammeters show that at any instant the current is the same at all points in

non-branching circuits. For example, in figure 7.2, the readings of ammeters M and N are the same. In branching circuits the current is the same *at any cross-section* of the circuit: in the figure the total current at the cross-section X-X is the same, at any instant, as the current at the cross-section Y-Y, whatever components are placed in the two branches K and L: $I_2 + I_3 = I_1$. The current divides to pass through the branches and then recombines.

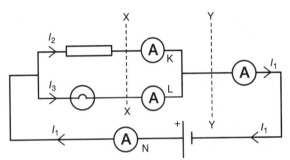

FIGURE 7.2

Experiments show that where we can measure the amount of charge Q passing through a circuit in time t (e.g. in an electrolysis experiment) we find that the current I is equal to the rate of passage of charge, so

$$I = \frac{Q}{t} \quad \text{or} \quad Q = It.$$

The ampere

The unit of electric current is called the **ampere** (A). It is one of the base units of the SI, and is defined in terms of the magnetic force between two current-carrying conductors as we shall see later. The equation $Q = It$ shows that if I and t are measured in amperes (A) and seconds (s) then the unit of electric charge must be the ampere second (A s). But it is convenient to have a special name for the unit, and it is called the **coulomb** (C). So

$$1\,C = 1\,A\,s$$
$$1\,A = 1\,C\,s^{-1}$$

Measurements show that the charge on one electron is $-1.60 \times 10^{-19}\,C$. Or we may say that $1\,C$ is the charge on 6.24×10^{18} electrons.

EXAMPLE

A car battery sends a current of 5.0 A through each of two headlamps and a current of 0.50 A through each of two sidelamps. In 20 minutes, how much charge passes through each headlamp, and how much passes through each sidelamp, and how much through the battery?

For one headlamp $Q = It = (5.0\,\mathrm{C\,s^{-1}})(20 \times 60\,\mathrm{s})$
$$= 6000\,\mathrm{C, \ and}$$
for one sidelamp $Q = It = (0.50\,\mathrm{C\,s^{-1}})(20 \times 60\,\mathrm{s})$
$$= 600\,\mathrm{C}.$$

The charge passing through the battery must be equal to the total charge passing through all four lamps, i.e.

$$2(6000\,\mathrm{C}) + 2(600\,\mathrm{C}) = 13\,200\,\mathrm{C}$$

7.2 Currents in solids

Metals

The most important class of conducting materials is the **metals**. In any solid the atoms are fixed at permanent sites in the material. But a small number (typically one per atom) of the electrons are shared with neighbouring atoms, providing the chemical bonds which bind the material together. In most non-metals these **valence electrons** are fixed within particular groups of atoms, but in metals the valence electrons are free to move through the crystal lattice, and it is this movement which consistutes the electric current in the metal.

In some materials there also exist unfilled vacancies (called **holes**) in the electronic structure of the material. It is possible for a *bound* electron (i.e. *not* a valence electron) to move into such a hole, and the hole is transferred to the atom from which the electron came. The hole behaves like a centre of positive charge, and the hole can move through the material. So although in many metals (e.g. copper, silver) the condution is by means of *negative* electrons (**n-type** conduction) in some metals (e.g. zinc) the conduction is mostly by means of *positive* holes (**p-type** conduction).

Semiconductors

Intermediate between the metals and the non-metals is a class of materials called the

semiconductors, such as germanium and silicon. In a pure semiconductor only about 1 in 10^{10} of the charge carriers (electrons or holes) are free to move (but the proportion increases with temperature). The conductivity can be increased by adding atoms of other elements: by a suitable choice of atom the material can be made p-type or n-type. Devices such as diodes, transistors and integrated circuits are made by creating crystals of silicon with regions of n- and p-type material within them.

The speed of the charge carriers

You might think that in a wire carrying a current, the electrons must be moving very fast: when you press switches, lamps light with no apparent delay.

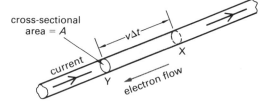

FIGURE 7.3

We can work out how fast the electrons are moving in a wire.

Consider a length of conductor of cross-sectional area A containing n free electrons per unit volume, as shown in figure 7.3. The number of electrons in the volume remains constant. Suppose the average speed of the electrons is v. Then the number of electrons passing any point such as Y in a time Δt is the number contained in a length $v(\Delta t)$. So

volume between X and Y = $Av(\Delta t)$
number of electrons between X and Y = $nAv(\Delta t)$
total charge ΔQ passing Y in time Δt is $enAv(\Delta t)$

where e is the charge on each electron.

But the current I = rate of flow of charge

$$= \frac{\Delta Q}{\Delta t} = enAv$$

so

$$v = \frac{I}{enA}.$$

The largest current which a copper fuse wire of diameter 0.50 mm can carry without melting is 5.0 A. What is the average speed of the electrons in it? Take the number density of free electrons to be $1.0 \times 10^{29}\,\mathrm{m}^{-3}$ and the charge on each electron to be $1.6 \times 10^{-19}\,\mathrm{C}$.

Using $I = enAv$

$$v = \frac{I}{enA} =$$

$$\frac{5.0\,\mathrm{A}}{(1.6 \times 10^{-19}\,\mathrm{C})(1.0 \times 10^{29}\,\mathrm{m}^{-3})\{\pi(0.25 \times 10^{-3}\,\mathrm{m})^2\}}$$

$$= 1.6 \times 10^{-3}\,\mathrm{m\,s}^{-1} = 1.6\,\mathrm{mm\,s}^{-1}$$

The speed calculated in this example is probably much smaller than you expected. This *is* a typical speed for electrons, but the electrical forces which set them in motion are sent round the circuit at very great speed – in fact with very nearly the speed of light $(3 \times 10^{8}\,\mathrm{m\,s}^{-1})$. It is like what happens when a stationary column of soldiers is told to march. The message travels with the speed of sound and reaches all of them (almost) simultaneously, so that they start marching simultaneously. The slow speed with which they start marching has no connection with the high speed with which the message reaches them.

In a simple circuit consisting only of copper conducting wire, the current is the same everywhere, but the speed of the electrons will depend on the cross-sectional area of the wire: where the area is smaller, the speed will be greater. In a fuse wire, which is made of copper but is narrower than the other wires in the circuit, the speed will be greater and there will be more kinetic energy to be converted to internal energy. This is part of the reason why the fuse wire melts before the other wire in the circuit.

A good conducting material is one for which n is large: for metals it is usually between 10^{28} and 10^{29}, but for semi-conducting materials it is about 10^{20} (but depends on temperature). You can deduce that the speeds of charge carriers in semi-conducting materials must therefore be *much* greater than they are in conductors.

7.3 Electrical energy

A current flowing in a circuit may be lighting a lamp or producing useful mechanical work (if it is passing through a motor). Electrical energy is being converted into some other form. It is obvious that these processes cannot happen unless there is, somewhere else in the circuit, some device which is converting a non-electrical form of energy into electrical energy at exactly the same rate as it is being converted in the motor or lamp. We cannot get something for nothing. There are several devices which can convert non-electrical forms of energy into electrical potential energy: two of the commonest are the cell and the dynamo. In both cases non-electrical forces separate electric charges at the terminals of the device and give them electric potential energy: if the terminals are then connected by a wire, the free electrons in the wire are given kinetic energy which is then converted into internal energy in the wire.

At this stage it may help you to use an *analogy* – one which will be useful later, too. The analogy is between the charged particles in electric circuits, which we have just been thinking about, and people moving up to the top of a hill on a ski-lift, and down again. You will see the analogy illustrated in figure 7.4.

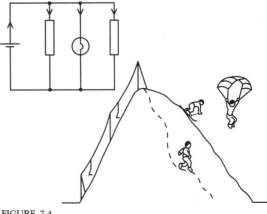

FIGURE 7.4

◆ *In electric circuits:*
(1) The cell pushes electric charge to the 'top' of the cell.
(2) The charge then has electric p.e.
(3) The charge can return from the 'top' to the 'bottom' of the cell. However this happens, the electric p.e. is converted into kinetic energy and then into internal energy.

◆ *In ski-lifts:*
(1) The ski-lift pushes people to the top of the hill.
(2) The people then have gravitational p.e.
(3) The people can return from the top to the bottom of the hill. However this happens, the gravitational p.e. is converted into kinetic energy and then into internal energy.

E.m.f.

We say that the cell or dynamo produces an **electromotive force**. This is an unfortunate form of words, since it if not a force at all, so we shall always use the abbreviation **e.m.f.** The cell or dynamo separates electric charge: if when a charge Q is separated, it is given electric potential energy W, then the e.m.f. $\mathscr{E}$ is defined by the equation

$$\mathscr{E} = \frac{W}{Q}$$

The unit of e.m.f. is therefore the JC^{-1} but for convenience we call it the **volt** (V). So

$$1\,V = 1\,JC^{-1}$$

For e.m.f. we shall use the symbol $\mathscr{E}$ to avoid confusion with the electric field strength E which you will meet later on.

Potential difference

An electric cell separates charges of opposite sign between its terminals and these regions of separated charge cause electrical forces on charged particles nearby. We say that there is an *electric field* between the terminals, in and around the cell. If there is a conducting path between the terminals the electric field acts on the charge carriers and sets them in motion. The electric forces therefore do work on the charge carriers, accelerating them

until they collide with the fixed atoms in the material, when they stop momentarily before being accelerated again. Their motion is rather like the motion of the balls in a pin-ball machine. So the electrical potential energy is converted into kinetic energy and then into internal energy.

We need some way of describing the amount of electrical potential energy which can be given to the charge as it moves between two points in an electric field, and we define the **potential difference (p.d.)**, V, between two points by the equation

$$V = \frac{W}{Q}$$

where W is the electrical potential energy converted to some other form when a charge Q passes between the points. The unit of electrical potential difference is therefore the same as the unit of e.m.f., i.e. the JC^{-1} or volt (V). Although the units of e.m.f. and p.d. are the same, they are not the same quantity. The e.m.f. tells us how much energy is converted from non-electrical forms (e.g. chemical, mechanical) into electrical potential energy in the device where the conversion takes place. The p.d. tells us how much electrical potential energy is converted into other forms (e.g. internal, mechanical) in the circuit outside the device. For example, the e.m.f. of a battery might be 12 V: i.e., each coulomb passing through it is given 12 J of electrical potential energy. But if some of that energy is converted in the device itself, the energy available for the external circuit will be less than 12 J per coulomb, so the p.d. will be less than 12 V.

EXAMPLE

An immersion heater is placed in a beaker of water and from measurements of the temperature rise it is found that in 5 minutes the energy supplied to the water is 15 kJ. If the current in the immersion heater was 4.3 A, what was the p.d. between the terminals of the heater?

Charge passed
$$Q = It = (4.3\,A)(5 \times 60\,s) = 1290\,C$$

$$\text{P.d. } V = \frac{W}{Q} = \frac{15000\,\text{J}}{1290\,\text{C}} = 11.6\,\text{V}$$

The e.m.f. of the battery was probably more than 11.6 V.

Measuring potential difference

Potential differences are measured using **voltmeters** which, like ammeters, may be analogue or digital instruments. In fact, as we shall see later, the same basic instrument may be used as either, though in practice a meter is usually adapted for use as one or the other. Figure 7.5(a) shows a photograph of a circuit in which voltmeters are being used to measure the p.d. V_r between the ends of a resistor, V_b between the ends of a bulb, and V_c between the ends of a cell. Figure 7.5(b) shows the circuit diagram. You can see that

$$V_c = 1.33\,\text{V}, \ V_r = 0.44\,\text{V and } V_b = 0.89\,\text{V}$$
so that
$$V_c = V_r + V_b$$

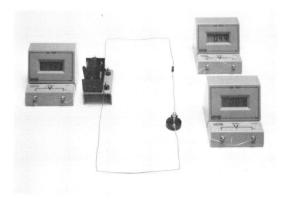

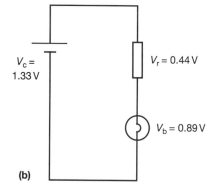

$V_c = 1.33\,\text{V}$

$V_r = 0.44\,\text{V}$

$V_b = 0.89\,\text{V}$

(b)

FIGURE 7.5

This is what you should expect, since what the voltmeters are doing is telling us how much energy is converted between the points where they are connected. Very little energy is converted in the wires themselves, so if, for each coulomb passing, 0.44 J of energy is converted in the resistor, and 0.89 J is converted in the bulb, and almost none in the wires, then *of course* the total (in resistor, bulb and wires) is 1.33 J!

The circuit was then disconnected, and the voltmeter connected across the same cell, with the result shown in figure 7.6. Now the p.d. between the cell terminals is 1.52 V. Why has it changed?

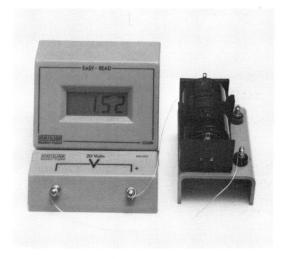

FIGURE 7.6

In the first circuit there was a current in the circuit, and therefore in the cell, and internal energy was being converted in the cell so that not all of the electrical energy was available for the external circuit. So we can see that for each coulomb the cell actually converted 1.52 J, but only 1.33 J was delivered to the circuit. The remaining 0.19 J was converted within the cell itself. So the e.m.f. of the cell is 1.52 V. In fact there must be *some* current passing through the cell when the voltmeter is connected or else the voltmeter would not work. However, most voltmeters, and especially digital voltmeters like the one shown, draw very little current from the cell and so the energy converted within the cell is negligible. Connecting a voltmeter across a cell when it is not delivering a current does effectively measure its e.m.f.

EXAMPLE

The diagram shows a circuit consisting of an ammeter, a bulb and a resistor connected in series to a cell, whose e.m.f. is known to be 1.20 V. A voltmeter connected between A and H reads 1.16 V, and a voltmeter connected between B and C reads 0.51 V.

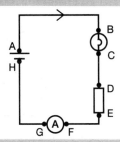

(a) What would you expect voltmeters to read if they were connected between A and B, C and D, D and E, E and F, F and G, G and H?
(b) When a current of 0.27 A flows for 10 minutes, what are the amounts of energy converted in this circuit?

(a) The voltmeters connected between A and B, C and D, E and F, F and G, G and H should all read *zero*, since there is negligible energy being converted in either the wires or the ammeter. Between A and F we know that the p.d. is 1.16 V, and that 0.51 V of this occurs between B and C, so there must be a p.d. of 1.16 V − 0.51 V (= 0.65 V) between D and E, since these are the only other points between which energy is converted.
(b) The charge Q which passes is given by $Q = It = (0.25\,A)(600\,s) = 150\,C$. So to calculate the energy converted in

◆ the cell, to electrical p.e.:
$W = \mathscr{E}Q = (1.20\,V)(150\,C) = 180\,J$.

◆ the bulb, to internal energy:
$W = VQ = (0.51\,V)(150\,C) = 76.5\,J$.

◆ the resistor, to internal energy:
$W = VQ = (0.65\,V)(150\,C) = 97.5\,J$.

The total amount of electrical p.e. converted to internal energy in the external circuit is 76.5 J + 97.5 J = 174 J, so the remaining 6.0 J of electrical p.e. must have been converted to internal energy within the cell itself.

Potentials

Let us make some more use of the analogy of the ski-lift and its passengers. Anyone walking down the hill will have a map on which *contour lines* are marked. These tell the walkers the heights at certain points on the map. But heights above what? Above some zero level which is an *arbitrary* choice (a different zero level would be chosen in a different country). But the walker does not worry about where the zero is: all that matters is the *difference* in level between certain points.

Just as maps are marked with contour lines so we sometimes find it helpful to talk about the **potentials** at points in an electric circuit, as well as the **potential differences between** points. But before we can do this we need to label some point in the circuit as having zero potential. It really does not matter which point we choose, but usually it is the negative terminal of the cell which is chosen to have zero potential. Figure 7.5(b) has been redrawn as figure 7.7 to show this.

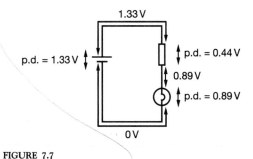

FIGURE 7.7

Remember that we assume that there is no p.d. between the ends of any of the connecting wires, so the whole length of each of these wires is at the same potential. Then these different lengths have potentials of 1.33 V, 0.89 V and 0 V. Where would be the point with a potential of 1.00 V? Somewhere in the resistor. Where would be the point with a potential of 0.50 V? Somewhere in the bulb.

P.d. in a branched circuit

Figure 7.8(a) shows a photograph of a branched circuit, and (b) shows the circuit diagram. Here we can see that the p.d.s between the ends of the components between B and C are the same, even

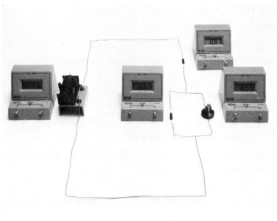

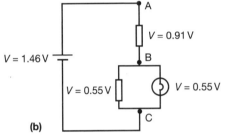

(b)

FIGURE 7.8

though they are clearly different components, and probably have different currents in them. These p.d.s *cannot* be different, however unlike the components are. Think of the walkers coming down the hill: suppose they all start by going down the same path, but then split up, taking different paths. However different their paths may be from then on, they *must* go down through the same vertical height between there and the bottom of the hill and *must* lose the same amount of gravitational potential energy. In the circuit the p.d.s *cannot* be different because whichever route the charge takes from B onwards, it must lose the same amount of electrical potential energy. The voltmeter reading across the cell terminals shows that there is 1.46 J of energy available (for each coulomb) at the start. The voltmeter reading between A and B shows that each coulomb passing between those points loses 0.91 J of electrical p.e., and so each coulomb has 0.55 J of electrical p.e. still to lose, whichever way it goes next. Probably more coulombs go one way than the other, but any one coulomb loses 0.91 J of its electrical p.e. between A and B and the remaining 0.55 J between B and C, whichever way it goes.

EXAMPLE

The p.d. between the terminals of the cell in the circuit shown in the diagram is measured to be 1.53 V and the p.d. across the bulb is 0.61 V. What are the p.d.s between the ends of resistors R_1 and R_2?

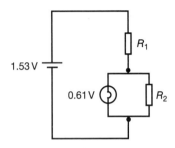

The p.d. between the ends of R_2 *must* be 0.61 V, since the bulb and R_2 are connected in parallel. The p.d. between the ends of R_1 must be 0.92 V.

The cell provides each coulomb with 1.52 J of electrical p.e., all of which must be converted before the coulomb returns to the cell. Each coulomb lost 0.92 J in R_1: some lose 0.61 J in the bulb and the others lose 0.61 J in R_2.

7.4 Electrical power

The *rate* at which energy is converted from one form to another in a mechanical or electrical device is called the **power** of the device so the power P is defined by the equation

$$P = \frac{W}{t}$$

where W is the energy converted in time t. The unit is the $J\,s^{-1}$ which for convenience is called the **watt** (W). So

$$1\,W = 1\,J\,s^{-1}$$

When charge Q passes through a device which has a p.d. V between its ends, the energy W is given by $W = VQ$. If this happens in time t,

$$P = \frac{W}{t} = \frac{VQ}{t} = V\left(\frac{Q}{t}\right) = VI$$

In the same way the power P_{tot} converted to electrical p.e. by a source of e.m.f. $\mathscr{E}$ is given by

$$P_{tot} = \mathscr{E}I$$

This is the *total* power converted to electrical form by the source of e.m.f.: part of this is lost as internal energy within the source, and the rest is delivered to the external circuit (where it can be measured using $P = VI$).

EXAMPLE

A battery of e.m.f. 12.00 V is connected to a bulb. The p.d. shown by a voltmeter connected across the battery is 11.63 V. An ammeter shows that the current in the circuit is 2.96 A. Calculate the rates of conversion of different kinds of energy in this circuit.

The rate of conversion, P_{tot}, of chemical energy to electrical p.e. in the battery is given by $P_{tot} = \mathscr{E}I$, so

$$P_{tot} = \mathscr{E}I = (12.00\,V)(2.96\,A) = 35.5\,W$$

The p.d. across the bulb must be the same as the p.d. across the battery, and the rate of conversion P of electrical p.e. to internal energy in the bulb is given by $P = VI$, so

$$P = VI = (11.63\,V)(2.96\,A) = 34.4\,W$$

The rate of production of internal energy in the battery must be the difference between P_{tot} and P, i.e. 1.1 W.

Lamp filaments

When there is an electric current in a wire, internal energy is produced in it and its temperature rises until the rate of loss of energy from its surface is equal to the rate of production of internal energy. The temperature then remains steady. For an electric lamp it is important to achieve the greatest luminous efficiency, i.e. to ensure that as much as possible of the energy converted appears as light. Luminous efficiency depends on temperature, so tungsten, which has a melting point of about 3700 K, is used. However, even tungsten cannot be used in a vacuum above 2400 K if serious evaporation is to be avoided. Most lamps are therefore filled with an inert gas. This reduces evaporation and enables temperatures of 2600 K or more to be used. Also the tungsten that does now evaporate is carried away by convection and deposited at the top of the bulb instead of producing a uniform blackening of the glass.

Halogen, or **quartz–iodine**, lamps run at an even higher temperature. The bulb is made of quartz and filled with iodine vapour: a chemical process enables the evaporated tungsten to be recycled to the filament. Thus the filament has a higher luminous efficiency because of the higher temperature, but the chemical process means that its life is prolonged. This type of filament is used in car headlamps and in projector lamps.

Fuses

The wires used for joining up an electric circuit must be thick enough so that the temperature which they reach is normally only very slightly above that of the surroundings. But an accidental short circuit would cause a large current which would heat the wires and cause damage. To avoid this happening in mains installations there is a **consumer unit** or **fuse-box** where the supply enters the house which either contains **circuit breakers** (which switch off the supply when the current becomes too great in one of the circuits) or **fuses**. In the consumer unit a fuse usually consists of a short length of relatively fine wire which is joined to the circuit by means of two screw-down terminals. In addition separate fuses, contained in small ceramic or glass tubes, are provided inside each plug used to connect an appliance. The diameter of the wire in the fuse is selected so that it quickly reaches its melting point and breaks the circuit if the current exceeds the maximum safe value.

Exercises on each section of this chapter may be found in the companion textbook, *Practice in Physics*.

SUMMARY

At the end of this chapter you should be able to:

◆ remember that the current is the rate of flow of electric charge and use the equation $Q = It$.

◆ remember that the total current flowing away from a junction is equal to the total current entering it.

◆ understand that in metals it is usually the (negative) electrons that are the charge carriers but that charge may be carried by (positive) holes in other materials.

◆ use the equation $I = enAv$ for the current in a conductor.

◆ remember that there is a difference between the drift speed of the charge carriers and the speed of transmission of the electric field.

◆ understand the meaning of e.m.f. $\mathscr{E}$ and use the equation $W = \mathscr{E}Q$.

◆ understand the meaning of potential difference V and use the equation $W = VQ$.

◆ understand that when two components are connected in series, the total p.d. is equal to the sum of the separate p.d.s.

◆ understand that when two components are connected in parallel, the p.d. between the ends of each of them is the same.

◆ understand that the p.d. between the terminals of a cell will be less than its e.m.f. when it is delivering a current.

◆ understand that we can assign values of potential to points in a circuit if we arbitrarily fix the value of the potential at some point (usually calling the cell's negative terminal zero).

◆ use the equation $W = Pt$.

◆ use the equations $P = \mathscr{E}I$ and $P = VI$ to calculate the power of a cell and the power of a resistor or other circuit component.

8 Electric Circuits

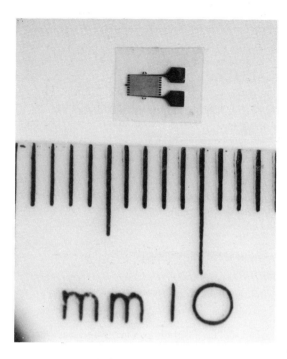

The photograph shows a strain gauge. It is a length of about 30 mm of very fine wire, doubled back on itself about a dozen times to give a device which measures about 2 mm by 1 mm. The ends are brought out to two metal pads to which connections can be made. The wire is fixed to a thin, flexible, plastic base, which in turn is glued to part of some large structure, such as a building or a bridge. Any strain in the structure is accompanied by the same strain in the strain gauge, so the wire becomes very slightly longer and very slightly narrower. The effect of both these changes is to increase the electrical resistance of the wire. As we shall see in this chapter, this change of resistance may be measured precisely, so the strain in the structure can be measured. In monitoring the effects of various kinds of loading on, say, a bridge, caused by traffic or wind forces, many such strain gauges would be fixed to different parts of the bridge, and wires from the strain gauges led back to a monitoring station where the variations in strain would be recorded.

8.1 Resistance

The current in a particular metallic conductor depends on a number of factors besides the p.d. across it. The most important of these is the temperature, but other physical conditions (elastic strain, illumination) may also affect the current. **Ohm's law** states that if all these are kept constant, the current I in a metallic conductor is directly proportional to the p.d. V across it, i.e.

$$\frac{V}{I} \text{ is constant}$$

for a given metallic conductor under steady physical conditions. The constant is known as the **resistance** R, so

$$R = \frac{V}{I}$$

Even when we are dealing with a circuit component to which Ohm's law does not apply, the quantity V/I is still defined as the resistance of the component. In this case the resistance is not constant. You can see that the unit of resistance must be the $V\,A^{-1}$, but for convenience this is called the **ohm** (symbol Ω, the Greek letter omega). Thus

$$1\,\Omega = 1\,V\,A^{-1}$$

The **conductance** G of a conductor is defined by the equation $G = 1/R$. The unit of conductance is the Ω^{-1} which is called the **siemens** (S). So for a metallic conductor obeying Ohm's law we have

$$G = \frac{1}{R} = \frac{I}{V}$$

You could investigate the relationship between current and p.d. for a wire, or any other component, with the circuit shown in figure 8.1(a). Using different numbers of cells, you would have different p.d.s (measured with the voltmeter) across the component, and different currents (measured with the ammeter) in it. You could also reverse the component in the circuit. This is equivalent to reversing the battery, so you could think of this process as applying negative p.d.s and so obtaining negative currents. If Ohm's law applies, the graph of current against p.d. will be a straight line passing through the origin, as shown in figure 8.1(b).

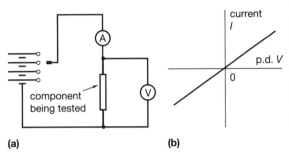

(a)   **(b)**

FIGURE 8.1

Fixed resistors

Resistors are commonly used circuit components, and can easily be obtained in a range of values from $0.03\,\Omega$ to $100\,\text{M}\Omega$. Most of them fall into two categories: **carbon** or **metal film**, and **wire-wound**.

The carbon or metal film is deposited on a ceramic *former* (a cylindrical supporting base). The carbon film resistors are cheap, but their **tolerance** is only $\pm5\%$, i.e. the manufacturer guarantees the resistance to within only $\pm5\%$ of the stated value. The tolerance of metal film resistors may be as low as $\pm0.1\%$, depending on the cost. In both cases a helical groove is cut in the film to adjust the resistance to the required value, as shown in figure

8.2(a). Then the whole device is coated with lacquer to protect it. The wire-wound resistors (figure 8.2(b)) consist of wire (usually the alloy **constantan**) wound on an insulating former: they are then coated with an epoxy resin, or an enamel coating. The advantage of these is that they have good **stability**: their resistance varies little with change in conditions, particularly temperature. The value of the resistance is often indicated by a series of coloured bands: the first three or four of these give the value in ohms, and the last indicates the tolerance. Increasingly the resistance is being shown in figures: e.g. 150R means $150\,\Omega$, and 6K8 means $6.8\,\text{k}\Omega$.

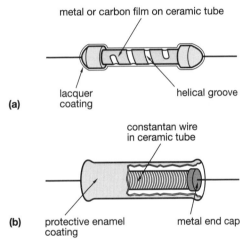

metal or carbon film on ceramic tube

(a)   lacquer coating   helical groove

constantan wire in ceramic tube

(b)   protective enamel coating   metal end cap

FIGURE 8.2

All resistors are available with a range of power ratings: e.g. you could ask for a $100\,\text{k}\Omega$, $0.5\,\text{W}$ resistor. This means that the maximum safe rate of conversion of energy in the resistor is $0.5\,\text{W}$. You can work out (using $P = VI$ and $V = IR$) that this means that the largest current which may be allowed to pass through it is $2.2\,\text{mA}$.

Variable resistors

Variable resistors can be constructed using wire wound on either a straight ceramic tube (figure 8.3(a)) or on a cylindrical insulating support (figure 8.3(b)). There are terminals at each end of the wire, and in addition a sliding contact which can touch the wire at different points along its length. The turns of wire are insulated from each other, but bared where the slider can touch them.

The straight version is usually called a **rheostat**: it is wound with relatively thick wire and can therefore carry quite large currents, perhaps as much as 10 A. The maximum current is specified on a label fixed to the rheostat: this should not be exceeded, otherwise the wire will overheat. The resistances of rheostats are usually in the range 5 Ω to 100 Ω.

(a)

(b)

FIGURE 8.3

The circular version is usually called a **potentiometer**: it will have many turns of relatively thin wire (or, in cheaper versions, just a carbon track). The resistance may be anything from 10 Ω to 1 MΩ. Some **multi-turn** potentiometers have a helical track, and 10 turns of the knob are needed to move the slider from one end of the track to the other. This makes them *sensitive*: a particular movement of the knob changes the resistance by only a small proportion of the total.

Characteristics

For the fixed and variable resistors considered so far, Ohm's law will apply. There are other types of circuit component for which it does not, and the variation of current with p.d. (the **characteristic** of

the component) can be investigated using the circuit of figure 8.1. Some of these components are listed below and the characteristics are shown in figure 8.4.

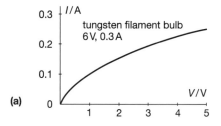

(a)

(*a*) *A metal lamp filament.* In normal use a filament becomes hot enough for its resistance to increase considerably. There is of course nothing special about the wire used to make the filament: its resistance *would* be constant for large values of V and I if the filament could be kept at a constant temperature. The graph is drawn for a typical pea bulb rated at 6 V, 0.3 A.

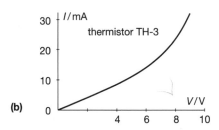

(b)

(*b*) *A thermistor.* A **thermistor** is a resistor made from some semi-conducting material such as silicon. Its resistance decreases with temperature, because the material then gains enough energy to enable additional charge carriers to become free to take part in the conduction process. So when there is a current in the thermistor the temperature rises and the resistance decreases. As with the lamp filament, the resistance would be constant if the thermistor could be kept at a constant temperature. The measurements for the graph were obtained by passing a current through the thermistor (type TH-3) and waiting for several minutes until its temperature became steady.

(*c*) *A semiconducting diode.* The **diode** consists of a silicon crystal which has been treated so that it contains n-type and p-type silicon. The effect is as shown in figure 8.4(c) for a particular diode

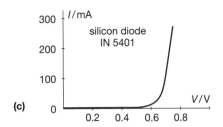

(1N5401). There is little current, even in the forward direction, until the p.d. rises to about 0.5 V (the **turn-on** p.d.). Then there is a rapid increase in current for very little increase in p.d. until at about 0.8 V its resistance is only about 1 Ω and is probably negligible compared with that of the other components in the circuit. In the reverse direction the resistance is not constant but it is very large (about 100 MΩ). It is also possible to construct diodes from germanium: the turn-on p.d. for these is about 0.2 V.

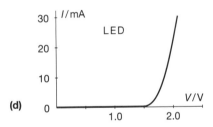

FIGURE 8.4

(d) *A light-emitting diode*. The **LED**, which is made from gallium phosphide or gallium arsenide phosphate, is a diode with the additional property that it emits light when a current passes through it in the forward direction. Different kinds of LED can produce blue, green, yellow or red light, or infra-red radiation: the colour depends on the material used. The graph shows that for one particular LED no current flows until there is a p.d. of about 1.50 V when the LED begins to glow dimly), and then the current grows rapidly to 30 mA at just over 2.0 V. This is a typical maximum current for an LED. They are often used as indicators in logic circuits, where the supply p.d. is 5 or 6 V, so a current-limiting resistor of about 100 Ω then needs to be connected in series with the LED so that the p.d. across the LED is only about 2 V. LEDs are also familiar in 7-segment numerical displays in which seven bar-shaped LEDs are used to construct the digits.

(e) *A light-dependent resistor*. The **LDR** is made from cadmium sulphide. Light falling on it liberates more charge carriers and typically its resistance falls from about 10 MΩ in the dark to about 100 Ω in bright light.

Energy conversion in resistors

The electrical power P converted in a resistor is given by $P = VI$: also, the resistance R is given by $R = V/I$.

Substituting for V we have

$$P = I^2 R$$

and substituting for I we have

$$P = \frac{V^2}{R}$$

These equations tell us that for constant resistance, the power is proportional to the *square* of the current or p.d.

EXAMPLE

A bulb is labelled '6 V, 5 W'. What would be its power if it were connected to a battery which provides a p.d. of 3.0 V?

$P = V^2/R$ so, for constant resistance, if the p.d. is halved the power will become one-quarter of what it was, i.e. 1.25 W. But at this lower power the temperature will be lower and the resistance less, so in fact the power will be rather more than 1.25 W.

We may use the equations $P = I^2R$ and $P = V^2/R$ only when calculating the rate of conversion of electrical potential energy *to internal energy in resistors*: The equation $P = VI$ may be used to calculate the rate of conversion to *any form of energy* where the p.d. is V and the current I.

EXAMPLE

An electric motor is connected to a power supply: when the p.d. across the terminals is 20 V, the current flowing is 0.50 A. The resistance of the coils of the motor is known to be 12 Ω. Calculate the rate of conversion of internal energy in the coils, and the rate of production of mechanical energy by the motor.

For the motor as a whole, the power supplied
$$= VI = (20\,V)(0.50\,A) = 10\,W$$

For the coils,
$$P = I^2R = (0.50\,A)^2(12\,\Omega) = 3.0\,W$$
so the rate of production of mechanical energy
$$= 10\,W - 3.0\,W = 7.0\,W.$$

Combinations of resistors

Many circuits contain resistors connected together and you need to be able to calculate the effective or total resistance of these combinations. However complicated the networks may appear to be, the problem of finding their total resistance can nearly always be reduced to that of finding either the total resistance of two resistors *in series* or the total resistance of two resistors *in parallel*. For example, in figure 8.5 you would:

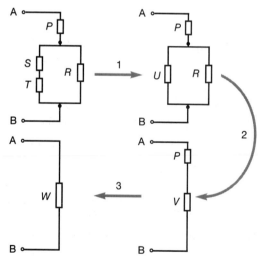

FIGURE 8.5

1 find the total resistance of S and T (in series): call this U
2 find the total resistance of U and R (in parallel): call this V
3 find the total resistance of P and V (in series): call this W. W is the total resistance of the resistors between the points A and B.

So you need rules for finding the total resistance of two resistors (a) in series, (b) in parallel.

Resistors in series

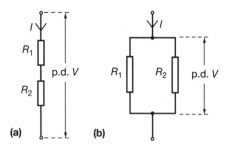

FIGURE 8.6

In figure 8.6(a) there is the same current I in each of the resistors R_1 and R_2. The total resistance R_{ser} is given by

$$R_{ser} = \frac{V}{I} \quad \text{or} \quad V = IR_{ser}$$

where V is the p.d. between A and B. V is the sum of the separate p.d.s across the resistors, so

$$V = IR_1 + IR_2$$
$$IR_{ser} = I(R_1 + R_2)$$
$$R_{ser} = R_1 + R_2$$

Resistors in parallel

In figure 8.6(b) there is the same p.d. V across each resistor. The total resistance R_{par} is given by

$$R_{par} = \frac{V}{I} \quad \text{or} \quad I = \frac{V}{R_{par}}$$

where I is the total current through the combination. This is the sum of the separate currents in the resistors, so

$$I = \frac{V}{R_1} + \frac{V}{R_2} = V\left(\frac{1}{R_1} + \frac{1}{R_2}\right)$$

so
$$\frac{V}{R_{par}} = V\left(\frac{1}{R_1} + \frac{1}{R_2}\right) \quad \Rightarrow \quad \frac{1}{R_{par}} = \frac{1}{R_1} + \frac{1}{R_2}$$

These results can be extended to combinations of any number of resistors: e.g. for three resistors in series, $R_{ser} = R_1 + R_2 + R_3$. In one particular case the result may be simplified, making calculations quicker. For n equal resistances, R, in parallel

$$R_{par} = \frac{R}{n}$$

EXAMPLE

Calculate the combined resistance of each of the arrangements of resistance shown in the figure:

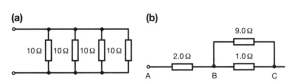

(a) In this case the resistances are equal, so

$$R_{par} = \frac{10\,\Omega}{4} = 2.5\,\Omega$$

(b) Between B and C there are two resistances in parallel:

$$\frac{1}{R_{par}} = \frac{1}{9.0\,\Omega} + \frac{1}{1.0\,\Omega} \quad \Rightarrow \quad R_{par} = 0.90\,\Omega.$$

This resistance is in series with $2.0\,\Omega$, and so the total resistance $R_{AC} = 2.0\,\Omega + 0.90\,\Omega = 2.9\,\Omega$.

Notice that in (a) of this example the total resistance of the four $10\,\Omega$ resistors is only $2.5\,\Omega$ (less than $10\,\Omega$). This should not surprise you if you realise that adding resistors in parallel makes it easier for the current to flow. It is like using a thick wire instead of a thin wire. In (b) there is another example of this idea. The $1.0\,\Omega$ resistor has a $9.0\,\Omega$ resistor connected in parallel, and the total resistance becomes less than $1\,\Omega$ (not much less, because the resistance of the $9.0\,\Omega$ resistor was

relatively large). Whenever two resistors are connected in parallel, the total resistance is less than the resistance of the smaller of the two.

8.2 Resistivity

Simple experiments show that for a uniform conductor the resistance is directly proportional to its length l and inversely proportional to its cross-sectional area A. The resistance R of such a conductor can therefore be written

$$R = \frac{\rho l}{A}$$

which is an equation which defines ρ, a property of the material of the conductor called its **resistivity**. You can see that the unit of resistivity is the $\Omega\,m$.

EXAMPLE

The 'extra-flexible' wire which is sometimes used to make connecting leads in laboratories consists of 55 strands of copper wire, each of diameter 0.10 mm. If the resistivity of copper is $1.7 \times 10^{-8}\,\Omega\,m$, calculate the resistance of a connecting lead made from 0.50 m of this wire.

The total cross-sectional area A of this wire is

$$55 \times \pi(0.05 \times 10^{-3}\,m)^2 \text{ so}$$

$$R = \frac{\rho l}{A} = \frac{(1.7 \times 10^{-8}\,\Omega\,m)(0.50\,m)}{55 \times \pi(0.05 \times 10^{-3}\,m)^2} = 0.020\,\Omega.$$

We could have calculated the resistance of a single strand of wire and then divided by 55, since the wire consists of 55 strands in parallel.

The **conductivity** σ of a material is defined by the equation

$$\sigma = \frac{1}{\rho}$$

and we can write $R = \dfrac{l}{\sigma A}$

Values of conductivity and resistivity

A conducting solid contains free charge carriers, either electrons or positive holes, which are free to wander through it in any direction. When there is no electric field in the material they are in rapid random motion. They form a kind of 'electron gas' confined within the crystal structure. When a p.d. is applied to the conductor, and produces an electric field in it, a slow drift of the electron gas takes place throughout the material. It is this that constitutes the electric current.

The conductivity of a material is proportional to n, the number of charge carriers per unit volume. Metals all have large values of n (about $10^{29}\,\text{m}^{-3}$) and are therefore good conductors. Silver is the best conductor, closely followed by copper: aluminium has a conductivity about half that of copper. But the density of copper is about three times that of aluminium, so a conductor of given resistance is much lighter if made of aluminium rather than copper. For this reason the overhead power lines of the electricity grid are made of aluminium (twisted round a steel core which gives the cables additional strength). Other metals have lower conductivities, but of the same order of magnitude.

Non-metallic substances have much lower conductivities, and greater resistivities, almost all of them more than $10^{4}\,\Omega\,\text{m}$. For example, the resistivities of different types of glass range between $10^{6}\,\Omega\,\text{m}$ and $10^{11}\,\Omega\,\text{m}$. Materials such as polythene have resistivities of about $10^{14}\,\Omega\,\text{m}$.

Intermediate between these extremes comes the class known as semiconductors. Only a small proportion of the valence electrons is free to move through the material: at 293K the resistivity of silicon is $4 \times 10^{3}\,\Omega\,\text{m}$ and the resistivity of germanium is $0.7\,\Omega\,\text{m}$. But minute traces of impurities may increase the number of charge carriers considerably and therefore reduce the resistivity. Impurities in a metal do not much affect the number of free charge carriers, but do affect their freedom of movement, so the resistivity of metals increases with increasing impurity.

Temperature effects

When the temperature of a metal is raised the thermal vibration of the atoms increases. This increases the interaction of the electrons with the crystal lattice, and the electrons cannot move as freely. The resistivity of the material therefore increases with temperature. The same effect happens in semiconductors, but a more important effect is that the greater thermal vibration sets free more charge carriers, i.e. n increases. The conductivity increases approximately *exponentially* with temperature: this is the explanation of the characteristic of the thermistor shown in figure 8.4(b). The resistance decreases by a factor of between 20 and 30 between 0°C and 100°C.

The resistance of a pure metal increases nearly linearly with temperature so we can write

$$R_\theta = R_0(1 + \alpha\theta)$$

where R_θ is the resistance at the Celsius temperature θ, R_0 is the resistance at 0°C and α is the **temperature coefficient of resistivity**. All metals have similar values of α: they are all about $4 \times 10^{-3}\,\text{K}^{-1}$. Semiconductors have large *negative* values of α, usually about $-0.06\,\text{K}^{-1}$, but their resistance varies with temperature in a much less regular manner.

Superconductivity

For metals a graph of resistivity against kelvin temperature T is close to a straight line passing through the origin, as shown in figure 8.7 for a typical pure metal. At very low temperatures there is still some resistance. For some metals, however, the resistance falls to *zero* at temperatures near absolute zero (e.g. at 1.18K for aluminium). Below this **transition temperature** the metal is said to be **superconducting**. In this state an electric current, once established, flows indefinitely without the

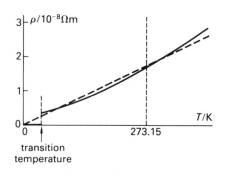

FIGURE 8.7

need for a source of e.m.f., since there is no conversion of the kinetic energy of the electrons to internal energy.

For many years after the discovery of superconductivity (by Kamerlingh Onnes in 1911) the effect was known to occur only at relatively low temperatures (below 20 K), which limited its usefulness. But in the last few years alloys (copper oxides containing elements such as tin, barium, yttrium, lanthanum, etc.) have been discovered which are superconducting at temperatures as high as 125 K, which is above the boiling point of nitrogen. Since liquid nitrogen is a relatively cheap substance, this increases the range of uses for superconducting substances. Using a superconducting material would be helpful in any situation where we want to avoid the wasteful or possibly harmful conversion of electrical energy to internal energy, but it does not offer the prospect of 'something for nothing': as soon as the energy of the current is used to do work, its energy is, of course, converted into some other form.

8.3 Circuit calculations

Here is a summary of the rules we have so far established for calculating values of current, p.d., resistance and power in electric circuits:

- charge passed = current × time ($Q = It$).

- in a series circuit the current is the same at all points.

- if a circuit has a junction, the current flowing into the junction is equal to the current flowing out of it.

- for components in series, the p.d. across all of them is equal to the sum of the p.d.s across each of them.

- for components in parallel, the p.d. across each of them is the same.

- resistance = $\dfrac{\text{p.d.}}{\text{current}}$

$$\left(R = \frac{V}{I} \quad \text{or} \quad V = IR \quad \text{or} \quad I = \frac{V}{R} \right).$$

- power P of cell = $\mathscr{E}I$ (converting chemical energy to electrical potential energy).

- power P of any circuit component = VI (converting electrical potential energy to some other form).

- power P of resistor = $VI = I^2R = \dfrac{V^2}{R}$.

We shall now use these rules in some examples. The examples have been chosen to make some important points, and should therefore be treated as part of the text, especially as there are comments after each example.

EXAMPLE 1

Using **V = IR** in a series circuit
A series circuit contains a cell, an ammeter, a 5.0 Ω resistor and a bulb. The cell has a terminal p.d. of 6.0 V.

(a) If the ammeter reads 0.40 A, what is the p.d. across the resistor?
(b) If a voltmeter connected across the bulb reads 4.0 V, what is the resistance of the bulb?

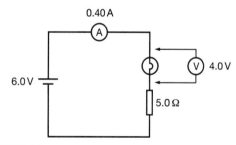

FIGURE 8.8

(a) For the resistor
$V = IR = (0.40\,\text{A})(5.0\,\Omega) = 2.0\,\text{V}$

(b) For the bulb $R = \dfrac{V}{I} = \dfrac{4.0\,\text{A}}{0.40\,\text{A}} = 10\,\Omega$

Notice that in this series circuit the p.d.s across the components are proportional to their resistances. This must be so, since for each of them $V = IR$, and I is the same for both. So once we knew that the p.d. across the bulb was twice the p.d. across the resistor we could have said that the resistance of the bulb must be twice 5 Ω, i.e. 10 Ω. This is a useful short cut.

EXAMPLE 2

Using $V = IR$ in a parallel circuit

In figure 8.9 ammeter A_1 reads 0.63 A, and ammeter A_2 reads 0.24 A. A voltmeter connected between A and C reads 6.0 V and the resistance of resistor Y is known to be 15 Ω.

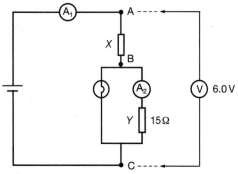

FIGURE 8.9

(a) What is the current in the bulb?
(b) What is the resistance of the bulb?
(c) What is the resistance of resistor X?

(a) The current in the bulb must be

$$0.63\,A - 0.24\,A = 0.39\,A$$

(b) For the bulb, we now know the current but not the p.d. But the p.d. across the bulb is the same as the p.d. across resistor Y, which we can calculate:

For resistor Y, $V = IR = (0.24\,A)(15\,\Omega)$
$= 3.6\,V\ (=\text{p.d. across bulb})$

For the bulb, $R = \dfrac{V}{I} = \dfrac{3.6\,V}{0.39\,A} = 9.2\,\Omega$

(c) $(\text{p.d.})_{AB} = (\text{p.d.})_{AC} - (\text{p.d.})_{BC}$
$= 6.0\,V - 3.6\,V = 2.4\,V.$

For resistor X, $R = \dfrac{V}{I} = \dfrac{2.4\,V}{0.63\,A} = 3.8\,\Omega$

EXAMPLE 3

Using a rheostat as a current limiter

You have available a battery of e.m.f. 6.0 V and negligible resistance, a 50 Ω rheostat and a 100 Ω resistor. What are the maximum and minimum currents which can be made to pass through the resistor?

Assuming that the circuit is connected as shown in figure 8.10, the maximum current is obtained with the slider at the bottom of the rheostat. Then the rheostat provides zero resistance, and the resistor is connected directly to the 6.0 V supply, so the current is given by

$$I = \frac{V}{R} = \frac{6.0\,V}{100\,\Omega} = 0.060\,A = 60\,mA$$

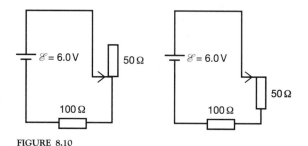

FIGURE 8.10

The minimum current is obtained with the slider at the top of the rheostat. Then total resistance of circuit = 100 Ω + 50 Ω = 150 Ω.

$$\text{Current } I = \frac{V}{R} = \frac{6.0\,V}{150\,\Omega} = 0.040\,A = 40\,mA.$$

So the rheostat can vary the current between 40 mA and 60 mA.

This is not a large variation. Making the rheostat's resistance larger would make it possible to achieve a smaller minimum current, but the current could never be reduced to zero. And the larger the rheostat's resistance, the smaller the movement of the slider for a particular change in resistance, so it would be difficult to adjust the current by small amounts. Used in this way the rheostat is said to be a **current limiter**. A better way of using the rheostat to control current is described after Example 4.

EXAMPLE 4

A potential divider

The battery maintains a constant p.d. of 12.0 V across the two resistors shown in figure 8.11.

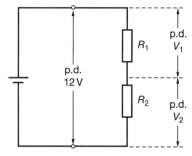

FIGURE 8.11

What are the p.d.s V_1 and V_2 across the resistors R_1 and R_2 when

(a) $R_1 = 50\,\Omega$, $R_2 = 50\,\Omega$?
(b) $R_1 = 50\,\Omega$, $R_2 = 100\,\Omega$?
(c) $R_1 = 500\,\Omega$, $R_2 = 1000\,\Omega$?

(a) The p.d. is proportional to the resistance in a series circuit, so $V_1 = V_2 = 6.0\,\text{V}$.
(b) Ratio of resistances $= 1:2$, i.e. $\frac{1}{3}$ of the p.d. is across R_1 and $\frac{2}{3}$ across R_2, so $V_1 = 4.0\,\text{V}$ and $V_2 = 8.0\,\text{V}$.
(c) Ratio of resistances is the same as in (b), so again $V_1 = 4.0\,\text{V}$, $V_2 = 8.0\,\text{V}$.

What *is* different in (c) is the current: this is one-tenth of what it was in (b). This circuit is the basis of a **potential divider**. The battery p.d. is *divided* into two parts. Any fraction of the battery p.d. is available between X and Y in figure 8.11 if the values of R_1 and R_2 are suitably chosen. For example, if the battery p.d. is 6.0 V, and a p.d. of 1.0 V is required, then there must be 5.0 V across R_1 and 1.0 V across R_2. The ratio of the resistances of R_1 and R_2 needs to be $5:1$, so the resistances could be $500\,\Omega$ and $100\,\Omega$. The battery p.d. could be thought of as the input p.d. and the p.d. between X and Y as the output p.d. – figure 8.12

also shows the circuit relabelled in this way. Then

$$V_{\text{out}} = V_{\text{in}} \times \frac{R_2}{R_1 + R_2}$$

In practice a rheostat or potentiometer would be used instead of fixed resistors, since then the ratio is infinitely variable. Look back at Example 3: this use of a rheostat as a *potential divider* would enable the current in the $100\,\Omega$ resistor to be varied from 60 mA to *zero*.

Sensor circuits

You know that a change in temperature will alter the resistance of a thermistor and that a change in light intensity will alter the resistance of an l.d.r. The component is then said to be acting as a **transducer** or **sensor**. The sensor and a fixed (or variable) resistor may be connected as shown in figure 8.13(a): when the temperature of the thermistor changes, V_{out} changes. It is common to connect the point X to the input of a transistor or operational amplifier in order to switch the potential of some other point from a high to a low potential, or vice versa, so that it can be used in a logic circuit. The next example illustrates this.

EXAMPLE 5

A sensor circuit

The graph in figure 8.13(a) shows the variation of resistance with temperature for a thermistor. The thermistor is connected in series with a fixed resistor of resistance $150\,\Omega$, as in figure 8.12(b). The potentials of A and B are 5.0 V and zero respectively. Point X is connected to the base of a transistor which will 'switch' when the potential of X rises to 1.2 V. Estimate the

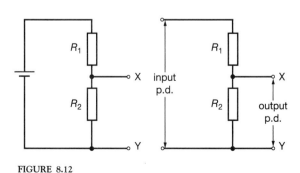

FIGURE 8.12

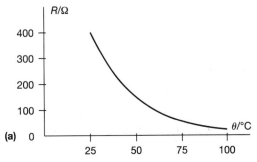

(a)

105

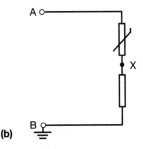

(b)

FIGURE 8.13

temperature at which this happens.

Using the potential divider equation

$$\frac{1.2\,V}{5.0\,V} = \frac{150\,\Omega}{150\,\Omega + R_{th}}$$

which gives $R_{th} = 47\,\Omega$. From the graph we can see that the thermistor has this resistance at a temperature of about 78°C.

EXAMPLE 6

Power in series and parallel circuits

A battery of e.m.f. 6.0 V and negligible resistance is connected to a $5\,\Omega$ resistor (X) and a $10\,\Omega$ resistor (Y) (a) in series, (b) in parallel. What are the powers of the resistors?

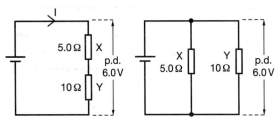

FIGURE 8.14

(a) There is the same current I in each resistor (the p.d.s are different): it is given by

$$I = \frac{V}{R} = \frac{6.0\,V}{15\,\Omega} = 0.40\,A$$

$$P_X = I^2 R_X = (0.40\,A)^2(5.0\,\Omega) = 0.80\,W$$
$$P_Y = I^2 R_Y = (0.40\,A)^2(10.0\,\Omega) = 1.60\,W$$

(b) The p.d. V across each resistor is the same (now the currents are different):

$$P_X = \frac{V^2}{R_X} = \frac{(6.0\,V)^2}{5.0\,\Omega} = 7.2\,W$$

and

$$P_Y = \frac{V^2}{R_Y} = \frac{(6.0\,V)^2}{10.0\,\Omega} = 3.6\,W$$

This example illustrates the important point that when we connect resistors *in series*, the power is *directly* proportional to the resistance, i.e. the smaller the resistance the smaller the power. This is what the equation $P = I^2 R$ tells us: that when I is constant, P is less when R is less. (We could have used the equations $P = VI$ or $P = V^2/R$ to solve this problem but it would have been more trouble: we should have had to calculate V. Also, the way in which P depends on R is not as obvious: you might think that the equation $P = V^2/R$ tells you that the greater the resistance, the *smaller* the power, but you need to remember that V is not the same for the two resistors.)

But when we connect different resistors *in parallel* the power is *inversely* proportional to the resistance, i.e. the smaller the resistance the greater the power. This is what the equation $P = V^2/R$ tells us: that when V is constant, P is greater when R is less. (We could have used the equations $P = VI$ or $P = I^2 R$ to solve this problem but it would have been more trouble: we should have had to calculate I. You might think that the equation $P = I^2 R$ tells you that the greater the resistance, the *greater* the power, but you need to remember that I is not the same for the two resistors.)

8.4 Circuits for measurement

Current and p.d. are measured using ammeters and voltmeters. How do we measure resistance and power? Since $R = V/I$ and $P = VI$ then in both cases we need to measure the p.d. across the resistor and the current flowing through it.

The circuit is shown in figure 8.15. In fact the ammeter actually measures the current passing through the resistor *and* the voltmeter, and the

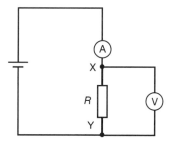

FIGURE 8.15

connected to the joulemeter. At the same time the energy supplied is recorded on a digital display.

The Wheatstone bridge

Resistances can be compared using the circuit shown in figure 8.16. This is called a **Wheatstone bridge network**. The battery maintains the same p.d. across each pair (P, Q and R, S) of resistors and each pair acts as a potential divider.

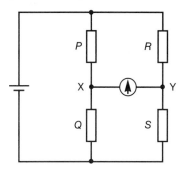

FIGURE 8.16

If the meter indicates zero current the potentials at X and Y must be the same, so P and Q, and R and S, must be dividing the p.d. in the same ratio. So the ratios of resistances must be the same, i.e.

$$\frac{P}{Q} = \frac{R}{S}$$

So if we want to find the value of an unknown resistor R we use three adjustable resistors P, Q and S and alter their values until there is zero current. P, Q and S would probably be **resistance boxes**, which are made in a variety of designs: nowadays rotary switches are used to bring into use a series of coils so that we can choose any value of resistance that we wish. Figure 8.17(a) shows a circuit diagram, and figure 8.17(b) shows a photograph of a typical box, indicating that a value of $236\,\Omega$ has been chosen.

presence of the voltmeter also disturbs the circuit because the resistance between X and Y is now the total resistance of the resistor and the voltmeter connected in parallel, which will be less than the resistance of the resistor alone. How much these two effects matter depends on how the resistance of the voltmeter compares with the resistance of the resistor. For example, if the resistor's resistance is about $1000\,\Omega$, and the voltmeter's resistance is $10\,M\Omega$ there is very little effect. But if the resistor's resistance is $10\,M\Omega$, then the voltmeter will need to have a higher resistance still. In this case we could use an **electrometer** (which is essentially an operational amplifier used as a buffer) which will have an input resistance of about $10^{12}\,\Omega$. When both the p.d. and the current have been measured, the resistance and the power can be calculated.

Multimeters and joulemeters

A **multimeter** is an instrument which can measure several electrical quantities. It may be analogue or digital. All multimeters can measure current, p.d. and resistance, and some can also measure frequency and capacitance, and test components such as diodes and transistors. They can measure resistance because they include a battery which passes a current through the resistor which is being measured: the current measured by the multimeter is an indication of the resistance of the resistor and interpreted as such on one of the meter's scales. The multimeter therefore gives a direct measurement of resistance.

Joulemeters are instruments containing electronic circuits which can be placed between the power supply and the resistor, effectively measuring and then multiplying the values of p.d. and current to display the power on a voltmeter

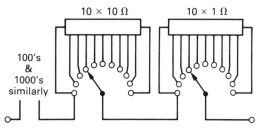

FIGURE 8.17(a)

FIGURE 8.17(b)

EXAMPLE

A Wheatstone bridge network is used to measure the resistance of a wire, which is represented by R in figure 8.14. The adjustable resistors are chosen so that $P = 100.00\,\Omega$, $Q = 1000.0\,\Omega$, $S = 1318\,\Omega$, and the meter indicates zero current. What is the resistance of the wire?

$$\frac{R}{S} = \frac{P}{Q} \Rightarrow \frac{R}{1318\,\Omega} = \frac{100.00\,\Omega}{10000\,\Omega} \Rightarrow R = 131.8\,\Omega$$

Notice that the values of P, Q and S are known with considerable precision, and how we can obtain values of resistance correct to the nearest $0.1\,\Omega$ even though the resistance box does not contain multiples of $0.1\,\Omega$.

The availability of high-resistance voltmeters and multimeters has largely made obsolete the technique described above, but the principle of the circuit is still used. One such use is with a **strain gauge**, which is used by structural engineers, and is illustrated at the start of this chapter. Its resistance is usually between about $100\,\Omega$ and $150\,\Omega$.

Suppose a strain gauge is represented by resistor R in figure 8.14 and the bridge is balanced (i.e. the p.d. between X and Y is zero). If there is a strain in the girder there is a change ΔR in the resistance R: the bridge becomes unbalanced and there is a small p.d. between X and Y. If a second strain gauge (Q in the circuit) is placed on the girder alongside the

first it will have the same change ΔR, and the p.d. between X and Y will be doubled, so this increases the sensitivity. Other strain gauges (P and S in the circuit) are placed at right angles to the first two (as shown in figure 8.16) so there is no change in their length or reistance. The purpose of P and S is to compensate for resistance change which might occur because of temperature change: any temperature changes the resistance of all the strain gauges by the same amount and the balance of the bridge is not affected.

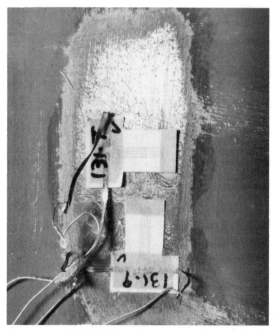

FIGURE 8.18

In this use of the Wheatstone bridge network no attempt is made to balance the bridge. If the out-of-balance p.d. between X and Y is small, it is proportional to the change in resistance, and therefore to the strain. It is this p.d. which is amplified and recorded by the engineers who are investigating the structure.

Potentiometer circuits

Figure 8.17 shows a battery maintaining a constant p.d. of about 3 V across the ends of a 1-metre length of resistance wire, which has a rule running alongside it. In this situation this cell is called the

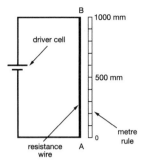 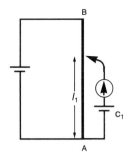 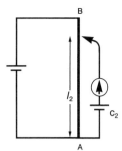

FIGURE 8.19

driver cell. Imagine that we do not have a voltmeter available so we do not know the exact value of the p.d. between the ends of the wire. Suppose the lower end of the wire is at zero potential. The potentials between the ends B and A on the wire range from about 3 V down to zero. We now bring along another cell C_1, whose e.m.f. is about 1.5 V, and connect its negative terminal to A, and its positive terminal through a meter to a **jockey** (which is essentially a terminal with a spade-like end which can make precise contact at any chosen point on the wire). The potential of the jockey will be about 1.5 V. There *must* be some point on the wire which is at the same potential. We can find where it is, because if we touch the jockey on the wire at that point, the meter will read zero, since because the wire and the jockey are at the same potential, no current will flow from one to the other. Suppose we find that the point is 523 mm from A. If we disconnect cell C_1 and put another cell C_2 in its place we can repeat the experiment. If the balance point is now 506 mm from A, we know that the cell C_2 does not provide such a large p.d. as C_1, and we can work out the ratio of the p.d.s: 506/523. Generally, if the two p.d.s are V_1 and V_2 and the balance lengths are l_1 and l_2

$$\frac{V_1}{V_2} = \frac{l_1}{l_2}$$

For this measurement to be precise (a) the p.d. across the wire must not change while the two measurements are being made, and (b) the wire must be uniform, so that the resistances of different lengths are proportional to those lengths.

The advantage of this method is that when the balance point has been found, no current flows through the cell, and the p.d. between its terminals is equal to *its e.m.f.* So using this method we are actually comparing e.m.f.s, and not just the terminal p.d.s.

This is called a **potentiometric** method of measuring p.d. We have described only a simple form of it. One refinement would be to replace the wire by a multi-turn linear potentiometer (shown in figure 8.18) fitted with a dial to indicate the position of the slider along its wire, and in its most refined form this method is capable of measuring with an uncertainty of 1 part in 10^6. It is inherently a precise method of measuring e.m.f. or p.d., *since it does not disturb or load the circuit* in which the measurements are being made. As described above, the method can only compare two p.d.s, but by making comparison with a standard cell (e.g. the **Weston** cell of e.m.f. 1.01859 V at 20°C) other p.d.s may be *measured* in terms of this practical laboratory standard.

FIGURE 8.20

EXAMPLE

A potentiometer is used with a lead–acid cell, whose e.m.f. is 2.00 V, to measure the e.m.f. of a dry cell. A standard cell (e.m.f. 1.0186 V) is available. When the cells are connected in the normal way, the balance lengths are found to be 736 mm and 505 mm respectively. Calculate the e.m.f. $\mathscr{E}$ of the dry cell.

$$\frac{\mathscr{E}}{1.0186\,\text{V}} = \frac{736\,\text{mm}}{505\,\text{mm}} \Rightarrow \mathscr{E} = 1.48\,\text{V}$$

It is not necessary to know the e.m.f. of the driver cell, and knowing it does not help us to find the p.d. across the potentiometer wire, since we do not know enough about the internal resistance of the cell, the contact resistances where the wires make connections, etc. The internal resistances of the dry cell and the standard cell are also irrelevant, since at balance the current through them is zero.

Exercises on each section of this chapter may be found in the companion textbook, *Practice in Physics*.

SUMMARY

At the end of this chapter you should be able to:

◆ use the equation $R = V/I$.

◆ understand that Ohm's law states that the resistance of a metallic conductor is constant if physical conditions are constant.

◆ describe how to check on whether Ohm's law is obeyed for a particular circuit component.

◆ describe how to plot characteristics (i.e. graphs of current I against p.d. V) for circuit components.

◆ draw the characteristics for a lamp filament, a thermistor and a silicon diode.

◆ use the equations $P = I^2R$ and $P = V^2/R$ to calculate the power of a resistor.

◆ use the equations $\quad R_{\text{ser}} = R_1 + R_2 \quad$ and $\quad \dfrac{1}{R_{\text{par}}} = \dfrac{1}{R_1} + \dfrac{1}{R_2}.$

◆ explain why the resistance of two resistors in parallel is always less than the resistance of the smaller of the two.

◆ use the equation $R = \rho l/A$ which defines resistivity ρ.

◆ remember that the resistivity of metals increases with temperature but the resistivity of semiconductors decreases with temperature.

◆ remember that the resistivity of all metals and alloys falls to zero at their transition temperature, and they are then superconducting.

◆ understand how a rheostat can be used as a current limiter, using two terminals.

◆ understand how a rheostat can be used as a potential divider, using three terminals, and what the advantage of this is.

◆ describe how to measure the resistance and power of a resistor.

◆ draw a Wheatstone bridge circuit and state the relationship between the resistance values when the bridge is balanced.

◆ draw a potentiometer circuit and explain how it can be used to compare the e.m.f.s of two cells.

9 Cells and Meters

We all rely on being able to carry with us portable stores of energy in cells and batteries. You may have a portable cassette deck or CD player, or you may have a modern camera in which several functions are battery-powered – and you almost certainly have a calculator. Cars and motorbikes need batteries so that the engine can be started and continue to run, and so that the lights work. At home there may be battery-powered electric clocks or smoke detectors, and you will certainly use a torch from time to time. In the building trade there is an increasing number of battery-powered tools: 'cordless' drills and 'cordless' screwdrivers. You probably use some battery-powered meters in your school or college laboratory. Large computers have batteries to provide power in an emergency when the mains supply fails. Your younger brothers and sisters have battery-powered toys: older people may use hearing aids. The photograph shows some of the enormous range of cells and batteries that are available: the range is needed to provide p.d.s of various sizes, and so that we can have batteries which can deliver currents of different sizes, and can have batteries which last for different lengths of time.

9.1 Electric cells

Rods of two different metals dipped in a tank of any electrolyte (e.g. copper and zinc placed in dilute sulphuric acid) act as a source of e.m.f. If an electric connection is made between the metals, a current flows. The energy of the source of e.m.f. comes from the chemical changes which take place between the metals and the electrolyte.

Primary cells

In a primary cell the chemical reaction is not reversible, and the cell has to be disposed of when all the chemicals have been used. Most primary cells are based on the Leclanché cell, which in one form or another has been in use for well over a hundred years. Most manufacturers make three versions of it, as shown, for example, in figure 9.1. These are 'dry' cells, i.e. they are sealed so that the moisture in the electrolyte cannot evaporate or leak. All use zinc as one of the electrodes and carbon (with manganese dioxide in close contact) as the other. The electrolyte may be ammonium chloride, zinc chloride or potassium hydroxide. Table 9.1 shows the properties of these cells.

FIGURE 9.1

TABLE 9.1 Properties of primary cells

Electrolyte	Relative life	Current in normal use
ammonium chloride	1	intermittent up to 0.5 A
zinc chloride	1.5	intermittent up to 0.5 A
potassium hydroxide	3	intermittent or continuous: up to 2 A

The last two are leak-proof, since the zinc case does not take part in the chemical reaction. The third one is described as an **alkaline** cell.

Other primary cells include the button type cells used in watches, small calculators, hearing aids, etc. These have mercuric oxide and zinc electrodes (e.m.f. 1.35 V), silver oxide and zinc electrodes (e.m.f. 1.55 V) or zinc and oxygen electrodes (e.m.f. 1.25 V). The first of these is being superseded by the third. All maintain an almost constant e.m.f. until the ends of their lives; unlike the zinc–carbon cells, where the e.m.f. falls continuously with use. One primary cell found only in laboratories is the Weston cell (already mentioned in the last chapter). Its e.m.f. of 1.01859 V (at 20°C) varies little with temperature and will remain constant to within 1 part in 10^6 for several years if the current taken from it is kept less than $10\,\mu\text{A}$.

Secondary cells

In these the chemical reaction is reversible. For many years almost the only secondary cell in common use was the **lead–acid** cell found in car batteries. In that situation the chemical reaction needs to be reversible so that a dynamo (driven by the car engine) can replace energy that has been supplied by the battery in starting the engine and running lamps, windscreen wipers and the many other electrical devices now found in cars. The lead–acid cell has an almost constant e.m.f. of 2.0 V until near the end of its life. It is now available in sealed maintenance-free forms. A similar cell, sometimes found in laboratories, is the **NiFe** (nickel–iron) cell. This has an e.m.f. of only 1.2 V (which falls in use) but is lighter than the lead–acid cell and is particularly robust. Both cells can supply far larger currents than the primary cells.

A secondary cell which is becoming more and more widely used in the nickel–cadmium (**nicad**) cell. It is the dry cell found in rechargeable devices such as calculators, toys, domestic tools (e.g. can openers, hand-held vacuum cleaners), power tools (e.g. drills, screwdrivers) etc. It has an almost constant e.m.f. of 1.2 V throughout its life, and as the list of uses implies, it can supply large currents.

Batteries

Electric batteries consist of several electric cells joined together. They can be joined in series or in parallel, as shown in figure 9.2.

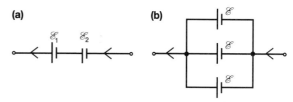

FIGURE 9.2

Cells in series – In this case the same charge Q must pass through each cell (whose e.m.f.s are $\mathscr{E}_1$ and $\mathscr{E}_2$) as shown in figure 9.2(a). The total energy given to the circuit is the sum of the energies provided by each cell, so

$$\text{total energy provided} = \mathscr{E}_1 Q + \mathscr{E}_2 Q$$

If the total e.m.f. is $\mathscr{E}$, then $\mathscr{E}Q = \mathscr{E}_1 Q + \mathscr{E}_2 Q$
so
$$\mathscr{E} = \mathscr{E}_1 + \mathscr{E}_2$$

Most cells have e.m.f.s between 1 V and 2 V. Batteries of larger e.m.f. may be made only by joining together cells in this way. For example, a battery labelled 9 V must have inside it six 1.5 V cells joined in series.

Cells in parallel – Suppose we have three cells of equal e.m.f. $\mathscr{E}$ joined in parallel as in figure 9.2(b). Each cell maintains an equal part of the total current passing through the battery. The resultant e.m.f. is $\mathscr{E}$, the same as the e.m.f. of one of the cells, but the battery has less resistance than a single cell (because the cells' resistance is arranged in parallel) and can provide a larger current than a single cell.

9.2 Internal resistance

You know that in the situation shown in figure 9.3 the voltmeter will record a **terminal p.d.** V which is less than the e.m.f. $\mathscr{E}$ of the cell.

We saw earlier that this happens because not all the electrical potential energy is delivered to the external circuit. Some of it is converted to internal energy within the cell. We say that the cell has

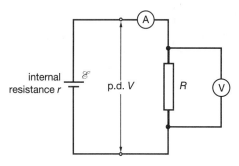

FIGURE 9.3

internal resistance r. We can now work out how $\mathscr{E}$, V and r are related to the current in the circuit.

If the current in the circuit is I, the power of the cell is $\mathscr{E}I$: this is the rate of production of electrical potential energy. We also know that the rate of conversion of electrical potential energy in the external circuit is VI, and if the internal resistance is r, the rate of conversion of electrical potential energy to internal energy within the cell is I^2r. So we can write

$$\mathscr{E}I = VI + I^2r$$

Since $V = IR$ we can write this equation as

$$\mathscr{E}I = I^2R + I^2r = I^2(R + r)$$

which makes it clear that the rate of working of the cell is equal to the combined rate of working of the external and internal resistors. These energy conversions are illustrated in figure 9.4, with the rate of working shown against the arrows.

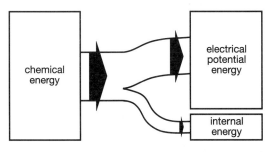

FIGURE 9.4

Dividing both sides of these equations by I we have

$$\mathscr{E} = V + Ir \quad \text{and} \quad \mathscr{E} = I(R + r)$$

The second of these is sometimes called the **circuit equation**, since it lets us work out the current in the most general circuit situation, i.e. when we have to consider the internal resistance of the cell as well as the resistance of the circuit.

If you write the first of these as

$$\mathscr{E} - V = Ir$$

you can see that the drop in p.d. ($\mathscr{E} - V$) depends on both I and r, so the smaller I is, the nearer the terminal p.d. will be to the e.m.f. Only when $I = 0$ does $\mathscr{E} = V$, but if we want to measure the e.m.f. of a cell and connect a high-resistance voltmeter across its terminals the current will be very nearly zero and so we shall have a very good approximation to the e.m.f.

The graph of terminal p.d. V against current I, in figure 9.5, is a straight line which illustrates the equation $V = \mathscr{E} - Ir$. The intercept on the V-axis is $\mathscr{E}$ and the slope of the graph is $-r$. However, the internal resistance is not constant for a particular cell: it depends on the life of the cell, and also on the current, if the current is large. The line has therefore not been drawn as a firm line to remind you of this fact.

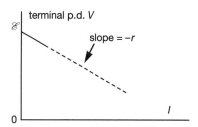

FIGURE 9.5

Measuring internal resistance

The internal resistance of a cell can be measured with the circuit shown in figure 9.3. The terminal p.d. V and current I are measured for different values of resistance. In practice the p.d. should be measured immediately after the circuit has been connected, since an effect called **polarisation** (in which the nature of the electrodes is affected by the current in the circuit) reduces the e.m.f. as time goes on. If a graph is plotted (as shown in figure 9.5) the slope is $-r$, so the internal resistance can be calculated.

EXAMPLE

A dry cell of e.m.f. 1.50 V and internal resistance $0.6\,\Omega$ is connected to a resistor of resistance $4.4\,\Omega$. Calculate (a) the current which flows, (b) the p.d. between the ends of the resistor, (c) the p.d. between the cell terminals.

(a) Using $\mathscr{E} = I(R + r)$ gives

$$I = \frac{\mathscr{E}}{R + r} = \frac{1.50\,\text{V}}{5.0\,\Omega} = 0.30\,\text{A}.$$

(b) For the resistor,
$V = IR = (0.30\,\text{A})(4.4\,\Omega) = 1.32\,\text{V}.$
(c) The p.d. between the cell terminals is (therefore) also 1.32 V.

Notice that we can *write down* the p.d. across the cell terminals *without any further calculation* because the p.d. across the cell *must* be the same as the p.d. across the resistor. We *cannot* use $V = Ir$ for the cell, since it is a source of e.m.f. If we needed to calculate V, we would have to use

$$V = \mathscr{E} - Ir = 1.50\,\text{V} - (0.30\,\text{A})(0.6\,\Omega)$$
$$= 1.50\,\text{V} - 0.18\,\text{V} = 1.32\,\text{V}.$$

If there had been no internal resistance, the current would have been

$$I = \frac{\mathscr{E}}{R} = \frac{1.50\,\text{V}}{0.60\,\Omega} = 0.34\,\text{A}.$$

Choice of battery

The fact that the terminal p.d. is less than the e.m.f. by the amount Ir means that the internal resistance must be taken into account when choosing a cell for a particular application. It is not just the e.m.f. that matters. The internal resistance of a cell depends upon:

♦ the dimensions of the cell: roughly, if all the dimensions of a cell are doubled, its internal resistance will be halved.

♦ the mobility of the ions in the electrolyte, so a wet cell (e.g. a lead–acid cell) has a much smaller internal resistance than any dry cell.

The maximum current which a cell can supply occurs when it is short-circuited. Since the only

resistance in the circuit is then the cell's internal resistance, the short-circuit current I is given by

$$I = \frac{\mathscr{E}}{r}$$

For a small dry cell, with $\mathscr{E} = 1.5\,\text{V}$ and $r = 0.75\,\Omega$, I might be 2 A, for a larger dry cell it might be 5 A, but for a lead–acid cell it might be several hundred amperes. Of course, short-circuiting a cell will convert its chemical energy into internal energy extremely quickly and wastefully. There may also be a rapid and dangerous rise in temperature in the cell, and damage to its interior. But the short-circuit current, though never used in practice, gives an indication of the uses to which the different types of cell might be put. A torch bulb, of resistance $5\,\Omega$, needing a current of about 0.2 A, could be powered by a dry cell; but two car headlamps, needing a total current of 10 A, could not. They would need the lead–acid battery with which the car is provided. The demands on a car battery are extreme: the current needed to turn the starter motor might be as much as 300 A.

EXAMPLE

A motorist is sitting in her car with the headlamps switched on. She then switches on the starter motor, and finds that the headlamps dim while the motor is turning. If the battery has an e.m.f. of 12 V, and the internal resistance of each of the cells is $0.005\,\Omega$, and the current in the starter motor is 160 A, calculate the p.d. across the battery terminals while the motor is turning, and explain why the headlamps dim.

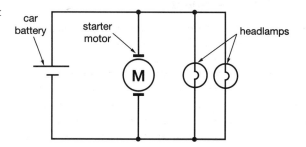

The current through the headlamps is small compared with the current in the motor, and can be ignored. So the current in the battery is

160 A, and the total internal resistance is 0.03 Ω.

$$V = \mathcal{E} - Ir = 12\,\text{V} - (160\,\text{A})(0.03\,\Omega)$$
$$= 12\,\text{V} - 4.8\,\text{V} = 7.2\,\text{V}.$$

The p.d. across the headlamps also falls to 7.2 V, and as the power $P = V^2/R$, P depends on V^2, so the power falls to the fraction $(7.2/12)^2$ of the original power, i.e. to only 36%.

It is essential to draw a circuit diagram before starting any question about electric circuits: in this question you have to understand that the headlamps and the starter motor are all connected in parallel across the battery.

The internal resistance of a lead–acid cell increases as the temperature falls, because the ions are not as mobile, so the p.d. available at the terminals of the starter motor will be less. This is one reason why a battery is less able to start a car engine in cold weather.

The total resistance of a battery of cells may be calculated using the normal rules for calculating total resistance for resistors in series and in parallel. For example, four cells which each have an e.m.f. of 1.50 V and an internal resistance of 0.8 Ω will have:

- *in series*, an e.m.f. of 6.0 V and an internal resistance of 3.2 Ω.

- *in parallel*, an e.m.f. of 1.50 V and an internal resistance of 0.2 Ω. This is why connecting cells in parallel enables them to provide larger currents.

Power supplies

Most laboratories use **power supply units**, as well as batteries, to deliver currents. These are described in detail later, but for the time being you should note that these too have internal resistance, though in this case it is usually called internal **impedance** (impedance is a term which includes other electrical properties, besides resistance, which tend to decrease the current). The nominal values of output p.d. stated on the case of the power supply unit are not likely to be achieved when a current is being drawn from the power supply.

One particularly relevant case to consider is the so-called **e.h.t** (extra high tension) power supplies which provide p.d.s of up to 6 kV. However, they are not dangerous because the internal impedance is very high. It may also be possible to include a resistor of 50 MΩ to increase the impedance further, for safety reasons. Then the short-circuit current (given $I = \mathcal{E}/r$) is about 3 mA.

Maximum power

How can we get the greatest power delivered to a resistor by a battery or any other supply which has some internal resistance? Consider once again the circuit in figure 9.3 and look at two extreme cases. Suppose the resistance of R is very large, so that the current I is small. Then the power $(P = VI)$ is small because I is small. Now suppose the resistance of R is very small: then the current will be large but the p.d. across the cell $(V = \mathcal{E} - Ir)$ will be small so again the power $(P = VI)$ will be small. The power will be greater than it was in either of these two cases if R is neither too large nor too small. In fact the **maximum power** occurs when $R = r$. This is sometimes known as the **maximum power theorem**. This is shown in the graph in figure 9.6, which is drawn for a cell of e.m.f. 1.50 V and internal resistance 0.5 Ω. This is an example of **impedance matching**. For example, the output of an amplifier in a hi-fi system might have an internal impedance of 4 Ω. A loudspeaker connected to it should therefore have an impedance of 4 Ω if the power delivered to it is to be at maximum. Also, a signal generator often has two alternative outputs: one with high and one with low impedance. The first would be used when connecting it to a high-impedance device like an oscilloscope, the second when connecting it to a low-impedance device like a loudspeaker.

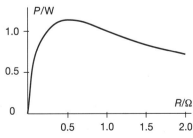

FIGURE 9.6

Recharging batteries

Cells in which the action, if reversible, can have the chemical energy restored to them by passing a current through them in the direction opposite to that in which it flows when they are delivering a current. This process is called **recharging** (though it should be called **re-energising**). Both lead–acid and nicad cells have very low internal resistance so some care must be taken to prevent the charging current from being too high. For nicad cells an electronic circuit is used which maintains a constant current until the battery is fully charged. Lead–acid battery chargers usually have a resistor in series to restrict the current to a safe value. A typical circuit is shown in figure 9.7, with a choice of resistors so that there can be either a rapid recharging in an emergency, or a slow 'trickle' charge overnight.

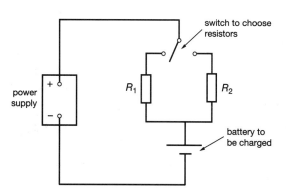

FIGURE 9.7

Notice that you need to take account of the different signs of the e.m.f.s.

Capacity

The **capacity** of a battery is the amount of charge which can pass through it before all its chemical energy has been converted. In practice the performance of a battery will tail off towards the end of its life, so the stated value of capacity may tell us the charge which can pass until the p.d. falls to some specified fraction of the original p.d. The unit of capacity could be the coulomb, but in practice it is more convenient to use the **ampere hour** (Ah):

$$1\,Ah = (1\,Cs^{-1})(3600\,s) = 3600\,C.$$

A car battery might have a capacity of 60 Ah, i.e. it could provide a current of 1 A for 60 h, or a current of 2.0 A for 30 h, etc. – but this sort of 'ideal' performance will not be obtained for large currents (e.g. you will not be able to obtain a current for 60 A for 1 h). On the other hand, a torch battery might have a capacity of 8 Ah, and a battery for a watch, a capacity of 100 mAh. The energy W stored in a battery can be calculated by multiplying the capacity Q (in C) by the e.m.f. $\mathcal{E}$: $W = \mathcal{E}Q$.

EXAMPLE

In a circuit like that shown in figure 9.7 used to recharge a car battery, the supply e.m.f. is 15.0 V, the car battery has an e.m.f. of 12.0 V, the internal resistance of the supply is 0.40 Ω, and the internal resistance of the battery is so small that it can be neglected. Calculate the values of the resistances R_1 and R_2 if there are to be two possible charging currents: 5.0 A and 0.50 A.

(i)
$$\mathcal{E} = I(R + r)$$
$$\Rightarrow \quad 15.0\,V - 12.0\,V = (5.0\,A)(R_1 + 0.40\,\Omega)$$
so
$$R_1 = 0.20\,\Omega$$
(ii) Similarly $R_2 = 5.6\,\Omega$.

EXAMPLE

A catalogue describes a cell as '1.5 V, 1.2 Ah'. Calculate the total energy which it can supply.

The total charge Q which the cell can pass is given by

$$Q = (1.2\,Cs^{-1})(3600\,s) = 4320\,C$$

The energy W stored $= \mathcal{E}Q = (1.5\,JC^{-1})(4320\,C)$
$$= 6480\,J = 6.5\,kJ$$

Energy stored in dry cells is very expensive, as you probably know if you run portable radio cassette decks from batteries. It is about 10 000 times more expensive than electrical energy supplied by the mains; you pay for the convenience of having a portable store of energy.

9.3 Meters

Moving-coil instruments

A typical moving-coil meter may be designed to give **full-scale deflection** (f.s.d.) for a current of only 100 μA. To be as sensitive as this, the coil needs to have many turns of fine wire, so its resistance may be as high as 1000 Ω. In its basic state it may be treated as an *ammeter* measuring currents of up to 100 μA, but it can also be thought of as a *voltmeter*, since when there is a current of 100 μA through its resistance of 1000 Ω there must be a p.d. V between its terminals given by

$$V = IR = (100 \times 10^{-6}\,\text{A})(1000\,\Omega)$$
$$= 1.0 \times 10^{-1}\,\text{V} = 100\,\text{mV}.$$

So full-scale deflection means both:

♦ a current of 100 μA passing through it; and

♦ a p.d. of 100 mV between its terminals.

Any ammeter is simultaneously a voltmeter and vice versa. The word **galvanometer** is used to mean any meter which detects electric current: it may refer to either an ammeter or a voltmeter, but particularly to an instrument used to make **null** measurements, i.e. when all that is needed is to detect whether or not there is any current.

One of the meter's terminals will be labelled '+' or coloured red. This is the terminal into which the current should pass. Otherwise the needle will deflect in the opposite direction. With some **centre-zero** instruments this does not matter. Other instruments may be able to record negative readings to some extent, but the pointer will probably come to rest against a stop, with an unknown, and possibly large, current flowing through the meter. It is clearly better to spend a moment thinking about the correct way to connect the meter.

Ammeters

To measure currents larger than 100 μA the basic meter needs to be fitted with a **shunt**. This is a resistor which is connected in parallel with the meter. We can calculate the resistance which a shunt needs to have. Figure 9.8(a) shows the circuit and (b) shows a photograph of a typical

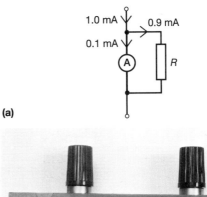

(a)

(b)

FIGURE 9.8

shunt. Suppose you want to be able to measure currents of up to 1 mA, and want the meter to give f.s.d. for this current. Then when the current is 1 mA the meter must have 100 μA (i.e. 0.1 mA) passing through it, so 0.9 mA must pass through the shunt. The shunt resistance must be chosen so that this happens. The shunt must have 9 times more current passing through it than passes through the meter, so its resistance must be one-ninth of the meter's resistance, i.e. $\frac{1}{9} \times 1000\,\Omega = 111\,\Omega$. The total resistance R of the meter as a whole, including the shunt, is given by

$$\frac{1}{R} = \frac{1}{1000\,\Omega} + \frac{1}{111\,\Omega} \quad \Rightarrow \quad R = 100\,\Omega$$

If larger currents are to be measured, shunts of smaller resistance are used to divert greater proportions of the current past the meter (since only 100 μA may pass through the meter itself). Table 9.2 gives values of shunt resistance and total resistance for various f.s.d.s, starting with a 1000 Ω meter with f.s.d. 100 μA. Notice that the total resistance is always less than the shunt resistance because the meter and the shunt are connected in parallel.

Because ammeters are placed *in* a circuit, their resistance should be small enough to be negligible.

TABLE 9.2 Shunt resistance and total resistance at various f.s.d.s

f.s.d	100 μA	1 mA	10 mA	100 mA	1 A	10 A
shunt resistance/Ω		111	11.1	1.11	0.111	0.0111
total resistance/Ω	1000	100	10	1	0.1	0.01

The table shows that the resistance *is* small when the meter is adapted to measure large currents. However, you may wonder whether the basic meter, with its large resistance of 1000 Ω, would make a suitable ammeter. The answer is that it would be used only where the current is small (<0.1 mA) and if the currents are as small as that, the resistance of the rest of the circuit must be much larger than the resistance of the ammeter. Of course, the p.d. across it must be less than 100 mV, so that confirms that it will be sensible to use it.

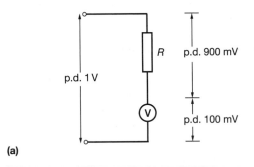

(a)

Voltmeters

We have already seen that a moving-coil meter may be thought of as a voltmeter. In its basic form the meter we have so far been considering will measure p.d.s of up to 100 mV. To measure larger p.d.s we need to connect resistors (called **multipliers**) in series with it. Suppose we want to measure p.d.s of up to 1 V. Figure 9.9(a) shows the arrangement and (b) shows a photograph of a typical multiplier. There must be only 100 mV across the meter, so the rest of the 1 V, i.e. 900 mV, must be across the multiplier. Thinking of this as a potential divider you can see that the multiplier must have a resistance of 9000 Ω. Multipliers of larger resistance would be used to enable the meter to measure larger p.d.s. Table 9.3 gives the values of multiplier resistance and total resistance, starting with our 1000 Ω meter which has an f.s.d. of 100 mV.

As with the ammeter, we need to consider the effect of the meter on the circuit. Look at the circuit in figure 9.10(a), where a battery provides a p.d. of 12.0 V across two resistors, each of resistance 100 kΩ, connected in series. The p.d. across each must be 6.0 V. Suppose a basic meter is now adapted to give it an f.s.d. of 10 V. When it is connected across R_1 you expect it to read 6.0 V, but

(b)

FIGURE 9.9

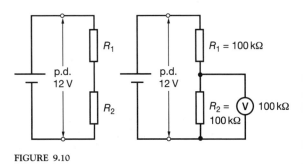

FIGURE 9.10

TABLE 9.3 Multiplier resistance and meter resistance at various f.s.d.s

f.s.d.	100 mV	1.0 V	10 V	100 V
multiplier resistance/Ω	0	9000	99 000	999 000
meter resistance/Ω	1000	10 000	100 000	1 000 000

it reads much less. If you look at the second circuit diagram you will see why. The voltmeter is actually a resistor (Table 9.3 shows that it has a resistance of $100\,k\Omega$) connected in parallel with R_2, so the lower part of the circuit consists of two $100\,k\Omega$ resistors in parallel, which have a total resistance of only $50\,k\Omega$. The proportions of p.d. are now therefore $2:1$, i.e. $8.0\,V$ between the ends of R_1, and $4.0\,V$ between the ends of both R_2 and the voltmeter. So the voltmeter measures a p.d. of $4.0\,V$. It *is* correctly measuring the p.d. that is there now, but it has disturbed the circuit. We always need to choose a voltmeter which has a much higher resistance than the resistance of the component across which we are connecting it.

Shunts and multipliers are usually simple plug-in attachments, labelled with the f.s.d. which they provide. But the meter itself has a scale which is marked from, say, 0 to 10, and this cannot (usually) be altered. So a pointer reading has to be *interpreted*. If the pointer points to 7.3 (out of 10) it will mean $7.3\,mA$ if the f.s.d. is $10\,mA$, but $73\,mA$ if the f.s.d. is $100\,mA$.

EXAMPLE

A meter has a resistance of $1000\,\Omega$ and an f.s.d. of $100\,\mu A$. How can it be adapted to give f.s.d.s of (a) $2.0\,A$, (b) $5.0\,V$?

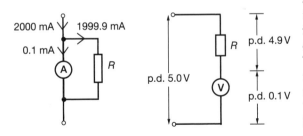

(a) A resistor must be provided in parallel, as shown in the figure. Working in mA for convenience, $2.0\,A = 2000\,mA$, and $100\,\mu A = 0.1\,mA$. The circuit diagram shows that only $0.1\,mA$ may pass through the meter, so $1999.9\,mA$ must pass through the shunt, so the shunt resistance must be only

$$\frac{0.1\,mA}{1999.9\,mA} \times 1000\,\Omega = 0.050\,\Omega$$

(b) A resistor must be provided in series, as shown in the figure. The maximum p.d. V across the meter itself is given by

$$V = IR = (100 \times 10^{-6}\,A)(1000\,\Omega) = 0.10\,V,$$

so the p.d. acrosss the resistor must be $4.90\,V$. So its resistance R is given by

$$R = \frac{4.90\,V}{0.10\,V} \times 1000\,\Omega = 49\,k\Omega$$

Light-beam meters

Most moving-coil meters have pivoted coils and the coil comes to rest when the torque of the magnetic forces is balanced by the opposing torque of two **hair springs**. These hair springs are relatively stiff and this is one of the factors which limits the sensitivity of a **pivoted-coil** meter. There are also **suspended-coil** meters; in modern versions the coil is supported in a vertical plane by two taut fine vertical wires, one above and one below the coil: these wires lead the current into and out of the coil and themselves provide the necessary resisting torque. As these wires are not stiff, the meter is very *sensitive*: i.e. a small change in p.d. leads to a relatively large movement of the coil. Another feature which increases the sensitivity still further is that an **optical pointer** is used instead of a metal pointer. A lamp enclosed in the case of the instrument provides a beam of light which is focused and reflected from a mirror fixed to the coil; as the coil rotates, the beam of light rotates. The length of the beam may be doubled by reflection from another mirror inside the case so that the pointer is effectively nearly $500\,mm$ long. The beam produces a spot of light which has a hairline shadow down its centre: this moves over a millimetre scale and in a typical meter a current of $1\,\mu A$ may produce a deflection of $20\,mm$, so currents of $0.05\,\mu A$ ($= 50$ nanoamperes) may be detected. The meter usually has some less sensitive ranges ($0.1\times$, $0.01\times$ and $0.001\times$) which can be selected using a rotary switch which connects the necessary shunts.

Digital instruments

Integrated circuits called **analogue-to-digital converters** are available which, if supplied with a p.d. will produce a digital reading of the p.d. on a display. Typically in a school or college laboratory the meter can display numbers up to 1999 and will have an input resistance of $100\,k\Omega$. The f.s.d. might be given by a current of $2\,\mu A$. We can work out what p.d. this corresponds to:

$$V = IR = (2 \times 10^{-6}\,A)(100 \times 10^{3}\,\Omega)$$
$$= 0.2\,V = 200\,mV$$

Fitting the meter with shunts or multipliers will enable it to measure larger currents and p.d.s. For example, the meter will have a resistance of only $20\,m\Omega$ when it has been adapted as an ammeter with an f.s.d. for $10\,A$, and will have resistance of $10\,M\Omega$ when it has been adapted as a voltmeter with an f.s.d. for $20\,V$.

The digital meters have several advantages over analogue meters.

♦ It is easier (and therefore quicker) to read them.

♦ For measurements near their f.s.d. (e.g. $13.68\,V$, $8.76\,A$) they are more sensitive than analogue meters.

♦ As ammeters they may have smaller resistances, and as voltmeters they may have larger resistances, so that they disturb the circuit less.

♦ There is not the problem of having to interpret scales, as with simple instruments there is often only one range available, and in more sophisticated instruments the decimal point may be moved automatically to the correct position.

But the digital meter samples the p.d. or current only at intervals, so although the updating of the display occurs quite frequently, it is not as useful as an analogue meter for measuring changing currents.

Multimeters

A **multimeter** is a meter on which different ranges of current and p.d. may be selected by means of rotary or push-button switches. It may be an analogue or a digital meter. It will also be able to measure resistances, as we have already explained. A typical multimeter would have current ranges

from $200\,\mu A$ to $10\,A$ (both d.c. and a.c.), p.d. ranges from $200\,mV$ to $1000\,V$ (both d.c. and a.c.) and resistance ranges from $200\,\Omega$ to $20\,M\Omega$. Some digital meters are also auto-ranging: the user selects the quantity to be measured, and the necessary shunt or multiplier is brought into use automatically. More sophisticated multimeters can also measure capacitance, frequency and (with a suitable probe) temperature.

Data loggers

It is also possible to use digital meters which not only take *but also store* measurements of p.d. These **data loggers** are usually used for measurements of p.d. which vary with time. The measurements are made at chosen fixed time intervals, which may be as short as $50\,\mu s$ or as long as 15 minutes. A typical school laboratory data logger would be able to take and store 4000 successive measurements or (because it has four channels to each of which a p.d. may be connected) 1000 successive measurements of four different p.d.s. Since almost any physical quantity (e.g. temperature, pressure, light intensity, wind speed, sound intensity) can, by use of a transducer, be represented by a p.d., such data loggers are extremely versatile: their use goes far beyond the measurement of p.d. in electric circuits. (Their use as timing devices has already been described in Chapter 1.) So, for example, a data logger with a suitable adaptor could be used to measure the current in a light bulb during the first few milliseconds after switching on, or to measure the temperature in a pond over a period of several days. Graphs of the variation of any of these

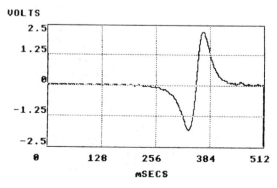

FIGURE 9.11

quantities against time may be displayed on an oscilloscope screen, or drawn on paper. Figure 9.11 shows how the e.m.f. induced in a coil varied as a magnet fell though it. You can see that with a data logger events may be captured which last just a few tens of milliseconds.

Electronic instruments

During the last 20–30 years the development of transistors which can be driven by currents as small as 10 picoamperes ($10\,pA = 10 \times 10^{-12}\,A$) has led to the use of instruments called **d.c. amplifiers** or **electrometers**. These are instruments capable of amplifying very small currents (of the order of picoamperes) so that they are large enough to give readings on an ordinary analogue or digital meter. The power required for the amplification is supplied by a battery. Early versions of these instruments were quite difficult to use, involving the connection and disconnection of resistors of various values, but the latest versions are straightforward, and indeed appear to the user as

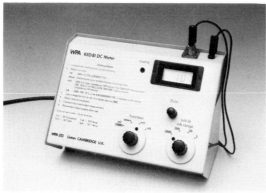

FIGURE 9.12

just very sensitive analogue or digital meters (as shown in figure 9.12). There are instruments which can measure currents as small as 1 pA (which makes them suitable for measuring the ionisation currents in radioactivity experiments) and instruments capable of measuring p.d.s as small as $1\,\mu V$ (which makes them capable of measuring the e.m.f. induced in a single loop of wire). It seems certain that the continuing development of electronics will make precise measurement easier and cheaper as time goes on.

9.4 The cathode ray oscilloscope

The oscilloscope (c.r.o.) consists of an evacuated tube, at one end of which a heated filament F emits electrons, as shown in figure 9.13 on the next page. These electrons are accelerated in an electric field created by a potential difference of several thousand volts between the filament and a series of anodes A_1, A_2 and A_3. After leaving the last anode the electrons have a very high speed: up to $3.0 \times 10^7\,m\,s^{-1}$ (one-tenth of the speed of light). The **focus** control on the front panel of the c.r.o. adjusts the p.d.s of these anodes so that the beam is focused to a fine point on a fluorescent screen S at the other end of the tube.

The screen is marked out with a grid, called a **graticule**, of lines at intervals of 10 mm, with other markings at 2 mm intervals. The spacings between these lines are often referred to as divisions, so that the height of a shape on the screen is measured in 'divisions' rather than mm. The screen glows where it is struck by the electrons; the glow persists for a few tenths of a second: this is called the **afterglow**. The electrons which have struck the screen are conducted back to the anode by means of a graphite coating on the inside of the tube. The number of electrons reaching the screen per second, and hence the brightness of the spot, is controlled by a cylindrical electrode G, called the **grid**, through which the electrons pass soon after leaving the filament. The **brilliance** control on the front panel of the oscilloscope controls the grid p.d.

Just beyond the anodes are two sets of **deflector plates**, to each of which a p.d. can be connected.

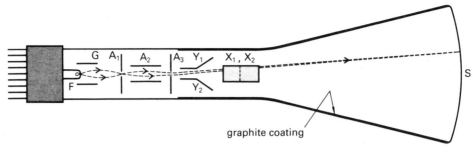

FIGURE 9.13

The p.d. creates an electric field between the plates, and while in this field the electron beam is deflected. The **Y-plates** are horizontal and push the beam up or down, depending on the polarity of the plates, and the **X-plates**, which are vertical, push the beam to one side or the other. P.d.s are available from within the oscilloscope which can be used to move and beam about: the controls are called the **Y-shift** and the **X-shift**.

EXAMPLE

(a) If the flow of electrons in a cathode ray oscilloscope provides a current of 40 mA, how many electrons reach the screen each second? The electronic charge is -1.6×10^{-19} C.
(b) If the screen is 200 mm from the last anode, and their speed is $2.5 \times 10^7 \, \text{m s}^{-1}$, how long do they take to travel from the anode to the screen?
(c) How many electrons are 'in flight' between these points at any time?

(a) The number N of electrons reaching the screen each second is given by

$$N = \frac{I}{e} = \frac{40 \times 10^{-3} \, \text{C s}^{-1}}{1.6 \times 10^{-19} \, \text{C}} = 2.5 \times 10^{17} \, \text{s}^{-1}$$

(b) The time taken t is given by

$$t = \frac{s}{v} = \frac{200 \times 10^{-3} \, \text{m}}{2.5 \times 10^7 \, \text{m s}^{-1}} = 8.0 \times 10^{-9} \, \text{s}$$

(c) Each second 2.5×10^{17} electrons leave the last anode, so in 8.0×10^{-9} s,
$$(2.5 \times 10^{17} \, \text{s}^{-1})(8.0 \times 10^{-9} \, \text{s})$$
electrons must leave, i.e. 2.0×10^9, so this is the number that are in flight at any time.

Measuring p.d.s

The Y-plates are connected, through an amplifier, to two sockets (coloured red and black), or a single coaxial socket, on the outside of the oscilloscope. The black socket, or the outer part of the coaxial socket, is connected to Earth through the oscilloscope's mains lead. The movement of the spot on the screen can be controlled by the p.d. applied to the sockets and the amount of amplification which the amplifier provides: most oscilloscopes have a switched range of different amplifications labelled 5 V/div, 1 V/div, etc. The rotary switch which controls the amount of amplification is called the **sensitivity** or **gain** control. You can see this, marked 'volts/div' near the right-hand end of the front panel of the oscilloscope shown in figure 9.14. The distance the spot moves (in divisions of the graticule on the screen) can be used to work out the size of the p.d. connected. *The oscilloscope therefore acts as a voltmeter*, capable of measuring a range of p.d.s from millivolts to volts. It has a high resistance: typically $10 \, \text{M}\Omega$. It is not, however, very precise, since the movement of the spot is never more than

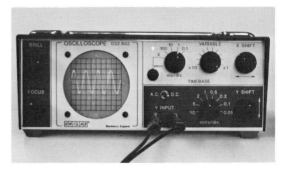

FIGURE 9.14

a few tens of mm, and this cannot be determined to better than about 1 mm.

A steady p.d. connected to the sockets displaces the spot to a new position: an alternating p.d. sweeps the spot up and down, and if the frequency is high enough (more than about 10 Hz) leaves a vertical line on the screen where the electrons have been, because of the afterglow. The beam is, of course, striking the screen at only one point at any moment. To see how the p.d. varies with time, however, it is necessary for a **time base** to be connected to the X-plates.

The time base p.d.

The time base consists of a p.d. which rises steadily and then falls abruptly, and repeats this periodic behaviour indefinitely: a graph of how this p.d. changes with time is shown in figure 9.15. It is called a **saw-tooth** waveform. The X-plates are vertical, so that the time base p.d. moves the electron beam steadily across the screen in a horizontal direction, at a speed which can be varied. The time taken for the spot to move across the screen can be chosen from a range which may vary from 1 s/div to 1 μs/div. The effect of having this time base connected is to draw a graph which shows how the p.d. connected to the Y-plates varies with time. The electrons spend so little time between the Y-plates, and have so little mass, that they can faithfully record the variation of the supply p.d. even up to frequencies of 100 MHz. A moving-coil meter cannot possibly do this because of the mass of its moving parts: in an oscilloscope the only moving parts are the electrons. In figure 9.14 you can see a typical trace.

Oscilloscopes can also be used to measure current, if they are used to measure the p.d. across a resistor of known value: then the current is calculated using $I = V/R$.

EXAMPLE

Find (a) the peak value of the alternating p.d. connected to the oscilloscope shown in figure 9.14, and (b) find its frequency.

(a) The vertical distance from peak to peak of the trace is 4.8 divisions, and since the sensitivity is set at 0.5 V/div the peak p.d. is 2.4 V.

(b) The switched time base is set to 1 ms/div (and the variable time base is set to ×1). There are 10 divisions across the screen, so the spot took 10 ms to travel across it. Since there are four complete cycles on the screen, each cycle took 2.5 ms, so the frequency f of the alternating supply is given by

$$f = \frac{1}{T} = \frac{1}{2.5 \times 10^{-3}\,\text{s}} = 400\,\text{Hz}$$

Synchronisation

We have already said that the glow left by the electron beam fades quite quickly, so to see a permanent trace on the screen, as in figure 9.14, it is necessary that the line should be 'painted' again and again on the screen in exactly the same place as before. In some oscilloscopes (such as the one illustrated) the time base p.d. is controlled so that the spot is not allowed to start moving across the screen until the p.d. connected to the Y-plates has reached some particular (small) value of p.d., so the start of the horizontal movement of the electrons is *synchronised* with the same or a similar point in the alternating cycle of the p.d. which is being observed. On more sophisticated oscilloscopes the size of the p.d. needed to start the spot moving may be varied (or be chosen to be negative or positive) and these oscilloscopes have additional **trigger** controls.

You will realise that simple oscilloscopes can

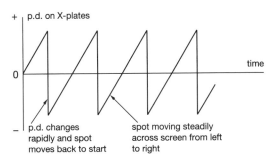

+ p.d. on X-plates

0

time

p.d. changes rapidly and spot moves back to start

spot moving steadily across screen from left to right

FIGURE 9.15

measure only periodic changes of p.d., since the trace has to be painted on the screen again and again if you are to be able to see it. They cannot record *transient* events, like the change of p.d. across a resistor when a current is first switched on. There are also, however, **storage oscilloscopes**, which will record a changing p.d. which may last for only a fraction of a second, and does not repeat itself.

Double-beam oscilloscopes

Oscilloscopes which can provide more than one trace are common. To provide the two traces a single filament provides an electron beam which passes through deflector plates to which is connected a very rapidly alternating p.d. which sends the beam first through one set of Y-plates and then through a second set of Y-plates. Effectively there are then two separate electron beams, so we can see simultaneously graphs of how two different p.d.s are changing. You will find later that in alternating circuits the applied p.d. and the current are not always in phase. A double-beam oscilloscope can be used to observe the variation of these two quantities.

Capacitative input

Figure 9.16(a) shows the apparently steady direct output from a laboratory power supply: the sensitivity was set at 5 V/div, so assuming that the trace was in the centre of the screen when no p.d. was connected, you can see that the output is 20 V. But this output has been produced from an alternating supply, and there may be a small alternating p.d., or **ripple**, superimposed on the steady p.d. If we wanted to examine the trace in more detail to see if the output really was steady, it would be natural to think of increasing the amplification. But increasing the amplification would amplify the p.d. of 20 V as well as the ripple, and the height of the trace would probably be so great that it would be off the top of the screen, and there would not be enough p.d. available from the Y-shift control to bring the trace back on to the screen. In this situation we can connect a capacitor between the input sockets and the amplifier. As you will see later, this blocks the direct part of the p.d. and leaves the alternating ripple, which can then, of course, be amplified considerably. This

(a)

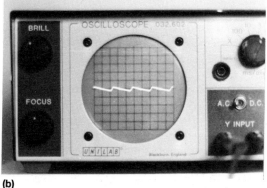

(b)

FIGURE 9.16

connection is made by a switch on the front of the oscilloscope: it may be labelled AC/DC, as in figure 9.14. In figure 9.16(b) the p.d. has been amplified much more than in figure 9.16(a) (the sensitivity setting was 0.5 V/div), and the switch is in the AC position, and you can see that there *is* a ripple with a peak-to-peak value of 0.5 V.

The switch may be in either position for an alternating p.d. like that shown in figure 9.14, but it *must* be in the 'DC' position for measuring steady p.d.s. If you tried to measure the p.d. across a battery with the switch in the DC position the battery p.d. would be across the capacitor and not across the oscilloscope's Y-plates, so no p.d. would be registered.

The X-input

We can use an oscilloscope to perform other tasks apart from measuring how a p.d. varies with time. When the time base p.d. is switched off (fully anti-clockwise in figure 9.14) a different p.d. can be connected to the X-plates: with the type of

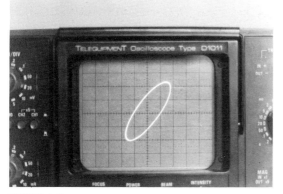

FIGURE 9.17

'out of step'. When the p.d.s are *in phase* (i.e. each produces a positive peak value at the same time) the trace is a straight diagonal line with a positive slope (figure 9.17(a)); if they were *in antiphase* (one has a positive peak value when the other has a negative peak value) the line is again straight but has a negative slope. If there is any other value of phase difference the trace will be an ellipse, as shown in figure 9.17(b). If the two sources do not have *exactly* the same frequency the trace changes continuously from a line through an ellipse to a line. Figure 9.18 is a time exposure, lasting about 5 s, taken for two sources which did not have *quite* the same frequency.

The trace changes from a broader ellipse to a narrower ellipse as the phase difference between the two sources changes. If we are trying to adjust two sources to give exactly the same frequency we need to get the trace stationary: this is a very sensitive technique for matching two frequencies.

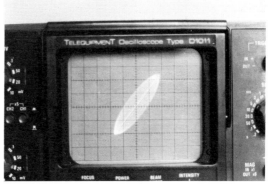

FIGURE 9.18

oscilloscope shown in figure 9.14 one lead goes to the X-input socket, and the other to the black (Earth) socket. If two alternating p.d.s *of the same frequency* are connected to the Y-plates and the X-plates the trace on the screen will be stationary. The nature of the trace shows the **phase difference** of the two p.d.s, i.e. the extent to which they are

---

Exercises on each section of this chapter may be found in the companion textbook, *Practice in Physics*.

SUMMARY

At the end of this chapter you should be able to:

◆ understand that in primary cells the chemical action is not reversible but that in secondary cells it is.

◆ understand that for cells in series the total e.m.f. is the sum of the separate e.m.f.s.

◆ understand that for cells of e.m.f. $\mathscr{E}$ in parallel the total e.m.f. is $\mathscr{E}$ (but that the cells can now supply a larger current).

◆ use the equations $\mathscr{E} = I(R + r)$ and $\mathscr{E} = V + Ir$ to analyse circuits in which the internal resistance r of the cell has to be taken into account.

◆ describe how to measure the internal resistance of a cell.

◆ explain that a cell delivers maximum power when the resistance of the external circuit is equal to the cell's internal resistance.

◆ draw a circuit diagram to show how a secondary cell may be recharged.

◆ understand what is meant by the capacity of a battery.

◆ calculate the value of shunt resistance needed to give an ammeter a higher f.s.d.

◆ calculate the value of multiplier resistance needed to give a voltmeter a higher f.s.d.

◆ understand why an ammeter must have a much lower resistance than the other components in the circuit in which it is placed.

◆ understand why a voltmeter must have a much higher resistance than the component across which it is measuring the p.d.

◆ understand the advantages and disadvantages of light-beam meters, digital meters, data loggers and electronic meters.

◆ describe how to use a c.r.o. to measure p.d.s and how the sensitivity (or gain) control is used.

◆ describe how to use a c.r.o. to measure the frequency of an alternating p.d. and how the time-base control is used.

10 Heating Solids and Liquids

We all heat water: we heat it for cooking, we heat it for hot drinks, we heat it for baths and showers, we heat it for washing clothes, we heat it for washing the house and the car, we heat it for central heating, we heat it for swimming pools, we heat it to make steam engines and steam turbines work. And we cool it to make ice. Knowing how much energy is needed to heat water and other substances is vitally important, since energy is expensive. For example, it is worth knowing that the cost of a shower is about 20% of the cost of a bath. This chapter will explain what happens when we heat all solids and liquids and what determines how much energy it takes and how much it costs.

10.1 Measuring temperature

We all have an idea of what is meant by hot and cold, but our impressions are only qualitative and they are also subjective. We need an instrument to measure hotness *quantitatively*: such an instrument is called a **thermometer**. The thermometer is placed in contact with the body whose temperature

we wish to find and left there until the temperature reading is steady. The thermometer and the body are then *in thermal equilibrium* and there will be no net exchange of energy between them. To make it easy to use, a thermometer:

- should not absorb too much energy from the body whose temperature is being measured, or else the temperature of the body will be affected.

- should allow energy to be conducted easily from the body to the thermometer, or else it will take a long time to reach thermal equilibrium.

Defining a temperature scale

When you think of a thermometer you probably think of a mercury thermometer, in which the expansion of mercury is used to measure the temperature. But many different types of thermometer are possible. The following properties are all currently used to measure temperature:

- the length of a column of liquid in a tube.

- the resistance of a piece of metal.

- the e.m.f. of a thermocouple.

- the pressure of some gas kept at constant volume.

Whichever property we use we need to:

- choose two **fixed points** (easily reproducible temperatures) and give them numbers. These points are often the temperature at which ice and water coexist in equilibrium, called 0° (the **ice-point**) and 100° (the **steam-point**) on the centigrade scale. In both cases the water must be pure, and the pressure nust equal standard atmospheric pressure.

- we need to find the value X of the property at these two fixed points. The scale of a centigrade ('100 steps') thermometer is then divided into 100 equal parts between these fixed points. Then if we want to measure a temperature θ, we measure the value of the property X_θ when it is at that temperature. If the values of the property at the ice-point and steam-point are X_0 and X_{100}, then we can work out, by proportion, what the temperature is in degrees on the **centigrade** scale. The formula is

$$\theta = \frac{X_\theta - X_0}{X_{100} - X_0} \times 100°C$$

Instead of using the formula we could deduce the temperature from a graph, as shown in figure 10.1, and the graph may help you to understand what is being done, even if the formula is more convenient in practice.

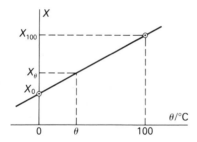

FIGURE 10.1

The symbol °C is also used for **'degrees Celsius'**: the difference between the two scales is a matter of definition and is not important at this stage. We shall use °C to mean degree centigrade or degree Celsius.

EXAMPLE

The resistance of a piece of platinum is measured when it is at 0°C, 100°C and some unknown temperature θ. It is found to be $100.1\,\Omega$, $136.9\,\Omega$ and $125.3\,\Omega$ at these temperatures. What is the unknown temperature?

$$\theta = \frac{125.3\,\Omega - 100.1\,\Omega}{136.9\,\Omega - 100.1\,\Omega} \times 100°C = 68.5°C$$

Thermometers in practice

(a) The liquid-in-glass thermometer

This type, shown in figure 10.2(a), is what most people think of as a thermometer. It has the advantage of being portable. It can be made *sensitive* by using a tube of small diameter, and a bulb which holds a large amount of liquid. The mass of liquid should not be too large, however, or else the time taken for the thermometer to come to equilibrium will be long (but this time can be reduced by making the walls of the bulb thin). Mercury is a liquid in the range −39°C to 357°C (under atmospheric pressure) so it has a convenient, though small, range. Ethanol (usually coloured red) can also be used as the liquid where precision is not important. It is cheaper, and is liquid in the range −117°C to 78°C, which enables it to measure lower temperatures.

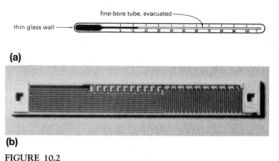

(a)

(b)

FIGURE 10.2

(b) Resistance thermometers

The electrical resistance of a metal (particularly platinum, because it can be obtained to a high degree of purity) may be used to measure temperature. The platinum may be in the form of a thin film (figure 10.2(b)) which is protected in a ceramic or metal sheath. This might be about 30 mm long and a few mm wide. Temperatures between −50°C and 500°C can be measured. In practice the platinum is often put in one arm of a Wheatstone bridge circuit. When the temperature, and therefore the resistance, changes, the bridge is no longer balanced. With suitable circuitry, for limited ranges of temperature, the out-of-balance p.d. can be used to give a direct reading in °C on a digital display. This is how some hand-held thermometer probes work. The resistance of semiconducting materials can also be used: their resistance changes more rapidly at low temperatures, so that they are particularly suitable below −150°C. **Thermistors** are used more as temperature *sensors* than thermometers, but there are other semi-conducting devices which *are* used as thermometers: e.g. a particular integrated circuit (LM35) will give an output of 10 mV per °C for the range −40°C to 110°C.

(c) Thermocouple thermometers

A thermocouple can be made by taking wire of two different metals (or alloys) A and B and joining them as shown in figure 10.3(a). The wires may be simply twisted together to form the two junctions.

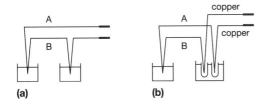

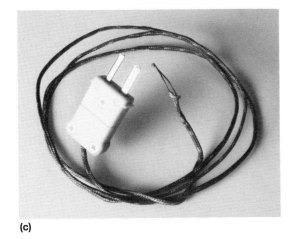

(c)

FIGURE 10.3

If the two junctions are at different temperatures θ_1 and θ_2, a **thermo-electric** e.m.f. is produced. As the size of the e.m.f. depends on the difference in temperature between the two junctions, the thermocouple can be used to measure temperature. One of the junctions has to be kept at a steady temperature (the *reference* temperature): the other is then placed where the temperature is to be measured. The e.m.f. may be amplified using an operational amplifier circuit. There is the possibility of additional thermo-electric e.m.f.s arising where the wires of the thermocouple materials join the copper wires in the rest of the circuit, so to avoid this the arrangement shown in figure 10.3(b) is used, with both junctions with copper being kept at the reference temperature.

In practice commercially-available thermocouple thermometers have circuits which compensate automatically for the reference junction not being kept at a steady temperature, and usually the thermocouple is arranged to give a digital reading in °C. For measuring different ranges of temperature, different pairs of metals or alloys are used. But all thermocouples have the advantage that they respond quickly to changes of temperature (perhaps in 20 ms). A length of thermocouple wire is shown in figure 10.3(c): note the small size of the junction.

(d) Pyrometers

The thermometers so far described must all be placed in contact with the substance whose temperature is to be measured. None of them therefore, can be used to measure temperatures of much more than 2000°C. In this region **pyrometers** are used to analyse the visible radiation coming from the substance, and to deduce its temperature.

(e) Constant-volume gas thermometer

In this instrument the pressure of a gas, kept at constant volume, is used as the thermometric property. This thermometer is large and cumbersome, and is totally impracticable for the everyday measurement of temperature. But it is the standard thermometer and is used in standards laboratories (such as the National Physical Laboratory at Teddington, England) to find precise values of temperatures which are needed as reference temperatures (e.g. the freezing-point of gold is measured to be 1064.43°C). Such a measurement might take months but once made it will be universally accepted as the freezing-point of gold.

The reason for this thermometer being accepted as the standard is that all gas thermometers, whatever gas they contain, give similar readings, and that even the small differences between them become smaller when the pressure in them is reduced. This is because as the pressure decreases the gases are losing their individual characteristics and are behaving increasingly like an **ideal gas** and so a temperature measured in this way is known as an **ideal gas scale temperature**.

The thermodynamic scale

A constant-volume gas thermometer indicates that there is a temperature at which, if the gas had not already liquefied or solidified, the pressure would be zero. As we cannot imagine a gas exerting a negative pressure, this temperature is taken to be the lowest possible temperature – an **absolute zero**. (There is much other evidence for thinking that this temperature is a genuine zero of temperature.) It would make sense to define another temperature scale such that this absolute zero was the lower fixed point. Our choice for the upper fixed point is the **triple-point** of water (the only temperature at which water can exist in equilibrium in the gaseous, liquid and solid phases). We could give this temperature any number we chose, but to make the degrees on this scale the same size as Celsius degrees we call the triple-point of water 273.16 kelvin, where the **kelvin** (K) is the unit of temperature for this ideal-gas scale. On this scale a temperature T is defined by the equation

$$T = \frac{p_T}{p_{tr}} \times 273.16\,K$$

where p_T is the pressure at temperature T, and p_{tr} is the pressure at the triple-point. (This is similar to the previous equation, but simpler because the pressure at the lower fixed point is zero.) This called the **thermodynamic** or **kelvin** scale of temperature, since it is based on the thermodynamic properties of heat engines. It can be shown to be identical with the ideal-gas scale.

The triple-point of water is 0.01°C, so that 0.01°C = 273.16 K, but for most purposes the lack of precision of our laboratory temperature measurements does not justify our using so many significant figures, so we usually simply add 273, not 273.16, to the centigrade (or Celsius) figure to obtain the temperature in kelvin: e.g.

$$20°C = 293\,K$$
$$100°C = 373\,K.$$

10.2 Internal energy

Internal energy

The graphs in figure 10.4 will remind you of how the k.e. and electric p.e. of a molecule in a solid are continually changing: as the molecule moves from A to O its electric p.e. decreases and its k.e. increases. As it moves from O to B, its electric p.e. increases and its k.e. decreases. This oscillation continues indefinitely. If the solid is heated the molecules are given more energy: the result is that the maximum amounts of k.e. and electric p.e. are greater, and the molecule oscillates over a greater distance, as shown in figure 10.4(b). Molecules in a liquid oscillate in a similar way, but on average have more electric p.e. since they are on average further apart than the molecules in a solid.

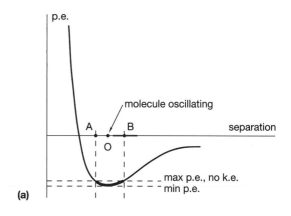

(a)

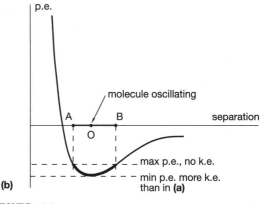

(b)

FIGURE 10.4

Molecules in a gas have much more electric p.e. than molecules in a solid or a liquid (because they are much further apart) and much more k.e. too (because they are at a higher temperature). We give the name **internal energy** to the total of the kinetic and potential energies of the molecules of a substance. So:

♦ 500 g of iron at 67°C will have more internal energy than 500 g of iron at 63°C.

♦ 500 g of iron at 56°C will have more internal energy than 300 g of iron at 56°C.

♦ 300 g of water at 0°C will have more internal energy than 300 g of ice at 0°C.

♦ 200 g of water vapour at 100°C will have more internal energy than 200 g of water at 100°C.

The symbol for internal energy is U and changes in it are represented by ΔU. Internal energy can be changed in two ways: by **heating** and **working**.

Heating

Heating is a process which occurs *because of a temperature difference*. For example, a gas flame *heats* some water because it is hotter than the water. The symbol for this kind of energy transfer is ΔQ. Notice that we have used the word 'heat' as a verb, not a noun, because heating is a *process*: we cannot talk about the *amount of* heat in a body (that would be its internal energy). You should not say that the flame 'gives heat to' the water: it gives internal energy to the water and therefore heats it. Unfortunately the way in which the word heat has been used in the past has led to its appearing (as a noun) in the quantities specific heat capacity and specific latent heat. These really mean 'specific internal energy capacity' and 'specific latent internal energy' but you can see that it would be awkward to use these phrases. So we shall call these quantities s.h.c. and s.l.h. to avoid using the word 'heat' as a noun (just as e.m.f. is used to avoid saying electromotive force).

Working

We could also change the internal energy of a body by *working* on it, as we do, for example, when we pump up a bicycle tyre and it becomes hotter, or the brake pad sliding on the disc of a car's disc brakes raises the internal energy of the pad and the disc. The symbol for this kind of energy transfer is ΔW. In this kind of process the internal energy changes because *forces* are used: ordered forms of energy become disordered forms.

Processes

Notice that the internal energy U of a body describes what it *has* – its *content*: heating and working are *processes* – methods by which the internal energy is changed. We cannot talk about the heat Q or the work W in a body – only about the changes ΔQ and ΔW. Some examples of the two processes, heating and working, are given in Table 10.1.

TABLE 10.1 Some examples of heating and working

Situation	Process by which energy is transferred	Type and site of energy	
		Initially	finally
electric bar radiant heater in a room	heating	internal energy of element	internal energy of room
person turning drill in block of wood	working	chemical energy of person	internal energy of drill and wood
car being stopped by brakes	working	k.e. of car	internal energy of brakes and surroundings
person walking out of warm room	heating	internal energy of person	internal energy of surroundings
electric element in kettle	working	electric p.e. of mains supply	internal energy of water and kettle

Notice that where there is *heating*, the initial and final forms are both internal energy, which is transferred because one body is hotter than the other. When there is *working*, the (disordered) internal energy appears at the expense of some ordered form of energy.

First law of thermodynamics

This law expresses the idea of **conservation of energy** in this particular situation. It states simply what you would expect: that the change in a body's internal energy is equal to the internal energy transferred by the two processes heating and working. For example, in a steam engine 63 kJ of internal energy might be supplied to the steam in a cylinder by heating, but at the same time the cylinder might push the piston out and do 25 kJ of work on the surroundings, so the net gain in internal energy would be $63\,\text{kJ} - 25\,\text{kJ} = 38\,\text{kJ}$.

Unfortunately examination boards do not agree about the way in which the first law should be written down, so we have given both alternatives here. You should use whichever your board uses.

First version

$$\Delta U = \Delta Q + \Delta W$$

where
ΔU is the gain of internal energy
ΔQ is the energy transferred to the body by heating
ΔW is the energy transferred *to* the body by working

In the steam engine example, therefore,

$$\Delta Q = +63\,\text{kJ}, \Delta W = -25\,\text{kJ}$$
$$\text{so} \quad \Delta U = (+63\,\text{kJ}) + (-25\,\text{kJ})$$
$$= +38\,\text{kJ}$$

the positive sign indicating a gain of internal energy.

Second version

$$\Delta Q = \Delta U + \Delta W$$

where
ΔU is the gain of internal energy
ΔQ is the energy transferred to the body by heating
ΔW is the energy transferred *from* the body by working

In the steam engine example, therefore,

$$\Delta Q = +63\,\text{kJ}, \Delta W = +25\,\text{kJ}$$
$$\text{so} \quad (+63\,\text{kJ}) = \Delta U + (+25\,\text{kJ})$$
$$\text{so} \quad \Delta U = +38\,\text{kJ}$$

the positive sign indicating a gain of internal energy.

The *result* is the same, whichever form of equation is used.

10.3 Heat capacity

If you are an engineer designing a washing machine one thing you will certainly want to know is how much energy it needs to heat up the water and other parts of the machine which will inevitably become hotter also. The more energy it needs, the longer any of the washing cycles will take. You would want to know how much energy it took to raise the temperature of the whole machine, including the water, by one centigrade degree. This quantity is called the **heat capacity** (h.c.) of the machine and the water. The heat capacity C of a body is defined by the equation

$$C = \frac{\Delta Q}{\Delta \theta}$$

where ΔQ is the energy supplied and $\Delta \theta$ is its temperature rise. You can see that its unit is the J K^{-1} or $\text{J}\,°\text{C}^{-1}$. We can use K or °C interchangeably here, since they are the same size, but K is preferred, since it is easier to write or say. The equation is often used in the form $\Delta Q = C\,(\Delta \theta)$ – when we need to know how much energy needs to be given to a body to raise its temperature. In this situation ΔQ is taken to mean the energy supplied by any means – heating or working.

EXAMPLE

A washing machine has a heater of power 2.5 kW. Water enters it at 12°C and it has to be heated to 60°C for one of the cycles. If this takes 31 minutes, what is the heat capacity of the machine with the water in it?

Energy supplied
$$= (2.5 \times 10^3 \, \text{J s}^{-1})(31 \times 60 \, \text{s}) = 4650 \, \text{J}$$
Temperature rise = 48 K

so heat capacity $C = \dfrac{\Delta Q}{\Delta \theta} = \dfrac{4650 \, \text{J}}{48 \, \text{K}} = 96.9 \, \text{J K}^{-1}$

Specific heat capacity

As a washing machine engineer you might want to redesign the machine to reduce the amount of water used for each cycle (to save water, and to save energy). Then you would want to know how much energy it needs to warm up each kilogram of water by each degree: this is called the **specific heat capacity** (s.h.c.) of water. The specific heat capacity c of a substance is defined by the equation

$$c = \frac{\Delta Q}{m(\Delta \theta)}$$

where ΔQ is the energy supplied, m is the mass of the substance and $\Delta \theta$ is its temperature rise. The equation is often used in the form $\Delta Q = cm(\Delta \theta)$. There is a simple connection between h.c. C, mass m and s.h.c. c:

$$C = cm.$$

EXAMPLE

If the s.h.c. of water is $4200 \, \text{J kg}^{-1} \, \text{K}^{-1}$, and 2.0 litres less water was used in the washing machine in the previous example, what would the new heating time become?

Heat capacity of 2.0 litres (mass = 2.0 kg) of water

$$= mc = (2.0 \, \text{kg})(4200 \, \text{J kg}^{-1} \text{K}^{-1}) = 8.4 \, \text{kJ K}^{-1}$$

so the heat capacity of the machine + water is reduced from $96.9 \, \text{J K}^{-1}$ to $88.5 \, \text{kJ K}^{-1}$ and the energy needed will be reduced in the ratio 88.5/96.9. Therefore the time taken will be reduced in this ratio also. The new heating time

$$= 31 \, \text{minutes} \times \frac{88.5}{96.9} = 28.3 \, \text{minutes}$$

Measuring s.h.c.

You could use a metal block fitted with an immersion heater and a thermometer, as shown in figure 10.5(a). You would measure the mass of the block. The block would need to be lagged to reduce energy losses during the heating: it could stand on a mat made of expanded polystyrene, and be wrapped in cotton wool. The immersion heater would be connected to a power supply (the electrical circuit is shown in figure 10.5(b)) and then the heater switched on and a stopclock started at the same time. You would measure the p.d. V across the heater, and the current I in it.

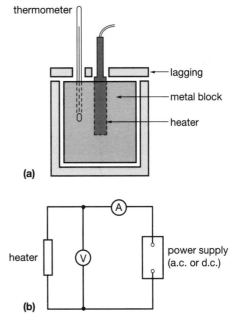

FIGURE 10.5

The temperature would be recorded at regular intervals, and a graph like that shown in figure 10.6 would be obtained. If you select a straight part of the graph you could measure the temperature rise $\Delta \theta$ which occurred in a certain time Δt. Then the value of the s.h.c. of the metal of the block is given by

$$c = \frac{VI(\Delta t)}{m(\Delta \theta)}.$$

There is a curve at the start of the heating process because it takes time for the energy to reach the thermometer from the heater. The curve at the top occurs because the block is losing energy more rapidly there. This method of measuring the s.h.c. by *direct heating* could be adapted for liquids and gases, though there the heat capacity of the container (which would inevitably also be heated) would have to be measured and allowed for.

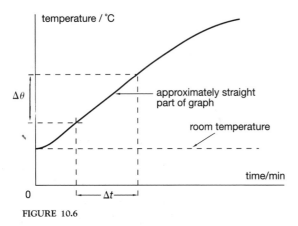

FIGURE 10.6

Energy losses

The graph in figure 10.6 shows that the temperature rises more and more slowly as time goes on. This happens because the block is losing energy to the surroundings at a greater and greater rate as the temperature difference itself and the surroundings increases. *The rate of loss of energy depends on the temperature difference between the block and its surroundings.* There must eventually come a time when the temperature stops rising because it is losing energy to the surroundings exactly as fast as it is being given it from the heater. It is not that the body 'has reached its limit': it is just that it is getting rid of the energy as fast as it is getting it. This is a most important idea and applies in many situations. For example, why does the filament of an electric light bulb reach a steady temperature? Because at just that temperature, it is hot enough to be able to get rid of the energy as fast as it is getting it. Why does a room, with an electric fire switched on, not become hotter and hotter indefinitely? Because the warmer the room gets, the greater the rate at which energy leaks out through windows and doors: the room reaches a temperature where the rate of loss of energy is equal to the rate at which the energy is being supplied.

EXAMPLE

A copper block of mass 1.0 kg was fitted with an immersion heater, and the p.d. and current were 10.3 V and 3.68 A respectively. When a graph of temperature against time was plotted, the temperature was found to rise steadily from 26.5°C to 41.9°C in 3.0 minutes. What value do these measurements give for the s.h.c. of copper?

$$\text{Energy supplied} = VI(\Delta t)$$
$$= (10.3\,\text{V})(3.68\,\text{A})(180\,\text{s})$$
$$= 6823\,\text{J}$$

Temperature rise $= 16.4\,\text{K}$

So $c = \dfrac{VI(\Delta t)}{m(\Delta\theta)} = \dfrac{6823\,\text{J}}{(1.0\,\text{kg})(16.4\,\text{K})} = 416\,\text{J}\,\text{K}^{-1}$.

The accepted value is $381\,\text{J}\,\text{kg}^{-1}\,\text{K}^{-1}$. Two sources of error, both of which would explain the difference, are that no allowance has been made for the heat capacity of the heater or for energy losses to the surroundings.

Correcting for energy losses

You need to be able to make a correction for the energy lost to the surroundings in an experiment like the one where a metal block is heated. An example will explain the idea.

Suppose a kettle is filled with water at 12.0°C and takes 3 minutes to reach a temperature of 96.0°C: at this point it is switched off. It is allowed to cool: its temperature falls by 1.2 K in $1\frac{1}{2}$ minutes. At the moment the kettle was switched off, it was 84 K above its surroundings, so while heating the *average* excess temperature was therefore 42 K. In the $1\frac{1}{2}$ minutes after it was switched off the excess temperature was (roughly) 84 K, so during this time it was losing energy twice as fast as it was (on average) while it was warming up, so in $1\frac{1}{2}$ minutes it loses as much energy as it did while it was warming up for 3 minutes. But in the second case (while cooling) we know what the loss of temperature was: 1.2 K. This must have been the temperature 'lost' while heating, so the corrected final temperature should be 96.0°C + 1.2 K $= 97.2$°C.

This is quicker to do than it is to explain, and the graph in figure 10.7 should help you to see what you have to do. You need to draw first an idealised form of the graph.

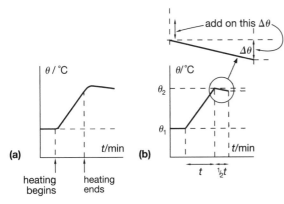

(a) heating begins / heating ends

FIGURE 10.7

An alternative way of correcting for energy losses, which is possible in some types of experiment, is to start with a body a few degrees below room temperature, and then to heat it until it is as many degrees above room temperature. Then the loss of energy while the body is above room temperature is exactly balanced by the gain of energy while it is below room temperature.

Rate equations

If we take the equation $\Delta Q = cm\,(\Delta \theta)$ and divide both sides by the time Δt in which the heating occurs, we have

$$\frac{\Delta Q}{\Delta t} = cm\,\frac{\Delta \theta}{\Delta t}$$

We can write this as $\dfrac{dQ}{dt} = cm\,\dfrac{d\theta}{dt}$

where dQ/dt is a symbol meaning the rate of supply energy (which would be measured in $J\,s^{-1}$ or W) and $d\theta/dt$ is a symbol meaning the rate of rise in temperature (in $K\,s^{-1}$). This form of the s.h.c. equation is useful when we want to find rates of rise of temperature.

EXAMPLE

The filament of an electric light bulb is made of tungsten and has a mass of 0.10 g. When it is first switched on the p.d. is 12.0 V and the current is 15 A. If the s.h.c. of tungsten is $142\,J\,kg^{-1}\,K^{-1}$, find the initial rate of rise of temperature.

The power $= (12\,V)(15\,A) = 180\,W$ so

$$\frac{dQ}{dt} = cm\,\frac{d\theta}{dt} \Rightarrow \quad 150\,W$$

$$= (142\,J\,kg^{-1}\,K^{-1})(0.10 \times 10^{-3}\,kg)\frac{d\theta}{dt}$$

so $\dfrac{d\theta}{dt} = 10.6\,kK\,s^{-1}$ (i.e. $10\,600\,K\,s^{-1}$)

This is a very high rate of rise, but of course it will start to lose energy very rapidly almost immediately: the final temperature of the filament might be about 2600 K, which would be reached in a few tenths of a second.

Heating a continuous stream of water

When immersion heaters in water tanks and kettles heat water, the whole mass of water has its temperature steadily raised. In 'instant' gas and electric water heaters the heating process is different: the water passes over a heating filament a little at a time, and each small mass (Δm) of water has its temperature raised by a large amount θ in a very short time.

The equation $\Delta Q = cm\,(\Delta \theta)$ should now be written $\Delta Q = c\,(\Delta m)\,\theta$, which as a rate equation, becomes

$$\frac{dQ}{dt} = c\left(\frac{dm}{dt}\right)\theta$$

EXAMPLE

An electric shower has a maximum power of 7.0 kW. If water enters the heater at a temperature of 10°C and leaves it at a temperature of 41°C, what is the maximum rate of flow of water? If you were content with an output temperature of 37°C, what rate of flow could you have? (S.h.c. of water = 4200 J kg⁻¹ K⁻¹.)

$$\frac{dQ}{dt} = c\left(\frac{dm}{dt}\right)\theta$$

$$7.0 \times 10^3 \, \text{W} = (4200 \, \text{J kg}^{-1} \text{K}^{-1})\left(\frac{dm}{dt}\right)(41 \, \text{K})$$

so
$$\frac{dm}{dt} = 0.041 \, \text{kg s}^{-1} = 2.4 \, \text{kg min}^{-1}$$

$$= 2.4 \, \text{litres min}^{-1}.$$

If θ was reduced from 41 K to 37 K, the rate of flow could rise in the ratio 41/37, i.e. to

$$2.4 \, \text{litres min}^{-1} \times \frac{41}{37} = 2.7 \, \text{litres min}^{-1}.$$

Values of s.h.c.

Most solid substances have s.h.c.s in the range 100 J kg⁻¹ K⁻¹ to 1000 J kg⁻¹ K⁻¹. Generally, for elements, the higher the density, the lower the s.h.c. For example, dense elements like lead and mercury have s.h.c.s of 120 J kg⁻¹ K⁻¹ and 130 J kg⁻¹ K⁻¹ and at the other extreme magnesium and sodium have s.h.c.s of 980 J kg⁻¹ K⁻¹ and 1190 J kg⁻¹ K⁻¹. Most common organic liquids have s.h.c.s of about 2000 J kg⁻¹ K⁻¹, but the s.h.c. of water (4200 J kg⁻¹ K⁻¹) is unusually high. The s.h.c.s of gases vary greatly with the gas and the conditions under which they are heated: it takes less energy to heat a gas kept at constant volume than it does to heat the gas which is allowed to expand while it is being heated (because work has to be done to push back the atmosphere). For air at constant volume, the s.h.c. is about 1000 J kg⁻¹ K⁻¹.

The high value of the s.h.c. of such a common substance as water has many consequences in everyday life. On a global scale, the oceans change temperature far more slowly than the land does, so that the temperature of places on the Earth depends partly on whether they are surrounded by sea or land. Take, for example, London (England), on an island surrounded by sea, and Orenburg (Russia), in the middle of a large continent, far from the sea. They are both at the same latitude and have similar heights above sea level, yet the monthly average temperature varies from +2°C to +19°C in London but from −18°C to +22°C in Orenburg.

On a smaller scale, the high s.h.c. of water is responsible for the relative coolness in summer of coastal areas, and their relative mildness in winter. On a still smaller scale, the high s.h.c. of water means that it may take a long time to raise the temperature of some water, but that the water will not fall in temperature quickly, either. This has obvious significance for people taking baths, or drinking coffee.

10.4 Latent heat

If you put a thermometer and a heater in a block of ice which is well below 0°C, and switch the heater on, the temperature of the ice will rise, as we would expect. However, after a time the temperature will stop rising and will remain steady. If we look at the ice we see that it is melting. When the ice has completely melted (to water) the temperature will start to rise again, until once again the temperature remains steady for a time. The water is then boiling and the temperature will remain steady until all the water has evaporated. Figure 10.8 is a graph which shows this behaviour.

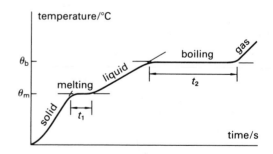

FIGURE 10.8

All substances behave similarly, unless the pressure is low enough for the substance to change from solid to gas directly, i.e. to sublime, without passing through the liquid phase.

Specific latent heat

The energy which we supply to change the phase of the substance (e.g. change ice to water) is called **latent heat** L. 'Latent' means 'hidden', and the term is used because the energy seems to disappear: it does not produce the expected temperature rise. We can define the **specific latent heat** (s.l.h.) l by the equation

$$l = \frac{\Delta Q}{m}$$

were ΔQ is the energy needed to change the phase of a mass m of the substance. You can see that the unit of l is the $J\,kg^{-1}$. We can refer to the s.l.h. of melting (or fusion), or the s.l.h. of vaporisation, whichever is appropriate. For any normal substance the latent heat is large compared with the energy needed to change the temperature of the substance by, for example, 100K. For example, for copper, the energy needed to raise the temperature of 10g of copper by 100K is, using $\Delta Q = cm(\Delta\theta)$,

$$\Delta Q = (400\,J\,kg^{-1}\,K^{-1})(10 \times 10^{-3}\,kg)(100\,K) = 400\,J$$

but the energy needed to melt 10g of copper is 2060J and the energy needed to evaporate 10g of molten copper is 48000J.

Freezing and condensing

When a gas condenses to form a liquid the latent heat is given back again. Although water at 100°C is dangerous, the same mass of water vapour at 100°C is even more dangerous: it would deliver about 8 times as much energy to your hand. Fortunately the mass of vapour is likely to be a lot less than the mass of water. Again, when a liquid freezes to a solid, latent heat is given back. This is why there is a pause when a liquid (like naphthalene or its safer alternative, hexadecanol) cools from about 80°C to room temperature: the substance continues to give energy to the surroundings, but the latent heat maintains its temperature until the substances has completely frozen.

Here is another example. Water (e.g. a puddle on the ground) does not completely freeze as soon as the air or ground temperature reaches freezing-point: it takes some time for the latent heat (in the water) to be transferred from the water to the colder surroundings, and some of the water will still be liquid for some time after the temperature of the surroundings has fallen below 0°C.

EXAMPLE

A kettle has a power of 2750W and contains 1.7kg of water at 16°C. Its h.c. is $500\,J\,K^{-1}$. If the water boils at 100°C, how long will it take to reach boiling point, and how much longer will it take for all the water to evaporate? S.h.c. of water = $4200\,J\,kg^{-1}\,K^{-1}$; s.l.h. of vaporisation of water = $2.3\,MJ\,kg^{-1}$.

Energy ΔQ needed to raise temperature

$$= (C + cm)(\Delta\theta)$$
$$= \{500\,J\,K^{-1} + (4200\,J\,kg^{-1}\,K^{-1})(1.7\,kg)\}(84\,K)$$
$$= 0.64\,MJ$$

$$\text{Time taken} = \frac{\Delta Q}{P} = \frac{0.64 \times 10^6\,J}{2750\,J\,s^{-1}}$$
$$= 233\,s \approx 4\,minutes.$$

Energy ΔQ needed to evaporate water

$$= ml = (1.7\,kg)(2.3 \times 10^6\,J\,kg^{-1}) = 3.91\,MJ$$

$$\text{Time taken} = \frac{\Delta Q}{P} = \frac{3.91 \times 10^6\,J}{2750\,J\,s^{-1}}$$
$$= 1422\,s \approx 23\tfrac{1}{2}\,minutes.$$

Notice that we do not need to know the h.c. of the kettle or the s.h.c. of the water for the second calculation, since then no temperature changes occur. The time taken is far longer than the time taken to bring the water to the boil.

Measuring the s.l.h. of fusion (melting)

We shall describe a method which you could use in your own laboratory. Ice is the most convenient solid to use. Take lumps of ice and add them to a known mass of water in a plastic beaker. The temperature of the water will fall, because energy is taken from it to melt the ice and warm the melted ice. So

energy lost by water = energy gained by ice *and melted ice*,

Using these symbols:
m_w is the original mass of water
m_i is the mass of ice added
$\Delta\theta_w$ is the temperature of the water

$\Delta\theta_i$ is the temperature rise of the melted ice
c is the s.h.c. of water
l is the s.l.h. of fusion of ice

$$m_w(\Delta\theta_w) = m_i l + m_i c(\Delta\theta_i).$$

Find m_i by subtracting m_w from the final mass. If c is known, and $\Delta\theta_w$ and $\Delta\theta_i$ measured, l can be calculated.

The sources of uncertainty in this experiment are:

- the h.c. of the plastic beaker is ignored, but its mass is so small (2–3 g?) that this is justified.

- it is assumed that all the ice is at 0°C; but if the lumps are small, and the surface is melting, this is reasonable.

- the lumps of ice will be wet before they are added to the water, so the added mass is not all ice, but partly water.

- there will be energy gained by the water from the surroundings as the water cools down. However, we could start with the temperature of the water slightly above room temperature, and add ice until the water has cooled down to the same amount below room temperature, so that the loss and gain of energy would cancel out.

Measuring the s.l.h. of vaporisation

Take about 250 g of water, heated to boiling point, and put it into a vacuum flask. Fit the flask with a rubber bung through which passes a tube which allows vapour to escape, and also two leads to an immersion heater (which could be a simple coil of nichrome wire whose resistance is about 4 Ω) as shown in figure 10.9(a). Place the flask on a top-pan balance. The electrical circuit is shown in figure 10.9(b): measure the p.d. across the heater and the current in it in the usual way.

Switch on the power supply and wait until vapour is seen to escape from the flask. Tare (i.e. set to zero) the top-pan balance and at the same time start a clock. Record the balance reading at regular time intervals. When the balance reading is falling at a steady rate you will know that all the energy supplied is being used to evaporate the water (and not to heat up the flask). Record the mass Δm evaporated in a time Δt. Then the s.l.h. of vaporisation l is given by

$$l = \frac{VI(\Delta t)}{\Delta m}$$

Since we are using a vacuum flask, the loss of energy is likely to be negligible, and the experiment, as described, is probably adequately precise. If it is suspected that the loss of energy is not negligible, a second experiment can be done using a different power. Then since the temperature of the flask must be the same in both experiments, the *rate* of loss of energy will be the same, and if the time for the two experiments is the same, the *loss* of energy will be the same. Call this loss of energy Q: then

$$V_1 I_1 t = m_1 l + Q$$
$$V_2 I_2 t = m_2 l + Q$$

and by subtraction we have

$$(V_1 I_1 - V_2 I_2)t = (m_1 - m_2)l$$

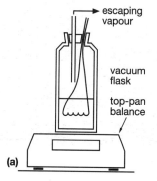

(a)

escaping vapour

vacuum flask

top-pan balance

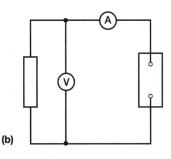

(b)

FIGURE 10.9

from which l can be calculated. Neither the individual masses m_1 and m_2, nor the difference in mass $(m_1 - m_2)$ should be small, or else there will be too much uncertainty in the value of l.

Molecular explanation

When the phase changes, the energy which we supply has two separate jobs to do. To melt a solid, some of the molecules have to be pulled away from their neighbours, and the structure no longer has any rigidity. To pull any molecules further apart, work must be done: we say that *the molecular bonds are broken*. Of course the bonds are not real, like chains or ropes or even springs, but it is convenient to use this phrase to describe the *electrical* forces holding the molecules together.

Once a liquid has been formed from a solid the number of broken bonds is constant, though it will not always be the same bonds that are broken. When a liquid evaporates, all of the remaining bonds must be broken. Since melting means the breaking of relatively few bonds we should expect the s.l.h. of melting to be less than the s.l.h. of vaporisation, and Table 10.2 shows that this is so: l_m is always much less then l_v. The relative sizes of l_m and l_v give some idea of the proportion of molecular bonds broken when the solid melts.

TABLE 10.2 l_m and l_v for various substances

	l_m/kJ kg^{-1}	l_v/kJ kg^{-1}
argon	30	163
carbon dioxide	189	932
copper	205	4790
mercury	11	296
sulphur	44	300
tungsten	192	4350
water	333	2260

Secondly there is nearly always expansion when a solid melts (ice is the only common exception) and there is always an expansion when a liquid evaporates. On expansion energy is needed to push back the atmosphere, but the amount of energy is small compared with the energy needed to break bonds, even for the liquid–gas phase change, when the volume change is greatest: e.g. water evaporating at 100°C requires only about 7% of the total latent heat to push back the atmosphere.

Evaporation

Some molecules in a liquid will have more than the average kinetic energy, and some will have less. At any temperature there will be some at the surface of the liquid which have enough energy to escape completely from the others. This escape from the *surface* of the liquid is called **evaporation**. (It is not the same as **boiling**, which occurs at a particular temperature: in the boiling process, liquid evaporates into air bubbles which are formed *throughout* the liquid.) Although this process happens at *any* temperature, it will occur at a greater rate at higher temperatures, since then there will be more molecules which have the energy necessary to escape.

Since it is the more energetic molecules which escape, the *average* kinetic energy of the remaining molecules falls, so the temperature falls. It will not continue to fall for long because as soon as it falls below the temperature of the surroundings, energy flows in from the surroundings – but any evaporating liquid (to which energy is not being supplied artificially) will be at a lower temperature than it would have been if it had not been evaporating. This is why you feel cold when you have just got out of a swimming pool; the water on your skin is evaporating.

The human body relies on this mechanism for the regulation of body temperature. When muscular activity raises body temperature, sweat

FIGURE 10.10

glands produce water on the surface of the body. While you are running the rate of loss of energy by this process balances the rate of production of energy, and your temperature is constant. When you stop running, your skin is still moist and the evaporation process continues and your body temperature could fall dangerously, especially if you are in a draught; this is why you see athletes wrapping up after events even if they feel warm (see figure 10.10). The 'chill factor' which is given in weather forecasts refers to the cooling effect of a wind: a wind makes you feel colder because it blows away the layer of warm air next to your skin and also because it increases the evaporation rate.

Exercises on each section of this chapter may be found in the companion textbook, **Practice in Physics**.

SUMMARY

At the end of this chapter you should be able to:

◆ understand the principle of measuring a temperature on the centigrade or Celsius scale.

◆ describe several types of thermometer and explain the advantages and disadvantages of each.

◆ understand the advantages of the kelvin scale of temperature.

◆ remember that $0\,\text{K} = -273°\text{C}$.

◆ understand what is meant by internal energy.

◆ understand the difference between heating and working.

◆ use the first law of thermodynamics with whatever sign convention your examination board uses.

◆ use the equations $\Delta Q = C\,(\Delta\theta)$ and $\Delta Q = cm\,(\Delta\theta)$, which define heat capacity (h.c.) C and specific heat capacity (s.h.c.) c.

◆ describe how to measure the s.h.c. of a metal in the form of a cylindrical block.

◆ understand that the rate of loss of energy from a body depends on the temperature difference between itself and the surroundings, and on its surface area and the nature of the surface.

◆ understand that a heated body must eventually reach a steady equilibrium temperature when it is losing exactly as much energy as it is gaining.

◆ make simple corrections for the loss of energy from a heated body.

◆ use the equation $\Delta Q = cm\,(\Delta\theta)$ in the forms

$$\frac{dQ}{dt} = cm\frac{d\theta}{dt} \quad \text{and} \quad \frac{dQ}{dt} = c\left(\frac{dm}{dt}\right)\theta.$$

◆ understand the significance of water having a particularly high s.h.c.

◆ use the equation $\Delta Q = ml$, which defines specific latent heat l.

◆ describe how to measure the s.l.h. of melting of ice.

◆ describe how to measure the s.l.h. of vaporisation of water.

◆ explain in molecular terms why energy is needed to melt a solid or evaporate a liquid.

◆ understand the importance of evaporation in regulating human body temperature.

11 Energy Transfer

Transfers of energy are going on naturally all around us. The most important is the daily delivery of energy from the Sun to the Earth: we would not survive long without it. But in the past we have captured very little of it: apart from photosynthesis by plants, most has been radiated away again at night. The depletion of our reserves of fossil fuels is making us look harder at ways of making use of renewable sources of energy: hydroelectric energy, wind energy, wave energy and solar energy come under this heading. The photograph shows a successful 'alternative technology' method of making use of what the Sun supplies.

In the last chapter we noticed that evaporation is one way in which energy can be transferred from a warm body to cooler surroundings. This chapter is about other methods of energy transfer: convection, conduction and radiation. These three processes, which all occur because of temperature differences, can legitimately be described as heating processes.

11.1 Convection

Because a substance generally expands when its temperature rises, its density decreases. If the fluid in a container is heated at the bottom, the fluid there will become less dense and will rise to the top, and the cooler, denser fluid at the top will fall to the bottom. This is why electric kettles and immersion heaters always have their heaters at the bottom (as shown in figure 11.1(a)) and why refrigerators have the freezing compartments at the top (figure 11.1(b)). This movement of a fluid, which transfers energy, is called **convection**.

(a)

(b)

FIGURE 11.1

On a much larger scale local heating of the Earth by the Sun warms the air in contact with it and causes the air to rise. Such movements of the atmosphere are the cause of most weather changes. Clearly convection can happen only in a fluid.

Cooling by convection

We know that a warm body, such as a mug of coffee placed on a table, cools down. How does the energy transfer take place? Partly by evaporation, and partly by conduction and radiation, but mainly by convection. The mug heats the air next to it (by conduction) and this warmed air rises, so that cool air takes its place and the process of cooling by convection continues. The same thing happens to an unclothed human body standing still; the body loses energy at a rate of about 100 W in that situation, and 30% of this occurs by convection. Figure 11.2 shows a photograph, taken using a special technique, of the convection currents streaming upwards from a human face. Wearing clothes on the rest of the body largely eliminates this loss, by preventing the layers of air from moving. Central heating radiators also lose energy

FIGURE 11.2

primarily by convection. The more powerful radiators have additional fins on the back of their panels to increase the surface area, and hence the amount of convection. Storage heaters also release their energy mainly by convection, sometimes assisted by a fan. Heat sinks (figure 11.3) which are fixed to electronic components to let them get rid of their excess internal energy, have fins: the fins should be placed vertically, as shown in the photograph, to enable the convection currents to flow easily.

As you would expect, the rate of flow of energy dQ/dt by *natural* convection depends on the surface area A, the nature of the surface (e.g. whether it is rough or smooth) and on the temperature difference $\Delta\theta$ between the surface and the surroundings. The relationship is

$$\frac{dQ}{dt} \propto A\,(\Delta\theta)^{5/4}$$

FIGURE 11.3

When there is **forced convection** (e.g. when the body is in a strong draught, perhaps created by a fan) a simpler relationship, known as **Newton's Law of Cooling**, applies:

$$\frac{dQ}{dt} \propto A\,(\Delta\theta).$$

11.2 Conduction

Conduction is the transfer of energy through a substance *without* the bodily movement of any of the substance. It is this process which, for example, makes the lid of a saucepan hot when it is only the base which is being given energy.

To study this effect we need to consider a simple situation: a bar of material, of constant cross-sectional area, heated at one end and cooled at the other, and surrounded by lagging so that energy does not flow out from the sides, as shown in figure 11.4(a). The energy supplied at one end will flow along the bar and out at the other end. Although the energy which first enters the bar will raise its temperature at that end, the temperatures at different parts of the bar eventually become constant, decreasing steadily from one end to the other, as shown in figure 11.4(b). Then, since the temperature of the bar is no longer rising, energy is leaving the bar at the cooler end at exactly the same rate at which it is entering it at the hotter end. Although the heat capacity of the bar will

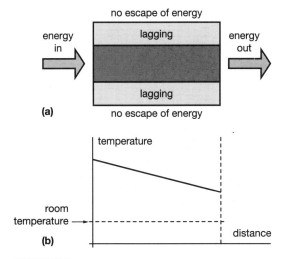

FIGURE 11.4

affect the time taken from the bar to reach a steady state, it has no effect on it once the steady state has been reached.

Thermal conductivity

There is an analogy between energy flow in a bar and the flow of electric charge in a conductor.

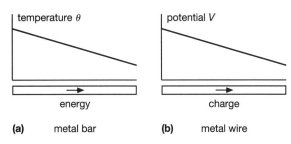

(a)   metal bar   (b)   metal wire

FIGURE 11.5

Figure 11.5 shows a bar along which energy flows and a wire along which charge flows. Each has cross-sectional area A and length l. You know that for the wire the equations

$$I = \frac{V}{R} \quad \text{and} \quad R = \frac{l}{\sigma A}$$

can be put together to give

$$I = \frac{\sigma A V}{l}$$

which we can write

$$\frac{dQ}{dt} = \frac{\sigma A (\Delta V)}{l}$$

We might expect the rate of flow of energy dQ/dt along the bar to depend in the same way on the temperature difference $\Delta \theta$, the length l, the cross-sectional area A. The existence of an *analogy* like this between what happens in two different branches of Physics does not of course prove that this is what the rate of flow of energy will depend on, but *experiments* show that this is indeed what happens:

$$\frac{dQ}{dt} = \frac{kA (\Delta \theta)}{l}$$

where k is the **thermal conductivity** of the substance, and is defined by this equation. You can see from the equation that its unit is the $\mathrm{W\,m^{-1}\,K^{-1}}$. Metals are the best thermal conductors: table 11.1 gives values for some typical materials (at 273 K).

TABLE 11.1   Thermal conductivity of various materials

Material	$k/\mathrm{W\,m^{-1}\,K^{-1}}$
copper	403
aluminium	236
iron	84
brick	≈ 1
glass	≈ 1
water	0.56
polythene	≈ 0.4
air	0.24
plaster	0.13
expanded polystyrene	≈ 0.01

EXAMPLE

A central heating radiator consists of two flat steel surfaces measuring 2240 mm × 600 mm. The steel is 3.0 mm thick and the thermal conductivity of the steel is $60\,\mathrm{W\,m^{-1}\,K^{-1}}$. If energy passes through the radiator panels at a rate of 1.8 kW, calculate the difference in temperature between the inner and outer surfaces of the steel.

$$\frac{dQ}{dt} = \frac{kA (\Delta \theta)}{l} \quad \Rightarrow \quad 1800\,\mathrm{W} \quad \Rightarrow \quad =$$

$$\frac{(60\,\mathrm{W\,m^{-1}K^{-1}})(2\times2.24\,\mathrm{m}\times0.60\,\mathrm{m})(\Delta\theta)}{3.0\times10^{-3}\,\mathrm{m}}$$

$$\theta = 0.033\,\mathrm{K}.$$

In the last example you saw that the outside surface of a radiator is at very nearly the same temperature as the inside. A common mistake is to think that a surface must be at the same temperature as the air surrounding it. You know it cannot be, since radiators feel hot! There *is* a large temperature difference between the radiator surface and the air in the room, nearly all of it occurring across a layer (a few mm thick) of air which is stationary or slow moving. There *must* be this temperature difference, in order to get the energy to flow into the room through the air, which is a bad conductor. Figure 11.6 shows how the temperature varies along a cross-section through the radiator.

FIGURE 11.7

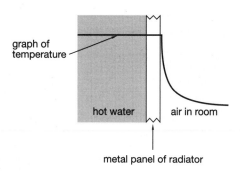

FIGURE 11.6

Insulation of houses

The increasing cost of energy and the decrease in the stocks of fossil fuels make it more and more important to insulate houses properly. Nowadays architects are required to design houses so that the walls, roof, floor, etc., are constructed in such a way that they insulate well. This is why new houses are never built with walls consisting of a single thickness of brick: there is usually a layer of brick on the outside, a layer of aerated concrete block on the inside, and a gap between them which may be filled with glass wool or

polyurethane foam. The whole construction is called a **cavity wall**. Figure 11.7 shows one method by which the rate of loss of energy from buildings may be reduced. In the building industry the ability of a surface to transfer energy is measured by its U-value, which is defined by the equation

$$\text{U-value} = \frac{k}{l}$$

and its unit is the $\mathrm{W\,m^{-2}K^{-1}}$ so it is a measure of the rate of transfer of energy through a slab of material per square metre per unit temperature difference. The thermal conductivity equation could be rewritten as

$$\frac{dQ}{dt} = UA\,(\Delta\theta)$$

which shows that you simply have to multiply the U-value by the cross-sectional area and the temperature difference to find the rate of flow of energy. In the United Kingdom the Building Regulations which govern the construction of new buildings lay down maximum U-values for walls, roofs and floors. For example, the U-value for house walls must not exceed $0.60\,\mathrm{W\,m^{-2}K^{-1}}$. A related quantity is the **thermal resistance** R of a slab of material defined by $R = 1/U$ so $R = l/k$.

EXAMPLE

Calculate (a) U (b) R for a single-thickness brick wall. Thickness of brick = 115 mm, k for brick = $0.78\,\mathrm{W\,m^{-1}K^{-1}}$.

(a) $U = \dfrac{k}{l} = \dfrac{0.78\,\mathrm{W\,m^{-1}K^{-1}}}{0.115\,\mathrm{m}} = 6.8\,\mathrm{W\,m^{-2}K^{-1}}$

(b) $R = \dfrac{1}{U} = 0.15\,\mathrm{m^2 K\,W^{-1}}$.

How would you calculate the U-value for a more complex wall, e.g. a cavity wall consisting of one brick layer, an air gap, and a layer of aerated concrete faced with plasterboard? This is where the idea of thermal resistance is useful. The various layers are thermal resistors *in series*, so the total thermal resistance is simply the sum of the thermal resistances. When we have calculated that, we find U from $U = 1/R$.

Surface layers of air

Before you calculate the U-value for a real wall you need to realise that several of the surfaces of the wall will have the still layers of air which were mentioned in the example of the central heating radiator. These layers of air exist on the inside and outside surfaces of walls, windows, etc., and in the case of windows, at least, are largely responsible for the insulation. For example, consider a window which is 1 metre square, made from glass which is 4 mm thick, and suppose that the temperature difference between the inside and outside air is 16 K. If the temperature difference between the two glass surfaces was the same as this, i.e. 16 K, the window would allow energy to pass at a rate of 4 kW! Fortunately the layers of air next to the glass cut down this rate of flow dramatically.

The values of these surface resistances are known for different conditions: e.g. the value of R for an internal and an external vertical surface are $0.13\,\mathrm{m^2 K\,W^{-1}}$ and $0.06\,\mathrm{m^2 K\,W^{-1}}$ respectively (the difference occurring because there is likely to be more movement of air outdoors than indoors, so that the layer of still air is unlikely to be as thick). The situation in a cavity wall is more complicated.

Energy transfer occurs not only by conduction but also by convection and radiation, so it is not possible to think of a cavity as simply providing two extra surfaces. Instead we are given values of **cavity resistance**, which depend on the cavity thickness and on whether the cavity is vertical or horizontal. For example, the cavity resistance for a vertical gap of 20 mm (or more) between brick walls is $0.18\,\mathrm{m^2 K\,W^{-1}}$. This includes an allowance for the air, so the thermal resistance of the air need not be separately calculated.

EXAMPLE

Find the U-value for a cavity wall consisting of the following layers: brick (115 mm), air (50 mm), aerated concrete (200 mm), plasterboard (10 mm). Values of k, in $\mathrm{W\,m^{-1}K^{-1}}$: brick, 0.78; aerated concrete, 0.19; plasterboard, 0.18. Cavity resistance = $0.18\,\mathrm{m^2 K\,W^{-1}}$; surface resistances, interior and exterior = 0.13 and $0.06\,\mathrm{m^2 K\,W^{-1}}$.

Total thermal resistance

$$= R_{\text{brick}} + R_{\text{concrete}} + R_{\text{plasterboard}} + R_{\text{cavity}} + R_{\text{int}} + R_{\text{ext}}$$

$$= \left(\frac{0.115}{0.78} + \frac{0.200}{0.19} + \frac{0.010}{0.18} + 0.18 + 0.23 + 0.06\right)\mathrm{m^2 K\,W^{-1}}$$

$$= 1.63\,\mathrm{m^2 K\,W^{-1}}$$

So $U = \dfrac{1}{R} = 0.64\,\mathrm{W\,m^{-2}K^{-1}}$

The wall still has too high a U-value, so the cavity would need to be filled with an insulating material such as polyurethane foam. The thermal conductivity k for this is only $0.026\,\mathrm{W\,m^{-1}K^{-1}}$, and so a thickness of 50 mm will have a thermal resistance of $1.92\,\mathrm{m^2 K\,W^{-1}}$. Adding this, and subtracting the cavity resistance, gives

$R = 3.37\,\mathrm{m^2 K\,W^{-1}}$ and
$U\,(= 1/R) = 0.30\,\mathrm{W\,m^{-2}K^{-1}}$.

EXAMPLE

Water is boiling in an aluminium saucepan on a gas hob: measurements of the rate of evaporation of the water show that energy is reaching it at a rate of 1000 W. The bottom of the saucepan has an area of $0.030\,m^2$ and is 3.0 mm thick. (k for aluminium = $236\,W\,m^{-1}\,K^{-1}$.)

(a) Find the temperature difference across the base of the saucepan, assuming that energy flow up the sides of the saucepan is negligible.
(b) If the inside surface of the saucepan is coated with a layer of Teflon (to give a non-stick surface) which is 10 μm thick, find the new rate of flow of energy into the saucepan, assuming that the temperature difference remains the same. (k for Teflon = $0.25\,W\,m^{-1}\,K^{-1}$.)

(a) $$\frac{dQ}{dt} = \frac{kA(\Delta\theta)}{l}$$

$$1000\,W = \frac{(236\,W\,m^{-1}\,K^{-1})(0.030\,m^2)(\Delta\theta)}{3.0 \times 10^{-3}\,m}$$

$$\Rightarrow \quad \Delta\theta = 0.42\,K$$

(b) Thermal resistance of aluminium base

$$= \frac{l}{k} = \frac{3.0 \times 10^{-3}\,m}{236\,W\,m^{-1}\,K^{-1}}$$

$$= 1.27 \times 10^{-5}\,m^2\,K\,W^{-1}$$

Thermal resistance of Teflon layer

$$= \frac{l}{k} = \frac{10 \times 10^{-6}\,m}{0.25\,W\,m^{-1}\,K^{-1}}$$

$$= 4.0 \times 10^{-5}\,m^2\,K\,W^{-1}$$

so total thermal resistance
$= 5.27 \times 10^{-5}\,m^2\,K\,W^{-1}$,
i.e. 5.27/1.27 times greater than it was, so the rate of flow of energy will be reduced to

$$1000\,W \times \frac{1.27}{5.27} = 241\,W = 0.24\,kW$$

We could also use the idea of thermal resistance to calculate how the temperature difference of 0.42 K was split between the aluminium and the Teflon, since it must split in the ratio of the thermal resistances (just as a p.d. splits in the ratio of the electrical resistances in a series circuit). So there is

$$\frac{1.27}{5.27} \times 0.42\,K = 0.10\,K \text{ across the aluminium}$$

and therefore 0.32 K across the Teflon. The layers need temperature differences in proportion to their thermal resistances so that the rate of flow of energy through each may be the same.

Measuring thermal conductivity

To measure the thermal conductivity k we need to set up an experimental arrangement so that we can find the values of all the other quantities in the equation

$$\frac{dQ}{dt} = \frac{kA(\Delta\theta)}{l}$$

but we must ensure that:

◆ at any point in the specimen the temperature is constant. We can do this by waiting for a time long enough for this to happen, testing the temperature at different points from time to time.

◆ the flow of energy is parallel to the length of the conductor. This can be achieved by lagging the sides of the conductor to prevent energy escaping to the surroundings (though lagging will be unnecessary when we deal with a poor conductor, which will have to be very short to get a reasonable flow of energy through it).

Good conductors

In the usual apparatus found in school and college laboratories the material is in the form of a solid copper bar (figure 11.8) whose length is about 300 mm and whose diameter is about 100 mm. It is referred to as **Searle's bar**. The bar is heated at one

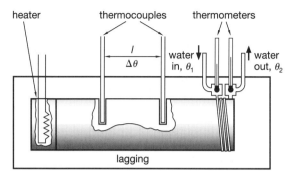

FIGURE 11.8

end, either electrically, or by passing steam through a cylindrical container fixed to that end of the bar. At the other end thin copper tubes, soldered to the bar to make good thermal contact, are supplied with cold water (from a constant-head apparatus to give a steady flow). The rate of flow is measured by collecting the mass m which emerges in a time t. The temperatures θ_1 and θ_2 of the water before and after it has passed through the tubes are measured with mercury thermometers. This supply of cold water creates the necessary temperature difference between the two ends of the bar: without it, the energy would not flow. At two points a distance l (about 100 mm) apart fine holes are drilled radially into the bar, and thermocouples measure the temperature difference $\Delta\theta$ across the length l. If these holes are narrow (as they can be, if thermocouples are used) they form a small proportion of the cross-sectional area of the bar and do not greatly disturb the straight-line energy flow. The current in the heater is switched on, the water in the cooling tubes is started, and the thermometer readings taken at regular intervals until each is constant. If the s.h.c. of water is c, the rate of removal of energy dQ/dt from the cooler end is

$$\frac{dQ}{dt} = \frac{cm(\theta_2 - \theta_1)}{l}$$

and assuming that the lagging is completely effective, this will be the rate of flow of energy through any cross-section of the bar. (If electrical heating is used, the power *supplied* could be measured by the usual method, using an ammeter and voltmeter. The average of this power, and that obtained from the cooling water, would then be

used to give the value of dQ/dt.) So we know dQ/dt in the thermal conductivity equation, and we can measure l, A (by calculating it from the diameter measured with vernier calipers) and $\Delta\theta$, so k can be calculated.

Poor conductors

The table earlier in the section showed that some materials have thermal conductivities which may be 1000 times smaller than those of good conductors. So how would you redesign the apparatus shown in figure 11.8? In the thermal conductivity equation you probably cannot change dQ/dt, the power of the heater, very much, or $\Delta\theta$, so if k is much smaller you would have to make A larger or l smaller, or both. The specimen would therefore have to be thin and wide, i.e. a thin disc rather than a narrow bar. You might like to try to design a piece of apparatus which could be used to measure the thermal conductivity of a poor conductor.

11.3 Radiation

Some bodies obviously radiate energy: the Sun, electric light bulbs, candle flames. We know this because we can see the light from them. What we cannot see is that the light from the Sun is a mixture of colours – which our eyes cannot distinguish until the 'white' light is split up by a prism or a diffraction grating, so that the different colours go in different directions. Then a spectrum is formed. Examining a spectrum (with suitable apparatus) we find that there is also *invisible radiation*: **ultra-violet** beyond the violet end, and **infra-red** beyond the red end, as shown in figure 11.9. Our eyes (and other detectors, like photographic film) see different colours because the light consists of waves of different wavelengths: the wavelengths of the visible spectrum range from about 400 nm at the violet end to about 750 nm at the red end.

Distribution of energy among wavelengths

All bodies emit infra-red radiation, even if they are cold. We are not usually aware of this, because the

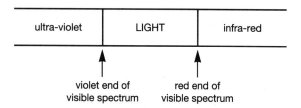

FIGURE 11.9

amounts emitted by cold bodies are relatively small. But the warmer they are, the more they emit, and the proportion that they emit in the visible and ultra-violet regions increases. The graph drawn in figure 11.10 shows this: at higher temperatures the peaks are higher, *and* the peaks occur at shorter wavelengths. For example, at 1500 K the peak is at a wavelength of 1930 nm; but at 2000 K the peak (now much higher) is at 1450 nm. Because the amount of radiation produced increases so greatly with temperature, it is impossible to show the graphs for higher temperatures on the same axes, so figure 11.11 is a second graph, with just the peaks shown for temperatures between 3000 K and 6000 K. There is a simple law connecting the temperature and the

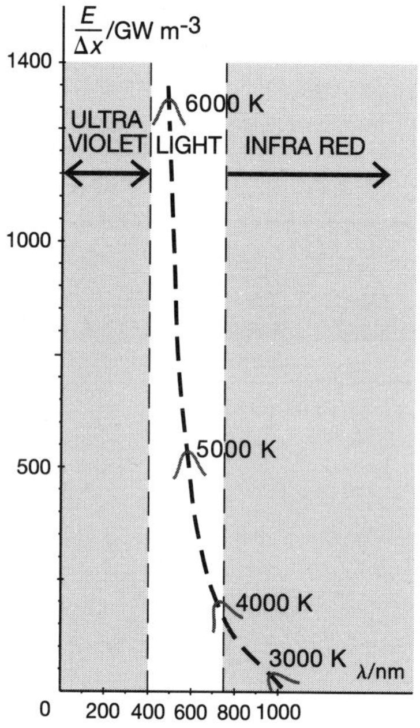

FIGURE 11.11

wavelength at which most radiation occurs: it is $\lambda T = 2.90 \times 10^{-3}\,\text{m K}$. You can see from the first graph that the visible radiation produced at 2000 K will be biased towards the red end of the spectrum, but the second graph shows that at 6000 K (the temperature of the surface of the Sun) there will be a good spread of energy across the visible spectrum, thus giving what we call 'white' light.

The graphs actually show the rate of emission of radiation within a band of 1 nm around the wavelength shown on the x-axis. For example, at 560 nm, the y-axis shows the rate of emission per m² between 559.5 nm and 560.5 nm. The graphs are therefore a sort of histogram. The graphs apply to a **black body**, which by definition has a surface which is a perfect emitter of radiation. For other surfaces the rate of emission will be less, but the curves will have a similar shape.

Filament temperatures

An ordinary tungsten filament bulb reaches a temperature of about 2600 K. At this temperature the peak occurs at 1120 nm, so there will be more red light than violet light, and the light will appear yellowish compared with daylight (from the Sun). Professional photographers use lamps which run at

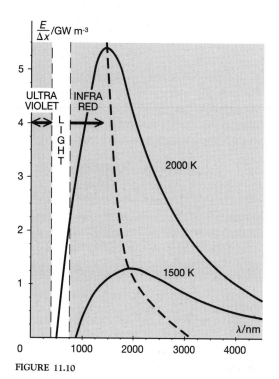

FIGURE 11.10

a temperature just below the melting-point of tungsten (i.e. at 3200–3400 K) because this gives light more like daylight; film for indoor use in studios is designed to give a correct rendering of colour when these lamps are used. Most film is sensitised to give a correct rendering for outdoor use. All filament lamps are very inefficient at producing light: you can judge from the graph of figure 11.10 what a high percentage (perhaps 95%) of the radiation is invisible.

Dependence on temperature

You can see from the graphs how much the rate of radiation depends on temperature. We had to use two different graphs to show the radiation in the ranges 1500–2000 K and 3000–6000 K. In fact, where T is the absolute temperature, the rate of emission is proportional to T^4. For example, if the temperature rises from 500 K to 1000 K, the total rate of emission of energy becomes 16 times greater. From 500 K to 3000 K the rate of emission is nearly 1300 times greater.

The rate of transfer of energy by radiation depends also on the area of the surface, of course, so that structures which need to lose energy by radiation often have fins built on to them (e.g. heat sinks, air-cooled motor-cycle engines). The nature of the surface also matters. Dark, matt, rough surfaces emit well: polished metallic surfaces emit badly. It would be better to paint central heating radiators with a dull black paint, but of course that would not be acceptable in houses and in any case, since the surface temperature is low, the 'radiators' transfer most of their energy by convection. You know that dark, matt, rough surfaces also absorb energy well (this is why you cannot see black objects!): polished metallic objects reflect well. So good emitters are also good absorbers: bad emitters are bad absorbers.

Detecting radiation

Radiation can be detected in the laboratory using a **bolometer**: the basis of this is simply a piece of material whose resistance changes when radiation falls on it. A more sensitive device is a **thermopile**, which consists of many thermocouples arranged in series. Figure 11.12(a) shows the electrical

(a)

(b)

FIGURE 11.12

arrangement: in practice there are about 100 thermocouples and they are packed closely together. Alternate junctions are shielded from the radiation and are therefore at a steady temperature, while the others have the radiation falling on them. These are usually at the narrow end of a cone, as shown in figure 11.12(b). The inside of the cone is silvered, so that the radiation received at the wide end of the cone is reflected on to the thermocouples. The sum of the e.m.f.s produced by the thermocouples can be detected by a sensitive voltmeter. The radiation coming from your hand, at a distance of about 200 mm, might create an

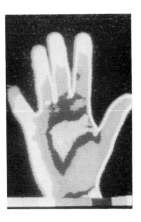

FIGURE 11.13

149

e.m.f. of about $50\,\mu V$. Imaging systems have been developed which record the emission of energy from the surfaces of buildings (to control wasteful energy losses) or from the bodies of human beings (to detect malfunctions in organs). Figure 11.13 shows the radiation from a normal hand. The 'spectrum' at the bottom of the photograph is the key to the surface temperatures, and covers a range of $3.5\,K$; the technique is very sensitive.

The greenhouse

It is instructive, when performing the thermopile experiment, to see the effect of placing a sheet of glass between the thermopile and different sources of energy. The glass has almost no effect on the radiation from a filament lamp, but it almost completely stops the radiation from your hand. We see that glass transmits high-temperature (short-wavelength) radiation but absorbs low-temperature (long-wavelength) radiation.

This is why, when we sit in a car on a sunny winter day, it can become surprisingly warm. On a summer day it can become unbearably hot (and dangerous for animals left without ventilation). This is because (short-wavelength) light and near infra-red radiation pass through the glass of the car windows and are absorbed by the materials and bodies inside the car, warming them up. These

bodies radiate energy again, but because their temperature is low compared with the Sun's temperature, the (long-wavelength) radiation they emit is in the far infra-red region, and this cannot pass through the glass of the car's windows. **Greenhouses** are designed to use this principle, as shown in figure 11.14: in a typical greenhouse, the glass acts like a one-way energy valve. It lets about two-thirds of the radiation in. About one-seventh is re-radiated, but little of this passes through the glass. The greenhouse has lent its name to the greenhouse effect.

The greenhouse effect

At night the temperature of the Earth's surface falls as it radiates energy away into space but when the Sun rises the next morning the Earth warms up again. This cycle is repeated, seemingly endlessly, with seasonal variations. But the balance is delicate: at present we are in a interglacial age, which started about $10\,000$ years ago. Before that, because the average temperature was just 2 or 3 degrees lower, we were in an Ice Age, when most of Europe and North America was covered with ice. We are not sure what causes these changes: it could be, for example, a small change in the output of energy by the Sun, or a change in the nature of its radiation. Apparently small changes in the energy balance can have catastrophic effects on the Earth's climate. What is worrying now is that the amount of carbon dioxide in the atmosphere is gradually but steadily increasing, largely as a result of the burning of fossil fuels. The carbon dioxide behaves like the glass in the greenhouse; it lets through the radiation from the Sun, but absorbs the radiation from the Earth, so the atmosphere becomes warmer. A warmer climate sounds attractive, but it would melt more of the polar ice cap and cause flooding. It would also have unforeseeable effects on crop production: it is possible that farmers in Britain, for example, would have to start planting new strains of the crops they currently grow, or even change crops altogether, to adapt to the higher temperatures. Much research is being carried out to determine what will be possible.

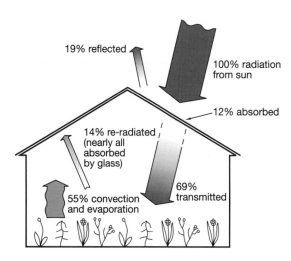

FIGURE 11.14

Exercises on each section of this chapter may be found in the companion textbook, ***Practice in Physics***.

SUMMARY

At the end of this chapter you should be able to:

◆ understand the mechanism of convection and conduction.

◆ use the equation $\dfrac{dQ}{dt} = \dfrac{kA\,(\Delta\theta)}{l}$, which defines thermal conductivity k.

◆ describe the most important methods of insulating houses.

◆ understand the importance of surface layers of air in reducing energy transfer.

◆ remember which materials are good conductors, and which are bad conductors.

◆ describe how to measure thermal conductivity for a good conductor.

◆ remember that infra-red radiation is part of the electromagnetic spectrum, with a wavelength greater than that of red light.

◆ understand that radiation can pass through a vacuum and that its speed in a vacuum is the same as the speed of light.

◆ understand that all bodies radiate, whatever their temperature.

◆ understand that the rate of transfer by radiation increases very rapidly with rise in temperature.

◆ remember that dull, dark, matt surfaces emit and absorb well, but shiny, light, polished surfaces emit and absorb badly.

◆ understand the significance of short- and long-wavelength radiation in the greenhouse effect.

12 The Ideal Gas

The photograph shows a typical power station scene: the cooling towers are the way we usually recognise a power station for what it is. The cooling towers are designed so that the power station can give energy to the surroundings, and are evidence of the energy which is wasted. If a power station does not have cooling towers, it must be near a river or the sea, so that it has an alternative way of giving energy to the surroundings. In fact all power stations waste more energy than they supply, and since a typical power station is capable of generating energy at a rate of 2 GW (i.e. 2×10^9 W) that is a lot of energy. Must they waste all that energy? Is this just carelessness or bad design? What are the laws of Physics that govern this situation? If losing this much energy is inevitable, are there ways in which some use may be made of it? Can anything be done with the energy which is wasted? These questions are answered in this chapter.

12.1 The ideal gas law

Robert Boyle, experimenting on what he called the 'spring of air' in 1660, discovered that the pressure p and volume V of a fixed mass of gas were inversely proportional to each other, or

$$p \propto \frac{1}{V} \quad \text{or} \quad V \propto \frac{1}{p} \quad \text{or} \quad pV = \text{constant}$$

provided that the temperature of the gas was kept constant.

We can use the apparatus in figure 12.1 to test this relationship (**Boyle's law**) in the laboratory, though unfortunately it does not allow us to test a very wide range of pressures (perhaps from 100 kPa to 250 kPa). The air is trapped in the left-hand tube, above some oil. The length l of this air column enables us to calculate the volume V of the air if we know the cross-sectional area of the tube

(a)

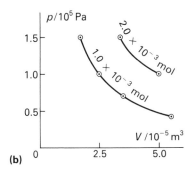

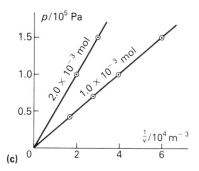

FIGURE 12.1

(which is usually $1.0\,cm^2$). The Bourdon gauge gives the pressure of the air in the space above the oil tank on the right. This is less than the pressure of the air in the tube by an amount equal to the difference in levels, but as this difference is never more than about 30 cm ($\Delta p < 3\,Pa$) it can be ignored. If we plot p against V we obtain a hyperbola (figure 12.1(b)). It is more useful to obtain a straight line: for this we plot p against $1/V$ (figure 12.1(c)).

It would be difficult to use any gas other than air in the apparatus shown, and also difficult to change the amount of gas in it – but experiments with other apparatus show that the results for other gases are exactly the same, if the same amount of substance (i.e. 1.0×10^{-3} mol) is used. If we change the amount of substance n, then for the same values of V we get pressures which are in proportion to the amount of substance (i.e. $p \propto n$ for constant V) so we can write Boyle's law in the form

$$pV \propto n \quad (T \text{ constant})$$

for any gas, though you should realise that Boyle's law is an idealisation of the way in which gases

behave. It holds very well for air at room temperature, and for oxygen and nitrogen separately, and for other gases like hydrogen or helium. But at lower temperatures, or high pressures, the law does not hold even for these gases. Boyle's law is like Hooke's law or Ohm's law: it is a statement which is useful as a working rule over a limited range of conditions.

The ideal gas

We *define* an ideal gas as one for which Boyle's law is exactly true, for all temperatures and pressures. How does the product pV for an ideal gas depend on its temperature? We must be careful here. By 'temperature' we mean the ideal-gas scale temperature, so *by definition p is proportional to T* at constant volume. We cannot 'discover' by experiment whether p is proportional to T, but the *definition* of temperature enables us to write

$$pV \propto nT \quad \text{or} \quad \frac{pV}{nT} = \text{constant}.$$

Gas constants

The constant in the previous equation could be measured experimentally – but you cannot use an ideal gas for this experiment! You have to use a real gas and measure the values of p, V, n and T at lower and lower pressures. You would find that pV/nT approached a particular value as you reduced the pressure more and more, as shown in figure 12.2. This value is $8.31\,J\,K^{-1}\,mol^{-1}$. It is labelled R and called the **molar gas constant**, so

$$\frac{pV}{nT} = R \quad \text{or} \quad pV = nRT$$

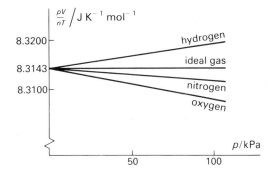

FIGURE 12.2

153

These measurements suggest that a real gas behaves increasingly like an ideal gas as the pressure approaches zero. But even at atmospheric pressure many real gases behave almost indistinguishably from the ideal gas. Note carefully the scale on the graph in figure 12.2: it shows that the value of pV/nT for these three gases differs from R by less than 0.1% when the pressure is atmospheric.

We can rewrite $pV = nRT$ in three useful ways:

(i) $$pV = \frac{m}{M}RT$$

where m is the mass of gas, and M the molar mass;

(ii) $$pV = \frac{N}{N_A}RT$$

where N is the number of molecules and N_A the Avogadro constant;

(iii) $$pV = NkT$$

where $k = R/N_A$ and is called the **Boltzmann constant**.

The value of k can be calculated from our knowledge of R and N_A:

$$k = \frac{R}{N_A} = \frac{8.31\,\mathrm{J\,K^{-1}\,mol^{-1}}}{6.02 \times 10^{23}\,\mathrm{molecules\,mol^{-1}}}$$

$$= 1.38 \times 10^{-23}\,\mathrm{J\,K^{-1}\,molecule^{-1}}.$$

The unit molecule^{-1} is not usually included (i.e. $k = 1.38 \times 10^{-23}\,\mathrm{J\,K^{-1}}$) but we have added it here to point out that if R is the *molar* gas constant, k is the *molecular* gas constant (i.e. the gas constant for one molecule).

EXAMPLE

(a) On a day when atmospheric pressure is $1.05 \times 10^5\,\mathrm{Pa}$, air is pushed into a vehicle tyre until the pressure is $3.20 \times 10^5\,\mathrm{Pa}$. If the volume of the inside of the tyre is $0.15\,\mathrm{m^3}$, what volume of air at atmospheric pressure was pushed in, assuming that the volume of the inside of the tyre, and the temperature of the air, remain constant?

(b) Later the tyre warms up, and the temperature rises from 15°C to 32°C. What does

the pressure in the tyre become, assuming that the volume remains constant?

(c) How many moles of air were in the tyre, if the molar gas constant $R = 8.31\,\mathrm{J\,mol^{-1}\,K^{-1}}$, and what was the mass of air, if the average molar mass of air is $30\,\mathrm{g\,mol^{-1}}$?

(a) The temperature is constant, so $p_1 V_1 = p_2 V_2$, i.e.
$$(1.05 \times 10^5\,\mathrm{Pa})V_1 = (3.20 \times 10^5\,\mathrm{Pa})(0.15\,\mathrm{m^3})$$

so $$V_1 = \frac{(3.20 \times 10^5\,\mathrm{Pa})(0.15\,\mathrm{m^3})}{1.05 \times 10^5\,\mathrm{Pa}} = 0.46\,\mathrm{m^3}$$

(b) $$\frac{p_1}{T_1} = \frac{p_2}{T_2}$$

15°C = 288 K and 32°C = 305 K

$$\Rightarrow \quad \frac{3.20 \times 10^5\,\mathrm{Pa}}{288\,\mathrm{K}} = \frac{p_2}{305\,\mathrm{K}}$$

so $$p_2 = 3.39 \times 10^5\,\mathrm{Pa}$$

(c) Using $pV = nRT$ for the air in the tyre at 15°C, where n is the amount of substance, i.e. the number of moles of air,

$$pV = nRT \Rightarrow (3.20 \times 10^5\,\mathrm{Pa})(0.15\,\mathrm{m^3})$$
$$= n(8.31\,\mathrm{J\,mol^{-1}\,K^{-1}})(288\,\mathrm{K})$$

which gives $n = 20\,\mathrm{mol}$, so that the mass of air is 0.60 kg.

Explanations of gas behaviour

You should now be asking yourself *why* real gases obey Boyle's law (very nearly). You know that a gas consists of molecules in rapid, random motion. What is it about the behaviour of molecules which makes $pV/nT = $ constant, even though the behaviour of individual molecules is entirely chaotic and unpredictable? You already know why gases exert a pressure and it is easy to get a *quantitative* idea of what is happening.

If the volume of the container is reduced, the molecules collide with the walls more often, so that the force on the walls is greater. It is reasonable to think that if the volume were halved, the collisions would happen twice as often, so that the pressure would be doubled, so we would expect the product pV to be constant. And if the amount of substance n were doubled (i.e. there were twice as many molecules) we would expect the pressure to be doubled if the volume stayed the same: there would simply be twice as many collisions in the same time.

But how does the pressure depend on such things as the mass and the speed of the molecules? A molecule can be a very complex particle, made up of several different atoms, and the forces which they exert on each other could be complex too. To make things easier, assume that they behave like tiny, smooth elastic spheres, which exert forces on each other only when they collide. They will have translational k.e. (i.e. k.e. as a result of their motion from place to place). Although they may exchange energy with each other if and when they collide, the total of this translational k.e. will be constant. This is the internal energy U of this simplified ideal gas. The process of making a model of a gas, like this, is called the **kinetic theory of gases**. For the first people to think along these lines (in the mid-nineteenth century) it was very much a step in the dark, since the ideas of molecular motion were then relatively new.

12.2 The kinetic theory of gases

We want to find out how the pressure which a gas causes depends on the mass and the speed of its molecules. Although deriving the result looks long and complicated it is just a matter of using Newton's laws of motion. The derivation has been broken down into stages which may help you remember what to do. But first we set up the situation, imagining N molecules in a box which

measures $a \times b \times d$ but considering first just one molecule, moving with velocity c as shown in figure 12.3. The resolved parts of its velocity parallel to the sides of the box are c_x, c_y and c_z but we shall consider only the x-direction. Assuming that the collisions are elastic the molecule keeps moving at speed c_x in that direction, however many times it bounces off the walls of the box. The steps are these:

♦ calculate time t to reappear at face X:

$$t = \frac{\text{distance}}{\text{speed}} = \frac{2a}{c_x}$$

♦ calculate momentum change when it collides with face X:

p changes from $+mc_x$ to $-mc_x$ so
$$\Delta p = -2mc_x$$

♦ calculate rate of change of momentum:

$$\frac{\Delta p}{\Delta t} = \frac{-2mc_x}{(2a/c_x)} = \frac{-mc_x^2}{a}$$

♦ state that $(\Delta p/\Delta t)$ = force on molecule
= −force on wall

♦ allow for there being N molecules in the box:

force on wall = sum of N terms like $\dfrac{mc_x^2}{a}$

Different molecules have different speeds, but

$$F = \frac{Nm\langle c_x^2 \rangle}{a}$$

where $\langle c_x^2 \rangle$ = average of squares of speeds like c_x.

♦ state that $\langle c_x^2 \rangle = \langle c_y^2 \rangle = \langle c_z^2 \rangle = \tfrac{1}{3}\langle c^2 \rangle$ so

$$F = \frac{Nm}{3a}\langle c^2 \rangle$$

♦ calculate pressure $p = \dfrac{\text{force}}{\text{area}} = \dfrac{F}{bd} = \dfrac{Nm}{3abd}\langle c^2 \rangle$

$$= \frac{Nm}{3V}\langle c^2 \rangle$$

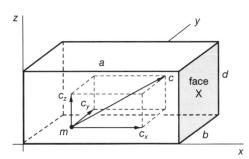

FIGURE 12.3

◆ rearrange to give $pV = \frac{1}{3}Nm\langle c^2\rangle$

$$\Rightarrow \quad p = \frac{1}{3}\frac{Nm}{V}\langle c^2\rangle = \frac{1}{3}\rho\langle c^2\rangle.$$

The $\langle c^2\rangle$ which occurs in this equation is called the **mean square speed**, since this is exactly what it is, the mean of the squares of the speeds. You may find different symbols used for $\langle c^2\rangle$ in this equation: you might find $\langle v^2\rangle$ or $\overline{c^2}$ or $\overline{v^2}$, e.g. $p = \frac{1}{3}\rho\overline{c^2}$. The term **root mean square** (r.m.s.) speed c_{rms} is also used, since this gives us an idea of the speeds of the molecules. Again, it is exactly what it says: the square root of the mean of the squares of the speeds. This is not the same as e.g. the mean of the square roots of the (speed)^2, as you can see if you take some simple numbers as an example: if the values of (speed)^2 are 16, 25, 36 and 49, the r.m.s. speed is $\sqrt{\frac{1}{4}(16 + 25 + 36 + 49)} = \sqrt{31.5} = 5.61$, whereas the average of the square roots is $\frac{1}{4}(4 + 5 + 6 + 7) = 5.50$. There is no short cut to avoid the complication of having to take the square root of the mean square speeds.

The assumptions

To get the result $p = \frac{1}{3}\rho\langle c^2\rangle$ we had to make some assumptions about what molecules were like. How justified were they?

◆ *We assumed* that molecules collide elastically: in a thermally insulated container the pressure of a gas does not decrease as time goes on (as it would if the collisions were not elastic).

◆ *We assumed* that we could treat the molecules like point objects with no size: this is certainly justified, since real gas molecules take up only about 0.01% of the volume of a container.

◆ *We assumed* that the molecules move about without affecting each other except when they collide: that is approximately true provided they are more than two or three diameters apart, which they are most of the time.

Incidentally, collisions between molecules do not

affect the argument, since in any collision any momentum which one molecule loses is gained by the other, so the rate of transfer of momentum across the box is the same, no matter which molecules are doing the transferring. Nor does it matter very much that real gas molecules accelerate downwards in a gravitational field: the speed of a molecule falling through a distance of about 1 m will increase by only about 0.005%. But you should understand that gas molecules *are* subject to gravitational forces, like any other body in a gravitational field.

It seems that the simple model is quite sensible, but the real test is how well it agrees with experiment, which we shall see in the next section. Meanwhile, in the next example, you will see how we can get some immediate results from the theory.

EXAMPLE

Estimate the root mean square speed of air molecules the atmosphere at sea level. Assume that the pressure is standard atmospheric (101 kPa) and that the density of air is 1.3 kg m^{-3}.

$$p = \frac{1}{3}\rho\langle c^2\rangle$$

$$\Rightarrow \langle c^2\rangle = \frac{3p}{\rho} = \frac{3(1.01 \times 10^5\,\mathrm{N\,m^{-2}})}{1.3\,\mathrm{kg\,m^{-3}}}$$

$$= 2.33 \times 10^5\,\mathrm{m^2\,s^{-2}}$$

$$c_{rms} = \sqrt{(2.33 \times 10^5\,\mathrm{m^2\,s^{-2}})}$$

$$= 4.8 \times 10^2\,\mathrm{m\,s^{-1}}.$$

In this example two *macroscopic* measurements (i.e. measurements of the bulk properties of materials, in this case pressure and density) have allowed us to calculate a *sub-microscopic* quantity (a property of the molecules which we could not have investigated even with a microscope), namely the r.m.s. speed of an air molecule. What we have calculated is a *typical* speed: we shall see later what decides the speeds of different molecules.

You may think this calculated speed is very high and therefore unlikely. But the speed of sound in air (340 m s^{-1}) must depend on, and be similar to, the speed with which the molecules of air move, so a value of 480 m s^{-1} for the speeds of the molecules is not unreasonable.

12.3 Consequences of kinetic theory

Temperature and the speeds of molecules

If we put together the equations

$$pV = \tfrac{1}{3}Nm\langle c^2 \rangle \text{ and } pV = NkT$$

we get

$$\tfrac{1}{3}Nm\langle c^2 \rangle = NkT$$

Dividing both sides by N, and mutiplying both sides by $\tfrac{3}{2}$ we get

$$\tfrac{1}{2}m\langle c^2 \rangle = \tfrac{3}{2}kT$$

i.e. the mean translational k.e. of each of the molecules of an ideal gas is $\tfrac{3}{2}kT$, and this statement tells us what temperature means in terms of molecular behaviour. The equation

mean translational k.e. of an ideal-gas molecule $= \tfrac{3}{2}kT$

tells us that the temperature of an ideal gas is proportional to the mean translational k.e. of its molecules: e.g. if the temperature of an ideal gas increases from 300 K to 600 K we should expect the mean k.e. of one of its molecules to double too. Figure 12.2 showed that in many conditions real gases behave very nearly like an ideal gas so we can apply this result to real gases provided that pressure is not too high and temperature not too low.

You may be wondering why the references are to the *translational* k.e. of the molecule. A complicated molecule could have other kinds of k.e., if it is rotating, or vibrating. The results we have derived apply only to the k.e. which the molecule has because it moves from place to place. It is 'the moving from place to place' which causes the pressure of a gas.

EXAMPLE

Find the mean translational k.e., and the r.m.s. speed, of a carbon dioxide molecule at a temperature of 290 K. For carbon dioxide, the relative molar mass is 44, and u = 1.66×10^{-27} kg.

$$\tfrac{1}{2}m\langle c^2 \rangle = \tfrac{3}{2}kT$$
$$= \tfrac{3}{2}(1.38 \times 10^{-23}\,\mathrm{J\,K^{-1}})(290\,\mathrm{K})$$
$$= 6.0 \times 10^{-21}\,\mathrm{J}$$

Since for carbon dioxide,
$m = 44 \times 1.66 \times 10^{-27}$ kg

$$\langle c^2 \rangle = \frac{2 \times 6.0 \times 10^{-21}\,\mathrm{J}}{44 \times 1.66 \times 10^{-27}\,\mathrm{kg}} = 1.64 \times 10^5\,\mathrm{m^2\,s^{-2}}$$

$$\Rightarrow c_{rms} = 4.1 \times 10^2\,\mathrm{m\,s^{-1}}.$$

The k.e. would be the mean translational k.e. of *any* molecule at 290 K, since the expression $\tfrac{3}{2}kT$ depends only on temperature. You could say straight away, for example, that the mean translational k.e. of a hydrogen molecule was also 6.0×10^{-21} J. However, the *speeds* of the two molecules are different, because they have different masses. The r.m.s. speed of a hydrogen molecule is given by

$$c_{rms} = \sqrt{\left(\frac{2 \times 6.0 \times 10^{-21}\,\mathrm{J}}{2 \times 1.66 \times 10^{-27}\,\mathrm{kg}} \right)}$$
$$= 1.9 \times 10^3\,\mathrm{m\,s^{-1}}.$$

The Avogadro and Dalton laws

(i) **Avogadro's law** states that samples of two different gases at the same volume, temperature and pressure contain the same number of molecules. For the two samples we can write

$$p_1 V_1 = \tfrac{1}{3}N_1 m_1 \langle c^2 \rangle_1 \text{ and } p_2 V_2 = \tfrac{1}{3}N_2 m_2 \langle c^2 \rangle_2$$

But $p_1 = p_2$, $V_1 = V_2$, and 'same temperature' implies 'same $\tfrac{1}{2}m\langle c^2 \rangle$' so we can see that $N_1 = N_2$, i.e. the two samples have the same number of molecules.

(ii) **Dalton's law of partial pressures** states that if two gases occupy the same container, the total pressure is equal to the sum of the pressures which each would exert if it were by itself. If there are N_1 and N_2 molecules of two different gases in the same container (necessarily at the same temperature) we can say that $p_1 V = N_1 kT$ and $p_2 V = N_2 kT$, and $pV = NkT$ where p is the total pressure. Since $N = N_1 + N_2$, $p = p_1 + p_2$, i.e. Dalton's law.

These two laws are both supported by experiment. The example which follows shows another way in which the ideas can be used. Suppose we want to estimate the mass of an ash particle used in a Brownian motion experiment, knowing that $k = 1.38 \times 10^{-23}\,\mathrm{J\,K^{-1}}$. Observation shows that an ash particle might appear to move about 1 mm in 1 second. But this 1 mm is from start to finish of a random zigzag path, so its mean speed would be more than this: let us guess that it might be 10 times greater, i.e. $10\,\mathrm{mm\,s^{-1}}$. Let us assume that c_{rms} is also $10\,\mathrm{mm\,s^{-1}}$, and that we can use $\frac{1}{2}m\langle c^2\rangle = \frac{3}{2}kT$ for any molecule or *small particle*. If room temperature $= 300\,\mathrm{K}$, and m is the mass of the ash particle, we can write

$$m = \frac{3kT}{\langle c^2\rangle}$$

$$= \frac{3(1.38 \times 10^{-23}\,\mathrm{J\,K^{-1}})(300\,\mathrm{K})}{(10 \times 10^{-3}\,\mathrm{m\,s^{-1}})^2} \approx 10^{-16}\,\mathrm{kg}.$$

This is a reasonable result. It is less than the mass ($\approx 10^{-14}\,\mathrm{kg}$) of a typical oil droplet in a Millikan experiment. There the Brownian motion of the droplet can just be observed, but it is not as noticeable as with the ash particles, which suggests that the ash particles are less massive.

The Maxwellian distribution

Before we go further let us stop and think what we are now doing. We have put together the ideal-gas definition $pV = nRT$ $(=NkT)$ and the kinetic theory result $pV = \frac{1}{3}Nm\langle c^2\rangle$. The result of this, and any further results, can be tested by experiment (for which Avogadro's law and Dalton's law suggest examples). One obvious experiment is to measure the speeds of the molecules of a gas at a particular temperature. They will have many different speeds, of course. The way in which the speeds are distributed is known as the **Maxwellian distribution** of speeds. Figure 12.4 shows the distribution for 1 000 000 oxygen molecules at temperatures of 300 K and 600 K. The meaning of the graph can be understood by looking at one point on it, e.g. the speed $750\,\mathrm{m\,s^{-1}}$. The graph shows that, at 300 K, 500 oxygen molecules (out of the total population of 1 000 000) have speeds between $749.5\,\mathrm{m\,s^{-1}}$ and $750.5\,\mathrm{m\,s^{-1}}$; at 600 K there are 1200 molecules within this speed range. This prediction depends on the ideas of the kinetic theory but it can be verified by direct measurement of the speeds of the molecules.

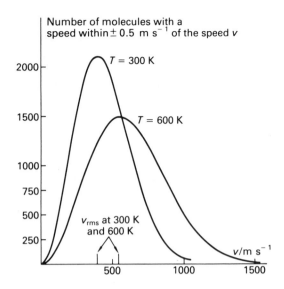

FIGURE 12.4

Measuring the speeds of molecules

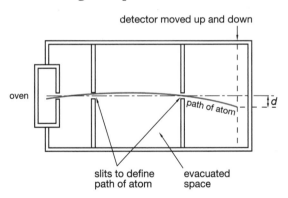

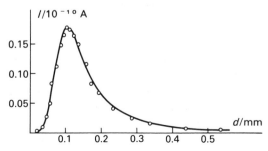

FIGURE 12.5

158

One method is to let the molecules escape from a hole in the side of a heated oven, as shown in figure 12.5(a): they fall in parabolic paths, the faster ones falling less far. A detector counts the number that fall different distances and hence the distribution of speeds can be found. Figure 12.5(b) (d is the distance fallen, and I is the current produced by the ionised molecules at different values of d) shows how the experimental results (the plotted points) agree with the predictions of the Maxwellian distribution (the smooth curve).

12.4 The internal energy of a gas

A **heat engine** is a device which converts internal energy into other forms of energy which are useful: e.g. a petrol or diesel internal combustion engine, a steam engine, a steam turbine. In all these heat engines gases are heated and, in expanding, do work: for example, the pistons in the cylinders of a petrol engine push the crankshaft round and hence the wheels of the car. In this section we shall therefore look at the various things that can happen to gas contained in a cylinder:

♦ a pressure change at constant volume.

♦ a volume change at constant pressure.

♦ an **isothermal** change, i.e. a change which takes place at constant temperature.

♦ an **adiabatic** change, i.e. a change which takes place without any energy exchange (between the gas and its surroundings) by heating.

Internal energy of a gas

The internal energy U of a substance is defined as the sum of the kinetic and potential energies of its molecules. When we deal with an ideal monatomic gas (or a real monatomic gas whose temperature is not too low, nor pressure too high) we assume that because the molecules are far apart we are dealing only with the k.e. of the molecules, so we can say simply

$$U = N(\tfrac{3}{2}kT) = \tfrac{3}{2}nRT$$

where N is the number of molecules and n the amount of substance. Notice that U does *not* depend on which monatomic gas it is: it depends simply on the number of molecules, and the temperature T.

Molar heat capacity

All gases behave in a very similar way: in particular, as we have just seen, at a specific temperature one mole of each monatomic gas has the same internal energy. This means that we can have just *one* heat capacity for one mole of *any* monatomic gas (whereas we have different specific heat capacities for each different substance in solid form). So for gases we usually use the **molar heat capacity** (m.h.c.), the heat capacity *per mole*. What *is* different for gases, compared with solids and liquids, is that it matters very much whether the gas is heated with its volume kept constant, or heated and allowed to expand. It takes more energy to heat a gas while it is expanding, because work is done in pushing back the atmosphere (whereas there is so little expansion with solids and liquids that this hardly matters). So there are *two* molar heat capacities: C_V for heating at constant volume and C_p for heating at constant pressure. The gas could be, and often is, heated in some other way (i.e. heated with both pressure and volume changing) and then the energy supplied would be different, but there is no need for other m.h.c.s in addition to C_V and C_p. Note that the m.h.c. symbol is C, not c. So these m.h.c.s are defined by the equations

$$C_V = \frac{\Delta Q}{n(\Delta T)} \text{ (constant volume)}$$

and

$$C_p = \frac{\Delta Q}{n(\Delta T)} \text{ (constant pressure)},$$

where ΔQ is the energy supplied by heating, n is the amount of substance and ΔT is the temperature change.

Values of m.h.c.

For a constant volume process there is no work done on or by the gas, so $\Delta Q = \Delta U$. Since

$U = \frac{3}{2}nRT$, we can write $\Delta U = \frac{3}{2}nR(\Delta T)$, so $\Delta Q = \frac{3}{2}nR(\Delta T)$.

Hence

$$C_V = \frac{\Delta Q}{n(\Delta T)} = \frac{\frac{3}{2}nR(\Delta T)}{n(\Delta T)} = \frac{3}{2}R = 12.47\,\mathrm{J\,mol^{-1}K^{-1}}.$$

The measured values of C_V for argon and helium, two real monatomic gases, are exactly this, i.e. $12.47\,\mathrm{J\,mol^{-1}K^{-1}}$, so we have excellent support for the kinetic theory ideas. The values of C_V for gases which are not monatomic are distinctly different, but the kinetic theory ideas can be extended to take account of this. You should realise that we now have the value of the m.h.c. for *any* monatomic gas: it will take $12.47\,\mathrm{J}$ to heat one mole of *any* monatomic gas through one degree (at constant volume). You can now see why we use the *molar* heat capacity for gases: many m.h.c.s are exactly the same, whereas the s.h.c.s would be very different.

There is a simple relationship between C_p and C_V (though you do not need to know how it is derived). It is

$$C_p = C_V + R$$

and this holds for all ideal gases (not just monatomic gases). For monatomic gases, therefore, for which $C_V = \frac{3}{2}R$, you can say at once that $C_p = \frac{5}{2}R \; (=20.78\,\mathrm{J\,mol^{-1}K^{-1}})$.

EXAMPLE

Calculate the s.h.c.s (at constant volume) of
(a) helium (mass of one mole = 4.0 g) and
(b) argon (mass of one mole = 40 g).
$C_V = 12.47\,\mathrm{J\,mol^{-1}K^{-1}}$ for both gases.

(a) $C_V = 12.47\,\mathrm{J\,mol^{-1}K^{-1}}$, so $12.47\,\mathrm{J}$ raise $4.0\,\mathrm{g}$ of helium through $1\,\mathrm{K}$.

$$\mathrm{s.h.c.} = \frac{\Delta Q}{m(\Delta T)} = \frac{12.47\,\mathrm{J}}{(4.0 \times 10^{-3}\,\mathrm{kg})(1\,\mathrm{K})}$$
$$= 3.1 \times 10^3\,\mathrm{J\,kg^{-1}K^{-1}}$$

(b) For argon, mass of one mole is 10 times greater, so s.h.c. is 10 times smaller, i.e. $3.1 \times 10^2\,\mathrm{J\,kg^{-1}K^{-1}}$.

Internal energy changes

Before looking more closely at the four processes mentioned at the start of this section (constant pressure, constant volume, isothermal, adiabatic), let us look at one or two general points. To fix our ideas on a concrete situation, we shall be thinking about gas contained in a cylinder fitted with a frictionless piston, as shown in figure 12.6. If the piston moves *inwards*, work is done *on* the gas; if it moves *outwards*, work is done *by* the gas. Outside the cylinder there is atmospheric pressure. If the piston is free to move, the pressure in the cylinder must also be atmospheric.

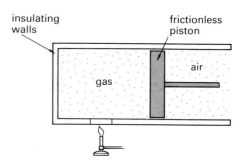

insulating walls — frictionless piston — air — gas

FIGURE 12.6

Constant volume processes

These are very easy to analyse. The piston does not move, so no work is done on or by the gas: $\Delta W = 0$, and $\Delta U = \Delta Q$. So since by definition

$$\Delta Q = nC_V(\Delta T), \; \Delta U = nC_V(\Delta T).$$

But there is one important point to be made. For any gas, monatomic or otherwise, U is proportional to T, so that ΔU is proportional to ΔT. There cannot be more than one relationship between ΔU and ΔT. Since we have such a relationship in $\Delta U = nC_V(\Delta T)$, this must be the relationship, and it must be true for all processes, whether or not they are constant volume processes. The internal energy change ΔU supplied in any circumstances, to produce a certain change in temperature ΔT, is equal to the internal energy which would have been supplied at constant volume, i.e.

$$\Delta U = nC_V(\Delta T) \text{ for } \textit{all} \text{ processes.}$$

EXAMPLE

0.050 mol of an ideal gas is enclosed in a cylinder fitted with a frictionless piston, and is heated at constant volume, so that the temperature rises from 290 K to 350 K. The constant-volume m.h.c. C_V for the gas is 21 J mol K^{-1}. (a) How much energy is supplied by heating, and (b) what is the increase in internal energy of the gas?

(a) $\quad \Delta Q = nC_V(\Delta T)$
$\quad\quad = (0.050\,\text{mol})(21\,\text{J mol}^{-1}\text{K}^{-1})(60\,\text{K})$
$\quad\quad = 63\,\text{J}.$

(b) $\quad \Delta W = 0$, so $\Delta U = \Delta Q = 63\,\text{J}.$

Constant pressure processes

This time, when the gas is heated, the piston moves outwards, as it must do to keep the pressure in the cylinder constant. As it does so, it does work in pushing back the atmosphere, and we need to know how to calculate the work done. We assume that the heating happens slowly enough for the pressure to be the same (atmospheric) on both sides of the piston. The process is shown in figure 12.7(a) and a graph of pressure against

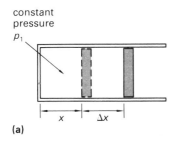

constant pressure
p_1

(a)

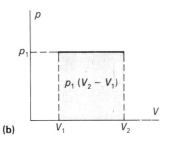

(b)

FIGURE 12.7

volume in figure 12.7(b). If this pressure is p_1, the area of the piston A, and the displacement Δx, the work done ΔW is given by

$$\Delta W = \text{force} \times \Delta x$$
$$= p_1 A (\Delta x)$$
$$= p_1 (\Delta V)$$
$$= p_1 (V_2 - V_1)$$

where V_1 and V_2 are the initial and final volumes. This work done is represented by the shaded area, beneath the line, in the p–V graph in the diagram.

EXAMPLE

Refer to the last example: the same gas is now heated at constant pressure, and the temperature rises by the same amount. Atmospheric pressure = 100 kPa; $R = 8.31\,\text{J mol}^{-1}\text{K}^{-1}$.

(a) What is the increase in internal energy?
(b) How much energy is supplied by heating?
(c) How much work does the gas do?

(a) Since the temperature rise is the same as before, ΔU must be the same as before, i.e. 63 J.
(b) $\Delta W = p(\Delta V)$ so we must find ΔV.
$\quad pV = nRT$
$$\Rightarrow V_1 = \frac{nRT_1}{p}$$

$$= \frac{(0.050\,\text{mol})(8.31\,\text{J mol}^{-1}\text{K}^{-1})(290\,\text{K})}{100 \times 10^3\,\text{N m}^{-2}}$$

$$= 1.20 \times 10^{-3}\,\text{m}^3$$

When T rises from 290 K to 350 K at constant pressure, V rises in the same proportion, i.e.

$$V_2 = \frac{350}{290} \times 1.20 \times 10^{-3}\,\text{m}^3 = 1.45 \times 10^{-3}\,\text{m}^3$$

so $\quad \Delta V = V_2 - V_1 = 0.25 \times 10^{-3}\,\text{m}^3$
so $\quad \Delta W = p(\Delta V)$
$\quad\quad\quad = 100 \times 10^3\,\text{N m}^2 \times 0.25 \times 10^{-3}\,\text{m}^3$
$\quad\quad\quad = 25\,\text{J}.$

(c) ΔW is work that the gas did on its surroundings, so the energy supplied by heating must have been enough to not only raise the internal energy of the gas but also to do this work, so $\Delta Q = 63\,\text{J} + 25\,\text{J} = 88\,\text{J}.$

In this example we have used the first law of thermodynamics but without using it in either of the forms (with their different sign conventions) mentioned in Chapter 10: it is easy to *think out* that ΔQ must be numerically greater than ΔU because the gas does work as well as having its temperature raised, and the *Physics* of the situation must be the same, whatever sign convention is used. However, we show here what signs would be attached to ΔU, ΔQ and ΔW in the two statements of the first law:

First version
$\Delta U = +63\,J$, $\Delta W = -25\,J$.
Since $\Delta U = \Delta Q + \Delta W$,
$\Delta Q = (+63\,J) - (-25\,J) = +88\,J$.

Second version

$\Delta U = +63\,J$, $\Delta W = +25\,J$.
Since $\Delta Q = \Delta U + \Delta W$,
$\Delta Q = (+63\,J) + (+25\,J) = +88\,J$.

Isothermal processes

Another important process is the isothermal process, one in which the temperature of the gas does not change. Then $pV = $ constant, so the p–V graph for this process will be identical with one of the Boyle's law p–V graphs, as shown in figure 12.8(a). Such a line is called an **isothermal**. The area beneath the line in p–V graph will represent the work done by the gas in the cylinder, since the area represents the sum of the small bits of work done as shown in figure 12.8(b): the narrower the strips are made, the closer their total area becomes to the area beneath the line. If you are studying mathematics you ought to be able to calculate the area beneath the graph exactly: otherwise you will

have to estimate it. You could use the 'trapezium rule' (as we shall in the next example) or you could draw the graph on graph paper and count the squares beneath the line and find the area that way.

You should see at once that in an isothermal process there can be no change in the internal energy of the gas: $\Delta T = 0 \Rightarrow \Delta U = 0$. So if we let some gas expand, and do work, it must at the same time be heated (by us or the surroundings) at an equal rate, or else its temperature would fall and there would be a change in its internal energy. The energy supplied by heating goes entirely to doing work.

EXAMPLE

0.050 mol of a gas is enclosed in a cylinder fitted with a frictionless piston. Its initial temperature is 290 K and it is in good thermal contact with the surroundings so that the temperature remains the same throughout. Initially the pressure in the cylinder is atmospheric ($= 100\,kPa$), but the piston is then pushed in slowly until the pressure rises to 200 kPa. How much energy does the gas supply to the surroundings by heating?

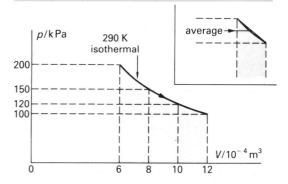

Energy removed from gas by heating = work done on gas, since the internal energy does not change. The diagram shows the p–V graph. W is given by the area beneath the graph. Using the trapezium rule, the area is divided into three strips. The area of each strip is calculated by multiplying the average height by the width. For the three strips shown, the three amounts of work done are 35 J, 27 J and 22 J (you should check this), giving a total of 84 J. So since 84 J

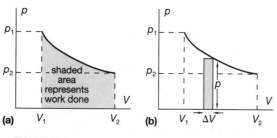

(a) shaded area represents work done

(b)

FIGURE 12.8

of energy are put into the gas by working, the gas must give 84 J of energy to the surroundings by heating.

This must be a slight over-estimate, since, as the inset shows, on this occasion this method treats the curve as if it were a series of three straight lines, each of which lies slightly *above* the actual curve. The calculated value, however, is 83.5 J, so the method gives a good approximation.

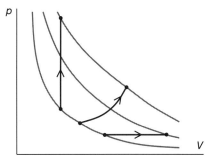

FIGURE 12.9

Adiabatic processes

One last important process is the kind in which there is no heating of the gas during the process: this is called an **adiabatic** process. This could happen because the container is well insulated, or because the change happens so quickly that there is no time for energy to flow in or out by heating. One example of the second situation occurs in sound waves: the compressions and rarefactions of the air happen so quickly that there is no time for energy to flow from place to place. A complete discussion of this process is outside the scope of this book, but you should realise that if $\Delta Q = 0$, then:

♦ in an adiabatic *expansion* the gas *cools* (since it does work on the surroundings, and the energy must come from the gas)

♦ in an adiabatic *compression* the gas *warms up* (since work is done on the gas, and so its internal energy must increase).

Summary of the four processes

Always draw some isothermals on a *p*–*V* diagram for the situation and remember that the lower *p*–*V* curves (i.e. the curves nearer the axes) are for lower constant temperatures, so all the moves shown by arrows in figure 12.9 result in rises of temperature, since the second temperature is higher than the first.

Figure 12.10 shows the four simple processes, with notes under each one. Statements about whether ΔU, ΔQ, ΔW are $>$ or $<$ 0 apply to the process shown, *if the process occurs in the direction of*

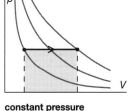

constant volume
so $\Delta W = 0$
Here $\Delta U > 0$ so $\Delta Q > 0$

constant pressure
ΔW calculated from area
Here $\Delta W < 0$. ‖ Here $\Delta W > 0$.
$\Delta U <$ or > 0, depending on ΔQ

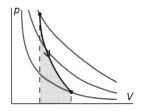

isothermal
so $\Delta T = 0$ so $\Delta U = 0$
ΔW calculated from area
Here $\Delta W < 0$ ‖ Here $\Delta W > 0$
so $\Delta Q > 0$ ‖ so $\Delta Q > 0$

adiabatic
so $\Delta Q = 0$
ΔW calculated from area
Here $\Delta W < 0$ ‖ Here $\Delta W > 0$
so $\Delta U < 0$ ‖ so $\Delta U < 0$

FIGURE 12.10

the arrow. Signs of ΔW depend on which version of the first law you are using: where it matters, vertical lines ' ‖ ' divides the alternatives: the signs that go with $\Delta U = \Delta Q + \Delta W$ are shown first, the signs that go with $\Delta Q = \Delta U + \Delta W$ are shown second (you might like to delete the alternative you are not using). On all the diagrams the three curves show isothermals.

12.5 Heat engines

It would be a pity to leave this study of gases without getting some idea of how the gas laws are *applied* to the machines which convert internal energy to mechanical energy. This is where we can answer some of the questions asked at the start of the chapter.

Heat engine principles

All heat engines have a **working substance**. All heat engines must perform a **cyclic** process, i.e. they must be able to repeat their operation for any number of cycles. The working substance must therefore be returned to its original state at the end of each cycle. If the substance is disposed of at the end of one cycle (as in an internal combustion engine) then we must start the new cycle with fresh working substance in the same state as the working substance at the beginning of the previous cycle.

In all heat engines the working substance is heated, usually by contact with a **source** which we shall call the **hot reservoir** (but sometimes through a direct injection of energy, as in the internal combustion engine). The working substance then expands and does work, so that heat engines convert internal energy into mechanical energy. At some stage all heat engines heat a **sink** which we shall call the **cold reservoir**: e.g. the exhaust of an internal combustion engine heats the surroundings. In that case the surroundings are the cold reservoir.

Thermal efficiency

The **thermal efficiency** η of a heat engine is defined by the equation

$$\eta = \frac{\text{work done by the engine}}{\text{energy absorbed by heating}}$$

although sometimes the efficiency is expressed as a percentage. This is a sort of 'ideal' efficiency for the engine, and does not take account of energy converted wastefully through frictional forces, electrical resistance heating, and unintentional heating of the surroundings. The **overall efficiency** of the heat engine will therefore always be less (sometimes much less) than the thermal efficiency.

A heat engine cycle

This example will take you through the principles which apply to heat engines. We shall use it to find the thermal efficiency of an imaginary heat engine.

Suppose we have a heat engine which consists of a cylinder, fitted with a piston, which contains 0.045 mol of a monatomic gas (so $C_V = \frac{3}{2}R$, $C_p = \frac{5}{2}R$) as its working substance, and that it performs the cycle shown in figure 12.11(a), where it is superimposed on some isothermals in a p–V diagram. The cycle follows the path *abcd*. Along *bc* the gas is being heated and its temperature rises; along *cd* it expands and does work; along *da* and *ab* it is cooled, as it has to be if it is to be a genuinely cyclic process. In a real heat engine the route taken on a p–V diagram would consist of a continuous curve, like that shown in figure 12.11(b), rather than this artificial set of straight lines: nevertheless, the principle is the same.

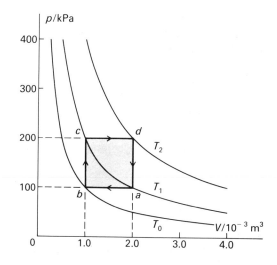

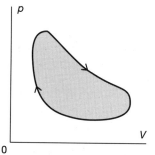

(b)

FIGURE 12.11

First, how much work is done by the gas? There is no work done along the paths bc and da. It does work when it expands along cd:

$$\Delta W = p(\Delta V) = (200 \times 10^3\,\mathrm{N\,m^{-2}})(1.0 \times 10^{-3}\,\mathrm{m^3})$$
$$= 200\,\mathrm{J}$$

and work is done on it when it is compressed along ab:

$$\Delta W = p(\Delta V) = (100 \times 10^3\,\mathrm{N\,m^{-2}})(1.0 \times 10^{-3}\,\mathrm{m^3})$$
$$= 100\,\mathrm{J}$$

so the net work done by the gas is $100\,\mathrm{J}$.

Secondly, how much energy is absorbed by the gas by heating? We need to use the equations $\Delta Q = nC_V(\Delta T)$ and $\Delta Q = nC_p(\Delta T)$ so we need to find the temperatures of each of the three isothermals. For the lowest temperature T_0 we have $p = 100\,\mathrm{kPa}$, $V = 1.0 \times 10^{-3}\,\mathrm{m^3}$, so

$$T_0 = \frac{pV}{nR} = \frac{(100 \times 10^3\,\mathrm{N\,m^{-2}})(1.0 \times 10^{-3}\,\mathrm{m^3})}{(0.045\,\mathrm{mol})(8.31\,\mathrm{J\,mol^{-1}\,K^{-1}})}$$
$$= 267\,\mathrm{K}$$

Along the path bc the temperature doubles since the pressure doubles (V constant) and along cd the temperature doubles again since the volume doubles (p constant). Hence $T_1 = 535\,\mathrm{K}$, and $T_2 = 1170\,\mathrm{K}$. Energy is absorbed during the constant volume process bc:

$$\Delta Q = nC_V(\Delta T)$$
$$= (0.045\,\mathrm{mol})(\tfrac{3}{2} \times 8.31\,\mathrm{J\,mol^{-1}\,K^{-1}})(535\,\mathrm{K}$$
$$- 267\,\mathrm{K})$$
$$= 150\,\mathrm{J}$$

and also during the constant pressure process cd

$$\Delta Q = nC_p(\Delta T)$$
$$= (0.045\,\mathrm{mol})(\tfrac{5}{2} \times 8.31\,\mathrm{J\,mol^{-1}\,K^{-1}})(1170\,\mathrm{K}$$
$$- 535\,\mathrm{K})$$
$$= 500\,\mathrm{J}$$

Along da and ab energy is given to the surroundings (because the working substance has to be returned to its original state so that the cycle can begin again): this energy does not need to be calculated using expressions for ΔQ, since it must be $550\,\mathrm{J}$, since a total of $650\,\mathrm{J}$ of energy is taken in by heating, and only $100\,\mathrm{J}$ of work done.

So thermal efficiency $\eta = \dfrac{100\,\mathrm{J}}{650\,\mathrm{J}} = 0.15$ or 15%

This is an artificial example and the efficiency of a real cycle would be higher. But what has to be done to make any engine more efficient? This is a question on which some engineers spend their whole working lives, so we cannot hope to answer it here! However, we can say that the thermal efficiency is a maximum for any cycle in which the energy is absorbed or rejected only at the highest and lowest temperatures T_h and T_c in the cycle. This condition is met by a cycle (called a **Carnot cycle**, and shown in figure 12.12) consisting of an isothermal compression, followed by an adiabatic compression, an isothermal expansion and an adiabatic expansion.

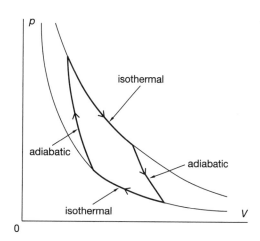

FIGURE 12.12

Then: thermal efficiency $= 1 - \dfrac{T_c}{T_h}$

so the efficiency will be large if T_c is small and T_h is large. An engine working so as to produce an efficiency equal to this maximum efficiency is called a **Carnot-efficiency** engine. In fact for most engines T_c is the temperature of the surroundings, so we cannot change this; we have to try to make T_h as large as possible, but this means using high pressures and increases the cost and complexity of the machinery.

In the steam turbines found in the most modern power stations, $T_h = 850\,\mathrm{K}$ and $T_c = 300\,\mathrm{K}$, so assuming that the cycle approximates to a Carnot cycle,

$$\text{thermal efficiency} = 1 - \frac{300\,K}{850\,K} = 0.65 \text{ or } 65\%.$$

You can see that it is the *maximum temperature* which governs the thermal efficiency of a heat engine like a steam turbine. Raising this is the key to making it more efficient, but you can imagine that it might be difficult to raise the temperature much above the 850 K which is the present maximum.

The *overall* efficiencies of modern power stations approach only 40%.

Heat engines in reverse

Figure 12.13(a) shows the energy flow diagram for a heat engine. Let us see what happens if a heat engine is operated in reverse, as shown in figure 12.13(b), i.e. so that a net amount of work is done *on* the working substance. In this mode of operation energy is being removed from the cold reservoir and given to the hot reservoir. The normal direction of flow of energy is from hot to cold, so work must be done to *pump* the energy up from cold to hot, and so such a device is called a **heat pump**. The heat pump is put to use in two apparently different but actually identical ways.

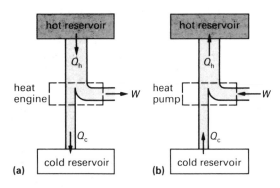

FIGURE 12.13

The refrigerator

Here the cold reservoir is the freezing compartment of the refrigerator, and the hot reservoir is the room in which the refrigerator is placed: at the back of any refrigerator is a grid of pipes by means of which energy is given to the room. The room becomes warmer, and the inside of the refrigerator

colder. All that is happening is that energy is being moved from one place to another, but because the direction (cold to hot) is abnormal, work must be done. In figure 12.14 the working substance needs to have a high s.l.h. of vaporisation and a low boiling point: e.g. CCl_2F (Freon 114) which has an s.l.h. of $0.17\,MJ\,kg^{-1}$ and a boiling point of $-30°C$.

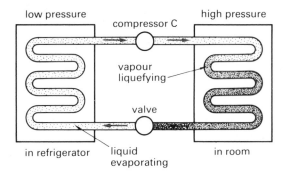

FIGURE 12.14

Most refrigerants which have been used in refrigerators are CFCs and are being replaced with other substances so as to avoid further damage to the ozone layer. The compressor, C, pumps the substance round the circuit. On the high-pressure side the vapour liquefies and gives its latent heat to the surroundings. The liquid then passes through a valve into a low-pressure region, where it evaporates again, taking the necessary latent heat from the surroundings.

The heat pump

Just as a refrigerator heats a room, at the expense of the inside of the refrigerator, so a heat pump heats a building, at the expense of some source of energy such as the surrounding air or the ground. The energy is taken from the air or ground by pipes. You would not normally think of the ground as containing energy, but that is only because the energy does not flow automatically into your house! There is plenty of energy there: it is just that it needs to be pumped up to a place where the temperature is higher. Figure 12.15 shows a commercially-available heat pump installed in a house. This is an excellent method of heating, since the energy from the surroundings is itself free, and the only running cost is the energy needed to

FIGURE 12.15

drive the pump, and this energy might be several times less than the energy delivered to the building. The owner of the house is paying for the use of a little energy to be able to make use of a lot of energy. The effectiveness of a heat pump or refrigerator is measured by a quantity η_r, which is called the **coefficient of performance**. If for a Carnot-efficiency engine (i.e. maximum-efficiency engine), working between the two extreme temperatures of the refrigerator, the efficiency was η, then for a Carnot-efficiency engine working in reverse

$$\text{coefficient of performance } \eta_r = \frac{1}{\eta} = \frac{1}{1 - (T_c/T_h)}$$

so that if the river temperature is 5°C (278 K) and the house temperature is 20°C (293 K), $\eta_r \approx 20$. For

a real pump η_r might be about 4, i.e. a heat pump supplied with energy at a rate of 1 kW will deliver 4 kW to the house. The principle on which heat pumps are based was known 100 years ago but they are not as common as they should be because until the 1970s fossil-fuel energy was cheap. The cost of installation involving, as it does, some system of piping to remove energy from an outside source, is certainly higher than for any other form of heating, and may make it too expensive for a single house, but for industrial use the heat pump offers clear advantages.

Energy costs

Figure 12.16 show parts of two domestic bills for electricity and gas supply. How does the cost of energy compare? Two prices per kWh are shown for electricity. The upper one (7.35 p) gives the standard rate; the lower one (2.74 p) the off-peak rate. For gas the price is 1.477 pence per kWh. Why is there such a great difference between the costs of energy by the two methods? What you have learnt in this chapter should answer this question for you: electrical energy is generated in power stations which are at most 40% efficient. So if a power station burns gas, we get $\frac{2}{5}$ of the energy: we have the pay $\frac{5}{2}$ times as much as if we bought the gas ourselves and used it in the home. Of course there are other factors influencing the cost as well: the costs of distribution and the efficiency of use in the home may be different (e.g. some energy supplied by gas heating goes wastefully into

(a)

Description	Meter Readings *		Units Supplied	Pence Per Unit	£
	Present	Previous			
ECONOMY SEVEN PRICES					
DAY UNITS TO 12 DEC	18676E	17496E	1180	7.35	86.73
NIGHT UNITS TO 12 DEC	5643E	5273E	370	2.68	9.92
QUARTERLY CHARGE					13.92
VAT 0.0% ON DOMESTIC USE OF			£110.57		0.00

(b)

Date of meter reading	Meter reading (see below for key ▬○)		Gas supplied 100's cubic			Charges
	Present	Previous	feet	kWh	Vat %	
28TH DEC	001456E	001127	329	10459	0	154.48
	METER NO	346				
STANDING CHARGE					0	9.40
CREDIT				1.477	Pence per kWh	
TARIFF						

If the present meter reading is an estimate (E) and you would like us to use your own meter reading, please write your reading on the back of the bill and send it to us as soon as possible. Or telephone and tell us your reading.   Thank you.

TOTAL £   163.88

FIGURE 12.16

flues), so the comparison needs to take into account more than just the overall efficiency of the power stations. We need electrical energy for motors (e.g. in washing machines) and to power TV sets and record players but using it for heating is clearly very wasteful when there are much cheaper alternatives available. This is one situation where what is good for us is good for the world as a whole: burning gas in our homes, instead of in power stations to produce electrical energy, saves fossil fuel.

Combined heat and power

Research is being done in this country into schemes which make use of the very hot water from power stations which goes to waste. The temperature of this pressurised water may be as high as 125°C. Unfortunately most of this country's power stations are situated far from towns so sending hot water along lengthy pipes may not be practical, since energy would be wasted. However, Rotterdam, in Holland, has made use of waste hot water for some time. The city was devastated during the Second World War, and during rebuilding the opportunity was taken to lay hot water mains running out from the city's power station. The overall efficiency of the power station is thereby raised to 55–60%. There are some snags, however: because houses have a hot water supply for heating and washing, there is no gas supply, which means that residents do not have a choice of fuels – they rely on electrical energy for cooking and any supplementary heating. Also, it is too expensive to meter the supply of hot water to individual houses, so residents pay an average calculated on the neighbourhood's use: there is no incentive for individuals to be economical. But undoubtedly the increasing cost and scarcity of energy will make combined heat and power schemes practical in this country before long.

Exercises on each section of this chapter may be found in the companion textbook, **Practice in Physics**.

SUMMARY

At the end of this chapter you should be able to:

♦ use the equations $pV = nRT$, $pV = \dfrac{m}{M}RT$ and $pV = \dfrac{N}{N_A}RT$.

♦ explain in molecular terms why pV = constant; why, for constant V and T, p is proportional to n; why, for constant n and V, p depends on T; why, for constant n and p, V depends on T.

♦ remember the assumptions made in the kinetic theory of gases.

♦ understand the steps in the derivation of the equation $p = \frac{1}{3}\rho\langle c^2\rangle$.

♦ understand what is meant by the root mean square speed of a collection of molecules.

♦ understand that the mean translational k.e. of the molecules of a gas is proportional to the kelvin scale temperature.

♦ understand why the m.h.c. at constant pressure is greater than the m.h.c. at constant volume.

♦ draw isothermals for different temperatures for an ideal gas.

♦ describe the changes in ΔQ, ΔW and ΔU for constant volume, constant pressure, isothermal and adiabatic changes.

♦ understand the principle of a heat engine.

♦ understand that the thermal efficiency is governed by the temperatures between which a heat engine works.

♦ understand that a heat pump is a heat engine working in reverse

13 Circular Motion

We all spend a great deal of our time going round in circles; but we don't notice the motion as the circles – round our latitude line on the Earth once a day, round the Sun once a year – are very big and the time to complete a single journey is very long. When you complete a circular journey very quickly, like the passengers on a fairground wheel, things are very different. You become acutely aware of the disorientation which comes from not living on a flat 'still' Earth. You also feel some very large forces which, if you think carefully about it next time you are at the fairground, always push you towards the centre of the circle in which you are whirling.

Forces of this kind, they are called centripetal forces, can be very much bigger than the weights of the bodies which are rotating. If the weight is mg, a centripetal force on the same mass might be 6 mg. We actually talk about the 'g-forces' which a fighter pilot experiences in coming out of a dive following the arc of a circle. Astronauts are trained to put up with huge g-forces by rotating them in a cabin on the end of a rotating arm; 15 mg is dangerously close to the safe maximum.

13.1 Describing circular motion

All the stars in figure 13.1, a long-exposure photograph in the night sky, have rotated through the same angle, the same fraction of a circle. If you divide the length of each star track (the arc, s) by its distance from the pole star (the radius, r) you get the same ratio for every track. This ratio is the angle through which each star has rotated, measured in **radians** (rad).

$$\theta = \frac{s}{r}$$

FIGURE 13.1

(The radian is not strictly a unit like, for example, the newton, as it is metres divided by metres. But we use it with angles measured in this way to remind ourselves that the angle is not in degrees.)

Since the total circumference of the circle is $2\pi r$, the total angle ($=360°$) round a point is

$$\frac{2\pi r}{r}\,\text{rad} = 2\pi\,\text{rad}$$

Hence the connection between the radian and the degree (the unit of angle you are more familiar with) is

$$2\pi \, \text{rad} = 360° \text{ or } 1 \, \text{rad} = 57.3°$$

and the conversion factor $\pi/180$ enables you to convert angles from degrees to radians and vice versa. Your calculator will do it for you as well. (You must be careful when you use a calculator to get sines and cosines: you need it to be in the degree mode or the radian mode depending on how the angle is expressed.)

EXAMPLE

A roll of sticky tape, of outside diameter 96 mm, is held in a dispenser. Through what angle is the roll rotated when a strip of tape 120 mm long is pulled off it?

The 120 mm of tape was part of a circle of radius 48 mm when it was on the roll.

$$\text{angle the roll moves} = \frac{120 \, \text{mm}}{48 \, \text{mm}} = 2.5 \, \text{rad}$$

$$\Rightarrow \text{angle} = 2.5 \times \frac{180}{\pi} \text{degrees} = 140°$$

Angular velocity

The average **angular velocity** ω_{av} of a rotating body is defined by the equation

$$\omega_{av} = \frac{\theta}{t}$$

For motion with a constant angular velocity ω, e.g. a circus roundabout, the angle θ rotated in a time t is given by

$$\theta = \omega t$$

The units of ω (the Greek letter omega) are radians per second, rad s^{-1}.

It is, however, quite common to quote angular speeds in revolutions per second or revolutions per minute (r.p.m.). You must always use radians in physics calculations so it is useful to remember that

$$1 \, \text{rev s}^{-1} = 2\pi \, \text{rad s}^{-1}$$

The time taken to complete one revolution is called the **period** T and is related to ω by

$$\omega = \frac{2\pi}{T}$$

The number of revolutions completed per second is called the rotational frequency n and is measured in rev s^{-1}.

Clearly

$$n = \frac{1}{T}$$

EXAMPLE

A skater spins at 2.5 revolutions per second. (a) Calculate the time period and the angular velocity of the motion. (b) If she stops in 1.2 s, through what angle does she spin as she comes to rest? Assume that she slows down uniformly.

(a) $2.5 \dfrac{\text{rev}}{\text{s}}$ means 1 rev every $\dfrac{1}{2.5}$ s,

i.e. time period $T = 0.40 \, \text{s}$

$$\text{angular velocity } \omega = \frac{2\pi}{T} = \frac{2\pi \, \text{rad}}{0.40 \, \text{s}}$$

$$= 16 \, \text{rad s}^{-1}$$

(b) $\theta = \omega_{av} t$
As she slows down uniformly

$$\omega_{av} = \frac{15.7 \, \text{rad s}^{-1}}{2}$$

so she rotates through

$$\theta = \frac{15.7 \, \text{rad}}{2} \cdot \frac{1}{\text{s}} \times 1.2 \, \text{s} = 9.4 \, \text{rad}$$

in coming to rest, i.e. about 1.5 revolutions or 540°.

Speed and angular velocity

In figure 13.2 suppose the radius OAB moves to OPQ in a time t. The point A covers a distance $r\theta$, and B covers a distance $2r\theta$. The speed of A is given by

$$v_A = r\theta/t = r\omega$$

and for B

$$v_B = 2r\theta/t = 2r\omega$$

i.e. the speed v of any point on a solid body is proportional to the distance r of that point from the axis of rotation.

$$v = r\omega$$

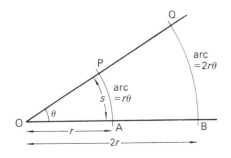

FIGURE 13.2

All the points on the body have the same angular velocity, but the speeds of points at different distances from the centre are different. The tracks of the stars in figure 13.1 (page 169) make this very obvious.

EXAMPLE

What are (i) the angular velocity and (ii) the speed of a point on the Earth's surface at a latitude of 60°N, e.g. the Shetland Isles or St. Petersburg? Take the Earth to be a sphere of radius 6400 km.

(i) All points on the Earth complete a circle once every 24 hours. Therefore St. Petersburg's angular velocity

$$\omega = \frac{2\pi}{T} = \frac{2\pi\,\text{rad}}{24 \times 3600\,\text{s}}$$
$$= 7.3 \times 10^{-5}\,\text{rad s}^{-1}$$

and this is independent of the latitude.
(ii) A point of latitude 60° is moving in a circle of radius $(6400\,\text{km})\cos 60°\ (=3200\,\text{km})$ about the Earth's axis of rotation. Therefore St. Petersburg's speed v is given by

$$v = r\omega = (3.2 \times 10^6\,\text{m})(7.3 \times 10^{-5}\,\text{rad s}^{-1})$$
$$= 230\,\text{m s}^{-1}$$

We could of course have found v by calculating the distance $(2\pi r)$ covered in a day and then calculating $2\pi r/T$, where T is 24 hours.

Measuring angular velocity

Measuring constant or steady angular velocities is straightforward. We make a mark on the rotating body and time how long it takes to make a counted number of revolutions. Alternatively, for high values of $n\ (= \omega/2\pi)$ we can use a calibrated flashing stroboscope to 'freeze' the rotating object and then read off the frequency f of the flashes: f is equal to n. There is, however, a chance that we have selected a flashing rate $f = n/2$, $n/3$ etc., for then the rotating object will still appear to be frozen. To check this we should increase f gradually and measure the *highest* f which gives a *single* frozen view of the object. Then $\omega = 2\pi f$.

Centripetal acceleration

A body, moving in a circle at a *constant* speed is nevertheless accelerating. How can this be? Because its *velocity is changing*. Velocity, a vector, changes when the direction of motion changes, so that in figure 13.3(a) though $v_P = v_Q$ in size (the same speed) v_P is not the same (velocity) as v_Q. You must add a velocity Δv to v_P to get v_Q as shown in (b). The average acceleration of the particle is thus $\Delta v/\Delta t$ in the direction of Δv. As Δt is made smaller the direction of Δv becomes more closely perpendicular to v_P, i.e. along PO.

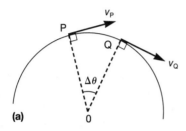

(a)

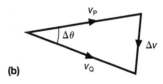

(b)

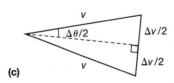

(c)

FIGURE 13.3

Therefore the instantaneous acceleration of the particle at any point on its circular path will be inward along a radius. It is said to be **centripetal**.

To establish the size of the average acceleration between P and Q refer to (c) in which $v_P = v_Q = v$.

$$\frac{\Delta v}{2} = v \sin \frac{\Delta \theta}{2}$$

But provided $\Delta \theta/2$ is small, we can write

$$\sin \Delta \theta/2 \approx \Delta \theta/2 \text{ (radians)}$$

(e.g. if $\Delta \theta = 10°$, then $\sin \Delta \theta/2 = 0.08716$ and $\Delta \theta/2 = 0.08727$. Try with other smaller values using your calculator; the difference between $\Delta \theta/2$ and $\sin \Delta \theta/2$ becomes even less.)

$$\therefore \qquad \Delta v = 2v \frac{\Delta \theta}{2} = v \Delta \theta$$

As

$$v = r \omega = r \frac{\Delta \theta}{\Delta t}$$

then

$$\Delta t = \frac{r \Delta \theta}{v}$$

so

$$a = \frac{\Delta v}{\Delta t}$$

$$\approx \frac{v \Delta \theta}{r \Delta \theta / v}$$

$$= \frac{v^2}{r}$$

(It might look as if this is only an approximate result, but calculus techniques can be used to show it is *exact*.)

$$a = \frac{v^2}{r}$$

i.e. a particle which is moving in a circle of radius r with a constant speed v has a centripetal acceleration which is of constant size v^2/r directed at any instant towards the centre of the circle.

Since $v = r \omega$ we can also write
$$a = r \omega^2$$

which is sometimes more useful.

EXAMPLE

A satellite takes 88.6 minutes to circle the Earth at a height of 200 km. What is its acceleration? (Radius of Earth = 6400 km.)

$$a = r \omega^2$$

$$= (6.6 \times 10^6 \, \text{m}) \left(\frac{2\pi}{88.6 \times 60 \, \text{s}} \right)^2$$

$$= 9.22 \, \text{ms}^{-2}$$

A satellite travelling very near the Earth's surface would be found to have an acceleration of $9.8 \, \text{ms}^{-2}$.

EXAMPLE

In an amusement park ride you are whirled in a circle of radius 4.5 m once every 3.2 s. Calculate your centripetal acceleration.

$$v = \frac{2\pi \times 5.4\,\text{m}}{3.2\,\text{s}} = 10.6\,\text{m\,s}^{-1}$$

$$\therefore a = \frac{v^2}{r} = \frac{(10.6\,\text{m\,s}^{-1})^2}{5.4\,\text{m}}$$

$$= 21\,\text{m\,s}^{-2}$$

i.e. just over 2g!

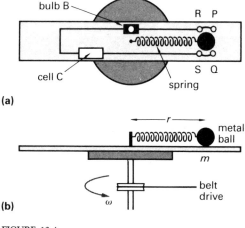

(a)

(b)

FIGURE 13.4

13.2 Centripetal forces

What causes a body to accelerate? A resultant or unbalanced force. The force must be in the same direction as the acceleration and of a size given by Newton's second law, $ma = F_{res}$. So for circular motion at a constant speed

$$m\frac{v^2}{r} = F_{res}$$

When you see objects moving in circles you can always always expect to find a force or forces which are pulling or pushing the object towards the centre of the circle. Some examples are the pull of the Earth on the Moon, the pull of a string on a conker, the push of the cage on the amusement park rider and the frictional push of the road on a car (think of what happens if the car hits an oily or an icy patch when cornering).

These are all centripetal forces: the word centripetal means only that the force is towards the centre – it is only telling us the direction of the force. It is *not* a new *kind* of force (like an electrical force or a frictional force).

Measuring centripetal forces

Figure 13.4 shows a piece of apparatus which enables F_{res}, m, r and ω to be measured separately and for each to be varied. The bulb B lights when a metal ball contacts either RS or PQ, two pairs of metal pegs which can be set in a number of places on a wooden board. The board is fixed to a

horizontal turntable which is driven at different speeds. The angular velocity $\omega = 2\pi n$ of the board is increased until the contact of the ball across RS is broken and the bulb goes out, and this speed is maintained while n is measured. If the speed rises too far the bulb comes on again through contact across PQ. The pull of the spring on the ball is the centripetal force F_{res}. With this arrangement we can measure F_{res} for a range of values of the variables m, r and ω and show that $F_{res} = mr\omega^2$.

The pull of the spring on the ball is a centripetal force; the pull of the ball on the spring is a centrifugal force (i.e. a force directed *outwards* from the centre) which is the same size (Newton's third law). The latter is seldom of any interest to us as it does not act on the object which is moving in a circle.

But we do feel centrifugal forces whenever we whirl something round in a circle with ourselves at the centre.

Suppose a small child of mass 15 kg is swung in a horizontal circle once every two seconds. Taking the radius of the circle in which the child moves to be 1.6 m, we can calculate the speed v of the child to be

$$v = \frac{2\pi(0.8\,\text{m})}{2.0\,\text{s}}$$

$$= 2.5\,\text{m\,s}^{-1}$$

so the size of the centripetal force is

$$(15\,\text{kg})\frac{(2.5\,\text{m\,s}^{-1})^2}{0.8\,\text{m}} = 120\,\text{N}$$

Therefore the centrifugal force on the whirler must also be 120 N.

The total force F acting between the child and the whirler is the resulant of the 180 N vertical force, the pull of the Earth on the child, and the 120 N horizontal centripetal force – the pull of the whirler on the child.

$$F^2 = (150\,\text{N})^2 + (120\,\text{N})^2$$
$$\Rightarrow \qquad F = 190\,\text{N}$$

This net force acts along the arms of the child and whirler which make an angle θ with the vertical, where

$$\tan\theta = \frac{120\,\text{N}}{150\,\text{N}}$$

which gives a value for θ of just below 40°.

EXAMPLE

A car of mass 900 kg is driven round a circle of radius 150 m on a horizontal track at steady speed of 20 m s^{-1}. What frictional force must be exerted by the ground on the car?

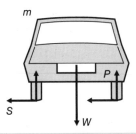

A free-body force diagram of the car shows the centripetal (sideways) frictional push S of the ground on the car. By Newton's second law

$$ma = S \text{ and as } a = v^2/r$$

$$S = (900\,\text{kg})\frac{(20\,\text{m s}^{-1})^2}{150\,\text{m}} = 2400\,\text{N}$$

In this problem friction supplies the centripetal force. Roads are often banked so that there is a horizontal resolved part of the contact force P acting towards the centre. This means that the frictional force F can be less, so that the car is less likely to slide.

EXAMPLE

An aeroplane of mass m is moving at a constant speed v in a horizontal circle of radius r. It does this by banking at an angle of θ to the horizontal. The diagram shows the force acting on the plane. Show that $v = \sqrt{(rg\tan\theta)}$.

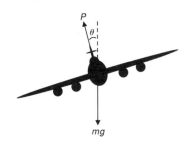

The aeroplane has no vertical acceleration so

$$mg = P\cos\theta$$

Horizontally it is accelerating centripetally so from Newton's second law:

$$m\frac{v^2}{r} = P\sin\theta$$

From these two equations

$$\frac{v^2}{r} = \frac{\sin\theta}{\cos\theta} = \tan\theta$$

$$\Rightarrow v = \sqrt{(rg\tan\theta)}$$

In both these examples the free-body force diagram is the key to a clear and quick solution.

13.3 Couples and torques

In order to produce rotation, to spin a washing machine tub for example, we use electric motors. Electric motors and internal combustion engines – in fact almost any device we use for producing mechanical energy – are designed to spin a shaft about a fixed axis. In dynamic terms their output is not a force which does work in moving along a line but a **torque** which does work in rotating through an angle.

To understand what torque is consider the book in figure 13.5. It is held by one corner and supported in equilibrium in a vertical plane. If the weight of the book is P then the person holding it must exert an upward force of the same size as P. The book is not, however, in equilibrium under the action of these two forces even though they add to produce no net vertical force. They do not act through the same point and so the sum of their moments is *not* zero. Something else produced by your finger and thumb (try it) is needed for equilibrium – see below.

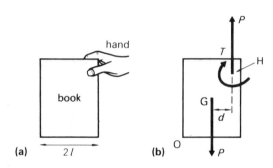

(a) $2l$ **(b)**

FIGURE 13.5

The sum of the moments of the two forces P is always dP anticlockwise – see page 25. It is easy to see that it is dP anticlockwise about G, since the moment of the downward force is zero, and the moment of the upwards force is simply dP anticlockwise. But you should also be able to see that the sum of the moments is still dP anticlockwise, even about other points such as H and O.

These two forces are said to form a **couple**. The turning effect of a couple is the sum of the moments of the two forces *about any point*.

$$\text{couple} = \begin{pmatrix} \text{size of} \\ \text{force} \end{pmatrix} \times \begin{pmatrix} \text{distance between} \\ \text{the two forces} \end{pmatrix}$$

In order to keep the book in equilibrium the person holding it must grip it and produce a turning action or **torque** T, such that

$$T \text{ (clockwise)} + dP \text{ (anticlockwise)} = 0$$
$$\text{i.e. } T - dP = 0$$

The finger and thumb torque is made up of millions of tiny frictional forces and, just as the pull of the Earth on a body is represented by a single force through its centre of gravity, you can replace a torque T by two equal and opposite forces, F, separated by d such that $T = dF$, that is, by a couple. Remember that a torque, like a couple, has units of N m, and produces only a turning effect. Other commonly met examples include:

◆ the torque produced when gripping a doorhandle or screwdriver or unscrewing lids of jars.

◆ the frictional torque on the bearings of any rotating shaft or the clutches which link gear boxes to engines in motor vehicles.

Torques and power

If the mechanical torque produced by a machine on a load is T, then the work done on the load in turning it through an angle θ is given by

$$W = T\theta = 2\pi NT$$

where N is the number of revolutions through which the shaft is driven; for instance, a torque of 12 N m produced by an electric motor might turn through 300 rev in one minute. The work done

$$= (2\pi \times 300)(12\,\text{N m})$$
$$= 22\,600\,\text{J}$$

and the power

$$= \frac{22\,600\,\text{J}}{60\,\text{s}} = 380\,\text{W}$$

This example illustrates that the power at which a torque operates, the rate at which it does work, is given by

$$P = \frac{W}{t} = \frac{T\theta}{t}$$

or

$$P = T\omega$$

where ω is the angular velocity. For a machine rotating at a steady rate of rotation n (revs per second), $\omega = 2\pi n$ and so

$$P = 2\pi nT$$

EXAMPLE

A person is working out on an exercise bicycle such as that found in a sports hall or a gymnasium. At two revolutions per second the tensions in the belt are $F_1 = 200\,N$ and $F_2 = 50\,N$ on a solid wheel of radius 0.20 m. What is the power delivered by the person to the shaft?

As the wheel is rotating at a constant rate the turning effect of the belt is equal to the frictional torque, T, which the rider can adjust to make it easier or harder to pedal.

$$T = (0.20\,m)(200\,N - 50\,N)$$
$$= 30\,N\,m$$
$$P = 2\pi n T$$
$$= 2\pi \times 2.0\,s^{-1} \times 0.20\,m \times 150\,N$$
$$= 380\,W$$

A fit athlete can keep up this power in pedalling for a minute or so.

Efficiency

The power output of an electric motor or car engine is *always* less than the power input. Energy is converted to 'wasted' internal energy.

The **efficiency** η of any energy converting device or transducer is defined by the equation

$$\eta = \frac{\text{useful energy output}}{\text{total energy input}}$$

$$\text{or } \eta = \frac{\text{useful power output}}{\text{total power input}}$$

Efficiency is often expressed as a percentage. The

efficiency of an electric motor or internal combustion engine varies with the conditions under which it is operated. An electric motor produces its highest torque T at low angular speeds ω. It is most efficient at high speeds when for a large motor η might rise to 85%. The internal combustion engine of a motor car has a low efficiency, typically 25% for a petrol engine and 40% for a diesel engine, over a limited range of speeds – figure 13.6(a).

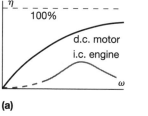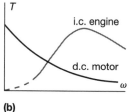

(a)   (b)

FIGURE 13.6

Thus a car requires a gear box to match this limited range of engine speeds to a much larger range of travelling speeds. Electrically driven vehicles (e.g. a milk float) do not need gear boxes. Figure 13.5(b) shows in a general way how T varies with ω for electric motors and internal combustion engines.

EXAMPLE

A screw jack of the type used to lift a car has a handle which moves in a circle of radius 150 mm and a screw with a pitch of 2.5 mm. It is 20% efficient. If a man can conveniently exert a force of 90 N in turning the handle, what is the weight of the heaviest car he can lift?

Consider one complete turn (2π rad) of the handle of the jack. It raises the side of the car a distance equal to the pitch of the screw.

If the man's push is F and the radius of the circle in which he moves his hand is r

$$\text{torque} = T = rF$$
$$\text{and energy input} = T\theta = (rF)(2\pi\,\text{rad})$$
$$= (150 \times 10^{-3})(90\,N)\,2\pi\,\text{rad}$$
$$= 85\,J$$

and energy output $= Ps = P(2.5 \times 10^{-3}\,\text{m})$

where P is the push of the jack on the car. As $\eta = 20\%$ then

$$0.2 = \frac{(2.5 \times 10^{-3}\,\text{m})P}{85\,\text{J}}$$

$\Rightarrow$ $\qquad\qquad P = 6800\,\text{N}$

The man could lift one side of a car which had a weight of 13 600 N, i.e. a mass of about 1400 kg.

Exercises on each section of this chapter may be found in the companion textbook, **Practice in Physics**.

SUMMARY

At the end of this chapter you should be able to:

- remember that to calculate an angle in radians the arc length is divided by the radius, $\theta = s/r$, and that 2π radians $= 360°$.

- use the equations:

$$s = \frac{\theta}{t} \quad \text{and} \quad \omega = \frac{v}{r} \quad \text{and} \quad \omega = \frac{2\pi}{T}$$

relating angular velocity ω to other quantities associated with circular motion.

- describe how to measure angular velocity.

- explain that a body moving in a circle at a constant speed is accelerating towards the centre of the circle.

- use the expressions

$$a = r\omega^2 \quad \text{and} \quad a = \frac{v^2}{r}$$

for centripetal acceleration.

- understand that the resultant force acting on a body moving in a circle at constant speed is directed towards the centre of the circle.

- draw free-body force diagrams and use Newton's second law in the form $mv^2/r = F_{res}$ for circular motion.

- remember that two equal forces P acting on a body in opposite directions but not along the same line form a couple and that the turning effect of the couple is dP, where d is the perpendicular distance between the lines of action of the forces.

- understand that a torque produces a pure turning effect which can be represented by a couple.

- use the equations:

$$W = T\theta \quad \text{and} \quad P = T\omega$$

for the work done by and power of a torque T.

14 Gravitational Fields

Gravity explains a great deal but remains basically mysterious. What goes up must come down. But does it have to? Tides are caused by the moon's gravity. Yet the high tide occurs hours after the moon has passed overhead. Nothing ever gets out of black holes. Then how do we know they exist? The Earth pulls a communications satellite towards it. Why does the satellite not therefore crash to the ground? We can even ask questions like 'how do gravitational forces arise?' What causes them?

With so many questions you must not expect this chapter to produce all the answers. What it does do is to show how a knowledge of the way in which gravitational forces vary, together with a basic understanding of mechanics can tell us exactly where to place a geosynchronous Earth satellite or how to predict to a fraction of a minute the moment when a deep space probe will be at its closest to one of the outer planets.

14.1 Newton's law of gravitation

Until the second half of the twentieth century our study of the solar system was based on observations taken from the Earth's surface. Because the Earth rotates around its own axis (daily) and moves around the Sun (yearly) it took a long time to understand the motions of the other planets which also, of course, are moving. Johannes Kepler (1571–1630) eventually established the following rules.

◆ The planets describe elliptical orbits, with the Sun at one focus (1609).

◆ The line drawn from the Sun to a planet sweeps out equal areas in equal time intervals (1609).

◆ The squares of the planets' periods of revolution are proportional to the cubes of their mean distances from the Sun, i.e. $T^2 \propto r^3$ (1619).

FIGURE 14.1

Apart from Pluto all the planets move in almost circular orbits and we usually assume that they do move in circles – the *most* elliptical, Mars, shown in figure 14.1, is so like a circle that if you drew its orbit accurately on a sheet of paper you would find it very hard to tell it was not circular.

Can we find a mechanism which correctly predicts these rules? The work of Galileo Galilei (1564–1642) and Isaac Newton (1642–1727) provided an answer. We must beware, however, of thinking that the role of the physicist is to aim to answer the question 'why?' Newton's law of gravitation does not explain *why* particles attract each other. It gives us a rule describing how strongly they do and so enables us to solve problems about satellites or the tides.

EXAMPLE

A falling apple is 6.39×10^6 m from the centre of the Earth. The Moon, which orbits the Earth every 27.3 days is 384×10^6 m from the centre of the Earth. (a) Compare the acceleration of the apple and the moon and (b) comment on their accelerations and their distances from the Earth's centre.

Moon

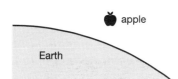
apple

Earth

(a) For the apple acceleration $= 9.8 \, \text{m s}^{-2}$

For the moon acceleration $= v^2/r$

and $v = \dfrac{2\pi r}{T} = \dfrac{2\pi \times 384 \times 10^6 \, \text{m}}{27.3 \times 24 \times 3600 \, \text{s}}$

$= 1020 \, \text{m s}^{-1}$

$\therefore a = \dfrac{(1020 \, \text{m s}^{-1})^2}{384 \times 10^6 \, \text{m}} = 2.7 \times 10^{-3} \, \text{m s}^{-2}$

As 9.8 divided by 2.7×10^{-3} is 3600, the apple's acceleration is 3600 times that of the Moon.

(b) The Moon's distance from the centre of the earth ÷ the distance of the apple is equal to 384 divided by 6.39 = 60.1 or almost exactly 60, and 60^2 is 3600, the ratio of the accelerations.

Newton made the calculations in the above example and realised that the Earth was pulling on both the apple and the Moon. It was from the ratio of 60 for the distances and 60^2 for the accelerations that he developed his law of gravitation.

The inverse square law

Newton's law of gravitation (1687) states that every particle in the universe attracts every other particle with a force F, where

$$F \propto \frac{m_1 m_2}{r^2}$$

and m_1 and m_2 are masses of the two particles and r is their separation. Predictions made with this inverse square law can be verified experimentally, the most convincing correlation between theory and practice coming from predictions about planetary motions rather than laboratory experiments. For example, the planets Neptune and Pluto were both first seen only after their existence had been predicted by analysing the motions of planets near them and assuming that gravitational forces were responsible for the variations in their motion. We can insert a constant and write

$$F = G \frac{m_1 m_2}{r^2}$$

where $G = 6.673 \times 10^{-11} \, \text{N m}^2 \, \text{kg}^{-2}$ and is called the universal constant of gravitation or 'big gee'. It is the most difficult of all the fundamental constants to measure precisely. Newton was not able to measure G himself but, by guessing the mean density of the Earth, was able to deduce a rough value for it.

Measuring G in the laboratory involves, in principle, measuring the force between two bodies of known mass, e.g. you and your neighbour. How big is this force? Knowing G is about $6 \times 10^{-11} \, \text{N m}^2 \, \text{kg}^{-2}$ and treating both of you as particles, it is roughly

$$F = G \frac{m_1 m_2}{r^2} = \left(6 \times 10^{-11} \frac{\text{N m}^2}{\text{kg}^2}\right) \frac{60 \, \text{kg} \times 60 \, \text{kg}}{(1 \, \text{m})^2}$$

$$\approx 2 \times 10^{-7} \, \text{N}$$

a force you will find very hard to detect. So you can see that laboratory experiments to measure G are difficult to perform as they all involve measuring the tiny attraction of one body for another.

EXAMPLE

If the mean density of the Earth is $5500\,\mathrm{kg\,m^{-3}}$ and its mean radius, r_E, is 6400 km, show that taking g to be $9.8\,\mathrm{m\,s^{-2}}$ at the Earth's surface leads to a value for G and find that value.

Applying Newton's second law to a lump mass m at the Earth's surface we have

$$mg = F$$

But also $\qquad F = G\dfrac{mm_E}{r_E^2}$

where m_E is the mass of the Earth

$$\therefore\ mg = G\frac{mm_E}{r_E^2} \quad \Rightarrow \quad G = \frac{gr_E^2}{m_E}$$

We have assumed that the Earth attracts as if it was a particle of mass m_E at its centre. Newton showed that this is true for spherical bodies like planets.

As $m_E = \frac{4}{3}\pi r_E^3 \rho$ for an Earth of average density ρ

$$G = \frac{3gr_E^2}{4\pi r_E^3 \rho} = \frac{3g}{4\pi r_E \rho}$$

$$= \frac{3 \times 9.8\,\mathrm{m\,s^{-2}}}{4\pi\,(6.4 \times 10^6\,\mathrm{m})(5500\,\mathrm{kg\,m^{-3}})}$$

$$= 6.6 \times 10^{-11}\,\mathrm{m^3\,kg^{-1}\,s^{-2}}$$

or $\qquad 6.6 \times 10^{-11}\,\mathrm{N\,m^2\,kg^{-2}}.$

The method of substituting numbers only when the final expression for G is established, is particularly useful in problems on gravitation, where the numbers are often very awkward.

14.2 Uniform gravitational fields

Newton's law of gravitation tells us that in the region near any lump of matter a mass will experience a gravitational force; the region is described as a gravitational field or a g-field. (To produce observably large forces, the lumps must be very large.) We measure the strength g of a gravitational field by the equation

$$g = \frac{F_g}{m}$$

where F_g is the gravitational force experienced by a body of mass m; g is measured in $\mathrm{N\,kg^{-1}}$. It is a vector quantity which has the same direction as the force. You have seen this equation before as $F = mg$, the pull of the Earth on a body of mass m in a state of free fall; g (little gee) was then $9.8\,\mathrm{m\,s^{-2}}$. You should check that the units $\mathrm{N\,kg^{-1}}$ and $\mathrm{m\,s^{-2}}$ are identical.

On or very near the Earth's surface the pull of the Earth on a body is constant in size and, over a limited area such as a town, can be assumed to be constant in direction. The gravitational field g is thus constant at about $9.8\,\mathrm{N\,kg^{-1}}$ within these limits. If we move away from the Earth's surface into 'space', g changes in size and, of course, it is different in direction above, for instance, London and Tokyo. We represent a uniform g-field by a set of parallel lines of force with arrows showing their direction – figure 14.2. To demonstrate that, for example, in a large room, g is uniform you need

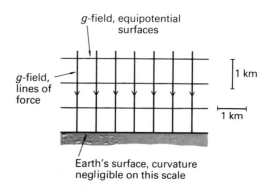

FIGURE 14.2

only take a test mass on a sensitive spring balance and show that the reading on the balance is independent of where we place it in the room, a trivial experiment but an important idea.

When a body is projected in a uniform g-field (it is then usually called a projectile) it follows a parabolic path in the absence of other forces, for example, air resistance.

A simple-pendulum determination of g

On page 282 it is shown that the period of oscillation T of a simple pendulum of length l is given by $T = 2\pi\sqrt{l/g}$. A graph of T^2 against l for a range of values of l will be a straight line, passing through the origin, with a gradient of $4\pi^2/g$, from which g can be found. See the Introduction, page 5. The theory assumes a rigid support, no air resistance and *small* amplitude swings and all these can be achieved with a small metal sphere supported by a cotton thread which swings through about 5° (figure 14.3).

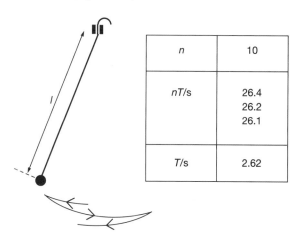

n	10
nT/s	26.4
	26.2
	26.1
T/s	2.62

FIGURE 14.3

Several measurements of nT (where $nT > 20s$) should be taken for each value of l. The number of oscillations n should be counted from the moment when the pendulum passes through the vertical position; and l must be properly defined at the support, perhaps by holding the thread between two flat pieces of wood. The length of the pendulum should be measured to the centre of the swinging sphere. A careful experiment will give g to better than ±1%.

Variation of g

If you go from sea level to the top of the Earth's highest mountain you find that g falls by about 0.3%; it is 1% below its sea level value at a height of about 32 km. The measured free fall acceleration g does, however, vary *at sea level* from 9.83 m s^{-2} at the poles to 9.78 m s^{-2} at the equator, a variation of 0.5%. There are two causes:

◆ 0.2% is the result of the Earth lacking symmetry; it is not a sphere but is like a sphere flattened slightly at the poles.

◆ 0.3% is the result of the Earth's present rotation (the 0.2% caused by its lack of symmetry, is mainly the result of its rotation in the distant past).

On a local scale tiny variations of g, of the order of one part in 10^7 or less, can be detected with gravimeters, very sensitive spring balances used by geologists to help them to predict what lies below the Earth's surface, e.g. oil, at that point.

Gravitational potential in a constant field

The change in gravitational potential energy ΔE_g of a body moving in a uniform g-field is defined as

$$\Delta E_g = \Delta(mgh)$$
$$= mg\Delta h$$

e.g. the first ten or so kilometres of the Earth's atmosphere. We would call ΔE_g the difference in gravitational potential *energy* difference for this body; $\Delta E_g/m$ is then called the **gravitational potential difference** ΔV_g. So

$$\Delta V_g = \frac{\Delta E_g}{m} = g\Delta h$$

and has units J kg^{-1}.

Why introduce ΔV_g? Firstly it is often useful to be able to consider surfaces on which all points have the same potential; some of these **equipotentials** were shown in figure 14.2 for a uniform field. Secondly, if ΔV_g is known, you can find the energy change for bodies of any mass as they move between the equipotential surfaces.

EXAMPLE

Taking the gravitational potential at the Earth's surface to be zero calculate the gravitational potential at heights of 2.0 m, 4.0 m and 6.0 m above the ground. Take g as $9.8\,N\,kg^{-1}$.

As $\Delta V_g = g\Delta h$, then between surfaces 2.0 m apart
$$\Delta V_g = 9.8\,N\,kg^{-1} \times 2.0\,m$$
$$= 19.6\,J\,kg^{-1}$$

$$\therefore \text{ at } 2.0\,m \; V_g = 19.6\,J\,kg^{-1}$$
$$\text{and at } 4.0\,m \; V_g = 39.2\,J\,kg^{-1}$$
$$\text{and at } 6.0\,m \; V_g = 58.8\,J\,kg^{-1}$$

Notice that the gradient of the potential ($\Delta V_g \div \Delta h$) is equal to g, the field. This is true in all kinds of field, e.g. in electrical and gravitational fields, and in both uniform and non-uniform fields.

14.3 Radial gravitational fields

The inverse square law of gravitation applies to *particles*. But provided the bodies we consider possess a spherical symmetry we can treat them, so far as they attract and are attracted by bodies

which lie beyond their surfaces, as particles with all their mass concentrated at their centres. Thus above the Earth's surface (i.e. at more than $r_E = 6400\,km$ above its centre) the Earth attracts gravitationally just as would a particle of mass $m_E = 6.0 \times 10^{24}\,kg$ placed at the Earth's centre.

By definition, a gravitational field g is measured by $g = F/m$. The pull of the Earth on the mass m at a distance r from the Earth's centre ($r > r_E$) is given by $F = Gmm_E/r^2$. Thus

$$g = \frac{Gm_E}{r^2}$$

Although the gravitational field is *symmetrical*, it is not *uniform*: as you know, it decreases with distance. A diagram of lines of force for the Earth ($r > r_E$) shows a pattern of radial lines with arrows towards the Earth's surface (figure 14.4). For a given value for r, g is the same all round the Earth so this field has a spherical symmetry.

A graph of g against r, for $r > r_E$, looks like the curved part of the graph of figure 14.5.

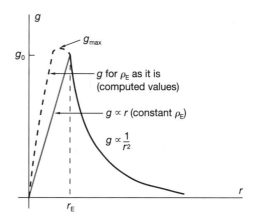

FIGURE 14.5

The straight grey part of the figure tells us how g would vary for $0 \leqslant r \leqslant r_E$ for an Earth of uniform density ρ_E; the dotted line is calculated from values of the Earth's density deduced from the study of seismic (earthquake) waves.

The maximum value g_0 of the gravitational field strength for a uniform planet occurs at its surface. For Mars $g_0 = 3.8\,N\,kg^{-1}$, for the moon $g_0 = 1.7\,N\,kg^{-1}$ (you can throw a cricket ball a long way!) and for Jupiter $g_0 = 25\,N\,kg^{-1}$.

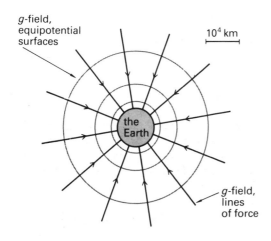

g-field, equipotential surfaces

10^4 km

the Earth

g-field, lines of force

FIGURE 14.4

EXAMPLE

If the Earth's radius is 6370 km and g at the equator is $9.78\,\mathrm{N\,kg^{-1}}$ calculate the value of g at a height of 200 km above the equator and find the percentage error in the value assuming that there was a uniform g-field up to that height.

$$\frac{g}{g} = G\frac{m_E}{r^2} \quad \text{i.e.} \quad g \propto \frac{1}{r^2}$$

and hence $\quad \dfrac{g'}{g} = \left(\dfrac{6370}{6570}\right)^2$

so that $\quad g' = 0.940 \times 9.78\,\mathrm{N\,kg^{-1}}$
$$= 9.19\,\mathrm{N\,kg^{-1}}$$

Assuming $g' = 9.78\,\mathrm{N\,kg^{-1}}$ at 200 km leads to an error, Δg, of $0.59\,\mathrm{N\,kg^{-1}}$

i.e. $\quad \dfrac{\Delta g}{g} = \dfrac{0.59}{9.78} = 0.060$ or 6%

Gravitational potential in a radial field

In a radial field V_g cannot change uniformly: there is a bigger energy change, and hence change of potential, in moving up 10 m *at* the Earth's surface than there is in moving 10 m away from the Earth when you are beyond the orbits of the outer planets.

How does V_g change with r, the distance from the centre of the Earth? It can be shown mathematically (using calculus) that if we choose V_g to be zero when r is infinity, then $V_g \propto 1/r$, and

the equation is

$$V_g = -G\frac{m_E}{r^2}$$

The gravitational potential energy, V_g thus depends *only* on position and has the same value a distance r from the Earth's centre in any direction. These spherical equipotential surfaces are shown as faint lines in figure 14.4.

Figure 14.6 shows a way of helping you to visualise changes in V_g near to the Earth. The surface is called a potential 'well' and is shaped so that its height above a flat disc (at its centre) for different values of r represents V_g at that place. The **zero of potential** is taken to be where the surface becomes flat, a very, very long way from the centre.

The gravitational potential energy, g.p.e., of a body of mass m at a point in the Earth's radial field is simply mV_g, i.e. $-Gmm_E/r$, and you can find *changes* in V_g or in g.p.e. by finding the difference between the two values of these expressions.

For example, the change in g.p.e. of a satellite of mass 5000 kg moving from 1000 km above the Earth's surface to 4000 km above it (remember that the radius of the Earth is 6400 km), is

$$\left(-\frac{Gmm_E}{10\,400 \times 10^3\,\mathrm{m}}\right) - \left(-\frac{Gmm_E}{7400 \times 10^3\,\mathrm{m}}\right)$$

So using $Gm_E = 4.0 \times 10^{14}\,\mathrm{N\,m^2\,kg^{-1}}$, the energy change

$$\Delta E_g = \frac{4.0 \times 10^{14}\,\mathrm{N\,m^2\,kg^{-1}}}{10^6\,\mathrm{m}} \times 5000\,\mathrm{kg}\left(\frac{1}{7.4} - \frac{1}{10.4}\right)$$

$$= 7.8 \times 10^{10}\,\mathrm{J}$$

The following example shows another way of finding this energy change from a knowledge of the gravitational force on the satellite.

EXAMPLE

The following table gives values for F, the pull of the Earth on a satellite of constant mass 5000 kg, as a function of its distance r from the centre of the Earth. The radius of the Earth is $6.4 \times 10^6\,\mathrm{m}$.

$F/10^4\,\mathrm{N}$	4.1	3.1	2.5	2.0	1.6
$r/10^6\,\mathrm{m}$	7.0	8.0	9.0	10.0	11.0

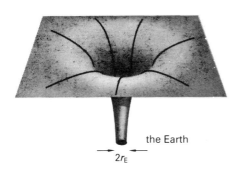

the Earth

$2r_E$

FIGURE 14.6

What is the work done by the pull of the Earth on the satellite, and hence its change in gravitational potential energy as it moves from 4000 km above the Earth's surface to 1000 km above its surface?

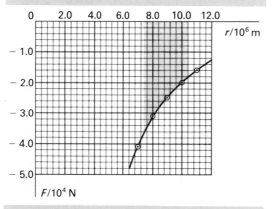

In this graph of F against r the shaded area represents the work done on the satellite as it moves between

$(4000 + 6400)\,km = 10.4 \times 10^6\,m$ and $(1000 + 6400)\,km = 7.4 \times 10^6\,m$ from the Earth's centre.

F is shown negative as the pull is inwards but r is measured outwards.

The shaded area = 98 small squares.
Each small square represents

$$(0.2 \times 10^4\,N)(0.4 \times 10^6\,m) = 8.0 \times 10^8\,J$$

Therefore work done $= 98 \times 8.0 \times 10^8\,J$
$$= 7.8 \times 10^{10}\,J$$

and so the loss of g.p.e. is $7.8 \times 10^{10}\,J$.

In the above example an approximate answer could have been obtained by saying that the 'average' force is roughly

$$\frac{(3.6 + 1.8)}{2} \times 10^4\,N = 2.7 \times 10^4\,N$$

and so the work done $= (2.7 \times 10^4\,N)(3 \times 10^6\,m)$
$$\approx 8 \times 10^{10}\,J$$

If there were no other forces acting on the satellite, the gain of k.e. must also be $8 \times 10^{10}\,J$.

Escape speed

A body of mass m at the Earth's surface has a gravitational potential energy $-Gmm_E/r_E$ and at infinity it would have zero g.p.e. The change in g.p.e. and hence the minimum kinetic energy it must be given is therefore equal to Gmm_E/r_E. The potential well of figure 14.6 can be used to illustrate this energy conservation. Imagine a small ball bearing placed at the edge of the surface. It would roll, slowly at first, towards the centre losing g.p.e. and gaining k.e. Similarly if the ball were projected from the centre it would reach the edge if it started with so much k.e. that it still had some left when it reached the edge. In a gravitational field the equivalent statement is, if it started with speed v, that

$$\tfrac{1}{2}mv^2 > \frac{Gmm_E}{r}$$

i.e. $$v > \sqrt{\frac{2Gm_E}{r_E}}$$

The value of v for which $v = \sqrt{(2Gm_E/r_E)}$ is called the **escape speed** v_e. If a body is projected with a greater speed than this, it will not return to the Earth.

EXAMPLE

With what speed must a body be projected from the Earth's surface if it is to 'escape' from the Earth's gravitational field? Comment on your calculation. Take $r_E = 6400\,km$ and $g_0 = 9.8\,m\,s^{-1}$

$$v_e = \sqrt{\frac{2Gm_E}{r_E}}$$

Beware of learning formulae like this one: it takes only two lines to derive.

At the Earth's surface

$$g_0 = G\frac{m_E}{r_E^2}$$

$\therefore \quad G\dfrac{m_E}{r_E} = g_0 r_E$

and so
$$\begin{aligned} v_e &> \sqrt{2g_0 r_E} \\ &> \sqrt{(2 \times 9.8\,m\,s^{-2})(6.4 \times 10^6\,m)} \\ &> 1.1 \times 10^4\,m\,s^{-1} \text{ or } 11\,km\,s^{-1} \end{aligned}$$

Notice that v_e is indeed an escape speed as it does not matter in what direction the body is projected.

Comments:

(i) The Earth's atmosphere has been ignored; such a simple expression is valid only once you are beyond the frictional drag of air.
(ii) It would be worth projecting the body in the sense in which the Earth rotates. Any body at the equator has a tangential speed of $r_E \omega \approx 0.5\,\text{km}\,\text{s}^{-1}$.
(iii) The escape speed is independent of the mass of the body. However, the escape *energy* $\frac{1}{2}mv_e^2$ is proportional to m.
(iv) The speeds of O_2 and N_2 molecules at atmospheric temperatures are much less than $11\,\text{km}\,\text{s}^{-1}$, so they do not escape, while H_2 molecules do.

14.4 Satellites

A body which moves under the action of one force which is the pull on it of the Earth or some other planet, star, etc. is in a state of **free fall**. Objects which remain in free fall for any length of time are called **satellites**. (A free-fall parachutist free-falls for only a few seconds before air resistance affects the motion. During most of the fall the parachutist is moving at a constant vertical velocity of about 125 m.p.h.) The Earth has only one natural satellite – our Moon – but there are thousands of satellites placed in orbit around the Earth: communications satellites, weather satellites, military satellites, scientific satellites, etc. Applying Newton's law of gravitation and his second law to any satellite which is moving in a circular orbit gives

$$m\frac{v^2}{r} = G\frac{mm_E}{r^2}$$

as

$$v = \frac{2\pi r}{T}$$

this leads to

$$r^3 = \frac{Gm_E}{4\pi^2}T^2$$

i.e.

$$r^3 \propto T^2$$

so that the further out the satellite, the longer the time it takes to orbit the Earth. (Incidentally, this is Kepler's third law mentioned at the beginning of this chapter.) The full expression also tells you that if G is known then measuring r and T for any one satellite lets us calculate the mass of the object it is orbiting. This is how we know the masses of the planets which have moons or of our own Moon around which we have put satellites.

Communications satellites

If you aim a dish aerial at the sky it will work only if the satellite at which it is pointing is always at the *same place* in the sky. As we, on the Earth's surface, rotate once a day, so must the satellite, and further thought will tell you it must rotate above the equator. Such satellites are said to be in **geosynchronous orbits**.

EXAMPLE

Calculate the height above the Earth's surface and the speed of the geosynchronous satellite Eutelstat. Take $Gm_E = 4.0 \times 10^{14}\,\text{N}\,\text{m}^2\,\text{kg}^{-1}$.

Using Newton's laws we get

$$mr\omega^2 = G\frac{mm_E}{r^2}$$

where r is the radius of the satellite's orbit and ω is equal to the angular velocity of the Earth.

$$\omega = 2\pi/(24 \times 3600)\,\text{rad}\,\text{s}^{-1} = 7.3 \times 10^{-5}\,\text{rad}\,\text{s}^{-1}$$

Therefore $r^3\omega^2 = Gm_E$, and as

$$Gm_E = 4.0 \times 10^{14}\,\text{N}\,\text{m}^2\,\text{kg}^{-1}$$

$$r^3 = \frac{4.0 \times 10^{14}\,\text{N}\,\text{m}^2\,\text{kg}^{-1}}{(7.3 \times 10^{-5}\,\text{rad}\,\text{s}^{-1})^2}$$

$$= 7.5 \times 10^{22}\,\text{m}^3$$

and $\quad r = 4.2 \times 10^7\,\text{m}$ or $42\,000\,\text{km}$

i.e. between $35\,000\,\text{km}$ and $36\,000\,\text{km}$ above the Earth's surface.

As both r and T are determined, the satellite must have a speed given by

$$v = 2\pi r/T = 3.1 \times 10^3\,\text{m}\,\text{s}^{-1}$$

All communications satellites are in circular orbits 42 000 km from the centre of the Earth. They do not bump into each other as they are all moving at the same speed of 3100 m s^{-1} and, anyway, there are international agreements about their spacing. Placing them in their geosynchronous orbits – parking them – thus involves giving them exactly the right speed at exactly the right height; not an easy task.

Orbits

Figure 14.7 summarises the possible paths for satellites projected tangentially at or near the Earth's surface (or the surface of any planet) with speed v. If v is zero the path is a straight line vertically downwards (not shown); on the other hand an infinitely large v (if that were possible) would give rise to a straight-line tangential path.

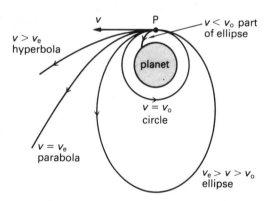

FIGURE 14.7

All other speeds give rise to paths which can be described as shown. Two interesting cases are:

◆ $v = v_e = \sqrt{(2g_0 r_E)}$, the escape speed with $g_0 = 9.8\,\text{N}\,\text{kg}^{-1}$, $v_e \approx 11\,\text{km}\,\text{s}^{-1}$. The path is a parabola.

◆ $v = v_0$, the speed sufficient for the satellite to have a circular orbit near the Earth's surface. The circle in figure 14.7 is assumed to be of radius 6.4×10^6 m. Since the centripetal acceleration is the free-fall acceleration:

$$\frac{v_0^2}{r_E} = g_0$$

$$
\begin{aligned}
v_0 &= \sqrt{g_0 r_E} \\
&= \sqrt{(9.8\,\text{m}\,\text{s}^{-2})(6.4 \times 10^6\,\text{m})} \\
&\approx 8\,\text{km}\,\text{s}^{-1}
\end{aligned}
$$

Many scientific satellites, including the space shuttles, move with this speed just above the Earth's atmosphere.

Weightlessness

The pull of the Earth on a body is called its weight. We can perceive our own weight only indirectly; it would be indeed surprising if we could feel the pull of the Earth on us just as it would be surprising if we could feel the air pressing against us. If a force of similar size to our weight acts on a small area of our bodies (e.g. the contact push of the floor on the soles of our feet), we do feel it. If we temporarily feel no such supporting force the brain mistakenly thinks we have no weight either. It is in this way that the idea of weightlessness arises. A person might *feel* weightless (and perhaps have a peculiar feeling in the stomach) when, for example, he or she has just stepped off a high dive board. An astronaut is in a state of *free fall* for much longer than a diver – figure 14.8. He feels no

FIGURE 14.8

supporting force when inside his spacecraft when its engines are off and so feels weightless. In this case it is *only* his weight which acts on him.

Mass and weight

There is sometimes confusion between mass and weight; the confusion is perhaps the result of both words being used loosely before Newton's second law is properly understood and the equation $mg = W$ for free fall clearly established. Note, however, that:

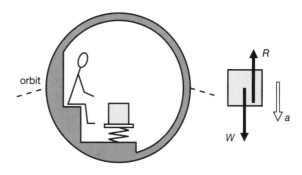

- mass is a scalar, weight (being a force) is a vector.

- mass is measured in kg, weight in N.

- mass is invariable, being associated with *how much matter* there is in a body. Weight depends on how far the body is from the Earth or some other object which attracts it gravitationally; weight can, of course, be zero in deep space.

EXAMPLE

The diagram shows a satellite orbiting the Earth. Inside the satellite a person has put a mass m on a top-pan balance. Explain why the balance reads zero.

If the local gravitational field is g, then a free-body force diagram of the mass would be as in the diagram where R is the contact push of the balance on the mass. Applying Newton's second law

$$ma = mg - R$$

But the acceleration of the whole satellite, including the mass m, is g towards the centre of the Earth, i.e. $a = g$, so substituting

$$mg = mg - R$$
$$\Rightarrow \qquad R = 0$$

and if the push of the balance on the mass is zero then so is the push of the mass on the balance (Newton's third law), i.e. the balance registers zero.

Exercises on each section of this chapter may be found in the companion textbook, **Practice in Physics**.

SUMMARY

At the end of this chapter you should be able to:

- use Newton's law of gravitation for two masses m_1 and m_2 separated by a distance r

$$F = G\frac{m_1 m_2}{r^2}$$

- explain that gravitational forces are only noticeable when one of the attracting bodies is at least of planetary size.

◆ use the equation $g = F_g/m$ for the gravitational field strength at a point where F_g is the gravitational force on a mass m placed at the point.

◆ remember that the gravitational field at the Earth's surface is about $9.8\,N\,kg^{-1}$.

◆ describe how to measure g using a simple pendulum.

◆ remember that the gravitational potential difference between two points is independent of the mass of a body moving between the points.

◆ draw diagrams to show gravitational field lines and equipotential surfaces.

◆ understand that the Earth's gravitational field is radial and is of size $g = Gm_E/r^2$, where m_E is the mass of the Earth.

◆ use the equation $V_g = -Gm_E/r$ for the gravitational potential at a point in the Earth's field and explain why V_g is zero at infinity.

◆ explain that there is a speed called the escape speed and that a body projected from the Earth's surface at any greater speed will not return.

◆ understand that a body is said to be in a state of free fall if the only force acting on it is the pull of the Earth.

◆ explain that a person feels weightless when he or she is in a state of free fall.

15 Storing Electric Charge

15.1 Capacitors

A capacitor is an arrangement of two conductors close together, but insulated from one another. Capacitors can be made in all sorts of ways but consider at first the simplest arrangement, which consists of just a pair of parallel metal plates – figure 15.1. The symbol for a capacitor is based on this arrangement whatever the actual form of construction in a particular case.

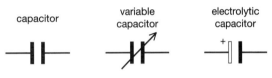

FIGURE 15.1

When you use a flash-bulb to help with indoor photography the power of the flash is high but the time for which it provides illumination is very short. The energy for the flash is provided initially by a cell but this could not provide the power required unless some of the chemical energy was first stored in a capacitor. This is a device which stores electric charge and can supply the energy associated with this very rapidly. Capacitors cannot be used to store large quantities of electrical energy as it would take a whole room full of them to supply a hundred watts for an hour or so.

Capacitors are also widely used in electrical and electronic time-delay circuits. Sometimes these provide repeated switching which leads to oscillations; there is such a circuit in your calculator and in a personal computer where the switching may occur millions of times a second. Or the delay may simply be to give time to get settled in front of the camera for a self-portrait; the time delay then being used to switch on the flash bulb circuit and simultaneously to open the camera shutter briefly.

A series circuit containing a capacitor and a d.c. source, e.g. a cell, is not a closed circuit, and there can be no continuous current in it. However, there is a momentary current when the capacitor is first connected to the d.c. supply. With a pair of plates

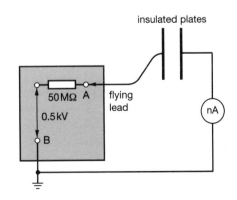

FIGURE 15.2

as in figure 15.2 placed several mm apart you need to use a supply of 1000 V or more and a very sensitive ammeter.

When the capacitor is first connected by joining the flying lead to the point A on the high-voltage supply, the meter registers a small pulse of current. If the flying lead is taken out and joined to the point B, the meter registers an equal negative pulse of current and it is clear that the same charge has flowed in the opposite direction in the capacitor leads, thus discharging it. With an electrolytic capacitor a 6 V supply may produce a detectable pulse on a milliammeter.

Capacitors and charge

How much charge flows in this experiment? Instead of discharging by connecting to B you can discharge the plates into a coulombmeter which measures up to 2000 nC. Now, for different values of the p.d. V, you will find that the charge Q that passes during charging or discharging is proportional to V. So

$$Q \propto V$$

For a given value of V you can vary the separation d of the plates; when d is increased, the charge Q is less. Larger values of Q are obtained when the plates are close together. This result suggests that

$$Q \propto \frac{1}{d}$$

By moving the plates sideways you can also vary the area of overlap A at constant separation; and this result suggests that

$$Q \propto A$$

Placing an insulator between the plates is also found to affect the charge Q that flows in this experiment. This may be demonstrated by putting a sheet of any good insulating material (such as polythene) between the plates; the charge Q is then much greater than with air between them for a given separation of the plates.

When a capacitor is connected to the d.c. supply in the above experiments, there is thus a momentary flow of charge in the circuit, as shown in figure 15.3(a). This leaves a surplus of positive charge on one plate of the capacitor and a surplus of negative charge on the other. The capacitor is

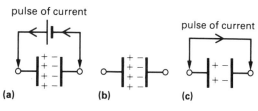

FIGURE 15.3

then said to be storing a charge Q; provided the insulation between the plates is good enough, this charge (+Q on one plate and −Q on the other) will remain isolated on them when the supply is disconnected (b). When the plates are subsequently connected together (c), a pulse of current passes in the circuit, carrying a charge Q from one plate to the other, so discharging the capacitor and returning it to its original state.

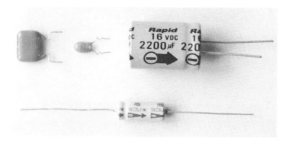

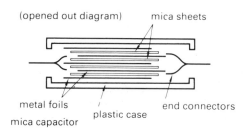

mica capacitor

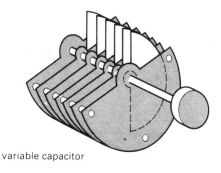

variable capacitor

FIGURE 15.4

Figure 15.4 shows some types of capacitor including a variable capacitor which you will find is what you are using when you tune a radio to different stations. To get the 'plates' *very* close together, and hence store a lot of charge, an electrolytic capacitor uses a film of oxide on the surface of aluminium foil as the insulator. A liquid electrolyte forms the other 'plate'. Because of this construction these capacitors have one terminal marked positive and must be used that way round. You should also check that you will not be exceeding the maximum safe voltage marked on any capacitor you use.

Storage capacitors

The use of semiconducting diodes to rectify a.c. is described on page 257. A single diode produces a half-wave rectified signal as in figure 15.5. The current in the circuit is then in one direction but is pulsating. A capacitor can be used to **smooth** the current.

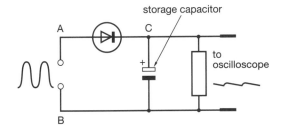

storage capacitor

A   C

to oscilloscope

B

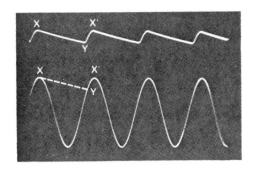

FIGURE 15.5

In bench-top 50 Hz units it will be an electrolytic capacitor. Figure 15.5 shows how the diode, capacitor and the load resistor are connected. (The load here means anything to which the rectifier is supplying current, e.g. a radio

or a Walkman.) We explain the smoothing as follows: during the first positive half-cycle the diode conducts, charging the capacitor with the upper plate positive. At the moment X (on the oscilloscope trace) the p.d. across the capacitor is equal to the *peak* p.d. of the supply. After this the potential of the point A in the circuit starts to fall, and the diode ceases to conduct. The capacitor and load are now effectively disconnected from the supply. The capacitor therefore starts to discharge through the load; this continues until the point A once more reaches a higher potential than C. There is then a forward p.d. across the diode, which starts conducting again. A pulse of current therefore passes through it between Y and X', recharging the capacitor to the peak p.d. of the supply. The average p.d. across the load is thus slightly less than the peak p.d. of the supply, and there is a slight a.c. **ripple** superimposed on it, as shown in the double-beam oscilloscope traces. You can apply an equivalent argument to the full-wave rectified supply from a bridge rectifier.

15.2 Capacitance

When a potential difference V is connected across the plates of a capacitor, a charge Q flows onto the plates. For a particular capacitor Q is proportional to V, so we can write

$$Q = CV$$

where C is a constant called the **capacitance** of the capacitor. You can see that the unit of capacitance is the $C V^{-1}$; but it is convenient to have a special name for this much-used unit, and it is called the **farad** (F) after Michael Faraday (1791–1867). The farad turns out to be a very large quantity, and capacitances are usually expressed in µF or even pF; $1 \, pF = 1 \times 10^{-12} \, F$. A variable air capacitor will have a maximum capacitance of 500 pF; the capacitor used in the flash unit of a camera might have a capacitance of 500 µF (a million times greater) and be the size of your thumb – it will be an electrolytic capacitor.

EXAMPLE

An electrolytic capacitor is used as shown in a camera flash-unit. It is marked $470\,\mu F$, $30\,V$.

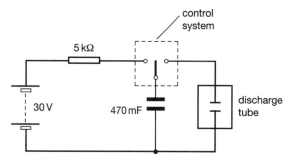

Calculate the maximum charge it can safely store and the average discharge current if it discharges in 0.20 ms.

$$Q = CV = (470 \times 10^{-6}\,F)(30\,V)$$
$$= 0.014\,C \text{ or } 14\,mC$$

$$\text{average current} = \frac{Q}{t} = \frac{14 \times 10^{-3}\,C}{0.20 \times 10^{-3}\,s}$$

$$= 70\,A$$

This is a large current but it lasts only for a very short time.

Measuring capacitance

Method 1

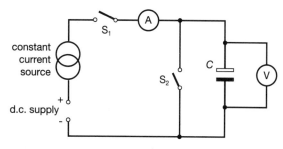

FIGURE 15.6

For large values of C the arrangement shown in figure 15.6 can be used. The **constant current source**, an electronic device which delivers a constant current I (perhaps about 0.5 mA), is switched on for a measured time t. The digital voltmeter, the resistance of which is so high that no measurable charge flows through it, records the p.d. V across the capacitor.

As
$$Q = It$$

then
$$C = \frac{Q}{V} = \frac{It}{V}$$

A series of values of t and V can be found so that a graph of V (up) against t (along) has a slope I/C, i.e.

$$\frac{\Delta V}{\Delta t} = \frac{I}{C}$$

The precision with which C can be found is then largely dependent on the accuracy of calibration of the milliammeter, probably not better than $\pm 2\%$.

A circuit very like this one can be used, with a capacitor of known C, to measure the time t for which S_1 is closed.

Method 2

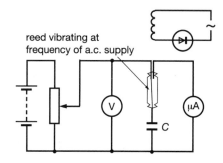

reed vibrating at frequency of a.c. supply

FIGURE 15.7

In this method the capacitor under test is charged to the p.d. V of the supply (up to 25 V, say), and is then discharged through a microammeter. This sequence of connections is performed many times a second by means of a metal strip (or reed) kept vibrating at a known frequency. One way of doing this is to employ a device known as a **reed-switch**. In this kind of switch the contacts to be joined are steel strips fixed in the ends of a glass tube. When the magnetic strips are magnetised by current in a coil around the tube, they are drawn rapidly together. The action is rapid enough to allow the switch to be opened and closed up to 1000 times a second. The diode in the driver circuit for the reed switch in figure 15.7 makes the number of discharges per second equal to the frequency f of its a.c. supply.

If the pulses of current through the ammeter follow one another at high enough frequency the meter shows a steady deflection which records the rate of flow of charge passing through it, i.e. the average current.

The capacitor gains a charge Q each time the switch operates, where $Q = CV$, so a charge CV passes through the meter at a frequency f. The average current I is thus given by $I = fCV$.

The method can be used for smaller values of capacitance than the previous method. For instance, if we take $f = 1000\,\text{Hz}$ and $V = 10\,\text{V}$, then $C = 1\,\mu\text{F}$ gives a current I of

$$I = (1000\,\text{Hz})(10^{-6}\,\text{F})(10\,\text{V})$$
$$= 10^{-2}\,\text{A} = 10\,\text{mA}$$

With a sufficiently sensitive meter, providing it is accurately calibrated, we can make precise measurements of capacitance down to $100\,\text{pF}$ or less.

Energy storage

The existence of a potential difference $(\text{J}\,\text{C}^{-1})$ between the plates of a capacitor implies that the charge stored there carries electrical potential energy. A camera flash unit uses this feature of a capacitor to store energy. Taking the numbers from the example at the beginning of this section, where $0.014\,\text{C}$ is discharged as the voltage falls from $30\,\text{V}$ to $0\,\text{V}$, we can see that the energy of the discharge is $(0.014\,\text{C})(15\,\text{J}\,\text{C}^{-1}) = 0.21\,\text{J}$, using the *average* p.d. of $15\,\text{V}$. This may not seem much energy but remember it is converted in $0.20\,\mu\text{s}$ so the average power is $1050\,\text{W}$, over a kilowatt!, and this nearly all becomes light in the xenon gas tube of the flash unit and not internal energy as it would in a filament floodlamp.

If the p.d. across a capacitor is V, and a small quantity of charge ΔQ is allowed to flow from one plate to the other, the energy converted from electrical to other forms (chiefly internal energy in the connecting wire) is $V\Delta Q$. This movement of charge partly discharges the capacitor, and the p.d. between the plates falls slightly. A graph of p.d. against total stored charge Q is shown in

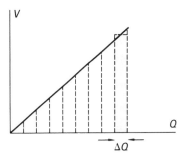

FIGURE 15.8

figure 15.8. You can see that the loss of energy $V\Delta Q$ is represented by the area of the shaded strip under the graph. As further quantities of charge ΔQ are allowed to flow from one plate to the other, the areas of the strips under the graph get steadily less as the p.d. falls. The total energy W_E stored in the capacitor is represented by the total area between the graph line and the axis of Q.

$$W_E = \tfrac{1}{2}QV$$

Since $Q = CV$, we may substitute for Q or V in this expression to obtain the three alternative formulae:

$$W_E = \tfrac{1}{2}QV = \tfrac{1}{2}CV^2 = \tfrac{1}{2}\frac{Q^2}{C}$$

The spring–capacitor analogy

Compare the following statements.

◆ A force exerted on a spring produces an extension in it

$$F = kx$$

◆ A p.d. produced across a capacitor produces a charge on it

$$V = \frac{1}{C}Q$$

In each case a *cause* (F or V) produces an *effect* (x or Q). You can use the analogy to help you think about capacitors. For a strong spring (high k), a large force produces only a small extension *and* for a small capacitor (low C, high $1/C$) a large p.d. produces only a small charge, etc. Such an analogy is more than a useful idea – you can predict the energy stored in a capacitor from a knowledge of the energy stored in a spring

$$W \text{ (mechanical)} = \tfrac{1}{2}kx^2$$

so

$$W \text{ (electrical)} = \tfrac{1}{2}\frac{1}{C}Q^2$$

by analogy. And it works!

Capacitors in parallel

The combined effective capacitance C_{par} of capacitors in parallel is equal to the sum of their separate capacitances. This may be shown as follows. In figure 15.9(a) the p.d. V is the same across the parallel capacitors C_1, C_2, and C_3, but the charges on them will be different. C_{par} is given by

$$Q = C_{par}V$$

where Q is the total charge on the capacitors, i.e. the sum of their separate charges.

So

$$Q = C_1V + C_2V + C_3V$$
$$= V(C_1 + C_2 + C_3)$$

and so

$$C_{par} = C_1 + C_2 + C_3$$

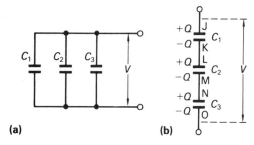

(a)　　　　　　　**(b)**

FIGURE 15.9

Capacitors in series

We need to consider first what happens when the series capacitors in figure 15.9(b) are charged by applying a p.d. V between the joints J and O. This will cause a charge $+Q$ to flow onto the plate J, and $-Q$ therefore onto the other plate K. The plates K and L and the connecting link between them form a single insulated conductor, whose total charge must remain zero throughout the process. Therefore if a charge $-Q$ appears on plate

K a charge $+Q$ must appear on plate L of the next capacitor, and so on through the sequence of capacitors. Thus, with capacitors in series the *same charge* is stored in each capacitor, $+Q$ on one of its plates and $-Q$ on the other. The combined effective capacitance C_{ser} is given by

$$Q = C_{ser}V$$

where V is the total p.d. across the combination.

Now

$$V = \frac{Q}{C_1} + \frac{Q}{C_2} + \frac{Q}{C_3}$$

$$= Q\left(\frac{1}{C_1} + \frac{1}{C_2} + \frac{1}{C_3}\right)$$

But

$$\frac{V}{Q} = \frac{1}{C_{ser}}$$

so that

$$\frac{1}{C_{ser}} = \frac{1}{C_1} + \frac{1}{C_2} + \frac{1}{C_3}$$

The above results may be tested very simply using either of the methods described for measuring capacitance.

Why should we ever want to use capacitors in series? The following example shows that a capacitor's maximum safe voltage may cause us to do so.

EXAMPLE

Four identical capacitors, marked $0.22\,\mu\text{F}$, $300\,\text{V}$, are connected in series across a $1.0\,\text{kV}$ d.c. supply. Deduce the charge stored from first principles.

In series the p.d. across each capacitor is $1000\,\text{V} \div 4 = 250\,\text{V}$. Each stores a charge of

$$\pm(0.22 \times 10^{-6}\,\text{C})(250\,\text{V}) = 55 \times 10^{-6}\,\text{C} \text{ or } 55\,\mu\text{C}$$

on its plates.

Of course, it is only the 55 μC on the end plates which is available if the charged capacitors are used in some way. The charges on the intermediate plates simply cancel out.

EXAMPLE

A capacitor of capacitance $8.0\,\mu F$ is charged to a p.d. of 400 V and then isolated from the supply. (i) What is the energy stored in it? (ii) If an identical capacitor (initially uncharged) is joined across it, what is the energy now stored in the pair of capacitors? Comment on the result.

(i) The initial energy W_E is given by

$$W_E = \tfrac{1}{2}CV^2$$
$$= \tfrac{1}{2}(8.0 \times 10^{-6}\,F)(400\,V)^2 = 0.64\,J$$

(ii) The two capacitors are now joined in parallel, and their combined capacitance is therefore 16.0 F. The total charge is unchanged (though it is now equally shared between the two capacitors). The p.d. across the capacitors therefore falls to half the initial value, i.e. it falls to 200 V. The final energy W_E' is therefore given by

$$W_E' = \tfrac{1}{2}(16.0 \times 10^{-6}\,F)(200\,V)^2$$
$$= 0.32\,J$$

Thus, half the energy (0.32 J) is apparently lost in this process of sharing the charge between the capacitors. Most of this energy is converted to internal energy in the conductors joining the two pairs of capacitor plates.

And what, in the above example, if the conductors are superconducting? The charge then oscillates back and forth between the capacitors and emits energy as electromagnetic waves: the same happens to some extent even with ordinary conductors. The frequency is very high – probably many MHz. Place a radio set nearby and you will hear a crackle as the capacitors are joined.

15.3 Meters

For many purposes we need a voltmeter with an exceptionally high resistance. An ordinary moving-coil voltmeter is often quite inadequate, e.g. for measuring the p.d. across a $1\,\mu F$ capacitor, as the current taken by the instrument would discharge the capacitor in milliseconds. For some purposes we can use an oscilloscope as a high-resistance

voltmeter, but even this takes a current, which although small is large enough to matter when we are dealing with very small quantities of electric charge. We then need voltmeters that take currents of 10 pA or less. Digital voltmeters using an operational amplifier **voltage follower** circuit (page 390) and an analogue to digital display work well. With added amplification of the voltage to be measured they are called electrometers rather than high resistance voltmeters.

Such meters may also be adapted to measure a very small current I by passing this through a known high resistance R joined across the input terminals – figure 15.10(a). The amplifier and meter measure the p.d. V across this; and we have

$$I = \frac{V}{R}$$

For instance, if the resistance R is $1.00 \times 10^7\,\Omega$ then the current range obtained is 0 to 200 nA (if the input p.d. is to be up to 0.20 V): the instrument is then a **nanoammeter**. With added amplification a picoammeter is obtained.

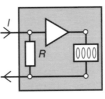
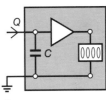

to measure current

to measure charge

(a) pico-ammeter   **(b)** coulombmeter

FIGURE 15.10

By joining a known capacitance C across the input terminals (figure 15.10(b)) the instrument may also be adapted to give the charge Q delivered to the terminals. If the capacitor is initially uncharged, the p.d. V across it is then given by

$$Q = CV$$

For instance suppose $C = 4.7\,\mu F$; then for input p.d.s up to 0.20 V the charge range of the instrument is 0 to 940 nC, and we have a **coulombmeter**. With suitable amplification instruments with maximum indications of 1999 nC are commonly used.

Measuring charge

A coulombmeter which measures nanocoulombs enables us to measure small quantities of charge collected on another insulated conductor or stored on the plates of a small capacitor. For instance, a small metal disc on an insulating handle may be used to collect a small sample of the charge on some other conductor, and then this may be measured by touching the disc on one terminal of the capacitor across the electrometer input. In practice almost the entire charge on the disc is given up to the capacitor.

Using the apparatus shown in figure 15.11 it is possible to take successive equal quantities of charge from a plate maintained at a high potential and to 'spoon' them into the coulombmeter capacitor. It will be found that the reading of the coulombmeter goes up in equal steps, confirming that the p.d. across the capacitor is directly proportional to the charge stored in it.

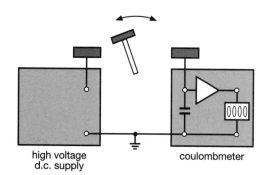

high voltage
d.c. supply   coulombmeter

FIGURE 15.11

The same thing may be done to measure the capacitance C' of a small capacitor (if it is much less than the coulombmeter capacitance C). The capacitor is charged to a suitable p.d. V (say 100 V); then it is disconnected from the supply and joined across the input capacitor of the electrometer, and gives up almost its entire charge Q to it. Then we have

$$C' = \frac{Q}{V}$$

The leaf electrometer

This consists of a rectangular piece of fine aluminium (or gold) leaf attached at its top edge to

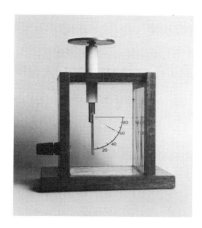

FIGURE 15.12

the side of a brass rod which is fixed in an insulating plug in the top of a conducting box with glass windows – figure 15.12. A small metal disc, called the cap of the electrometer or electroscope, is usually mounted at the top of the brass rod. When a p.d. is produced between the cap and the case, electrical forces draw the metal leaf out towards the case at an angle to the vertical, as shown. A scale may be mounted on one glass window to indicate the angle of deflection as in the photo, but often the instrument is used only for qualitative observations. Full-scale deflection requires a p.d. of about 1500 V. The leakage current through the insulating plug can be less than 10^{-13} A; and the capacitance of the instrument is only a few pF, so that a charge of about 10^{-9} C can be sufficient to produce maximum deflection.

EXAMPLE

A coulombmeter has an input capacitance of 10 nF. A student uses it to 'measure' the capacitance of a capacitor marked 0.22 μF. Discuss whether the experiment is worth while.

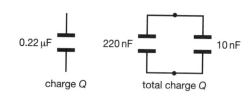

0.22 μF   220 nF   10 nF

charge Q   total charge Q

Before connecting to the coulombmeter suppose the $0.22\,\mu F$ ($= 220\,nF$) capacitor had a charge Q. This charge is shared after the connection. But most of it remains on the $220\,nF$ capacitor since the p.d. across each becomes the same and the $220\,nF$ capacitor has 22 times more capacitance. As the p.d.s across the capacitors are now the same, the charge on the $10\,nF$ capacitor is

$$\frac{10}{10+220}\times Q = 0.0435Q$$

The coulombmeter works by measuring the p.d. across its input capacitor. As only 4% of the charge is transferred to the coulombmeter, only 4% of the expected p.d. is measured so the experiment is quite hopeless.

A coulombmeter with a $4.7\,\mu F$ input capacitor would be able to measure capacitances of up to $0.1\,\mu F$ to within 2%.

15.4 Charging and discharging

When a charged capacitor is discharged through a resistor the charge on the capacitor decreases as shown in figure 15.13. This was produced by recording the p.d. across the capacitor at different times with a data-logging system, and then using $Q = CV$.

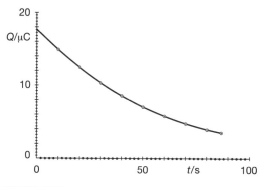

FIGURE 15.13

The shape of this curve is very special: its slope (or gradient) decreases as the value of Q decreases. So also do inverse and inverse square graphs but here, for **exponential** decay:

gradient of Q–t curve is proportional to Q

or   rate of change of $Q \propto Q$

or $$\frac{dQ}{dt} \propto Q$$

or $$\frac{dQ}{dt} = -(\text{constant})Q$$

the minus sign indicating that Q decreases as t *increases*.

To 'solve' this equation you need to have studied mathematics but the important idea is that when $dQ/dt = -(\text{constant})Q$, or $dV/dt = -(\text{constant})V$ etc. the solution looks like the curve of figure 15.13.

To predict the shape of this curve, suppose you have a $500\,\mu F$ capacitor charged to $20\,V$ and connected to a $200\,k\Omega$ resistor at $t = 0$. The initial charge on the capacitor is

$$Q = CV = (500 \times 10^{-6}F)(20\,V)$$
$$= 10\,000\,\mu C$$

and the current, at the instant the switch is closed, is

$$I = \frac{V}{R} = \frac{20\,V}{200\times 10^{3}\,\Omega}$$

$$= 1\times 10^{-4}A$$
or $100\,\mu A$, i.e. $100\,\mu C\,s^{-1}$

If the current continued at $100\,\mu A$, the capacitor would be completely discharged in $100\,s$ (since $100\,\mu C\,s^{-1} \times 100\,s = 100\,000\,\mu C$).

However, as the charge falls (we say *decays* by analogy with radioactivity) so does the p.d. and so the current falls in the same proportion. You should check that when the charge is $7500\,\mu C$ the p.d. is $15\,V$ and the current $75\,\mu A$; when the charge reaches $5000\,\mu C$ the p.d. is down to $10\,V$ and the current $50\,\mu A$. The sequence of calculations is an *iterative* process (it is repetitive) which can be represented by a flowchart or a computer program. The smaller the time interval chosen the more closely the predicted curve will follow the experimental curve of figure 15.13. Graphs of p.d. against time and of current against time have *exactly* the same shape as the Q–t graph. They too are exponential decay curves.

Exact solutions

Referring to the below circuit (figure 15.14(a)) which starts to discharge at $t = 0$, the current I in the resistor is $I = V/R$. Suppose charge ΔQ flows off the capacitor in a time Δt then $I = -\Delta Q/\Delta t$, since ΔQ is negative (a discharge). Thus

$$I = \frac{V}{R} = -\frac{\Delta Q}{\Delta t}$$

Since $Q = CV$, the fall in p.d. ΔV is given by $\Delta Q = C\Delta V$,

so
$$\frac{V}{R} = -C\frac{\Delta V}{\Delta t}$$

or
$$\frac{\Delta V}{\Delta t} = -\frac{1}{CR}V$$

In calculus form this is

$$\frac{dV}{dt} = -\frac{1}{CR}V$$

The solution to this *differential equation*, i.e. V expressed as a function of t, is written

$$V = V_0 e^{-t/CR}$$

where V_0 is the initial p.d. and e is 2.72, a number which emerges from the mathematics. It is known as 'ee' and when raised to a power, e^x for example, we talk of 'ee to the ex' as being an exponential function.

This is the equation of the curve in figure 15.14(b). The **time constant** of the decay process is CR; i.e. when $t = CR$, $V = V_0/e$ $= 0.37V_0$. Thus, in the case discussed above, we have

$$CR = (10\,000 \times 10^{-6}\,\text{F})(200 \times 10^3\,\Omega)$$
$$= 200\,\text{s}$$

This means that in the first 200 s the p.d. falls from 20 V to 0.37×20 V (= 7.4 V). In the next 200 s the p.d. again decreases in the same ratio to 0.37×7.4 V (= 2.7 V). The same happens in each successive interval CR. You can calculate that the p.d. falls to less than 2 V (i.e. 1%) in a time 5CR. The time for the p.d. to halve is 0.69CR or about $\frac{2}{3}$CR.

The equations which show how charge Q and current I change with time are similar to the

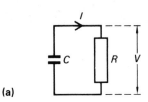

(a)

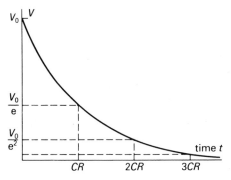

(b)

FIGURE 15.14

> **EXAMPLE**
>
> Show that the unit of CR is the second.
>
> The unit of CR is the FΩ, but the farad is equal to a coulomb per volt, and the ohm is equal to a volt per ampere
>
> so the unit of $CR = \dfrac{C}{V} \times \dfrac{V}{A} \times \dfrac{C}{A}$
>
> but a coloumb is an ampere $\times$ second
>
> unit of $CR = \dfrac{As}{A} = s$

equation for p.d. They are

$$Q = Q_0 e^{-t/CR} \text{ and } I = I_0 e^{-t/CR}$$

In a simple laboratory experiment I is the easiest variable to measure.

Time delay circuits

We often want a system which switches on for only a limited, usually a short, time and then switches off automatically. Or we want something to come on a short time after we close the switch, e.g. a remote camera shutter.

Figure 15.15 shows how to use a capacitor and a resistor as part of a time switch. When S is pressed the voltage at Y becomes 12 V so that $V_{YZ} = 12$ V.

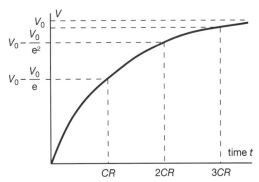

FIGURE 15.16

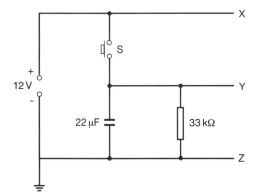

FIGURE 15.15

The capacitor charges immediately. On releasing S the capacitor discharges through the 33 kΩ resistor so V_{YZ} falls exponentially with a time constant $CR = 22\,\mu\text{F} \times 33\,\text{k}\Omega = 0.73$ s. During the same period V_{XY} rises from 0 as, $V_{XY} + V_{YZ} = 12$ V all the time. The falling voltage at Y can be used to switch something on (or off) with appropriate electronic circuitry – see page 393.

Charging

When a capacitor is repetitively charged and discharged through a resistor the discharge is as shown in figure 15.14(b) and the charging as in figure 15.16. The charging equation is

$$V = V_0(1 - e^{-t/CR})$$

It represents the rise of a voltage which approaches a value V_0; the 'gap' decreases exponentially. The time constant is still CR (and refers to the size of

this 'gap'). Figure 15.17 shows the charging and discharging process on an oscilloscope screen.

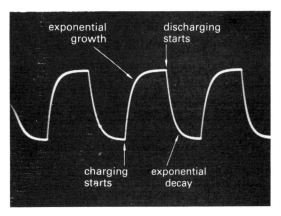

FIGURE 15.17

EXAMPLE

When the switch is closed in the circuit shown the p.d. across the resistor V_R, the p.d. across the capacitor V_C, the current in the circuit I and the charge on the capacitor Q vary with time as shown in the four graphs.

(a) Explain the relationships between the shapes of the graphs and (b) draw a similar set of graphs showing what happens when a charged capacitor is discharged through a resistor.

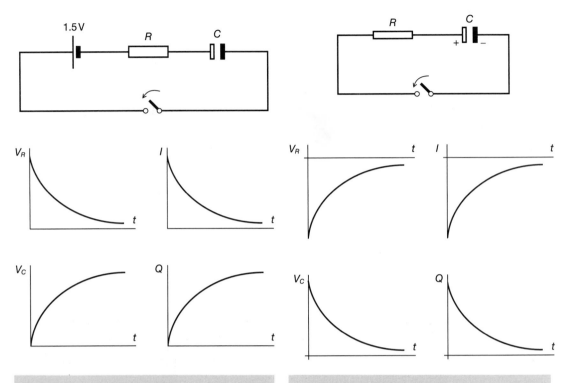

(a) Once the switch is closed

$$V_R + V_C = 1.5\,\text{V}$$

e.g. if $V_R = 0.9\,\text{V}$ at $t = 0.4\,\text{s}$ then $V_C = 0.6\,\text{V}$ at that instant. As $V_R = RI$, the graphs of V_R and I are the same shape, they show exponential decay. As $Q = CV_C$, the graphs of V_C and Q are the same shape, they rise to a maximum and the gap decreases exponentially.

(b) Once the switch is closed

$$V_R + V_C = 0$$

and so, if V_C is positive, V_R is negative.

Notice that in both (a) and (b) the slope of the Q–t graph gives values of I for the I–t graph.

Exercises on each section of this chapter may be found in the companion textbook, **Practice in Physics**.

SUMMARY

At the end of this chapter you should be able to:

◆ understand that a capacitor can store charge.

◆ use the equation $Q = CV$ which defines capacitance and remember that the unit of capacitance, the farad, is so large that capacitances are usually only μF or pF.

◆ explain that capacitors can store energy and use the equations:

$$W_E = \tfrac{1}{2}QV = \tfrac{1}{2}CV^2 = \tfrac{1}{2}\frac{Q^2}{C}$$

◆ use equations for the capacitance of capacitors in parallel and in series:

$$C_{par} = C_1 + C_2 + C_3 \quad \text{and} \quad \frac{1}{C_{ser}} = \frac{1}{C_1} + \frac{1}{C_2} + \frac{1}{C_3}$$

◆ describe how to measure capacitance for both small and large values of C.

◆ understand that a coulombmeter measures charge by detecting the potential difference across its input terminals when a charge is transferred to the capacitor connected between them.

◆ draw graphs showing the variation of charge and of potential difference against time for the discharging and the charging of a capacitance through a resistor.

◆ explain that

$$\frac{dQ}{dt} = -\frac{1}{CR}Q$$

is a differential equation the solution of which is called an exponential decay curve.

◆ remember that the product RC is called the time constant of the circuit and is measured in seconds.

◆ describe an experiment to measure the time constant for an RC circuit.

◆ use the charging and discharging equations:

$$Q = Q_0 e^{-t/CR} \quad \text{and} \quad Q = Q_0(1 - e^{-t/CR})$$

16 Electric Fields

Lightning results from the frictional charging of ice crystals which rise in the convection currents found in thunderclouds. As they rise they collide with large hailstones, the ice crystals becoming positively charged and the hailstones negatively charged. In this manner a huge charge accumulates at the centre and lower edge of the cloud and the sudden flow of the lower charge to earth typically carries an energy of 10 000 MJ. It is estimated that there are about 10 000 thunderstorms going on around the world at any given moment, many of them, of course, over the oceans. You can see that the Earth's atmosphere is an incredible electrical machine.

A flash of lightning or a spark produced in the laboratory by a Van der Graaff machine both happen when there is an electric field or potential gradient of more than 3000 V mm^{-1}. Such fields can be readily produced across a tiny gap in the automatic lighting devices you use with gas cookers or gas fires. Here the field is created by squeezing a crystal – it is called the piezoelectric effect – and the field and its resulting spark can be produced time and time again.

16.1 Electrical forces

When the surfaces of insulating bodies are placed in contact, they are found to become electrically charged, i.e. they attract or repel each other. Two surfaces 'in contact' actually touch only over a small fraction of the area involved because of the surface irregularities present. To increase the surface charge produced we need to rub the surfaces together. Charges produced by surface contacts are therefore usually called **frictional charges**. Polythene, for example, rubbed with a woollen duster acquires a *negative* charge; and two polythene strips so treated will be found to repel one another – figure 16.1. Perspex and cellulose acetate (the substance used for the base of a photographic film) acquire a *positive* charge when rubbed with a woollen duster; and these will be attracted towards a polythene strip that has been rubbed with wool – figure 16.1(b).

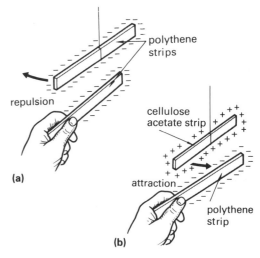

FIGURE 16.1

The frictional charge that appears on the surface of an insulator is small by other standards, probably not more than 10^{-10} C on the polythene strips mentioned above. But the difference of potential between the strip and the duster may be 10 000 V or more; you may have noticed sparks passing between frictionally charged objects – pulling off a nylon vest provides a familiar example!

EXAMPLE

If, in a thunderstorm, there is an average current from cloud to earth of 0.6 A and each lightning strike discharges 15 C of charge, calculate how many lightning strikes occur in 5 minutes.

0.6 A is $0.6\,Cs^{-1}$ so to get 15 C at this rate will take a time t, where

$$(0.6\,Cs^{-1})t = 15\,C$$

$$\Rightarrow \quad t = \frac{15\,C}{0.6\,Cs^{-1}} = 25\,s$$

In 5 minutes, i.e. 300 s, there are therefore $300 \div 25 = 12$ lightning strikes.

E-fields

Any space in which charged particles are acted on by electrical forces is said to contain an **electric field** or **E-field**. The deflection of the test strip in figure 16.2(a) shows that there is an electric field

between the plates. The field is produced by the charges on the surface of the plates.

The strength of the electric field, the E-field, at a point is measured by the force per unit electric charge that acts on a small positive charge placed there. Thus the force F_E acting on a small charge Q is given by

$$F_E = QE$$

(We have to assume that an electric charge Q introduced to measure the value of E is indeed small enough to produce negligible disturbance of the field.) The unit of electric field strength implied by the above definition of electrical field strength is the NC^{-1}; but any other equivalent combination of units may be used. In fact it is often convenient to use the Vm^{-1}, as explained in the next section.

The nature of an electric current is revealed in a striking way if a light metallised sphere is suspended on an insulating thread in the electric field between a pair of charged parallel plates – figure 16.2(b). As long as the sphere remains uncharged it experiences no force. But as soon as it touches one of the plates (say the positive one) it acquires from it some positive charge and is pushed towards the other plate; here it delivers up this charge, receiving a negative charge instead. It then returns to the positive plate, and so the cycle is repeated indefinitely at a rate of several oscillations per second. A meter joined in series with the plates and the high-voltage supply registers a steady deflection. Alternatively a coulombmeter can be used as the positive plate and the charge can be registered each time the ball hits it – providing the frequency is low.

Many substances that we ordinarily regard as insulators (at the low p.d.s of the batteries used in simple electric circuits) must be treated as poor conductors at the high p.d.s commonly used when we are studying electric fields. For instance, if the parallel plates of figure 16.2 are charged by connecting them for a moment to a high-voltage supply we may discharge them in about a second by joining the plates with a wooden ruler or a piece of paper, though in especially dry conditions these may behave as insulators. The only really good insulators are water-repellent substance such as nylon and other modern plastics.

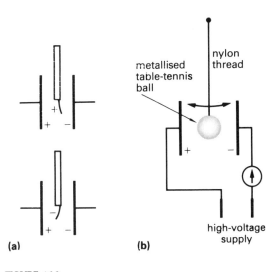

metallised table-tennis ball

nylon thread

high-voltage supply

(a)   **(b)**

FIGURE 16.2

Lines of force

Electric fields may be described by means of lines of force in much the same way as gravitational fields or the magnetic fields you may have met around magnets and current-carrying wires. The lines of the electric field indicate the direction of the force that would act on any charged particle placed in the field. The arrow on a line of force is always drawn to show the direction of the force that would act on a *positive* charged particle.

The lines of force of an electric field may be plotted in a rather similar way to that used with magnetic fields. It is found that a small particle or fibre tends to set itself with its axis parallel to a line of force when the fibre is floated on an insulating liquid in a small dish.

Figure 16.3 shows the pattern of lines of force found for a variety of arrangements of electrodes with the arrows inserted according to the usual rule. The simplest is that between a pair of parallel plates (a). Near the edges of the plates the field is non-uniform, varying in direction and magnitude from point to point; but in the central region between the plates the lines of force form a pattern of parallel lines across the gap between the plates; and here the field is a **uniform** one. By this we mean that the force acting on a small charge placed in the field is the same in size and direction at all points. In (c) there is a **radial** field between the central rod and the circular electrode.

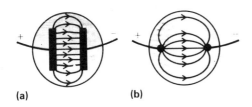

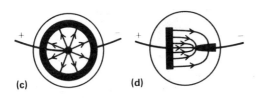

(a)   **(b)**

(c)   **(d)**

FIGURE 16.3

Suppose the gun gives each droplet a positive charge. As the drops approach the plant the positive charge attracts electrons from the ground to all parts of the plant – its leaves, stalk, etc. These are induced charges. There is now an electric field between the gun and the plant and the drops are pulled along the lines of force of this electric field. In so doing they end up on all parts of the plant, not only on top of the leaves but also on the underside. It is this effect which makes the process efficient. Not much spray is wasted because it fails to hit the plant.

16.2 Electric potential

Up to this point we have used *potential* as a way of describing the electrical situation in a *circuit*. Current passes (in the conventional sense) from points at high potential in a circuit to points at low potential, and the size of the current depends on the potential difference between these points. The p.d. V describes the energy converted from electrical to other forms and is defined by the equation

$$V = \frac{W}{Q}$$

where W is the energy converted when charge Q passes. This conversion of energy takes place under the action of the electrical forces acting in the circuit.

The same idea is useful for describing an electric field in an insulating medium such as the air. The electrical forces act on any charged particle that happens to be present, and if it is a positive charge they act to drive it in the direction of the arrow along a line of force. Just as in an electric circuit, the potential decreases as we pass along a line of force from the positive charge from which it starts towards the negative charge on which it ends; and the potential difference V between two points in the field is the energy converted by the forces of the electric field per unit charge that passes from one point to the other.

Earthing

As with electric circuits only *differences* of potential really concern us; but it is often convenient to have a conventional **zero of potential** with reference to which other potentials are measured. The Earth is normally taken as this. In practice this means that the walls, the floor, the bench top and the experimenter are at zero potential, unless special steps are taken to insulate one of them.

A knowledge of potentials at two points enables us to calculate the work done by the electrical forces in moving a charged particle from one point to the other, without having to work out the field

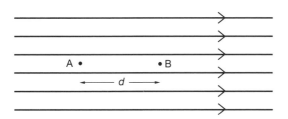

FIGURE 16.4

strength at intermediate points. Suppose a particle of charge Q is moved from a point at potential V_1 to another at a potential V_2. Then the work W done by the electrical forces is given by

$$W = Q(V_1 - V_2)$$

EXAMPLE

Suppose a tiny drop, of mass 2.0×10^{-13} kg, carries a charge of 3.2×10^{-19} C. Calculate how much energy it gains in moving from a place where the electric potential is 200 V to a place where it is 50 V and find an upper limit to its speed if it started from rest. Explain why it is an upper limit.

As $W = Q(V_1 - V_2)$ then

energy converted $= (3.2 \times 10^{-19}\,\text{C})(200\,\text{V} - 50\,\text{V})$
$$= 4.8 \times 10^{-17}\,\text{J}$$

If this were all to become kinetic energy $\frac{1}{2}mv^2$ then for this drop

$$\tfrac{1}{2}(2.0 \times 10^{-13}\,\text{kg})\,v^2 = 4.8 \times 10^{-17}\,\text{J}$$
$$\Rightarrow \qquad v = 0.022\,\text{m s}^{-1}$$

This speed of 2.2 cm s^{-1} is an upper limit because viscous (frictional) forces would act on the drop as it moved through the air. Gravitational forces have also been ignored.

The units of v^2 emerge from this calculation as $J\,kg^{-1}$ but

$$J\,kg^{-1} = (N\,m)\,kg^{-1}$$
$$= (kg\,m\,s^{-2}\,m)\,kg^{-1}$$
$$= m^2\,s^{-2}$$

so the units in the equation are consistent.

Potential gradient

A knowledge of the way in which the potential varies from point to point in a field enables us to calculate the value of the E-field. Figure 16.4 shows a uniform field in which the potential at A is V_1 and at B is V_2. A particle with charge Q is moved by electric forces from A to B, a distance d.

If the field strength is E, the force F_E acting on the particle is given by

$$F_E = QE$$

Since the field is uniform, this force remains constant as the particle moves. In moving it the distance d the work W done by the forces of the field is given by

$$W = F_E d = QEd$$

If the potentials at A and B are V_1 and V_2 respectively, we can also say

$$W = Q(V_1 - V_2)$$

and W is *independent* of the path taken by the charge in moving from A to B. (It is this fact which makes potential and p.d. such important and useful ideas.)

$$E = \frac{V_1 - V_2}{d}$$

The quantity on the right in this equation is called the **potential gradient** of the field, because it is telling us how the potential changes with distance.

For example, a uniform potential gradient of $200\,\mathrm{V\,m^{-1}}$ would mean that in $1.0\,\mathrm{m}$ the potential changes by $200\,\mathrm{V}$, in $0.5\,\mathrm{m}$ the potential changes by $100\,\mathrm{V}$, and so on. Potential gradient *is equal to the strength of* the E-field. The unit of potential gradient is the $\mathrm{V\,m^{-1}}$, and this is often the most convenient unit for recording electric field strengths. You should check for yourself that the $\mathrm{V\,m^{-1}}$ is the same as the $\mathrm{N\,C^{-1}}$.

How big are E-fields?

It is worth giving some estimate of the size of the E-field in different circumstances. For instance a typical dry cell (e.m.f. $1.5\,\mathrm{V}$) used in a torch is $60\,\mathrm{mm}$ long. The field in the space round the cell as it lies in a cupboard is not of course uniform; but if it was uniform from top to bottom, we should have

$$E = \frac{1.5\,\mathrm{V}}{60 \times 10^{-3}\,\mathrm{m}}$$

$$= 25\,\mathrm{V\,m^{-1}}$$

The electric field will actually be rather greater than this near the top where the outer case and central positive terminal are quite close together.

With a pair of parallel plates joined to a high-voltage supply, as in figure 16.2(b) (but with the sensitive meter removed), we can find what E-field

is sufficient to disrupt molecules of air in the gap; when this happens some of the air is ionised and a spark passes between the plates. For instance, with the plates set $1.5\,\mathrm{mm}$ apart a spark passes when the p.d. of the supply is increased to $4.5\,\mathrm{kV}$. Thus in the gap

$$E = \frac{4.5 \times 10^3\,\mathrm{V}}{1.5 \times 10^{-3}\,\mathrm{m}}$$

$$= 3.0\,\mathrm{MV\,m^{-1}}$$

At the surfaces of good insulators that have been charged by friction the E-field can well be close to this value; hence the small sparks observed in such circumstances.

Because of thunderstorms the upper atmosphere carries a permanent positive charge. The result is that in normal (fair-weather) conditions there is a uniform downward electric field at the Earth's surface of about $200\,\mathrm{V\,m^{-1}}$.

Equipotentials

We have two alternative ways of describing electric fields: (i) by means of lines of force and the electric field strength E; or (ii) by means of diagrams showing the electric potential at points in the field. Figure 16.5 shows both these forms of description for the field between a pair of parallel plates, one of which is earthed (zero potential), and the other at a potential of $6000\,\mathrm{V}$. The dashed lines connect points at a common potential. Seen

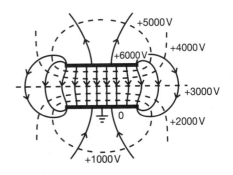

FIGURE 16.5

in three dimensions these lines would be parts of *surfaces* on each of which the potential has a given value. Such surfaces are referred to as **equipotential surfaces**. The lines of force are *always at right-angles to* the equipotential surfaces. In the central region of figure 16.5, where the field is uniform, the equipotentials are planes parallel to the plates, and the potential decreases uniformly through the air gap from one plate to the other.

If we wish to investigate the pattern of equipotential surfaces near two electrodes it is possible to make a scale model of the electrode system. In many cases it will be sufficient to study the nature of the field in two dimensions only; we can then use a special sort of paper coated with a uniform poorly-conducting layer. The positions of the electrodes can be marked in on this paper with paint which is a good conductor, and a suitable low p.d. applied between them – figure 16.6. By selecting a reading on the digital (very high resistance) voltmeter, e.g. 0.60 V, all the points on the conducting paper at 0.60 V above the lower electrode can be found and the 0.60 V equipotential lines drawn. Continuing with other values a full pattern of equipotentials can be located.

the smallest charge that could be measured by any other means.

In 1909 the American physicist R.A. Millikan found that the charges on such oil drops are always integral multiples of a certain smallest quantity of charge e; i.e. the charges obtained were e, $2e$, $3e$, ... and so on.

The charge was never (for instance) $0.6e$, $1.3e$, $4.7e$, or any other fractional multiple. The experiment demonstrates conclusively that electric charge exists in indivisible packets of size e; i.e. that charge is **quantised**. We now believe that charge is quantised at a deeper level in particles called quarks which combine to form protons and neutrons (but not electrons).

The present model is that quarks carry charges of $+\frac{2}{3}e$, $-\frac{1}{3}e$ and antiquarks carry charges of $-\frac{2}{3}e$ or $+\frac{1}{3}e$. As yet, however, no experiment has demonstrated the existence of free, isolated quarks but physicists continue to search for them.

A simplified form of Millikan's apparatus is shown in figure 16.7. The electric field is produced

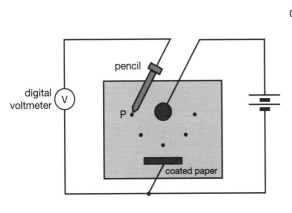

FIGURE 16.6

16.3 The electronic charge

By measuring the electrical force acting on a charged oil drop in a known electric field, it is possible to find the charge it carries. The drops used are of microscopic size, and the charges they carry are many orders of magnitude smaller than

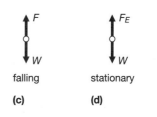

FIGURE 16.7

by a steady p.d. applied between two horizontal plates P and Q (a); these are held exactly parallel by an accurately made spacer ring of insulating material. The ring has two windows cut into it. The upper plate has a small hole in the centre through which oil drops are allowed to fall from a spray. Frictional effects in the nozzle of the spray result in at least some of the oil drops being charged. An optical fibre brings a beam through one window at the side into the space between the plates; and a low-power microscope at the other window is used to observe the drops by means of the light reflected from them – figure 16.7(b). They are seen as sharp points of light against a relatively dark background. There is a transparent scale fixed in the eyepiece by which the distances the drops move can be measured.

When the p.d. is connected to the plates, the motion of some of the slowly falling drops is reversed because of the electrical forces acting on them; a suitable drop is selected and by switching off and on alternately it may be held near the centre of the space until all the other drops have landed on one plate or the other. Measurements are now conducted on this single oil drop. The oil must be of the type used in vacuum apparatus; this has a very low vapour pressure so that the evaporation of the drop is slow, and its weight remains practically constant for a considerable time.

In order to measure the charge Q on the drop, the field is switched on in such a direction as to oppose the weight of the drop. The p.d. V between the plates is adjusted until the drop is held stationary. The drop is then again in equilibrium and the electrical force F_E is equal in magnitude to its weight (figure 16.7(d)). This will be possible only if the plates are exactly horizontal; otherwise the drop soon drifts out of the field of view. If the electric field strength is E when the drop is stationary, we have

$$F_E = W$$

But

$$F_E = QE = Q$$

Hence

$$\frac{QV}{d} = W$$

$\Rightarrow$

$$Q = \frac{Wd}{V}$$

The most difficult part of the experiment is to find the weight of the drop. For drops of a given density you can look up the weight of drops which take different times to fall 1.00 mm in air with the electric field switched off. Figure 16.7(c) shows the forces then acting on the drop as it moves down at a constant speed. In effect you are using a knowledge of the drop's terminal velocity and of the viscous properties of air to find the drop's weight. For example, for oil of density $920\,\text{kg m}^{-3}$, a drop which takes 15.0 s to fall 1.00 mm has a weight of $1.79 \times 10^{-14}\,\text{N}$. If for this drop the p.d. needed to hold it stationary was 670 V, with the top plate negative, and the plate separation d was 12.0 mm, then

$$Q = +\frac{(1.79 \times 10^{-14}\,\text{N})(12.0 \times 10^{-3}\,\text{m})}{670\,\text{V}}$$

$$= +3.21 \times 10^{-19}\,\text{C}$$

The charge on the drop can then be changed by holding a radioactive source nearby (a β-source is most effective); this slightly ionises the air between the plates. The drop will soon collide with one or more ions and its charge is then changed – and suddenly a new value of V is needed to hold the drop stationary.

The calculated charge on any drop is always found to be an integral multiple of the smallest charge obtained. This smallest charge is taken to be the electronic charge e and the presently accepted value is

$$e = 1.602 \times 10^{-19}\,\text{C}$$

16.4 Permittivity

The strength E of the electric field between the plates of a capacitor is equal to the potential gradient across the gap. Thus

$$E = \frac{V}{d}$$

where V is the p.d. between the plates, and d is their separation. We need now to discover how the field strength E depends on the *charges* on the surfaces of the plates that produce the field. In the last chapter we saw how the charge Q carried on the plates depends on their area A and separation d, giving

$$Q \propto \frac{AV}{d}$$

Rearranging, we can write this as

$$\frac{Q}{A} \propto \frac{V}{d}$$

The quantity Q/A is the charge per unit area on each plate, and is referred to as the surface density of charge on the plates; it is denoted by the symbol σ (Greek letter 'sigma'), and is measured in $C\,m^{-2}$. On the other side of the proportionality, V/d is equal to the electric field strength E. Therefore we can write this as

$$\sigma \propto E$$

i.e. electric field strength E is directly proportional to the charge density σ from which it arises. If there is empty space between the plates we may write

$$\sigma = \epsilon_0 E$$

where ϵ_0 (Greek letter 'epsilon') is a constant. It is called the **permittivity of free space** or simply the electric constant. Its value is fixed by the values of other physical constants, i.e. it cannot be measured (page 303) and, to 3 significant figures

$$\epsilon_0 = 8.85 \times 10^{-12}\,F\,m^{-1}$$

You should check that the unit $F\,m^{-1}$ emerges from the defining relationship $\sigma = \epsilon_0 E$.

EXAMPLE

Two conducting plates are placed close together and a power supply connected to them, giving them charges of $\pm Q$. The power supply is then removed. They are held using polythene gloves and are slowly moved apart. Discuss what happens as their separation increases.

Assuming that they are isolated (not connected to a source of p.d.), the charge on the plates cannot change, nor therefore does the charge density σ. This means that the electric field between the plates remains constant. To keep E constant while d increases means that the p.d. V between the plates must rise. If Q is fixed but V rises, the energy stored in the capacitor, $\frac{1}{2}QV$, also rises. The extra energy comes from the work done in pulling the oppositely charged plates apart.

The capacitance C of any capacitor is given by

$$C = \frac{Q}{V}$$

so, rewriting $\sigma = \epsilon_0 E$

as

$$\frac{Q}{A} = \epsilon_0 \frac{V}{d}$$

and rearranging you can see that the capacitance of a parallel plate capacitor can be expressed as

$$C = \epsilon_0 \frac{A}{d}$$

where A is the area of the plates and d their separation. This relationship helps us to understand the design of capacitors in the last chapter.

Relative permittivity

Generally a capacitor has a material insulator filling the space between its plates; the insulator affects the strength of the electric field in the gap, and is sometimes called a **dielectric**. A larger charge is now required to produce a given electric field in the gap. So we can write

$$C = \epsilon_0 \epsilon_r \frac{A}{d}$$

where ϵ_r is a dimensionless constant (with no units), referred to as the **relative permittivity** of the dielectric. It simply tells us how many times greater the capacitance is when the insulator is present. The relative permittivity of a vacuum is exactly 1 (by definition). For other materials is greater than 1: for air $\epsilon_r = 1.006$, for polythene 2.4, and mica 6.0: for substances with polar molecules it can be much higher, e.g. for water (in the purest possible state) $\epsilon_r = 80$.

EXAMPLE

A student makes a capacitor measuring 80 cm by 12 cm using a sheet of polythene 0.30 mm thick as the dielectric and using layers of kitchen foil as the conducting plates. Estimate

(a) the capacitance of capacitor, taking ϵ_r for polythene to be 2.4 and (b) the dimensions of the capacitor if a second layer of polythene was laid on top of one sheet of foil and then rolled up to form a cylinder.
Assume $\epsilon_0 = 9 \times 10^{-12}\,\mathrm{F\,m^{-1}}$.

(a) As $A = 0.80\,\mathrm{m} \times 0.12\,\mathrm{m} = 0.096\,\mathrm{m^2}$ and $d = 3.0 \times 10^{-4}\,\mathrm{m}$ then

$$C = \epsilon_0 \epsilon_r \frac{A}{d} = \frac{2.4(9 \times 10^{-12}\,\mathrm{F\,m^{-1}})(0.096\,\mathrm{m^2})}{3.0 \times 10^{-4}\,\mathrm{m}}$$

$$= 6.9 \times 10^{-9}\,\mathrm{F} \text{ which is about } 7\,\mathrm{nF}$$

(b) The volume of the two sheets of polythene and two sheets of foil – assume the foil is 0.1 mm thick – will be

$$V = (0.096\,\mathrm{m^2})(2 \times 0.3 \times 10^{-3}\,\mathrm{m}$$
$$+ 2 \times 0.1 \times 10^{-3}\,\mathrm{m})$$

$$= 7.7 \times 10^{-5}\,\mathrm{m^3}$$

If this is rolled into a cylinder of length of 0.12 m then the radius is given by r, where

$$\pi r^2 (0.12\,\mathrm{m}) = 7.7 \times 10^{-5}\,\mathrm{m^3}$$

$$r = 0.014\,\mathrm{m} \text{ or } 1.4\,\mathrm{cm}$$

It is necessary to add the second sheet of polythene in order to prevent the two 'plates' of the capacitor shorting when it is rolled up. This also has the effect of doubling the capacitance – can you see why? – so the capacitance of the tubular capacitor is about 14 nF.

Measuring ϵ_r

This can be done by comparing the capacitance C_0 of an air-filled (strictly vacuum) parallel plate capacitor and the capacitance C of the same plates with the material under investigation filling the space between the plates. As C_0 will be only a few hundred pF a vibrating reed method is required – see page 192. It is necessary only to set the vibrating reed to a fixed frequency and to read off the average discharge currents I_0 and I for the same charging voltage:

♦ firstly with the plates separated by a sheet of the insulating material of thickness d for which ϵ_r is required; and

♦ secondly with the plates separated by insulating spacers of the same thickness d and very small area.

Clearly

$$\epsilon_r = \frac{C}{C_0} = \frac{I}{I_0}$$

The value obtained for ϵ_r will have an uncertainty of up to $\pm 5\%$ because of the 'stray' capacitances between the non-earthed plate and earthed objects (clamp stands, etc.) round about.

16.5 Fields near conductors

In the absence of any d.c. supply producing a continuous supply of electrical energy, the charges in a system of conductors reach equilibrium under the action of the electric field in a very short time (less than $10^{-12}\,\mathrm{s}$ usually). When this has happened the field inside the material of a conductor must be zero everywhere. If not, the free charges in it will move under the action of the field until the distribution of charge is such as to make this so. Only if there is a current is the field inside a conductor non-zero. It follows that the potential inside the material of a conductor (with no current in it) is the same everywhere.

A maintenance worker on high-voltage wires wears clothing with a closely wound wire mesh layer. He, the clothing and the high-voltage wires are at the same potential and so no current can flow between them.

You should also realise that the field at the surface of a conductor must be everywhere perpendicular to the surface. If it was not, it would have a component parallel to the surface and would set in motion the free charges in the surface.

Close to a charged surface the charge density is proportional to the electric field strength. It is found that once again, the electric constant ϵ_0 connects the two quantities and $\sigma = \epsilon_0 E$ (for air $\epsilon_r = 1$ so it does not appear in the equation).

EXAMPLE

Estimate the charge density σ on the facing surfaces of a pair of parallel plates 20 mm apart across which there is a p.d. of 5.0 kV. Assume that the permittivity of air is the same as that of a vacuum. If the plates are 0.20 m square, estimate the total charge Q on one plate. Take $\epsilon_0 = 8.85 \times 10^{-12}$ Fm^{-1}.

The field strength E in the gap is given by

$$E = \frac{V}{d} = \frac{5.0 \times 10^3 \text{ V}}{20 \times 10^{-3} \text{ m}}$$

$$= 2.5 \times 10^5 \text{ V m}^{-1}$$

$$\Rightarrow \sigma = \epsilon_0 E = (8.85 \times 10^{-12} \text{ Fm}^{-1})(2.5 \times 10^5 \text{ V m}^{-1})$$

$$= 2.2 \times 10^{-6} \text{ C m}^{-2}$$

Assuming σ is constant over the surface of a plate we have

$$Q = (2.2 \times 10^{-6} \text{ C m}^{-2})(0.20 \text{ m})^2$$
$$= 8.8 \times 10^{-8} \text{ C} \approx 10^{-7} \text{ C}$$

Induced charges

When an object carries an electric charge, the electric field lines that start on it must end on an equal and opposite charge at some other place. Thus an insulated electric charge always causes an equal and opposite charge to flow onto other objects in its surroundings. The latter charges are referred to as induced charges; induced and inducing charges are *always* equal in size.

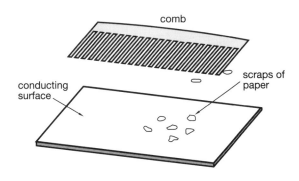

conducting surface

comb

scraps of paper

FIGURE 16.8

An understanding of induced charges helps us to explain many everyday electrostatic phenomena. When a negatively charged comb – figure 16.8 – is held above small pieces of paper, some of them jump up and stick to the comb (and are sometimes then pushed rapidly away). The reason is that the paper and the surface on which they are resting conduct charge. Thus electrons are pushed from the pieces of paper and these, as light positively charged objects, are then attracted to the comb. (When they reach it their charge is cancelled and some will receive an excess of negative charge which results in their repulsion.)

You should be able to explain why rubber balloons can be made to 'stick' to the walls of a room using similar reasoning. 'Dusting' plastic surfaces will, if anything, charge them and make them *attract* dust particles unless you wipe the surface with an anti-static cloth like the ones used on LP discs.

Charge density and surface curvature

On a charged conductor of irregular shape the charge density varies from point to point on its surface. A simple test with a proof-plane and a leaf electrometer (a coulombmeter is not sensitive enough here) shows that the charge density is always greatest at the most highly curved convex parts of the surface, i.e. the charge is found to concentrate chiefly near points and corners.

Point discharge

A consequence of this concentration of charge at the most curved parts of a conductor is that it is possible, without using very high p.d.s, to produce a very large field locally near a sharp point or fine wire, because the field strength E at the surface of the conductor is proportional to the charge density σ. Near a sharp point or fine wire it may even be sufficient to ionise the air there (which requires a field strength of 3 MV m^{-1}). Those ions that have the same sign of charge as the conductor are violently repelled from it, giving rise to an appreciable electric 'wind'. If you hold a lighted candle near a sharp needle raised to a high potential, the wind can be sufficient to blow the flame horizontal (figure 16.9(a)). It carries away the charge on the needle into the surrounding air, and eventually to earthed objects nearby. This is called a point discharge or corona discharge. Overhead power lines are often at a very high potential and the discharge from them to nearby trees and pylons can be observed.

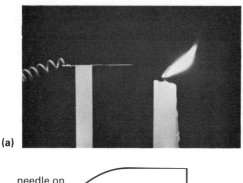

(a)

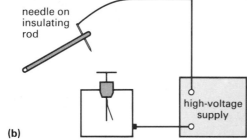

needle on insulating rod

high-voltage supply

(b)

FIGURE 16.9

Any object placed in the way of the 'wind' may collect some of the charge on its surface. We may show this by holding a sharp needle connected to a high-voltage source near the cap of an electrometer (figure 16.9(b)); we then observe a rapidly increasing deflection, as the charge blows onto the cap.

By placing thin high-voltage wires at the bottom of chimneys it is possible to use the resulting ions to charge smoke particles in the flue gases and then to attract them to charged plates within the chimney. Such **electrostatic charge precipitation** is highly developed at coal-fired power stations and the resulting ash, which accumulates by the tonne, is used for making breeze blocks and in civil engineering projects rather than entering the atmosphere as a pollutant.

16.6 Radial fields

An isolated charged particle produces a radial field where the electric field strength follows an inverse square law relationship – figure 16.10.

$$E \propto \frac{1}{r^2}$$

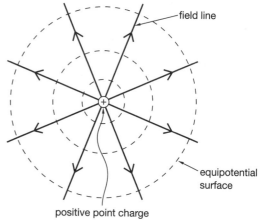

field line

equipotential surface

positive point charge

FIGURE 16.10

Coulomb's law

Two charged particles, carrying charges Q_1 and Q_2, placed a distance r apart in a vacuum (or air) each exert a force of size F on the other where

$$F = k\frac{Q_1 Q_2}{r^2} = \frac{1}{4\pi\epsilon_0}\frac{Q_1 Q_2}{r^2}$$

This is called Coulomb's law after the French physicist Charles Coulomb (1736–1806). Electric field strength E is defined as the force per unit charge placed in the field. You can see that this gives the radial field of the charged particle as

$$E = k\frac{Q}{r^2} = \frac{1}{4\pi\epsilon_0}\frac{Q}{r^2}$$

In both relationships the constant, $k = 1/4\pi\epsilon_0$, has the same function as G in the equations for gravitation, but charges, unlike masses, can both attract *and* repel. You should check that the value of $k = 1/4\pi\epsilon_0 = 9.0 \times 10^9\,\mathrm{N\,m^2\,C^{-2}}$ to 2 significant figures.

Although the law is stated for a vacuum it is effectively the same for air where ϵ_0 should be replaced by $\epsilon_0\epsilon_r$ but ϵ_r for air is so near to 1 that it is often omitted. It is forces which obey this law which 'make your hair stand on end' – figure 16.11 – or which enable smoke particles to be removed in electrostatic charge precipitation.

FIGURE 16.11

Testing Coulomb's law

You can check the law approximately by measuring the force F which two charged light conducting balls exert on each other at various distances apart. One of these is fixed on an insulating support, while the other is suspended from a pair of long nylon threads (figure 16.12). If the balls carry

charges of the *same* sign, the repulsion between them deflects the suspended ball; the sideways displacement of this is best measured by observing its shadow cast on a screen by a small light source. If the displacement is small compared with the length l of the support system, the force F is proportional to the *horizontal* movement s of the ball (figure 16.12(a)).

In figure 16.12(b) the three forces acting on the suspended ball are shown; T is the pull of the thread on the ball. Resolving the forces

$$F = T \sin \theta = T\frac{s}{l}$$

and $\qquad mg = T \cos \theta \approx T$ (if θ is small)

dividing $\qquad \dfrac{F}{mg} \approx \dfrac{s}{l}$

i.e. $F \propto s$ (if θ is small)

To check the law you need to show that the horizontal movement s is inversely proportional to the square of the distance apart r of the pith balls;

EXAMPLE

The diagram shows the electric field produced by an electric dipole formed by two charges $+4.0\,nC$ and $-4.0\,nC$. Calculate the strength of the E-field (i) at A and (ii) at B and (iii) explain how the direction of the field would be found at P. Take the distance a to be $0.10\,m$ ($10\,cm$), and $1/4\pi\epsilon_0 = 9.0 \times 10^9 \, N\,m^2\,C^{-2}$.

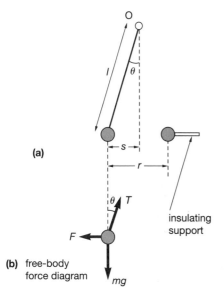

(a)

(b) free-body force diagram

FIGURE 16.12

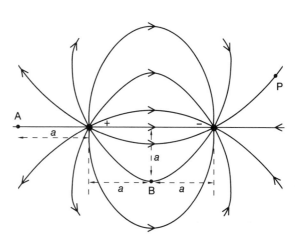

(i) A is 0.10 m from the +4.0 nC charge

$$E_+ = (9.0 \times 10^9 \, N \, m^2 \, C^{-2}) \frac{4.0 \times 10^{-9} \, C}{(0.10 \, m)^2}$$

$$= 3600 \, N \, C^{-1} \text{ to left}$$

but A is 0.30 m, i.e. 3 times further, from the −4.0 nC charge so

$$E_- = 3600 \, N \, C^{-1} \div 3^2$$
$$= 400 \, N \, C^{-1} \text{ to right}$$

It is only necessary to notice that A is 3 times as far from the −4.0 nC as from the +4.0 nC and then use the inverse square idea that the field is one-ninth.

$$E \text{ at } A = (3600 - 400) \, N \, C^{-1}$$
$$= 3200 \, N \, C^{-1} \text{ to left}$$

(ii) The distance of B from each charge is $a\sqrt{2}$, so the size of the E-field produced by each at B is

$$E = (9.0 \times 10^9 \, N \, m^2 \, C^{-2}) \frac{4.0 \times 10^{-9} \, C}{(\sqrt{2} \times 0.10 \, m)^2}$$

$$= 1800 \, N \, C^{-1}$$

Resolving this field into components parallel and perpendicular to the line between the charges, the perpendicular components cancel as that from A is down and that from B is up, but the parallel components add as each is to the right.

$$\therefore \quad E \text{ at } B = 2(1800 \, N \, C^{-1}) \cos 45°$$
$$= 2500 \, N \, C^{-1} \text{ to right}$$

In general, at points such as P, the components of the E-field from each end of the dipole must be added by vector addition. A scale diagram is often the easiest method.

and both these distances we find from the positions of the shadows on the screen. Leakage of charge from the pith balls produces uncertainties; but with care we can show that a graph of s against $1/r^2$ is approximately a straight line.

The hydrogen atom

In a hydrogen atom the average distance apart of the proton and the electron is 5.3×10^{-11} m. It is interesting to work out the strength of the E-field of the proton at this distance from it. The charge on the proton is the electronic charge $e \, (= 1.60 \times 10^{-19} \, C)$. Therefore

$$E = \frac{1.6 \times 10^{-19} \, C}{4\pi(8.9 \times 10^{-12} \, F \, m^{-1})(5.3 \times 10^{-11} \, m)^2}$$

$$= 5.1 \times 10^{11} \, N \, C^{-1} \text{ or } 5.1 \times 10^{11} \, V \, m^{-1}$$

This is an enormous field strength; compare it with the maximum E-field in air $(3 \times 10^6 \, V \, m^{-1})$.

The force F acting on the electron in this field is given by

$$F = (1.60 \times 10^{-19} \, C)(5.1 \times 10^{11} \, N \, C^{-1})$$
$$= 8.2 \times 10^{-8} \, N$$

This may not seem a very large force. But it is vastly greater than the gravitational force with which the proton attracts the electron. You should be able to show that the gravitational force at this separation of the particles is only $4.1 \times 10^{-47} \, N$, which is less than the electrical force by a factor of 2×10^{39}.

Bearing in mind the small mass of an electron $(9.1 \times 10^{-31} \, kg)$, the electrical force pulling it in a hydrogen atom is *very* large. The acceleration a of the electron in its orbit is given by Newton's second law as

$$a = \frac{F}{m_e}$$

$$= \frac{8.2 \times 10^{-8} \, N}{9.1 \times 10^{-31} \, kg}$$

$$= 9.0 \times 10^{22} \, m \, s^{-2} \approx 10^{-22} \, g$$

The electric potential near a point charge

In this case it is convenient to take our zero of potential at an *infinite distance* from the isolated point charge Q as we did when dealing with gravitational potential on a large scale around the Earth. In practical terms this means that we are assuming that the earthed walls and floor of the laboratory are at a sufficient distance for their effects to be ignored.

The potential V at a distance r (in air) from a particle with a charge Q may be derived from the expression for the electric field strength near the charge:

$$E = \frac{Q}{4\pi\epsilon_0 r^2}$$

It is necessary to find an expression for the potential V such that the potential gradient is equal at every point to E as given above. This may be done using calculus; but we shall just quote the result here:

$$V = \frac{Q}{4\pi\epsilon_0 r}$$

Thus the potential varies inversely with the distance r from the point charge. It falls to zero at an infinite distance from the point (see also figure 14.6).

The potential V at a distance of $5.3 \times 10^{-11}\,\text{m}$ from a proton, the nucleus of a hydrogen atom, is given by

$$V = \frac{1.6 \times 10^{-19}\,\text{C}}{4\pi(8.9 \times 10^{-12}\,\text{F m}^{-1})(5.3 \times 10^{-11}\,\text{m})}$$

$$= 27\,\text{V}$$

and the electric potential energy of an electron there is 27 eV. The energy needed to ionise a hydrogen atom is exactly half this, i.e. 13.5 eV. The kinetic energy of the orbital electron supplies the other half of the energy needed to remove the electron completely from the proton.

EXAMPLE

Referring to the previous example (page 213) calculate the electric potential (i) at A and (ii) at B.

(i) A is 0.10 m from the +4.0 nC charge

$$V_+ = (9.0 \times 10^9\,\text{N m}^2\text{C}^{-2})\frac{4.0 \times 10^{-9}\,\text{C}}{0.01\,\text{m}}$$

$$= 360\,\text{J C}^{-1}, \text{ i.e. } 360\,\text{V}$$

A is 0.30 m from the −4.0 nC charge, so the potential is one-third, and negative

$$V_- = -120\,\text{V}$$
$$V \text{ at A} = 360\,\text{V} - 120\,\text{V}$$
$$= 240\,\text{V}$$

(ii) The potential at B is zero as, whatever the value of V_+, the value of V_- will be the same size but of opposite sign.

B lies on a line perpendicular to the line joining the charges and mid-way between them. This is the zero-volt equipotential line; in fact it is part of a plane zero-volt equipotential surface.

To calculate V at other points, such as P, it is only necessary to find the values of V for each end of the dipole and to add them. Potential is a scalar quantity.

A conducting sphere

Just as for a point charge, the field has spherical symmetry. *Outside* the sphere the E-field is exactly the same as though the charge Q carried on its surface was concentrated at its centre. The field strength E at a distance r from the centre of the sphere (in air) is therefore given by $E = Q/4\pi\epsilon_0 r^2$ and the potential V is given by $V = Q/4\pi\epsilon_0 r$ (again taking the zero of potential at an infinite distance).

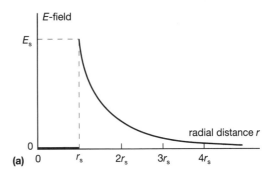

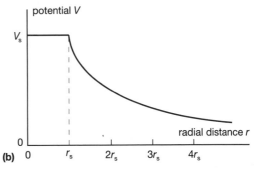

FIGURE 16.13

The similarity of the mathematics of radial electric fields and planetary gravitational fields mean that you can learn about one *by analogy* with the other. Indeed, whichever you find the easier is the one to use to grasp the difficult ideas of field and potential. The work on gravitation is dealt with in Section 14.3.

Inside the sphere (a hollow space) the E-field is zero. If the sphere is of radius r_s then at its surface the field E_s and potential V_s are given by

$$E_s = \frac{Q}{4\pi\epsilon_0 r_s^2} \text{ and } V_s = \frac{Q}{4\pi\epsilon_0 r_s}$$

The potential has the same value as this throughout the interior of the sphere, since there is no E-field inside the sphere, and therefore no potential gradient. Figure 16.13 shows how E and V vary with r near a charged conducting sphere.

EXAMPLE

A large Van der Graaff generator has a dome which is a sphere of diameter 4.0 m. When fully charged the potential of the dome is 9.0 MV above its surroundings. Calculate (a) the total charge on the dome and (b) the electric field at the surface of the dome.

(a) From $V = \dfrac{Q}{4\pi\epsilon_0 r}$, $Q = 4\pi\epsilon_0 rV$

$\therefore \quad Q = 4\pi(8.9 \times 10^{-12}\,\text{F m}^{-1})(2.0\,\text{m})(9.0 \times 10^6\,\text{V})$

$\quad = 0.0020\,\text{C or } 2.0\,\text{mC}$

(b) As $\qquad E = \dfrac{Q}{4\pi\epsilon_0 r^2} = \dfrac{Q}{4\pi\epsilon_0 r} \times \dfrac{1}{r}$

then $\qquad E = V \times \dfrac{1}{r} = \dfrac{9.0 \times 10^6\,\text{V}}{2.0\,\text{m}}$

$\qquad\qquad = 4.5 \times 10^6\,\text{V m}^{-1}$

This field is greater than the 3 MV m^{-1} needed to cause ionisation in air at atmospheric pressure, so it will cause sparking unless the air pressure is increased around the dome. This means that such a machine needs to be totally enclosed in a pressure vessel to operate at this potential.

Exercises on each section of this chapter may be found in the companion textbook, ***Practice in Physics***.

SUMMARY

At the end of this chapter you should be able to:

♦ understand that insulating bodies become charged when placed in contact.

♦ describe an experiment to demonstrate that current is a rate of flow charge.

♦ use the equation $E = F_E/Q$ for the electric field strength at a point where F_E is the electric force on a charge Q placed at the point.

♦ draw diagrams to show electric field lines and associated equipotential surfaces.

♦ understand the relationship between potential gradient and electric field strength in uniform fields and use the equation:

$$E = \frac{V_1 - V_2}{d}$$

♦ explain the equivalence of the units $\mathrm{N\,C^{-1}}$ and $\mathrm{V\,m^{-1}}$ for electric field strength.

♦ describe an experiment to study equipotential surfaces.

♦ understand that charge is quantised and explain the nature of the experimental evidence for this.

♦ remember that the electronic charge is $1.6 \times 10^{-19}\,\mathrm{C}$.

♦ use the equation $\sigma = \epsilon_0 E$ for charge density where ϵ_0 is the permittivity of free space.

♦ describe experiments to investigate the factors affecting the capacitance of a parallel plate capacitor.

♦ use the equations $C = \epsilon_0 A/d$ and $C = \epsilon_r \epsilon_0 A/d$ for the capacitance of parallel plate capacitors.

♦ explain how bodies can become charged by induction.

♦ use the following equations associated with point charges and radial fields

$$F = \frac{1}{4\pi\epsilon_0}\frac{Q_1 Q_2}{r^2}, \quad E = \frac{1}{4\pi\epsilon_0}\frac{Q}{r^2}, \quad V = \frac{1}{4\pi\epsilon_0}\frac{Q}{r}$$

♦ describe experiments to investigate the factors affecting the capacitance of a parallel plate capacitor.

17 Magnetic Fields and Forces

We have all had some experience of magnets. They are used in the home in magnetic door catches, to help keep refrigerator doors shut, in intruder alarms fitted to doors and windows, and outside in compasses for navigation. But much larger magnets, and electromagnets, are used in industrial applications: the photograph shows the 'people carrier' which is used to take passengers from Birmingham International Airport to the National Exhibition Centre. The track consists of

a continuous iron plate: electromagnets fixed to the carriages are pulled up by the magnetic force and this force supports the weight of the train. There is a gap of about 158 mm between the electromagnets and the track. This gap is monitored by sensors and kept at a constant value. This provides a frictionless transport system.

17.1 Magnetic fields

You will have used magnets in experiments at school, where you will have discovered that a freely-suspended bar magnet points north–south, that the magnetic effect seems to be concentrated at its ends (which are called poles), that the end which points north is called a north-seeking or north pole, and that like poles repel each other and that unlike poles attract each other. You will know that some materials are magnetic (iron, nickel, cobalt and specially designed alloys) and it is from such materials that most magnets are made. You may also have used other types of magnet: U-shaped or horseshoe magnets, or ceramic magnets which have their poles on their faces.

You will have used plotting compasses or iron filings to discover the direction of the magnetic field around some of these magnets: as a reminder, some of these fields are shown in figure 17.1,

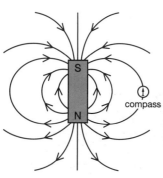

bar magnet

FIGURE 17.1

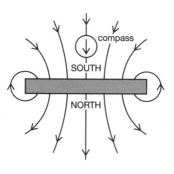

ceramic magnet

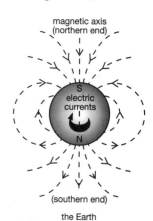

the Earth

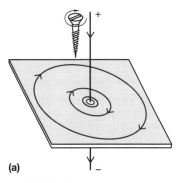

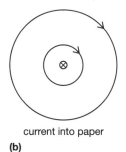

current into paper

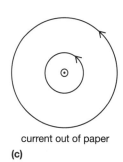

current out of paper

(a)　　　　　　　　　　　　　　　　**(b)**　　　　　　　　　　　　　　　　**(c)**

FIGURE 17.2

together with the magnetic field around the Earth. But remember that the fields are *three-dimensional*. In the diagrams you are looking at two-dimensional cross-sections of the fields.

The field of a current

You will probably also know that there is a magnetic field around a wire which carries an electric current and it is this, and its effects, that we shall be concentrating on in this chapter.

To investigate the magnetic field pattern around a straight wire you would need to have the wire arranged vertically, passing through a horizontal card, as shown in figure 17.2(a). Using plotting compasses you would find that the direction of the field is as shown, i.e., clockwise (seen from above), if the current is downwards, passing into the card. It is easier to draw a diagram which gives a two-dimensional view of the situation, as shown in (b) and (c). In these diagrams the cross means that the current is going away from you: the dot means that it is coming towards you (think of the cross as the tail end of a dart or an arrow, and the dot as its point). You need some way of remembering the direction of the field. You could use the **right-hand corkscrew** rule:

the direction of the lines of force round a current is that in which a corkscrew would be turned in order to move it forwards in the direction of the current

or the **right-hand grip** rule:

the direction of the lines of force round a current is that in which the fingers pass round the wire if the wire is gripped by a right hand with the thumb pointing in the direction of the current.

These rules are illustrated in figure 17.2(a): use whichever you find easier.

Once you know what the field around a straight wire is like, you can predict the field around a coil or a solenoid. By a **coil** we mean one or more turns of wire, nearly all of which are in the same plane: i.e., a coil is two-dimensional, as shown in figure 17.3(a). A **solenoid** is a coil which is 'drawn out' along its axis, as shown in (b). Check for yourself that the field pattern is as you would predict, using either of the rules given above. Notice that the field of a solenoid is similar to the field of a bar magnet, with 'poles' at the ends of the solenoid: the field of a coil is like that of a thin disc-shaped

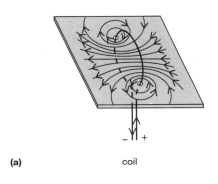

(a)　　　　　　　　coil

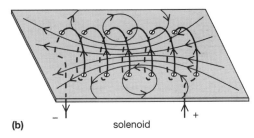

(b)　　　　　　　　solenoid

FIGURE 17.3

219

magnet with poles on its faces. The diagrams in figure 17.4 show a convenient way of remembering which end of a coil or solenoid is north or south: the arrows on the letters point the same way as the currents.

FIGURE 17.4

17.2 The force on a current-carrying conductor

When a straight wire carrying a current is placed at right angles to a magnetic field, a force acts on it in a direction which is at right angles to both the field and the current. The direction in which the force acts can be found using the **left-hand motor rule** which is illustrated in figure 17.5:

if the thumb and first two fingers of the left hand are put at right angles to each other, with the First finger pointing in the direction of the Field, and the seCond finger pointing in the direction of the Current, then the thumb gives the direction of the force.

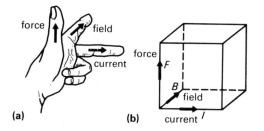

FIGURE 17.5

Because the three quantities B, I and F are at right angles to each other it is convenient to draw diagrams so that the direction of the magnetic field is either into or out of the paper (so that the current and the force will then be in the plane of the paper) as shown in figure 17.6 where the magnetic field has been represented by a grid of crosses, representing a field whose direction is *into* the paper. A grid of dots would represent a field whose direction is *out* of the paper. (Compare this with the single cross and the single dot which are used to indicate that an electric current is going away from us, or coming towards us.)

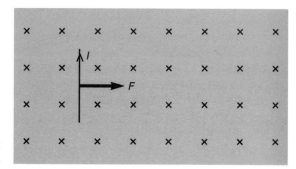

FIGURE 17.6

Measuring a magnetic force

You can measure the magnetic force on a wire carrying a current with the apparatus shown in figure 17.7. Place a U-magnet, consisting of two ceramic magnets mounted on an iron yoke, on a top-pan balance. The balance needs to have a sensitivity of 0.01 g. Support a stiff wire frame (you could make one yourself if one is not available) so that there is a horizontal portion, about 5 cm long, between the poles of the U-magnet, and the wire is at right angles to the magnetic field. Then *tare* the balance (i.e., set it to zero, so that the weight of the U-magnet is automatically allowed for). Pass a current of a few amperes through the wire. You will find that the balance gives a reading of about 0.50 g. It may be a positive or negative reading, depending on the direction of the force. If the balance reading is positive, there is a downward push on the balance, and by Newton's third law, an upward push on the wire of the same size. If the balance reading is negative, there is an upward pull on the balance, and a downward pull on the wire.

If you vary the current you will find that the magnetic force is proportional to it. If you remove one of the magnets so that the field is only half as

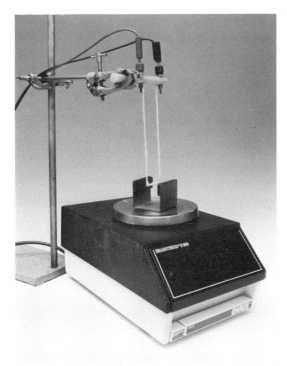

FIGURE 17.7

strong as it was, you will find that the magnetic force is half what it was. If you alter the length of the horizontal part of the frame, you will find that the force is proportional to the length (provided that the wire is not partly outside the almost uniform field between the magnets).

Magnetic flux density B

These simple experiments show that the magnetic force F is proportional to the strength of the field (which is called the **magnetic flux density** B), the current I, and the length l, and we choose to *define* B by the equation

$$F = BIl$$

The equation shows that the unit of magnetic flux density is the $N\,A^{-1}\,m^{-1}$, but for convenience this is called the **tesla** (T). (The reason for using the term 'magnetic flux density' will become clear later.) More sophisticated experiments also bear out these experimental results.

The strength of the magnetic field between the two magnets in such a system depends on the width of the gap: the narrower the gap, the greater the field. The largest magnets with a very narrow

gap can produce a flux density of about 0.1 T (100 mT).

With electromagnets (iron cores with current-carrying coils wound round them) the flux densities may reach values of up to 20 T in research laboratories. To achieve these values large currents must be used and either the materials must be superconducting or sophisticated water cooling processes must be used to get rid of the internal energy produced in the coils.

By contrast the flux density of the Earth's magnetic field is about 50 μT. In Great Britain the field is inclined downwards at about 70° to the horizontal, which provides a horizontal component of about 17 μT and a vertical component of about 47 μT.

EXAMPLE

Calculate the magnetic flux density in the gap between a pair of ceramic magnets on an iron yoke placed as shown in figure 17.7, given that the length l of wire between the magnets is 40 mm, the current is 3.2 A, and the balance reading is 0.43 g. Take $g = 9.8\,N\,kg^{-1}$.

The force on the balance is

$$(0.43 \times 10^{-3}\,kg)\left(9.8\,\frac{N}{kg}\right)$$

$$= 4.2 \times 10^{-3}\,N.$$

So
$$B = \frac{F}{Il} = \frac{4.2 \times 10^{-3}\,N}{(3.2\,A)(40 \times 10^{-3}\,m)}$$

$$= 0.033\,T = 33\,mT.$$

Wires at an angle to the field

In the experiment described above you would find that the force depends on the angle which the wire makes with the field. Figure 17.8 shows how to calculate the force in this situation. You need to find the resolved part of the flux density at right angles to the wire. We shall call this $B_\perp$. In the diagram you can see that $B_\perp$ is equal to $B\sin\theta$, but you should not try to remember this, as you will also have to remember which is the angle θ in the diagram. It is best to work it out from first principles each time.

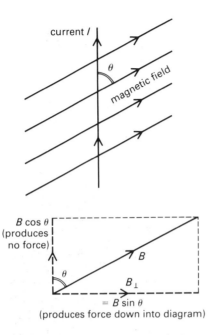

FIGURE 17.8

A power cable crosses a gap of 70 m between two pylons and carries a current of 2300 A in a direction making an angle of 73° with north. If the horizontal component of the Earth's magnetic field is 17 μT, and assuming that the wire is horizontal, calculate the vertical magnetic force on the wire.

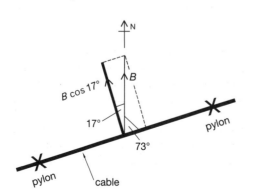

The flux density makes an angle of 73° with the current; its resolved part $B_\perp$ is $B\cos(17°)$

So $F = B_\perp Il$
$$= (17 \times 10^{-6}\,\text{T})(\cos 17°)(2300\,\text{A})(70\,\text{m})$$
$$= 2.6\,\text{N}.$$

The current in an actual power cable is alternating, so the direction of the force is alternately up and down. When the current is flowing (roughly) east, the left-hand motor rule shows that the force is upwards; when the current is flowing (roughly) west, the force is downwards. There is also a horizontal force on the wire, caused by the vertical component of the Earth's flux density.

The torque on a coil

If a rectangular coil, carrying a current, is placed in a uniform magnetic field, as shown in figure 17.9(a), there will be magnetic forces on the wires which form the sides of the coil. In the diagram the vertical forces (on the horizontal sides) balance each other, but the horizontal forces (on the vertical sides) are not acting in the same line, and therefore form a couple, which tends to turn the coil until its plane is at right angles to the field. You can also explain what happens by thinking of the coil as a flat magnet with poles on its faces: the couple on the coil then turns it so that the coil is aligned with the field, just as a freely-suspended bar magnet would be.

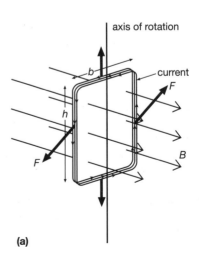

(a)

222

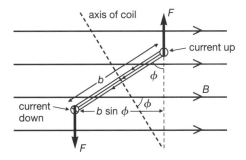

axis of coil

F

current up

b

φ

φ

B

current
down

b sin φ

φ

F

(b)   view looking down axis of rotation

FIGURE 17.9

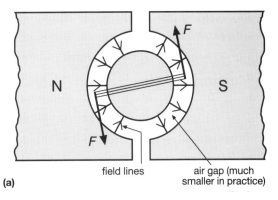

F

N

S

F

field lines

air gap (much
smaller in practice)

(a)

With the quantities shown on figure 17.9(b), and if the coil has N turns of wire, each force $F = N(BIh)$, and the distance apart of the forces F is $b \sin \phi$, so the torque T of the couple is

$$T = Fb \sin \phi = NBIhb \sin \phi = BANI \sin \phi$$

where $A \, (= bh)$ is the area of the coil.

The size of the torque therefore varies as the coil rotates: it has its maximum value ($BANI$) when $\phi = 90°$, i.e. when the plane of the coil is parallel to the field; and its minimum value, zero, when $\phi = 0°$, i.e. when the plane of the coil is at right angles to the field.

It is possible to arrange for the torque not to depend on the angle, but to have a value which depends on only B, A, N and I. This is achieved by arranging for the coil to move in a radial field, which can be created as shown in figure 17.10(a). Then the magnetic forces are always at right angles to the plane of the coil, and the couple always has the value $BANI$.

This is used in the **moving-coil meter**, as shown in the photograph of figure 17.10(b). The current to be measured passes through the coil, and the magnetic forces exert a couple on the coil. A spiral spring, designed to resist rotational motion, brings the coil to rest at an angle θ which is proportional to the size of the current. (It is rather like the spring in a spring balance, which extends until the pull of the spring is equal to the pull of the mass being hung from it.) This means that the meter will have *a linear scale*, which is obviously convenient.

This principle of arranging for there to be a torque on a coil is also used in some kinds of electric **motor**, but this will not be discussed until the next chapter.

(b)

FIGURE 17.10

17.3 The force on a moving charge

We say that there is a magnetic force on a wire carrying an electric current, but in fact the force is the total of all the forces acting on the individual electrons which are moving in the wire and providing the current. Figure 17.11 shows part of a conductor, of length l and cross-sectional area A, which is carrying a current I. The wire is at right angles to a magnetic field of flux density B. The number of charge carriers per unit volume is n: suppose they each carry a charge Q and have a speed v. The force F on the wire is given by

$$F = BIl$$

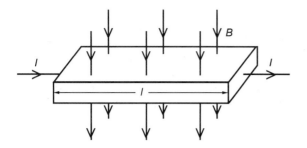

FIGURE 17.11

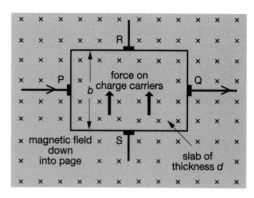

FIGURE 17.12

and since we know that $I = nAvQ$ we can write

$$F = B(nAvQ)l.$$

But for the length of conductor we are considering the volume is Al, and therefore the number of charge carriers is nAl, so that the force on each charge carrier is given by

$$F = BQv.$$

If the velocity v of the charge carriers is not at right angles to the magnetic field, the field has to be resolved so that we have the resolved part $B_\perp$ at right angles to v: then the force F is given by $F = B_\perp Qv$. The direction of the force is given by the left-hand motor rule if the charge carriers are positively charged, but if (as is often the case) they are electrons (with a negative charge) the current is in the opposite direction to the velocity of the charge carriers, and that must be taken into account.

The Hall effect

The sideways force on the charge carriers in figure 17.11 causes a sideways push on the conductor: it also means that inside the conductor the charge carriers will be pushed to one side, as shown in figure 17.12. With the direction of current and field shown, the left-hand motor rule shows that the magnetic force on the charge carriers is from S to R. (The direction of this force depends on the direction of the *current*: it does not matter what sign the charge carriers have.) The charged particles therefore drift from S to R, and create the p.d. V_H between R and S. If the charge carriers are positive, R will have a higher potential than S; if negative, R will have a lower potential than S. This **Hall effect** therefore allows us to discover the

sign of the charge carriers. For example, there are two types of silicon used in transistors: p-type silicon has mostly positive charge carriers, whereas n-type has mostly negative charge carriers and the two materials can be distinguished by the sign of the **Hall p.d.**

But the effect is quantitative too. The electric field E which the p.d. V_H creates is given by

$$E = \frac{V_H}{b} \quad \text{so} \quad V_H = Eb$$

where b is the width of the slab. The sideways drift of the charge carriers stops when the electric field has grown to such an extent that it exerts an electric force F_E on each charge carrier which is the same size as (but in the opposite direction to) the magnetic force F_B on the charge carrier. Then

$$F_E = F_B,$$

so that

$$Ee = Bev \quad \Rightarrow \quad E = Bv \quad \Rightarrow \quad V_H = Bvb$$

We also have $I = Avne$, where $A(= bd)$ is the cross-sectional area of the conductor, so

$$v = \frac{I}{bdne} \quad \Rightarrow \quad V_H = Bb\left(\frac{I}{bdne}\right) = \frac{BI}{dne}$$

You can see that V_H will be large if B and I are large, as you would have expected, but also that V_H is large if n is small, so you can expect there to be a large Hall p.d. for semi-conducting materials in which the number of charge carriers per unit volume is small.

Hall probes

The equation $V_H = BI/dne$ predicts that for a given piece of semi-conducting material, with a given current passing through it, the Hall p.d. is proportional to the magnetic flux density B. This result is used in the **Hall probe**, in which a piece of germanium (about 1 mm square) is mounted at the end of a plastic rod. A steady current is passed through the germanium, and when the slice is placed at right angles to a magnetic field, the p.d. generated is measured. Since, as we have just seen, the p.d. is directly proportional to the flux density, the strengths of two magnetic fields can be compared by comparing the Hall p.d.s which they produce. A **magnetic flux density meter** (shown in figure 17.13) is a device which gives direct readings of magnetic flux density. It consists of a Hall probe connected through an amplifier to a voltmeter: the amplification can be varied and the device is calibrated so that measurements of the B-field can be read directly from the voltmeter. Typical full-scale deflections are for flux densities of 10 mT, 50 mT, 100 mT and 1 T.

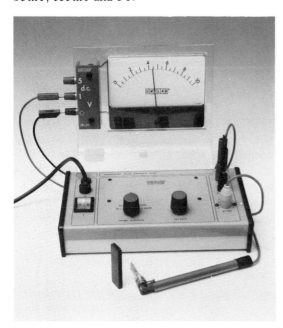

FIGURE 17.13

Electron beams

It is possible to produce streams of electrons (and other electrically charged particles) moving

through a vacuum at a high speed and this is usually the situation in which we use the equation $F = BQv$. Of course there is no magnetic force if the direction of travel of the charged particles is the same as, or opposite to, that of the field.

We shall be concerned only with the situation where the velocity is at right angles to the field.

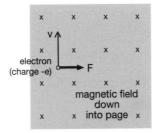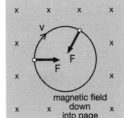

FIGURE 17.14

This is the situation shown in figure 17.14. Then, if the particle is an electron, the force will be in the direction shown. The force is at right angles to the velocity of the electron, so it does no work on it. The electron's speed therefore does not change (and so the force, given by $F = BQv$, also remains the same size). But the force *will* give the electron an acceleration which will change the *direction* of the electron's velocity. This acceleration will always be the same size, so the rate of change of velocity must be constant. This means that the rate of change of direction must be constant, and the electron must move in a *circular* path. So the magnetic field provides a force which gives the electron its centripetal acceleration, and we can write

$$\frac{mv^2}{r} = BQv$$

The equation can be rewritten as

$$r = \frac{mv}{BQ}$$

which shows how the radius of the circle depends on the magnetic flux density B, the electric charge Q and the momentum mv of the particle.

EXAMPLE

In a nuclear physics laboratory a beam of protons is travelling vertically downwards and the experimenters wish to turn them through an angle of 90° so that they move horizontally (and can strike a target in an experiment). If the momentum of each proton is 8.7×10^{-20} N s, the charge on the proton is 1.6×10^{-19} C and the maximum available magnetic flux density is 1.2 T, what is the radius of the circular arc in which the protons move as they turn through 90°?

Using $\quad r = \dfrac{mv}{Bq}$

$\Rightarrow \qquad r = \dfrac{8.7 \times 10^{-20}\,\text{N s}}{(1.2\,\text{T})(1.6 \times 10^{-19}\,\text{C})} = 0.46\,\text{m}.$

Specific charge

Electrolysis experiments allow us to measure the *average* charge on an electron but it is not easy to measure the charge on a single electron, or its mass. The first experiments were able to measure only the **specific charge** of an electron, that is the ratio

$$\frac{\text{charge of an electron}}{\text{mass of an electron}}$$

which is given the symbol e/m. You can make this measurement in a school or college laboratory using modern apparatus of the kind shown in figure 17.15(a). A heated cathode (at the left of the photograph) emits electrons, which are accelerated to the anode by a p.d. of a few kV. They then continue at a constant speed into a region where they can be subjected to magnetic and/or electric fields. A nearly uniform horizontal *magnetic* field is provided by two vertical coils, carrying a steady current, and a nearly uniform vertical *electric* field is provided by two horizontal metal plates. Both the magnetic force and the electric force on the electrons will then be vertical, and it can be arranged that these two forces are in opposite directions, as shown in figures 17.15(b,c,d).

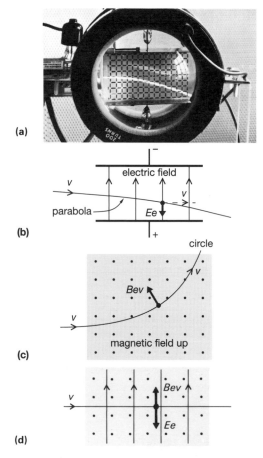

(a)

(b)

(c)

(d)

FIGURE 17.15

The magnetic flux density can be measured by removing the tube and placing a Hall probe at the centre of the space through which the beam was moving when undeflected. The strength E of the electric field can be found by measuring the p.d. V between the metal plates, and their separation d: then $E = V/d$.

The fields are adjusted until the magnetic force F_B and the electric force F_E are the same size, as shown by the electron beam being undeflected. Then $F_B = F_E$, so $Bev = Ee$, and

$$v = \frac{E}{B}$$

We can also say that the kinetic energy of an electron, $\frac{1}{2}mv^2$, is equal to the electric p.e. it lost in passing through the p.d. V_a between the cathode and the anode. Electric p.e. $W = QV$, so here the electric p.e. lost $= eV_a$, so

$$eV_a = \tfrac{1}{2}mv^2$$

Eliminating v between these two equations we have

$$eV_a = \tfrac{1}{2}m\left(\frac{E}{B}\right)^2$$

which gives

$$\frac{e}{m} = \frac{E^2}{2B^2V_a}$$

but it is more important to understand the two equations $Ee = Bev$ and $eV_a = \tfrac{1}{2}mv^2$ than to remember this result.

EXAMPLE

In an experiment to measure the specific charge e/m, the following measurements were made. With the beam undeflected, the magnetic flux density was 1.9 mT, the p.d. between the plates was 3.6 kV and they were 45 mm apart. The accelerating p.d. V_a was 5.0 kV. Calculate the value of e/m.

The electric field strength E between the plates is given by

$$E = \frac{V}{d} = \frac{3.6\times10^3\,\text{V}}{45\times10^{-3}\,\text{m}} = 8.0\times10^4\,\text{V m}^{-1}$$

For the undeflected beam $Bev = Ee$, so

$$v = \frac{E}{B} = \frac{8.0\times10^4\,\text{V m}^{-1}}{1.9\times10^{-3}\,\text{T}} = 4.2\times10^7\,\text{m s}^{-1}$$

For the accelerated electrons $eV_a = \tfrac{1}{2}mv^2$ so

$$\frac{e}{m} = \frac{v^2}{2V_a} = \frac{(4.2\times10^7\,\text{m s}^{-1})^2}{2(5.0\times10^3\,\text{V})}$$

$$= 1.8\times10^{11}\,\text{C kg}^{-1}.$$

The accepted value for e/m is $1.759\times10^{11}\,\text{C kg}^{-1}$. Notice the high speed which the electrons have after being accelerated through a p.d. of as little as 5 kV: it is about one-seventh of the speed of light. In the experiment the electrons are all deflected by the same amount: this shows that they all have the same speed and that they all have the same specific charge, i.e. they are all identical. Experiments (in other kinds of tube) with cathodes made from other metals, give the same value for e/m, which suggests that *all* electrons are the same, and that they are a universal constituent of all matter.

Measurements which enable us to find the value of e/m can also be made in another type of tube, called a **fine-beam tube**. In these tubes there is gas (usually helium) at a very low pressure. The electron beam starts *inside* the magnetic field and the strength of the field is adjusted so that the beam of electrons makes a complete circle, as shown in figure 17.16. In this type of tube the beam of electrons is visible because the electrons excite the helium atoms, which then emit a green light. This provides a very striking demonstration of the fact that charged particles, moving at right angles to a magnetic field, move in circular paths. To find the value of e/m we would need to use the equations $r = mv/Be$ and $\tfrac{1}{2}mv^2 = eV_a$ and measure the values of B, r and V_a.

FIGURE 17.16

Much of what we know about the structure of the atom and the particles of which it is made has come about through firing very fast moving particles at the nuclei of atoms. Most of these particles are charged, since, if they are charged, they can be accelerated using electric fields. One machine which can be used to accelerate charged particles is the Van de Graaff generator. You have probably seen a small version in your own laboratory, but they can be so large that they occupy several floors of a building. There are difficulties in making them able to produce p.d.s of more then 10 MV, so to accelerate charged particles through larger p.d.s an alternative is needed. In a **cyclotron** charged particles move in circular paths of ever-increasing radius and are accelerated many times by the *same* potential difference. The details need not concern us, but the construction of such machines is made much easier by the fact that the charged particles take the same time to complete one revolution, whatever their speed. You can see that this is true by looking again at the equation $r = mv/Be$. The time T for one revolution is given by

$$T = \frac{2\pi r}{v} \quad \text{and we know that} \quad \frac{r}{v} = \frac{m}{Be}$$

so $\quad T = \dfrac{2\pi m}{Be}$

and each of the terms on the right-hand side of the equation is constant, so T is constant. The physical reason for this is that as the speed increases, the radius increases, in direct proportion. The *angular velocity* $\omega(= v/r = Be/m)$ of the particles is constant.

17.4 Magnetic flux densities around coils and wires

If you move a Hall probe into a solenoid carrying a steady current you will find that the magnetic flux density increases at first and then reaches a steady value which remains constant until you approach the far end. Also, the flux density has a constant value over the whole cross-section of the solenoid. A solenoid therefore provides a way of producing a uniform magnetic field. A graph of how the flux density varies with distance along the axis is shown in figure 17.17: note that the flux density at each end of the solenoid is exactly half its value at the centre. The flux density outside the solenoid is

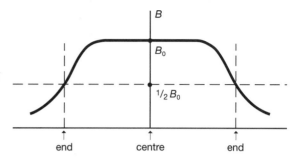

FIGURE 17.17

very small, except near the ends. The extent of the region in which the flux density is constant depends on the length and cross-sectional area of the solenoid: the longer it is, and the narrower it is, the larger the region in which the field is uniform. An 'ideal' solenoid would be infinitely long.

You could use a Hall probe to show that the flux density B near the centre of the solenoid is proportional to the current I in it, and proportional to how densely the coils of wire are wound on it, i.e. to the number of turns *per unit length*, n. The flux density also depends on the material inside and outside the solenoid. If the space is empty of all material we can write

$$B = \mu_0 nI$$

where μ_0 is the **permeability of free space** or **magnetic constant**. It is a constant which governs the size of magnetic forces, just as G governs the size of gravitational forces. In practice it makes very little difference whether the space is evacuated or filled with air, so for air-filled solenoids we use the equation as stated.

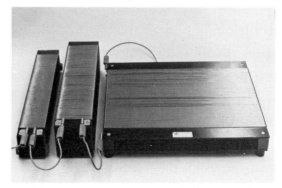

FIGURE 17.18

The equation makes no mention of the cross-sectional area of the solenoid. *The area does not matter, and nor does the shape.* Figure 17.18 shows three solenoids with the same number of turns per unit length, connected in series (so that they inevitably have the same current). In each of them the flux density is the same.

EXAMPLE

A rectangular wire frame, pivoted at its centre, as shown in the figure, is placed so that the edge AB is inside a flat solenoid like the one shown in figure 17.18. When a current passes through the solenoid the edge AB is pushed down, but the wire is restored to equilibrium in its original position by loading the opposite edge CD with a small piece of paper of mass 0.26 g. If the length of AB is 15 cm, the current in the solenoid is 5.4 A, the current in the frame is 2.2 A, there were 120 turns in a 10 cm length of the solenoid, and $g = 9.8 \, \text{N kg}^{-1}$, calculate the value of μ_0.

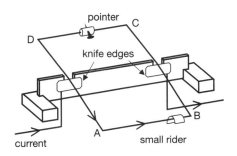

The weight of the paper

$$= (0.26 \times 10^{-3} \, \text{kg}) \left(9.8 \, \frac{\text{N}}{\text{kg}} \right)$$

$$= 2.5 \times 10^{-3} \, \text{N}$$

so the magnetic force $= 2.5 \times 10^{-3} \, \text{N}$

Using $F = BIl$, where B is the flux density in the solenoid, I is the current in the wire, and l is the length of the wire,

$$B = \frac{F}{Il} = \frac{2.5 \times 10^{-3} \, \text{N}}{(2.2 \, \text{A})(0.15 \, \text{m})} = 7.7 \times 10^{-3} \, \text{T}$$

For the solenoid, $B = \mu_0 n I$ where $n = 1200$ turns m^{-1}

so

$$\mu_0 = \frac{B}{nI} = \frac{7.7 \times 10^{-3} \, \text{T}}{(1200 \, \text{m}^{-1})(5.4 \, \text{A})}$$

$$= 1.2 \times 10^{-6} \, \text{N A}^{-2}.$$

As we shall see later, the correct value of μ_0 is specified by the definition of the ampere and is exactly $4\pi \times 10^{-7} \, \text{N A}^{-2}$, which is $1.26 \times 10^{-6} \, \text{N A}^{-2}$.

A flat circular coil

You could use a Hall probe to investigate the magnetic flux density at the centre of a flat circular coil. The field there is at right angles to the plane of the coil. As you would expect, you will find that the flux density is proportional to the current I in the coil, and its number of turns N. If you examine a coil of different radius, you will find that if you halve the radius, the field doubles (for the same values of I and N): the flux density is inversely proportional to the radius. So

$$B \propto \frac{NI}{r}$$

and it can be shown that $B = \mu_0 NI / 2r$.

A straight wire

The field lines run concentrically round a straight wire, and you could use a Hall probe to check this. At a particular distance from the wire, you will detect the maximum field if you hold the probe with its plane parallel to the wire, i.e. in the plane of the paper in figure 17.19. Again, at a particular distance from the wire, you will find that the flux density B is proportional to the current I in the wire. If you double the distance of the probe from the wire you will find that the flux density halves so for a particular current the flux density is inversely proportional to the distance r from the wire. So

$$B \propto \frac{I}{r}$$

and it can be shown that $B = \mu_0 I / 2\pi r$.

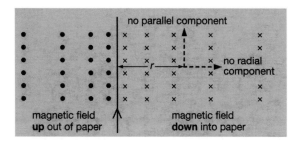

FIGURE 17.19

Coaxial cables

You would detect very little magnetic field near a mains cable, since there are two parallel conductors, very close together, carrying currents in opposite directions, and the resultant magnetic field is very nearly zero. But around the outside of a **coaxial** cable (e.g. the cable used to carry the signal from a TV aerial to the TV set), which consists of a central insulated wire (carrying a current in one direction) surrounded by a braided sheath which carries the return current there is *no* magnetic field. Nor is the central wire affected by magnetic fields from outside the wire.

The magnetic constant μ_0

Suppose two straight, parallel infinite wires carry currents in the same direction, as shown in figure 17.20. Then the left-hand wire produces a magnetic field as shown: up on the left, and down on the right. You can use the left-hand motor rule to show that the left-hand wire therefore attracts the right-hand wire. By approaching the situation from the point of view of the right-hand wire, or by using Newton's third law, you can show that the right-hand wire attracts the left-hand wire. If one of the currents was in the opposite direction to the other, the wires would repel each other.

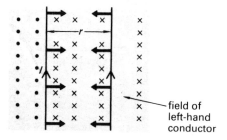

field of
left-hand
conductor

FIGURE 17.20

If the wires carry a current I and are a distance r apart, the right-hand wire is in the field B produced by the left-hand wire: $B = \mu_0 I/2\pi r$. The magnetic force on a length l of this wire is given by

$$F = BIl = \left(\frac{\mu_0 I}{2\pi r}\right)Il = \frac{\mu_0 I^2 l}{2\pi r}.$$

The **definition of the ampere** uses this expression for the force with which the wires attract each other:

one ampere is the current which, if flowing in each of two infinitely long parallel wires, causes one of them to exert a force of 2×10^{-7} N on a one-metre length of the other.

This implies a particular value for μ_0. If you substitute $I = 1$ A, $l = 1$ m, $r = 1$ m and $F = 2 \times 10^{-7}$ N in the equation you will find that $\mu_0 = 4\pi \times 10^{-7}$ N A^{-2}.

17.5 Magnetic materials

You know that there is a magnetic field around a coil which carries an electric current, so you will not be surprised to find that an atom, which has electrons in orbit round the nucleus, also produces a magnetic field. In addition the electrons have spin, which also causes magnetic effects. The overall effect is such that we can think of the atoms as being tiny 'atomic' magnets, which we call magnetic **dipoles**. For most substances, however, no effect is observed unless they are placed in a magnetic field: either the magnetic effects of a *single* atom cancel out, or the effects cancel out when there are *many* atoms. When placed in a magnetic field, there is some magnetic effect for *all* substances: the magnetic flux density is slightly increased or decreased. However, for a few substances (iron, nickel, cobalt, two other elements, and many alloys) the magnetic flux density may be increased by a factor of anything up to 10^5: these are called **ferromagnetic** materials. For these materials there may be an increase in magnetic flux density (depending on their previous history) even without an external magnetic field being applied. This very large increase in magnetic flux density comes about because in these materials it is possible for neighbouring atoms to affect each other in such a way that the magnetic effects are *aligned* in regions called **domains**. The sizes of domains vary widely, from about 0.01 mm to about 1 mm, so that they may contain anything from 10^{15} to 10^{21} atoms.

Domains

Figure 17.21 shows a photograph of the surface of a single crystal which contains some large domains.

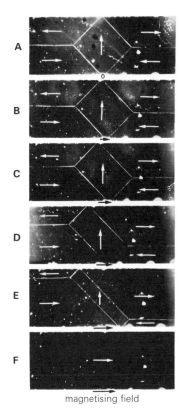

magnetising field

FIGURE 17.21

The domain boundaries are shown up by covering the surface with a liquid which carries a suspension of fine magnetic particles. These particles concentrate along the boundaries of the domains (rather as iron filings do at the boundary between two touching bar magnets). The arrows superimposed on the photograph show the direction of magnetisation of each domain. Under each photograph an arrow shows the size and

direction of the applied magnetic field.

In a freshly-formed crystal the magnetic fields of the domains cancel out, and almost no field can be detected. We say that the specimen is *unmagnetised*, although at the atomic level there is considerable *local* magnetisation. If it is now placed in a weak magnetic field the result is a general shifting of the domain boundaries in such a way that the proportion of atomic dipoles lined up with the field increases, as shown in the successive photographs in figure 17.21. The specimen now produces a magnetic field which is additional to the field which is magnetising it. To some extent this magnetisation is reversible: when the magnetising field is removed, the boundaries return more or less to their original position and the magnetisation disappears. However, in larger fields the axes of whole domains rotate abruptly until all the atomic dipoles are aligned with the field. No further magnetisation is then possible. These changes are usually not reversible, so the material retains its magnetisation when the field is removed, and so we are left with a **magnet**.

Thermal agitation tends to disrupt the alignment of the dipoles within the domains so materials cease to be ferromagnetic at high temperatures. The temperature at which this happens is called the **Curie temperature**: for iron it is 1040 K. Above this temperature ferromagnetic materials are no more magnetic than any other.

Ferromagnetic materials

Some consequences of the existence of domains are listed.

- Once the domain boundaries have been shifted so that all the atoms are aligned with the field, no further increase in magnetisation is possible, and the material is said to be **saturated**. We cannot indefinitely increase the magnetic flux density in a piece of magnetic material.

- If the magnetising field is reduced to zero the magnetic material retains its magnetism to some extent. To reduce the magnetisation to zero, a magnetic field in the opposite direction must be applied. If only a small field is required, the material is said to be a **soft** magnetic material; if a large field is required, the material is said to be magnetically **hard**. A soft material is used for electromagnets (in which most of the magnetisation must be lost when the current is switched off) or for transformer cores (in which the magnetisation must frequently be reversed). A hard material is used to make permanent magnets. Most magnetic materials are alloys specially developed for their purpose: their commercial names often tell us some of their constituents, e.g. **Alnico** (used for permanent magnets) contains iron, aluminium, nickel and cobalt). Not all magnetic materials are metallic, however: you will yourself have used ceramic magnets (commercial name **Magnadur**). These are **ferrites**, which are oxides of a number of materials (but always including iron).

◆ If a large magnetic field is required to change the magnetisation of a specimen, it usually also follows that a lot of energy is needed too. Just as the repeated loading and unloading of a rubber band will generate internal energy in it, so the repeated magnetisation and remagnetisation of a ferromagnetic material will also generate internal energy. In both cases we say there is **hysteresis**. For a typical soft material the energy generated is about 20 mJ per kg *in each cycle*. You can imagine that in a transformer operating at a frequency of 50 Hz this continual generation of internal energy will be wasteful and also make it necessary for some cooling system to be used.

Permeability

The effect of filling the core of a coil or solenoid with a ferromagnetic material is to greatly increase the flux density B for a given current. Consider a solenoid which has been bent into a ring shape so that the ends are joined together. This avoids the complication of there being ends at which the field is weaker, and is in any case similar to the real-life situation of the core of a transformer which has

wire wrapped round it. The magnetic flux density in such a solenoid is given by

$$B_0 = \mu_0 nI.$$

The effect of filling the ring with a magnetic material is to increase the magnetic flux density B. In practice the factor by which it is increased is not constant: it depends on the previous history of the specimen, and on the size of the current. We shall make the approximation that the factor *is* constant and denote it by μ_r, which is called its **relative permeability** of the material. Then

$$B = \mu_r B_0 = \mu_r \mu_0 nI$$

μ_r is a dimensionless quantity and therefore has no units. For a vacuum $\mu_r = 1$, by definition, and it is very little different from 1 for anything other than a ferromagnetic material. But for soft ferromagnetic materials its value is at least 1000.

In practice, when the core is filled with iron, it makes very little difference to the size of the magnetic flux density whether the coils are spread evenly or bunched up in one place. The effect of the iron is to concentrate the field lines within the core and to prevent them leaking through the air. The flux density is almost constant throughout the core.

EXAMPLE

A coil of 80 turns, carrying a current of 0.50 A, is wound on a ring-shaped iron core of average circumference 0.40 m. If the relative permeability μ_r of the iron is 2500, calculate the flux density in the iron, stating any assumptions made. Take $\mu_0 = 4\pi \times 10^{-7} \, \text{N A}^{-2}$.

The number of turns per unit length,

$$n = \frac{80}{0.40 \, \text{m}} = 200 \, \text{m}^{-1}$$

$$\Rightarrow B = \mu_r \mu_0 nI$$
$$= (2500)(4\pi \times 10^{-7} \, \text{N A}^{-2})(200 \, \text{m}^{-1})(0.50 \, \text{A})$$
$$= 0.31 \, \text{T}$$

The calculation assumes that the magnetic flux density increases in proportion to the current, i.e. that the iron does not become saturated.

Exercises on each section of this chapter may be found in the companion textbook, **Practice in Physics**.

SUMMARY

At the end of this chapter you should be able to:

◆ describe how to plot the magnetic field pattern around a magnet.

◆ draw the magnetic field pattern around a straight current-carrying wire and remember a rule for predicting the direction of the field lines and draw the magnetic field patterns around a coil and a solenoid.

◆ remember the rule for predicting the direction of the magnetic force on a current-carrying wire placed in a magnetic field.

◆ use the equation $F = BIl$, which defines magnetic flux density B.

◆ use the equation $F = BQv$.

◆ understand what is meant by the Hall effect.

◆ understand how an electron gun produces a beam of electrons.

◆ understand that a charged particle deflected by a magnetic field will move in a circular path.

◆ understand that a charged particle deflected by an electric field will move in a parabolic path.

◆ understand the principle of the measurement of the specific charge of the electron.

◆ use the equation $B = \mu_0 nI$ for a solenoid.

◆ use the equation $B = \mu_0 NI/2r$ for a coil, and $B = \mu_0 I/2\pi r$ for a straight wire.

◆ understand the principle of the definition of the ampere, and that it leads to an exact value for μ_0.

◆ understand, in terms of domains, how magnetic materials may be magnetised and why there is a limit to the magnetisation of a particular piece of magnetic material.

18 Electromagnetic Induction

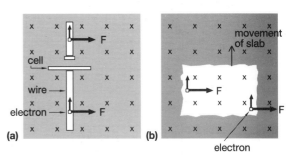

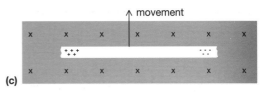

(c)

FIGURE 18.1

The photograph shows the blades of a steam turbine in a power station. Fuel is used to produce steam which strikes the blades which rotate the shaft and turn a dynamo which generates an alternating current. In the United Kingdom the dynamo rotates at 3000 r.p.m. (i.e. 50 revolutions per second) and this is why the mains supply alternates at a frequency of 50 Hz. These power stations, which deliver electrical energy to the grid system, are the source of most of the electrical energy we use. When you turn on a light, or switch on an electric kettle, or start a washing machine you are making use of energy provided by a dynamo. What is a dynamo and how does it work?

18.1 E.m.f.s induced by moving conductors

Figure 18.1(a) shows a wire which is part of a circuit. The cell is creating electric forces which are pushing electrons along the wire (up the page). The wire is in a magnetic field (into the paper) so

there is a force on the electrons which pushes them to the right. This is another example of the Hall effect. Now look at figure 18.1(b). It shows a slab of conducting material (whose atoms of course contain protons and electrons) being made to move up the page across a similar magnetic field. Again electrons are moving. Not this time because they are being pushed by electric forces from a cell, but because someone or something is pushing them mechanically. It does not matter whether the charges are pushed by electric or mechanical forces, the result is the same: just as for the wire, a magnetic force acts on these charged particles – because they are moving at right angles to a magnetic field.

The result is that negative charge will accumulate at one end of the conductor, leaving a surplus of positive charge at the other end. The *shape* of the conductor does not matter: for convenience we have drawn a wire in figure 18.1(c). So what you have seen is that a conductor moving across a magnetic field separates electric charge. In this way it acts just like an electric cell because what they both do is to cause a separation of electric charge. We therefore say that the moving conductor, like a cell, is a **source of e.m.f.** A cell, and a moving conductor, are drawn side by side in figure 18.2 so that you can see the similarity. If the cell, or the moving conductor, is connected to a complete circuit, a current flows.

You know that the energy provided by the cell comes from the chemicals contained within it.

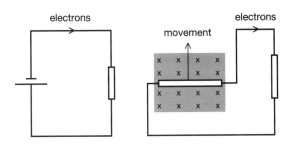

FIGURE 18.2

How is the energy provided in the circuit with the moving conductor? When the electrons are moving along the wire, there is a current in it, which is at right angles to the magnetic field. The left-hand motor rule shows that there is therefore a magnetic force pushing in the opposite direction to the way the wire is being pushed. Because there is this magnetic resisting force, work has to be done to push the wire: *whatever is pushing the wire* is providing the energy for the circuit. We should expect that, since the energy for the current has to be supplied from somewhere.

This leads to a very important conclusion: that whenever an e.m.f. is induced, its direction is such that if it were to produce a current, the effect would be *to oppose the change causing the e.m.f.* This is a fundamental law, since it is necessary if energy is to be conserved. It is known as **Lenz's law**. The generation of this sort of e.m.f. is called **electromagnetic induction**, and the e.m.f.s are called **induced e.m.f.s**.

Electromagnetic induction was discovered by Michael Faraday in a series of brilliant experiments which were difficult to design because the measuring instruments available to him were insensitive. However, you have probably already demonstrated them to yourself, or seen them

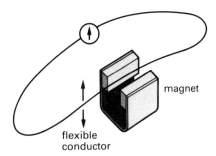

FIGURE 18.3

demonstrated, with simple apparatus like that shown in figure 18.3.

There is no effect if the conductor is held still: there is only an effect while the wire is 'cutting' the magnetic field lines (so nothing happens if the wire is moved parallel to the lines). The effects are the same if the conductor is held still and the magnet is moved: it is *relative* motion that matters. Figure 18.4(a) is a graph (obtained with a data logger) which shows the e.m.f. induced when a wire was pushed down quickly between the poles of a magnet, and figure 18.4(b) shows the result when the wire was pulled up, more slowly. You can see that the direction of the e.m.f. depends on the direction of movement, and that the e.m.f. is larger when the movement is faster. However, quantitative experiments are difficult to do, since the field of the magnets is not uniform, and any meter's movement is probably heavily damped.

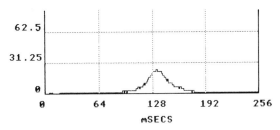

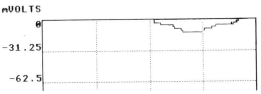

FIGURE 18.4

Dynamos

A machine which is designed to generate a direct e.m.f. (i.e. in one direction only) is called a **dynamo**. If you took the apparatus of figure 18.3 again and doubled up the wire as shown in figure 18.5(a) and rotated this loop about the axis shown, there would be an e.m.f. $\mathscr{E}$ induced in each wire, and the two e.m.f.s would add to give an e.m.f. of $2\mathscr{E}$. If you made more loops of the wire, as shown in figure 18.5(b), you would get much larger

shows how the e.m.f. would vary with time. In practice, therefore, there would be several coils, wound at an angle to each other, as shown in figure 18.6(a), so that the total induced e.m.f. had a steadier value. Figure 18.6(b) shows the separate e.m.f.s produced by two coils placed at right angles to each other, and (c) shows the sum of these e.m.f.s. In a commercially-produced dynamo the magnet which provides the magnetic field might be an electromagnet. The design and construction of dynamos is a highly technical matter and we need pursue it no further.

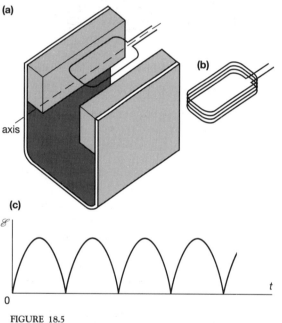

(a)

axis

(b)

(c)

FIGURE 18.5

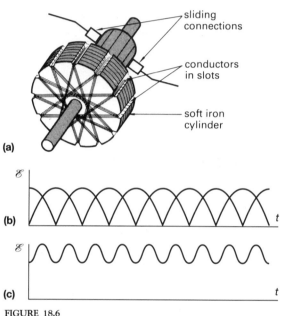

sliding connections

conductors in slots

soft iron cylinder

(a)

(b)

(c)

FIGURE 18.6

e.m.f.s. This is the basis of a dynamo: not only is the e.m.f. going to be much larger than would be possible with a single wire, but it is easier to rotate the loop continuously than to make it move up and down. In practice the ends of the wires are brought out to a device called a **split-ring commutator** in which, through sliding contacts, the rotating coil is connected to the external circuit in such a way that the current in that circuit flows only in one direction. In this simple arrangement the induced e.m.f. would vary considerably in size, since when the coil had rotated through 90° from the position shown in figure 18.5(b), it would be (momentarily) not cutting any lines and there would at that moment be no induced e.m.f. The graph in figure 18.5(c)

Alternators

The principle of this is very similar to that of the dynamo, except that the connection to the outside circuit is made through **slip rings**, which preserve the alternating nature of the current. Figure 18.7 shows photographs of two oscilloscope traces obtained by rotating a single-turn coil of wire between the poles of a U-magnet: in (b) the speed was higher than it was in (a). You can see that the size as well as the frequency of the induced e.m.f. are greater in (b).

In practice, however, especially in large machines, an electromagnet is used to provide the magnetic field, and because the current in this is

236

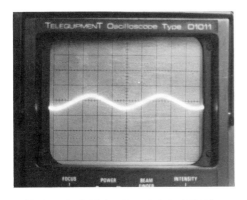

FIGURE 18.8

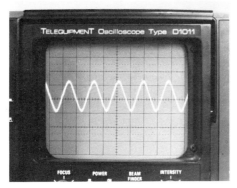

FIGURE 18.7

much smaller than the current flowing from the coils, it is usually arranged so that the coils should be stationary (and form what is called the **stator**) while the part providing the magnetic field rotates (and is called the **rotor**). This simplifies the task of making the sliding contacts: with this arrangement, they have to carry relatively small currents. The alternator is what is used in power stations to generate electrical energy on a vast scale. These alternators have a stator with three pairs of coils, equally spaced, so that three separate e.m.f.s are generated, each 120° out of phase with the next. In this way a power station generates a **three-phase** supply. Steam turbines drive these alternators at a rate of 3000 r.p.m., which means that they provide an alternating e.m.f. at a frequency of 50 Hz. At the other end of the scale you will, if you ride a bicycle, have almost certainly used an alternator in a bicycle lamp 'dynamo', as shown in figure 18.8: in this, the rotating magnet is a permanent magnet (thus needing no wire connections) and is driven by friction from the moving rim of the bicycle wheel.

What does the size of $\mathscr{E}$ depend on?

A dynamo provides us with a way of investigating quantitatively some of the factors on which the size of $\mathscr{E}$ depends. If you connect a voltmeter to its output, and drive the dynamo at various constant rates of rotation, you would find that the e.m.f. $\mathscr{E}$ is proportional to the rate of rotation, i.e. proportional to the speed v at which the wires in the coil are cutting field lines. If you vary the current to the field coils (but keep the current small enough to prevent magnetic saturation), you will find that the e.m.f. is proportional to the current and hence to the field: i.e. $\mathscr{E} \propto B$. To find out how the induced e.m.f. depends on the length of the wire we have to go back to the simple apparatus shown in figure 18.3, so that there is twice the length of wire in the magnetic field. If the magnets produce the same magnetic field, you would find that the induced e.m.f. is proportional to the length l. So now you would know that the induced e.m.f. $\mathscr{E}$ was directly proportional to B, l and v, i.e. $\mathscr{E} \propto Blv$. It turns out that $\mathscr{E}$ is actually *equal* to the product Blv, as we shall now show.

Showing that $\mathscr{E} = Blv$

Consider again a wire of length l at right angles to a magnetic field of flux density B and moving across it at a steady speed v (as in figure 18.9). When the conductor is first made to move, the magnetic forces push the electrons to the right. This continues until the electric field created by these electrons is large enough to provide an

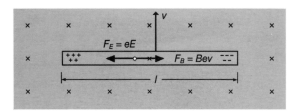

FIGURE 18.9

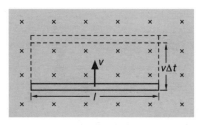

FIGURE 18.10

electric force on each electron which balances the magnetic force. Again, it is just another example of the Hall effect. The conductor is not connected to a circuit, so the p.d. across it is equal to the e.m.f. $\mathscr{E}$.

The electric field $\mathscr{E}$ in the wire is equal to the potential gradient, i.e.

$$E = \frac{\mathscr{E}}{l} \text{ (do not confuse field } E \text{ and e.m.f. } \mathscr{E})$$

so the electric force F_E on each electron is

$$F_E = eE = e\frac{\mathscr{E}}{l}$$

The magnetic force F_B on each electron is
$$F_B = Bev$$

so
$$e\frac{\mathscr{E}}{l} = Bev \implies \mathscr{E} = Blv.$$

If the magnetic field is not at right angles to the plane in which the wire is moving, we need to find the component $B_\perp$ at right angles to the plane, so generally $\mathscr{E} = B_\perp lv$.

In a time Δt the wire moves a distance $v(\Delta t)$, so the area ΔA swept out by the wire is, as shown in figure 18.10, given by

$$\Delta A = lv(\Delta t)$$

$$\implies \qquad \frac{\Delta A}{\Delta t} = lv$$

and we can write
$$\mathscr{E} = B_\perp \frac{dA}{dt}.$$

This equation gives the e.m.f. as the product of $B_\perp$ and the rate of sweeping out area dA/dt. This makes it possible to calculate $\mathscr{E}$ in other situations; for example, where a conductor like a disc is rotating.

EXAMPLE

You start to do an investigation to find out the size of the e.m.f. induced when a wire is pushed down between the poles of a large U-magnet in your laboratory. If the magnetic field between the poles of the magnet is about 0.1 T, and the length of the gap is about 30 mm, how sensitive a voltmeter would you need?
Estimate: maximum speed of pushing wire $= 2\,\mathrm{m\,s^{-1}}$.

using $\mathscr{E} = Blv$, we have
$$\mathscr{E} = Blv = (0.1\,\mathrm{T})(30 \times 10^{-3}\,\mathrm{m})(2\,\mathrm{m\,s^{-1}})$$
$$= 0.006\,\mathrm{V} = 6\,\mathrm{mV}$$

so a 10 mV voltmeter would do.

However the damping of the meter's coil would make it unlikely that the meter would read the full 6 mV before the e.m.f. dropped to zero again when the wire stopped.

You should check for yourself that the unit $T\,m^2\,s^{-1}$, which appears in the calculation, is the same as the volt.

A rotating coil

Figure 18.7 shows that the e.m.f. induced in a rotating coil varies sinusoidally with time. The peak value of the e.m.f. can be calculated as follows. The peak e.m.f. occurs at the moment when the plane of the coil is parallel to the field, as shown in (a), because at these moments the two sides BC and DA are moving at right-angles to the field. If these sides are of length l, the peak e.m.f. $\mathscr{E}$ induced by one of these conductors is given by $\mathscr{E} = Blv$ where B is the magnetic flux density and v is their speed. If the breadth of the coil is b, the sides BC and DA are moving in circular paths of diameter b, so that

$$v = \pi b n$$

where n is the number of revolutions per unit time. If the coil has N turns, the numbers of conductors like BC and DA is $2N$, so the total peak e.m.f. $\mathscr{E}_0$ is the sum of all the e.m.f.s, i.e.

$$\mathscr{E}_0 = 2NBlv$$

$$= 2NBl(\pi b n)$$

$$= 2\pi n BAN$$

where $A \, (= lb)$ is the area of the coil. The frequency f of the e.m.f. is equal to the rate n of turning of the coil. It can be shown that this result is true for a coil of area A of any shape.

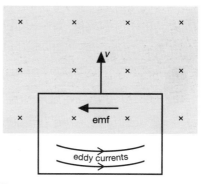

FIGURE 18.11

EXAMPLE

Calculate the peak e.m.f. induced in a circular coil of 100 turns of radius 0.10 m which is rotating at 50 revolutions per second about a diameter which is at right angles to a magnetic field of flux density 0.20 T.

$$\mathscr{E}_0 = 2\pi n BAN$$

$$= 2\pi(50\,s^{-1})(0.20\,T)\,(\pi)(0.10\,m)^2(100)$$

$$= 200\,V.$$

The frequency of the e.m.f. will be 50 Hz, since the coil rotates at 50 revolutions per second.

Eddy currents

We began this chapter by thinking about a slab of metal being pushed through a magnetic field, as shown in figure 18.11. Now suppose the slab is just starting to enter the magnetic field. An e.m.f. is induced in the shaded part which is already in the field, but not in the (unshaded) rest of the slab. But the unshaded part connects together the ends of the shaded part, so an induced current can flow, as shown. This sort of current, which flows, not in wires, but in the body of a conductor, is called an **eddy current**, since the current eddies round the

metal, rather like water running out of a sink. Its precise size and direction would be difficult to calculate, but we do know, from Lenz's law, that it must oppose the change, so in this case it will produce a magnetic field opposite to the original field. The e.m.f. may not be very large, but the resistance of the metal is low, so that large currents may pass.

The effect may be demonstrated by swinging a pendulum, whose 'bob' is a thick copper plate, between the poles of an electromagnet, as shown in figure 18.12. The plate swings freely when there is no current in the electromagnet coils. When the electromagnet is switched on, large eddy currents are induced in the plate, and the plate is quickly brought to a halt. The kinetic energy of the plate is converted, by the eddy currents, into internal energy in the plate. If a plate with slots cut in it is used instead, the braking effect is much reduced, because the eddy currents can now flow only inside the relatively narrow teeth left by the slots.

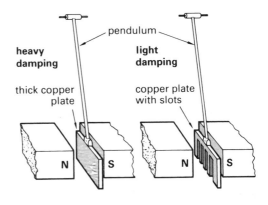

FIGURE 18.12

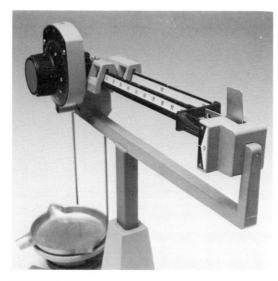

FIGURE 18.13

Figure 18.13 shows a balance which is damped by eddy currents. The aluminium vane at the end of the beam moves between two pieces of ceramic magnet which create a magnetic field. When the beam is loaded with weights it would oscillate for some time if its kinetic energy were not converted by the eddy currents into internal energy in the aluminium vane.

Eddy currents must be minimised in iron-cored apparatus (e.g. motors and dynamos) or else considerable losses of energy would occur. In these

cases the iron parts of the apparatus are built up from a stack of thin iron sheets, called **laminations**, as you can see in figure 18.14. The laminations are insulated from one another, sometimes by a thin layer of paper stuck on one side of each lamination, or by a coat of varnish.

18.2 Magnetic flux

We use the idea of magnetic field lines when we sketch the magnetic field patterns around magnets and wires. The convention is that there is a strong field where the lines are close together. In figure 18.15 the loop of wire has many lines passing through it, or linking it. If it is moved further away, the number of lines passing through the loop decreases, but could be increased again if the loop were made larger. We say that the amount of **flux** linking the loop is shown by the number of field lines passing through the loop. This will depend on the area of the loop and the strength of the field.

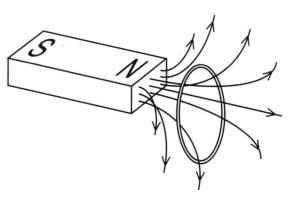

FIGURE 18.15

But of course the field lines do not actually exist, so we need a formal definition of magnetic flux as well. We define **magnetic flux** ϕ by the equation

$$\phi = B_\perp A$$

where $B_\perp$ is the component of magnetic flux density perpendicular to the loop and A is the area of the loop. Its unit must therefore be the $T\,m^2$, but

FIGURE 18.14

this is given the name **weber** (Wb). The equation can be rearranged to show that

$$B_\perp = \frac{\phi}{A}$$

and we can now see why B is called the magnetic *flux density*: it measures the *flux per unit area*, and this equation shows that the unit of magnetic flux density (the tesla, T) could be called the $Wb\,m^{-2}$.

EXAMPLE

Calculate the magnetic flux passing through a football pitch which measures 110 m by 70 m at a place where the magnetic flux density of the Earth's magnetic field is 47 μT in a direction making an angle of 66° with the horizontal.

Using $\phi = B_\perp A$
where $B_\perp = (47 \times 10^{-6}\,T)(\cos 24°)$
and $A = 7700\,m^2$,
        $\phi = (47 \times 10^{-6}\,T)(\cos 24°)(7700\,m^2)$
         $= 0.33\,Wb$

The continuity of magnetic flux

Magnetic flux is *continuous*: whenever flux enters a space, just as much leaves it at some other point. This applies not only to the fields produced by wires and coils, where the fact that the loops are closed is obvious, but also to the fields of magnets. Magnetic flux does not sprout from the ends of a magnet, but exists in closed loops: down the inside of a bar magnet from south pole to north pole, out of the north pole into the air, and back again through the air into the south pole – in a very similar manner to the flux produced by a solenoid.

18.3 Another way of inducing e.m.f.s

We saw in the first section of this chapter that an e.m.f. is induced when the B-field is constant, and the area through which it passes is changing: this was summed up in the equation $\mathscr{E} = B_\perp(dA/dt)$.

However, if $B_\perp$ is constant we could write

$$\mathscr{E} = B_\perp\left(\frac{dA}{dt}\right) \quad \text{as} \quad \mathscr{E} = \frac{d}{dt}(B_\perp A)$$

$$\text{or as} \quad \mathscr{E} = \frac{d\phi}{dt}$$

since we now know that we can think of the product $B_\perp A$ as the magnetic flux ϕ.

It also happens that an e.m.f. is induced when the area remains constant and the magnetic field changes, i.e. even when the conductor is not moving. For example, suppose two loops of wire are placed next to each other, as in figure 18.16: if the current in the right-hand loop changes, there is a change *in the flux passing through* the left-hand loop, and an e.m.f. is induced in this loop. *No conductors have moved* in this process, but an e.m.f. has nevertheless been induced. As you would expect, the size of the e.m.f. depends on the area of the left-hand loop and also on the rate at which the field is changing, i.e. this kind of e.m.f. is proportional to A and to dB/dt.

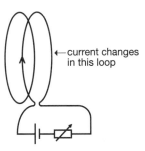

←current changes in this loop

FIGURE 18.16

It is possible to demonstrate these effects in a situation where an e.m.f. can be induced *either* by arranging for a conductor to cut field lines *or* by changing the magnetic field and leaving the conductor alone. In the first case A is being changed and in the second case B is being changed. A possible practical arrangement is shown in figure 18.17, where a wire frame is placed near the end of a solenoid which produces a magnetic field which passes through the area of the frame. The wire frame has one moveable side XY. There are two ways in which the magnetic flux passing through the wire frame can be reduced to

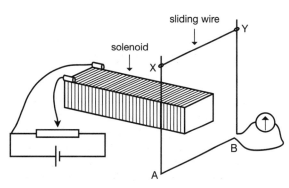

solenoid

sliding wire

X

Y

A

B

FIGURE 18.17

$$\mathscr{E} = A\frac{dB_\perp}{dt}$$

when the flux density in a coil of area A is changing

but realise that both are special cases of

$$\mathscr{E} = \frac{d}{dt}(B_\perp A) = \frac{d\phi}{dt}.$$

If a coil has more than one turn, the same e.m.f. will be induced in each turn, so that where a coil has N turns, the e.m.f. induced is given by

$$\mathscr{E} = N\left(\frac{d\phi}{dt}\right) \quad \text{or} \quad \mathscr{E} = \frac{d(N\phi)}{dt}$$

where $N\phi$ is called the **flux-linkage**.

This equation, which states that the induced e.m.f. is equal to the rate of change of flux-linkage is called **Faraday's law**.

In situations where there is not a constant rate of change of flux, we can calculate only the average e.m.f.: then

$$\mathscr{E} = \frac{\text{change of flux}}{\text{time}} = \frac{\Delta\phi}{\Delta t}$$

$$\left(\text{or } \frac{\Delta(N\phi)}{\Delta t} \text{ for a coil of } N \text{ turns}\right).$$

zero: either the wire XY can be slid down until the area ABYX is zero, or the current in the solenoid can be reduced to zero.

It is easy enough to show that in each case an e.m.f. is produced, but we really need also to measure the sizes of the two e.m.f.s. There is a complication because the solenoid's magnetic flux density is not uniform (so, for example, moving the wire at a steady speed will not produce a constant e.m.f.) but we can try to make both changes (i.e. moving the wire, and reducing the current) at whatever varying rate is necessary to produce a constant e.m.f. For example, when moving the wire, we would have to move the wire faster where the field is weaker.

When we do this we find that *if both changes of flux take the same time, the two e.m.f.s are the same size*. That is, no matter how the change of flux is brought about, when a certain change of flux occurs in a certain time the e.m.f. induced is the same. Either a conductor is cutting flux, or the magnetic flux density through the circuit is changing: they must just be different ways of thinking about the same thing. So the most general equation for calculating the size of an induced e.m.f. is $\mathscr{E} = d\phi/dt$, i.e. that the induced e.m.f. $\mathscr{E}$ is equal to the rate of change of flux ϕ in the circuit.

To sum up, in practice we shall use

$$\mathscr{E} = B_\perp\frac{dA}{dt} \text{ (often in the form } \mathscr{E} = B_\perp lv)$$

when a conductor is moving, and

EXAMPLE

A metal-framed window, measuring 70 cm by 40 cm, is hinged about a vertical side. When closed, the window lies in a north–south plane. If the horizontal component of the Earth's magnetic flux density is $18\,\mu\text{T}$, find the average e.m.f. induced when the window is opened through 90° in 1.5 s.

Initially the flux linking the window is zero; when the window is open the flux ϕ is given by

$$\phi = BA = (18 \times 10^{-6}\,\text{T})(0.70\,\text{m})(0.40\,\text{m})$$

$$= 5.0\,\mu\text{Wb}$$

so the average e.m.f.

$$\mathscr{E} = \frac{\Delta\phi}{\Delta t} = \frac{5.0 \times 10^{-6}\,\text{Wb}}{1.5\,\text{s}} = 3.4\,\mu\text{V}.$$

In this example we can find only the average e.m.f., since the rate of change of flux is not constant: it is greatest as the window starts to be opened. Notice how much simpler this is than calculating the e.m.f. by thinking of the individual parts of the frame cutting lines of force. If we had done this question that way, we should have had to consider the effect of the horizontal parts of the frame cutting the vertical component of the flux density.

Lenz's law

Earlier in the chapter, when we were dealing with conductors cutting magnetic field lines, we saw that Lenz's law enabled us to predict the direction of an induced e.m.f.: the direction of the induced e.m.f. was such as to oppose the change. We can also use Lenz's law when we deal with e.m.f.s induced by flux changes and say that the direction of the induced e.m.f. is such that, if it were to produce a current, the effect would be to oppose the change of flux.

EXAMPLE

A coil of 50 turns of wire is laid in a horizontal plane on a table. A vertical bar magnet, with its N pole nearest the coil, is placed above it, and (later) a second coil is placed above it, with a current flowing in a clockwise direction as seen from above the table. The flux emerging from the bar magnet is $1.0\,\mu\text{Wb}$, and when a current of $2.0\,\text{A}$ flows in the upper coil, the flux emerging from it is $4.0\,\text{mWb}$. All the flux emerging from the bar magnet and the upper coil passes through the first coil. Find the average size of the induced e.m.f. when

(i) The magnet is moved vertically away from the coil, in $0.050\,\text{s}$.
(ii) The current in the upper coil is reduced from $2.0\,\text{A}$ to zero in $2.0\,\text{s}$.
(iii) The current in the upper coil is increased from $2.0\,\text{A}$ to $3.0\,\text{A}$ in $1.0\,\text{s}$.

(i) The change of flux $\Delta\phi = 1.0\,\mu\text{Wb}$, and the change happens in a time $\Delta t = 0.050\,\text{s}$, so the average e.m.f. is given by

$$\mathscr{E} = N\frac{\Delta\phi}{\Delta t}$$

$$= \frac{50 \times 1.0 \times 10^{-6}\,\text{Wb}}{0.050\,\text{s}} = 1.0\,\text{V}$$

The coil in which the e.m.f. is induced forms a complete circuit, so a current will flow. The direction of the flux was downwards, so the current in the lower coil must flow in such a direction as to try to maintain this flux, i.e. to produce a downward flux, so the current must be anti-clockwise.

(ii) The change of flux $\Delta\phi = 4.0\,\text{mWb}$, and the change happens in a time $\Delta t = 2.0\,\text{s}$, so the average e.m.f. is given by

$$\mathscr{E} = N\frac{\Delta\phi}{\Delta t}$$

$$= \frac{50 \times 4.0 \times 10^{-3}\,\text{Wb}}{2.0\,\text{s}} = 0.10\,\text{V}.$$

The direction of the flux was upwards, so to try to oppose the change the current in the lower coil must produce upward flux, so the current in it must flow clockwise.

(iii) $\Delta\phi$ is half as much and Δt is also half as much, so the average induced e.m.f. will again be $0.10\,\text{V}$. But since the upward flux is increasing, the current in the lower coil must produce downward flux, so the current in it must flow anti-clockwise.

Eddy currents when the flux is changing

We saw earlier in the chapter that an induced e.m.f. can cause eddy currents to flow in a solid lump of metal when the metal is moving at right angles to a magnetic field. There will also be eddy currents when the magnetic flux through a lump or sheet of metal is changing; for example, when a bar of metal is placed inside a solenoid carrying alternating current. The bar rapidly becomes hot.

This conversion of electrical energy to internal energy, needs to be minimised in any iron cores or coils that carry alternating currents, and this is achieved by laminating the cores, just as with the iron parts of motors and dynamos. The higher the frequency, the larger the induced e.m.f.s, so the more the cores have to be subdivided to reduce the eddy currents. If thin laminations are not sufficient, a bundle of iron wires may be used, or even a core packed with iron dust. Another way of reducing the eddy currents is to use magnetic substances which have high resistivity, such as the iron compounds called ferrites. Many of these could be classed as electrical insulators, although their magnetic properties are similar to those of iron. Ferrites are widely used to make magnetic cores for the aerial coils in radio receivers.

18.4 Magnetically linked circuits

Mutual induction

We saw in the last section that a changing current in one coil can induce an e.m.f. in a coil which is nearby. This process is called **mutual induction**. Figure 18.18 shows this happening. The coil at the bottom of the steel retort stand has an alternating current in it: this magnetises the steel rod in alternate directions at the frequency of the supply.

The changing magnetic flux through the aluminium ring induces an alternating e.m.f. in it, which generates an alternating current. Whichever way the current is flowing in the coil, the current in the ring must be in the opposite direction, since the two repel each other, and the upward push of the coil on the ring balances the ring's weight. The reality of the current can be tested by feeling the aluminium ring: it will soon become hot. Also there is no lifting force when the continuous ring is replaced by a ring with a gap in it.

This demonstration does not work if the coil is supplied with a direct current. Then there is a momentary effect when the current is first switched on: the ring jumps up, but immediately falls back down again. The demonstration is effective mainly because of the large increase in magnetic flux caused by the presence of the iron. You are familiar with the effect in the transformer, where again the iron core efficiently transfers energy from the primary to the secondary side. A transformer is an excellent example of how mutual induction is put to use.

Let us start, however, with air-cored coils because these are simpler. Figure 18.19(a) shows an arrangement you could use to investigate the effects. Two solenoids are placed one inside the

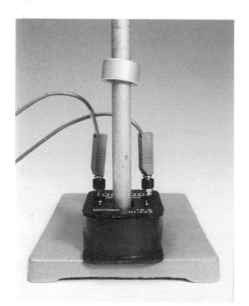

FIGURE 18.18

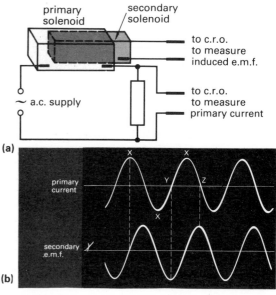

FIGURE 18.19

244

other. One, the primary solenoid, is supplied with alternating current, and a resistor is placed in series. This resistor can be connected to one channel of a double-beam oscilloscope to show the size of the current and how it varies with time. The terminals of the secondary solenoid are connected to the other channel of the oscilloscope. Figure 18.19(b) shows the traces you would see on the screen. At moments such as X the magnetic field produced by the primary solenoid has a maximum value, but is momentarily unchanging. At this moment the e.m.f. induced in the secondary solenoid is zero. However, at moments such as Y or Z, the magnetic field is momentarily zero, but is changing at its greatest rate. The induced e.m.f. therefore has its greatest value at these points. We can describe what happens by saying that the e.m.f. induced in the secondary *lags* a quarter of a cycle behind the variation of current in the primary.

If the frequency of the primary current is varied, while keeping the peak value constant, we find that the peak value of the induced e.m.f. is proportional to the frequency. We use a signal generator to do this, varying the frequency from about 10 Hz to about 100 kHz: a dramatic increase in induced e.m.f. is observed.

The a.c. search coil

By using many (perhaps 5000) turns of very fine wire we can make a small secondary coil to take the place of the secondary solenoid above. Figure 18.20 is a photograph of two such **search coils**, each mounted at the end of a strip of perspex which supports the wires leading to the coil. Such coils may be used to explore how the field varies from one part to another of the primary solenoid, but they can also be used to explore the magnetic field around any conductor which carries an alternating current.

The e.m.f. induced in a search coil is proportional to:

◆ the number of turns N it contains and its cross-sectional area A, which will be constant for a particular search coil. Since we want to use the search coil to examine the field at different points, the area must be small, but the very large number of turns means that the e.m.f. is of a measurable size.

◆ the size of the magnetic flux density in which it is placed, and it also depends on the orientation. The e.m.f. will be a maximum when the plane of the coil is perpendicular to a field, and zero when it is parallel to the field. It can therefore be used to tell the direction of the field at any point.

◆ the frequency of the magnetic flux density in which it is placed, since the higher the frequency the greater the rate of change of flux.

The search coil is usually used just to compare the values of magnetic flux density at different points in a magnetic field; the weaker the field, the higher the frequency that would need to be used, but once chosen, the frequency obviously needs to be kept constant. The e.m.f. generated by a search coil is sometimes displayed on an oscilloscope (with the time base switched off) but a multimeter with an a.c. millivolt range is more sensitive.

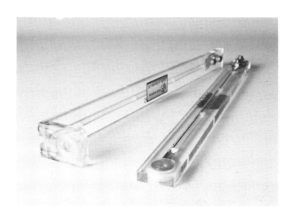

FIGURE 18.20

EXAMPLE

A student places a search coil 20 mm from a straight wire which carries an alternating current of peak value 1.0 A and frequency 50 Hz, and adjusts its orientation until he finds the maximum e.m.f. induced: this is 1.5 mV. Draw a diagram to show the wire, the magnetic field, and how he placed the search coil, and calculate the e.m.f. induced when the search coil is placed, (a) at a distance of 40 mm, with a current of the same size and frequency in the wire, (b) at a distance of 100 mm, with a current of 5.0 A at the same frequency, (c) at a distance of 500 mm, with a current of 5.0 A at a frequency of 500 Hz.

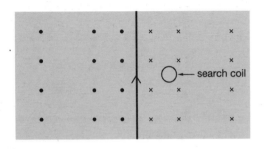

The diagram shows that the wire must be in the plane of the coil.

(a) The flux density is inversely proportional to distance from the wire, so at 40 mm (double the distance) the e.m.f. induced will be halved, i.e. 0.75 mV.

(b) At 100 mm from the wire the e.m.f. would be only a fifth the size, i.e. 0.30 mV, but the effect of increasing the current by a factor of 5 is to increase this to 1.5 mV.

(c) At 500 mm from the wire the e.m.f. would be only 1/5 of what it was at 100 mm, but the frequency is 10 times greater, so in total the e.m.f. induced is twice as great, i.e. 3.0 mV.

At a distance of 500 mm from the wire the peak magnetic flux density is only 2 μT when the current is 5.0 A, so you can see how sensitive a search coil is, especially when the frequency of the flux density is high. Incidentally it would not detect the Earth's B-field, since that is not alternating.

Mutual inductance

The e.m.f. induced in the secondary coil when the current in the primary coil is changing is proportional to the rate of change of flux linking it. Since the flux is proportional to the current in the primary coil, the e.m.f. is proportional to the rate of change of current in the primary coil: i.e., $\mathscr{E}$ is proportional to dI/dt, and we could write

$$\mathscr{E} = M\frac{dI}{dt}$$

where M is the **mutual inductance** of the two coils, and is a measure of how well the two coils are linked magnetically: the larger the mutual inductance, the larger the e.m.f. in the secondary for a certain rate of change of current dI/dt in the primary coil.

The size of M depends on such factors as the numbers of turns on the two coils, how close together they are, the nature of the material between them, and so on, i.e. on the construction of the coils. The size of M could, in theory, be calculated for a particular pair of coils, but in practice there are too many uncertainties for the result to be reliable. The unit of M must be the $V \div A/s$, i.e. the $V\,s\,A^{-1}$, but this is given the name **henry** (H).

The size of M can be measured, as the definition suggests, by finding the e.m.f. induced in the secondary when there is a measured rate of change of current in the primary. One way of doing this is to connect a **ramp generator**, which is an electronic device which produces a steadily rising current, to the primary coil: then the induced e.m.f. in the secondary can be measured. Figure 18.21 shows (a) the current which the ramp generator would produce in the primary coil or solenoid: (b) shows the induced e.m.f. which would be produced in the secondary. The advantage of using a ramp generator is that it causes a steady rate of change of current dI/dt, so that the induced e.m.f., which is equal to $M(dI/dt)$, is also steady and can easily be measured.

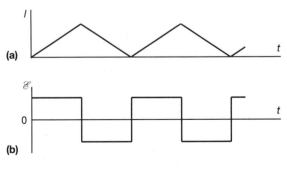

FIGURE 18.21

A ramp generator was used to give a rise of current of 0.70 A in 3.8 s in a solenoid. A second solenoid was placed inside the first one, and the e.m.f. induced in this was found to be 0.24 mV. Calculate the mutual inductance of the solenoids.

Rate of change of current

$$\frac{dI}{dt} = \frac{0.70\,A}{3.8\,s} = 0.18\,A\,s^{-1}.$$

$$\mathcal{E} = M\left(\frac{dI}{dt}\right) \Rightarrow M = \frac{0.24 \times 10^{-3}\,V}{0.18\,A\,s^{-1}}$$

$$= 1.3 \times 10^{-3}\,H = 1.3\,mH$$

18.5 Inductance

A single coil of wire produces its own magnetic flux in and around itself, so that if the current in it changes, there is an e.m.f. induced *in the coil itself*. This process is called **self-induction**. The induced e.m.f. $\mathcal{E}$ will be proportional to the rate of change of current dI/dt in the coil so we can write

$$\mathcal{E} = L\left(\frac{dI}{dt}\right)$$

where L is the self-inductance or just the **inductance** of the coil. This is obviously similar to the equation $\mathcal{E} = M\,(dI/dt)$ which is the definition of the mutual inductance M of a *pair* of coils. The unit of L, like the unit of M, is the henry (H). L is a property of the coil, and it depends on the size, number of turns and the geometry of the coil.

The effect is not very great in a coil consisting of just a single turn of wire, but the effect is four times greater in two turns because now there is twice as much flux, and it links twice as many coils. The self-inductive effect will increase roughly in proportion to the square of the number of turns so that the effect will be considerable when there are several hundred turns. The inductive effects may be further increased by a factor of several hundred times if the coil is wound on an iron core. A coil which is designed to have considerable inductance is called an **inductor**.

An iron-cored coil of 1100 turns of wire has a resistance of $1.6\,\Omega$. (a) What current flows when it is connected to a direct current supply which provides a p.d. of 12.0 V? (b) When an a.c. supply of frequency 50 Hz is used instead, the current in the circuit is found to be only 43 mA. Explain this, and calculate the e.m.f. induced in the coil.

(a) Using

$$I = \frac{V}{R} \quad \text{we have} \quad I = \frac{12.0\,V}{1.60\,\Omega} = 7.50\,A.$$

(b) The much smaller current is due to the self-induced e.m.f. $\mathcal{E}$ in the coil. If the supply p.d. is V, the effective p.d. is now $V - \mathcal{E}$ (since by Lenz's law the induced e.m.f. must oppose the change), so using $V = IR$ we have

$$12.0\,V - \mathcal{E} = (43 \times 10^{-3}\,A)(1.60\,\Omega)$$

$$\mathcal{E} = 11.9\,V$$

We see that the induced e.m.f. is nearly as great as the applied p.d. It would have been greater still if the alternating supply had had a higher frequency.

Direct current circuits

Inductors in d.c. circuits have an effect only when the current is switched on or off. To understand their behaviour you must realise that any inductor has both inductance and resistance. It is often convenient to calculate currents in circuits by treating an inductor as consisting separately of an inductor and a resistor, as shown in figure 18.22(a), but at any time the p.d. across it must consist of two parts:

(a) the p.d. needed to keep the current I flowing: this p.d. $= IR$; and (b) the p.d. needed to keep the current growing at a rate dI/dt: this p.d. $= L(dI/dt)$.

The sum of these two p.d.s must always be equal to the applied p.d., so as one increases the other decreases. That is why in a circuit like this the current varies with time as shown in Figure 18.22(b) (where as I increases, dI/dt decreases). The next example will help you understand what happens.

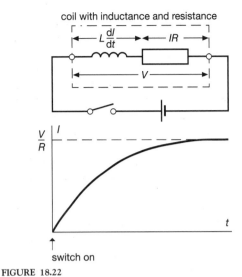

coil with inductance and resistance

switch on

FIGURE 18.22

EXAMPLE

In a circuit like that shown in figure 18.22(a) the battery maintains a p.d. of 6.00 V across the inductor, whose resistance is 2.00 Ω and inductance 4.00 H. When the switch is closed the current increases as time goes on. For the values of current given, calculate the values of the p.d. across the inductive and resistive parts of the resistor to fill the gaps in table 18.1 (* means 'after a long time'). Also calculate the rate at which the current is growing after 2 s.

When $I = 0$, $V_R = 0$, so $V_L = 6.00\,\text{V}$

When $I = 1.89\,\text{A}$,
$V_R = IR = (1.89\,\text{A})(2.00\,\Omega) = 3.78\,\text{V}$, so

$V_L = 6.00\,\text{V} - 3.78\,\text{V} = 2.22\,\text{V}$.

Similarly
when $I = 2.59\,\text{A}$,
$V_R = 5.18\,\text{V}$ and $V_L = 0.82\,\text{V}$,

and when $I = 2.85\,\text{A}$,
$V_R = 5.70\,\text{V}$ and $V_L = 0.30\,\text{V}$.

TABLE 18.1

t/s	I/A	V_R/V	V_L/V
0	0		
2	1.89		
4	2.59		
6	2.85		
*			

Eventually I reaches a steady value so that $dI/dt = 0$:

then $V_L = 0$ and $V_R = 6.00\,\text{V}$, so

$$I = \frac{V_R}{R} = \frac{6.00\,\text{V}}{2.00\,\Omega} = 3.00\,\text{A}.$$

$$V_L = L\left(\frac{dI}{dt}\right) \Rightarrow \frac{dI}{dt} = \frac{V_L}{L},$$

so after 2 s, $\quad \dfrac{dI}{dt} = \dfrac{2.22\,\text{V}}{4.00\,\text{H}} = 0.555\,\text{A\,s}^{-1}$

Switching off

When a current is switched off in an inductive circuit the current has to fall to zero very quickly, so the rate of change of current is very large, and the induced e.m.f. is therefore very large.

You can demonstrate this for yourself by switching off the current in a circuit which contains a 1.5 V dry cell and an iron-cored inductor with about 1000 turns. If you connect a neon lamp across the switch (figure 18.23) you will see it light when you open the switch: the neon lamp needs a p.d. of at least 100 V to make it light, so this shows that there is a large e.m.f. induced. If you use an old-fashioned 'knife' switch with visible contacts you will see a spark at the contacts. Do not try to do this experiment with more than one cell: it could be dangerous.

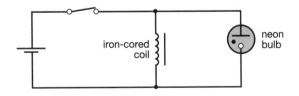

FIGURE 18.23

The ignition coil which provides the spark in a petrol internal combustion engine makes use of this effect: the p.d. which makes the current flow in the coil is only 12 V, but the induced e.m.f. when the circuit is broken is more than 10 kV.

Another everyday application occurs in the

design of the starter circuit for a fluorescent tube. The mains p.d. (about 240 V) is not enough to cause ionisation in the mercury vapour in the tube. When the lamp is first switched on a current flows through a small cylindrical tube (called the starter) which contains a bimetallic strip. After a fraction of a second the current in the bimetallic strip warms it up enough to cause it to bend and break the contact. But the circuit also contains an inductor, and the breaking of the current through the inductor induces a large e.m.f. which is enough to start ionisation in the mercury vapour. Once the current starts the electrodes at the ends of the tube warm up enough for the ionisation to continue and the tube stays alight.

Mechanical analogies

Most of us find mechanics easier to understand than electricity, and so a mechanical analogy of what is happening in an inductive circuit may be helpful. Connecting the battery in the circuit of figure 18.22(a) is rather like letting a stone fall in a viscous liquid. Initially a force (its weight) is applied to the stone, which then accelerates until it reaches a terminal speed when the weight is balanced by the resisting force. Using the usual notation, initially $F = ma$, but eventually $a = 0$ when $F = kv$. In this electrical circuit initially $V = L(dI/dt)$ but eventually $dI/dt = 0$ when $V = IR$. Table 18.2 shows the similarities.

TABLE 18.2   Comparison of mechanical and electrical properties

Mechanical	Electrical
velocity v	current I
acceleration dv/dt	rate of change of current dI/dt
force F	p.d. V
mass m	inductance L

For the time being the most important similarity is between mass m and inductance L. If a body has a large mass, it will resist changes to its motion: it will not gain speed quickly, but nor will it slow down quickly. Similarly it is the inductance of an inductor which means that the current grows slowly, and which equally prevents the current from decreasing quickly, as the last example showed. Inductance is the *inertial* property of a circuit.

You know that a charged capacitor stores electric potential energy in the electric field of the capacitor, and it might seem reasonable that an inductor with a current in it might store magnetic potential energy (m.p.e.) in its magnetic field. For example, the energy for the spark when a current is switched off in an inductive circuit must come from somewhere. We can use the mechanical–electrical analogy comparing mass to inductance to guess the expression for m.p.e. The kinetic energy of a mass m which has a speed v is given by $\frac{1}{2}mv^2$: this is the energy which a mass has because it is moving. If m and v in mechanics are analogous to L and I in electricity, then we might expect the energy of moving charge to be $\frac{1}{2}LI^2$, and this would be the m.p.e., since it is the moving charge which creates the magnetic field. Of course, the existence of an analogy does not prove that the magnetic energy $= \frac{1}{2}LI^2$, but the analogy helps us to understand it better. There are mathematical techniques for showing that the m.p.e. is indeed $\frac{1}{2}LI^2$.

18.6 Energy conversions

Figure 18.24 shows a U-shaped wire frame which has two parallel rails PQ and RS. A wire AB can be made to slide along the rails. There is a magnetic field, into the diagram, in the whole of the region. You already know that when the wire AB is slid along the frame, and cuts the flux, an e.m.f. $\mathscr{E}$ will be induced given by $\mathscr{E} = Blv$, where l is the length of AB and v its velocity.

We can also derive this result using the principle of conservation of energy. Suppose that the

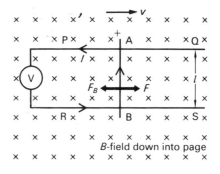

FIGURE 18.24

249

movement of the wire, at velocity v, induces an
e.m.f. $\mathscr{E}$ and an induced current I. The existence
of this current, which is at right angles to the flux
density B, means that there must be a magnetic
force F, given by $F = BIl$, on the wire AB. By
Lenz's law the direction of F will be opposite to
that of the velocity v. *To maintain the constant
velocity v of the wire the person moving the wire must
push the wire with a force of exactly this size (i.e. BIl).*
So his or her rate of working P is given by

$$P = Fv$$

$$= BIlv$$

What happens to this work done? It appears first as
electrical energy in the circuit and then as internal
energy. The principle of conservation of energy
tells us that the person's rate of working must be
equal to the rate of production of internal energy.
Since the rate of working of the e.m.f. $\mathscr{E}$ is $\mathscr{E}I$,

$$\mathscr{E}I = BIlv$$

or $$\mathscr{E} = Blv$$

as before.

The d.c. machine

There is a torque on a coil carrying a current in a
magnetic field. This torque tends to turn it until its
plane is at right angles to the field. To produce
continuous rotation there must be some device
which reverses the connections to the coil
whenever it passes through this equilibrium
position. The moving part (the **armature**) consists
of a set of coils sunk into slots spaced round an
iron core, as shown in figure 18.25. Each coil is
joined to a pair of contacts on a **commutator**
which is fixed to the shaft of the motor. Current is
led into and out of the commutator segments by a
pair of graphite blocks, called brushes, which are
arranged so that the connections to each coil are
automatically reversed at the right positions in
each revolution. The armature rotates between the
poles of an electromagnet whose coils draw current
from the same source that supplies the current to
the armature.

You can see that the construction of the d.c.
motor is identical with that of a d.c. dynamo
(compare figures 18.6(a) and 18.25). Any d.c.
machine may be used either as a dynamo or a

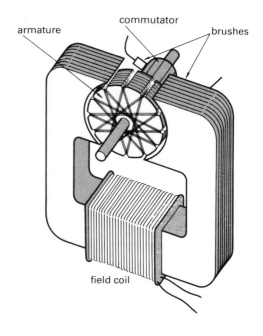

FIGURE 18.25

motor. When mechanical energy is supplied to the
machine, and converted to electrical energy, we
call the machine a *dynamo*: when electrical energy
is being converted into mechanical energy, we call
it a **motor**. The difference between the two is
merely a matter of which way round the energy
conversion is taking place.

*In fact the d.c. machine must always at the same
time behave as both a dynamo and a motor*, since a
current-carrying coil rotating in a magnetic field
must:

♦ because it carries a current, experience a
torque.

♦ because it is moving, generate an e.m.f.

This will become clearer when we analyse the
machines in more detail below.

The dynamo

Someone or something rotates the coil which
generates an e.m.f. $\mathscr{E}$. The agent therefore works at
a rate $\mathscr{E}I$ (compare the similar situation with the
sliding wire at the start of this chapter). The e.m.f.
drives a current I through an external resistor R, as
shown in figure 18.26(a).

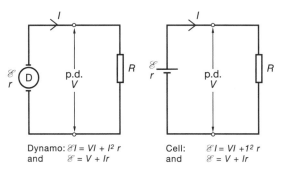

Dynamo: $\mathscr{E}I = VI + I^2 r$
and $\mathscr{E} = V + Ir$

Cell: $\mathscr{E}I = VI + I^2 r$
and $\mathscr{E} = V + Ir$

FIGURE 18.26

If the armature of the dynamo has internal resistance r the p.d. V at the terminals of the dynamo will be less than $\mathscr{E}$ (compare $V = \mathscr{E} - Ir$ for a cell). The power delivered to the external resistor will be VI, and the power delivered to the internal resistance will be I^2r, so using the principle of conservation of energy we have

$$\mathscr{E}I = VI + I^2 r$$

i.e. total power supplied = electrical power output
+ rate of heating armature.

The dynamo behaves like an electric cell in that in both the dynamo and the cell non-electrical forces are used to separate electric charge and push it round a circuit (if a circuit is connected). So it is not surprising that the power equation for the dynamo ($\mathscr{E}I = VI + I^2r$) is the same as that for the cell, where $\mathscr{E}$ = e.m.f. of the cell, V = terminal p.d. and r = internal resistance of the cell. The circuit for the cell is shown alongside the circuit for the dynamo in figure 18.26(b).

The motor

An external source of p.d. (e.g. a battery whose terminal p.d. is V) drives current I through the armature, as shown in figure 18.27(a): the rate of working of this source is therefore VI. The armature rotates, and an e.m.f. $\mathscr{E}$ is induced in it: it acts so as to oppose the change, so is often called a **back e.m.f.** Its rate of working is $\mathscr{E}I$. Also internal energy is generated in the armature coils at a rate I^2r. So again using the principle of conservation of energy we have

$$VI = \mathscr{E}I + I^2 r$$

The term I^2r represents the rate of production of wasteful internal energy, but what is the significance of the term $\mathscr{E}I$? We want it to be as large as possible a proportion of VI, since it represents the *mechanical power output* of the motor. So

electrical power = mechanical power output
supplied + rate of heating armature.

Just as the dynamo behaves like a cell which is maintaining a current in a circuit, so the motor behaves like a cell which is having energy put back into it. In both the motor and the cell electrical forces are used to convert electric potential energy into a non-electrical form: mechanical energy in the case of the motor, and chemical energy in the case of the cell. Again, not surprisingly, the power equation for the motor ($VI = \mathscr{E}I + I^2r$) is the same as that for the cell, where V = terminal p.d., $\mathscr{E}$ = e.m.f. of the cell, and r = internal resistance of the cell. The circuit for the cell is shown alongside the circuit for the motor in figure 18.27(b).

The relationships for both the dynamo and the motor are illustrated in figure 18.28. The d.c. machine is therefore a reversible energy-converter in much the same way as a rechargeable electric cell. A lead–acid cell is a device for the interconversion of electrical energy and chemical energy. The direction of the conversion depends on whether the direction of the current is the same as the direction of the e.m.f. A practical example will help to make this clear.

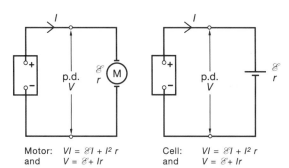

Motor: $VI = \mathscr{E}I + I^2 r$
and $V = \mathscr{E} + Ir$

Cell: $VI = \mathscr{E}I + I^2 r$
and $V = \mathscr{E} + Ir$

FIGURE 18.27

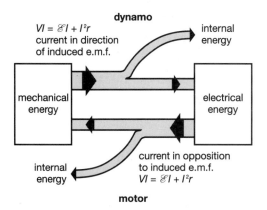

dynamo

$VI = \mathscr{E}I + I^2r$
current in direction
of induced e.m.f.

internal
energy

mechanical
energy

electrical
energy

internal
energy

current in opposition
to induced e.m.f.
$VI = \mathscr{E}I + I^2r$

motor

FIGURE 18.28

A practical example

Consider an electric delivery van (e.g. a milk-float)
which uses a 100 V lead–acid battery, climbing
over the brow of a hill. As it climbs the hill it
travels quite slowly, with the induced e.m.f.
generated in its motor proportional to the speed.
Since the 'motor' is at this point working as a
motor, this e.m.f. $\mathscr{E}$ is less than the battery e.m.f.
V: suppose it is 95 V. The difference between these
two e.m.f.s, i.e. 5 V, drives a large current through
the armature and provides the large torque
necessary to keep the float moving up the hill. The
net result is the conversion of chemical energy of
the battery into gravitational potential energy of
the float. Suppose the internal resistance r of the
armature is 0.1 Ω: then the current at this point is
given by $I = (V - \mathscr{E})/r = (5\,V)(0.1\,\Omega) = 50\,A$.
The electrical power P supplied by the battery is
given by

$$P = VI = (100\,V)(50\,A) = 5000\,W$$

and the rate of conversion P_m of electrical energy
to mechanical energy is given by

$$P_m = \mathscr{E}I = (95\,V)(50\,A) = 4750\,W$$

and the rate of conversion P_r of electrical energy to
internal energy is given by

$$P_r = I^2r = (50\,A)^2(0.1\,\Omega) = 250\,W$$

as expected from the principle of conservation of
energy: P must be equal to $P_m + P_r$.

As the milk-float approaches the brow of the
hill, and the slope becomes less, the vehicle starts
to go faster. The increased speed induces a larger
back e.m.f.: suppose it is now 99 V. The current
falls because the difference $V - \mathscr{E}$ is now smaller.
The difference has fallen to 1 V, so the armature
current falls to 10 A. The torque decreases, but this
is enough to maintain the vehicle's speed on the
smaller slope. The rate of conversion P_m has now
dropped to $(99\,V)(10\,A) = 990\,W$, which
corresponds to the required smaller rate of gain of
gravitational potential energy by the float.

As the float starts to descend the other side of
the hill it goes faster still, and at some point the
back e.m.f. is equal to the battery e.m.f. The motor
then draws no current from the battery: the energy
required to keep the float moving against resistance
forces is supplied by the loss of gravitational
potential energy. If the speed increases still further
the induced e.m.f. becomes greater than 100 V and
the machine behaves as a dynamo, driving current
in the reverse direction through the battery. This
current acts so as to oppose the change, and
therefore acts like a brake. But the descending
milk-float is losing gravitational potential energy
and is able to turn the armature against the
opposing torque which this current produces. In
this way mechanical energy is converted into (a
little) internal energy in the armature and
electrical energy (and most of this to chemical
energy) in the battery. So much of the
gravitational potential energy acquired by the
milk-float in climbing the hill is changed back to
chemical energy in the battery.

Starting a motor

At the moment of starting a motor its speed is zero,
so the induced e.m.f. is zero. If there is no other
resistance in the circuit, the starting current I_1 is
given by

$$I_1 = \frac{V}{R}$$

which, for the milk float of the example, gives a
current

$$I_1 = \frac{100\,V}{0.1\,\Omega} = 1000\,A$$

which would be a large and damaging value.
Suitable resistances therefore need to be included
in series with the armature when starting a motor

which is supplied with a large p.d. If it is required to limit the current to 50 A then the total resistance would need to be 2 Ω. As the motor speed increases the induced e.m.f. rises from zero and the starting resistance must be progressively reduced (by moving the rotating arm clockwise from position 1 in figure 18.29). Near full speed the induced e.m.f. is enough by itself to limit the speed of the motor and at this point the resistance can be reduced to zero: the arm is then in position 4 in the figure.

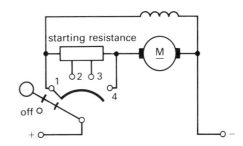

FIGURE 18.29

Reversing an electric motor

Figure 18.29 shows the normal connection of the electromagnet coils: they are connected in parallel with the armature. Reversing the polarity of the supply terminals therefore does not reverse the direction of rotation of the motor, since the direction of the field is reversed as well as the direction of the current, and the magnetic forces are in the same direction as before. The connections to only one of coils, field or armature, must be reversed. This may not apply to very small motors, which have permanent magnets: then reversing the polarity of the supply does reverse the direction of rotation.

Exercises on each section of this chapter may be found in the companion textbook, **Practice in Physics**.

SUMMARY

At the end of this chapter you should be able to:

♦ understand that in electromagnetic induction mechanical energy is converted into electrical potential energy.

♦ use the equation $\mathscr{E} = Blv$ for the e.m.f. $\mathscr{E}$ induced in a moving conductor.

♦ remember that the direction of the induced e.m.f. is such as to tend to oppose the change causing it (Lenz's law) and that this law is a consequence of the principle of conservation of energy.

♦ remember that eddy currents are generated in slabs of conducting material which are pushed through a magnetic field, and that they can be reduced by laminating the material.

♦ use the equation $\phi = B_\perp A$ which defines magnetic flux and understand what is meant by magnetic flux, and why B is called the magnetic flux density.

♦ understand that e.m.f.s may also be induced when the flux linking a loop of wire is changing, even if the conductors are not moving.

♦ understand that the two apparently different kinds of e.m.f. can be summed up in Faraday's law: the e.m.f. is equal to the rate of change of flux linkage.

◆ understand that eddy currents are induced in the cores of coils and solenoids and that this leads to the unwanted production of internal energy.

◆ use the equation $\mathscr{E} = M\,(dI/dt)$, which defines the mutual inductance M of two circuits.

◆ describe how to use an a.c. search coil to compare magnetic flux densities.

◆ use the equation $\mathscr{E} = L\,(dI/dt)$, which defines the self-inductance L of a conductor.

◆ understand that when the current changes in a d.c. circuit, an e.m.f. will be induced which will slow the rate of change of current.

◆ understand that a very large e.m.f. will be induced when the current is switched off in an inductive circuit.

◆ understand that any d.c. machine is acting simultaneously as a motor and a dynamo.

◆ remember that for a dynamo $\mathscr{E}I = VI + I^2r$ (and this is identical with the equation for a cell delivering a current).

◆ remember that for a motor $VI = \mathscr{E}I + I^2r$ (and this is identical with the equation for a cell which is having energy restored to it).

19 Alternating Currents

Power stations turn chemical or nuclear energy into electrical energy: thick cables carry the energy away from the power station to us, the users. The transmission system must be designed so that it is as economical as possible to build and run. The cables are necessarily long, so to keep their resistance low they need be made of very thick wire: of the metals which are cheap enough to be practicable, copper is the best conductor, but copper wires would mean heavy cables and either stronger pylons, or more pylons. So aluminium, which is not as dense, and quite a good conductor, is used instead. Even so, the amount of energy carried by these cables is so large that more must be done to keep losses of energy through the heating of the cables as low as possible: even 1% of such a large amount of energy would be considerable. It turns out that it is an advantage that the generators produce alternating *p.d.s, since this enables* **transformers** *to be used to raise and lower the p.d. at the start and end of transmission cables.*

19.1 Measurements

The potential difference provided by the electricity grid varies sinusoidally with time, at a frequency of 50 Hz (in Britain). This frequency is determined by the rate of rotation of the alternators in the power stations. These rotate at 3000 r.p.m., i.e. 50 times each second. The current that flows in a circuit which consists only of resistors (e.g. lamps, heaters, etc.) will be proportional to the p.d. V and at any instant we can use the equation $I = V/R$ to calculate the current I if the resistance of the circuit is R. The maximum values of V and I are called their **peak values** V_0 and I_0. When we were dealing with simple harmonic motion we saw that a sinusoidal oscillation of amplitude x_0 and frequency f has a displacement x which varies with time according to the equation

$$x = x_0 \sin 2\pi ft$$

and so for these sinusoidally varying p.d.s and currents we can write

$$V = V_0 \sin 2\pi ft \quad \text{and} \quad I = I_0 \sin 2\pi ft$$

where f is the frequency (in Hz, as usual) or

$$V = V_0 \sin \omega t \quad \text{and} \quad I = I_0 \sin \omega t$$

where $\omega = 2\pi f$: ω is measured in radians per second, and is the angular velocity of the alternators in the power stations.

The first two graphs in figure 19.1 show the variation with time of p.d. and current in a resistive circuit. We could calculate the power P at any instant by multiplying together the values of V and I at any instant. Looking at the graphs you can see that when V is positive, I is positive, and when V is negative, I is negative, so the product is always positive. The signs of V, I and P are tabulated below the graphs. The fact that the product is always positive means that power is always being delivered from the source to the circuit. You know this must be true, since alternating supplies are used for heating: there would be no overall heating effect if the energy was alternately supplied and then removed again! Its peak value P_0 is given by $P_0 = V_0 I_0$. The last graph in figure 19.1 shows how the power varies with time: it fluctuates sinusoidally between zero and P_0 at twice the frequency of the supply. You can see that its mean value $P_{av} = \frac{1}{2}P_0 = \frac{1}{2}V_0I_0$.

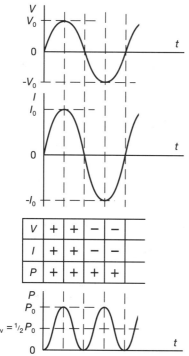

FIGURE 19.1

Root mean square values

An alternating supply whose peak p.d. is $10\,V$ is clearly not as effective as a direct supply which provides a constant p.d. of $10\,V$. We need to ask how effective, for heating purposes, an alternating supply of peak p.d. V_0 is: the answer will be some fraction q of V_0. We shall deal only with sinusoidally varying alternating p.d.s: for all of these the fraction q will be the same. Figure 19.2 shows the same resistor R connected to an alternating supply and a direct supply.

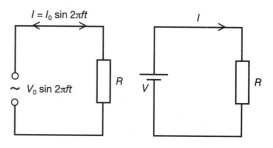

FIGURE 19.2

Because $P_{av} = \frac{1}{2}V_0I_0$ and $I_0 = V_0/R$ we can write

$$P_{av} = \frac{\frac{1}{2}V_0^2}{R}.$$

Suppose we now use a direct supply which provides a steady p.d. V which is enough to provide a power equal to the average power P_{av} provided by the alternating supply. Then

$$P = P_{av}$$

and since

$$P = \frac{V^2}{R} \text{ and } P_{av} = \frac{\frac{1}{2}V_0^2}{R},$$

$$\frac{V^2}{R} = \frac{\frac{1}{2}V_0^2}{R}$$

So

$$V^2 = \frac{1}{2}V_0^2 \text{ and } V = \frac{1}{\sqrt{2}}V_0.$$

So the fraction

$$q = \frac{1}{\sqrt{2}} (= 0.71).$$

For example, if an alternating supply has a peak p.d. of $50\,V$, it is, for heating purposes, equivalent to a direct supply whose p.d. V is given by

EXAMPLE

An alternating supply of peak p.d. $20\,V$ and frequency $50\,Hz$ is connected to a resistor of resistance $15\,\Omega$. What is the current in the resistor (a) $5.0\,ms$, and (b) $6.2\,ms$ after a cycle has begun?

(a) The period T of the supply is given by $T = 1/f = 1/(50\,Hz) = 20\,ms$, so $5.0\,ms$ is one-quarter of a period. The current is then at its maximum value. The peak current $I_0 = V_0/R = (20\,V)/(15\,\Omega) = 1.3\,A$.
(b) We have to use the equation $I = I_0\sin 2\pi ft$, so

$$I = (1.33\,A)\sin\{(2\pi)(50\,Hz)(6.2\times10^{-3}\,s)\}$$

$$= (1.33\,A)\sin(1.95\,rad)$$

$$= 1.24\,A$$

$$V = \frac{1}{\sqrt{2}} \times 50\,V = 35\,V.$$

This effective value, qV_0, is called the **root mean square** (r.m.s.) value, since it has been obtained by taking the square root of the mean square p.d. It is usually this value which is quoted when we refer to alternating p.d.s: e.g. when we say that in Britain the mains p.d. is 240 V, this is the r.m.s. value: the peak value is $\sqrt{2} \times 240\,V \approx 340\,V$.

We refer to r.m.s. values of *current* in a similar way. A fuse rated at 5 A should melt at either at steady current of 5 A or an alternating current of peak value $\sqrt{2} \times 5\,A \approx 7\,A$.

Rectification

For many purposes alternating current is entirely adequate. If we want to heat a filament in an electric lamp or an electric fire it does not matter which way the current flows. Many electric motors are designed to work with alternating current. Occasionally, however, a direct current is needed: for example, for power supplies for electronic circuits in computers and calculators, for electrolysis and for battery charging. Then it is convenient to be able to **rectify** an alternating supply, i.e. to modify it so that it passes a current in one direction only. For this purpose a **diode** is used.

The diode

A diode is a circuit component which allows current to pass easily in one direction through it. Nowadays these are almost always made of a semiconducting material such as germanium or silicon: two different forms (p-type and n-type) of one of these elements are created in a single crystal and such a crystal effectively allows current to pass in one direction only. The graph of current against p.d. for a silicon diode is shown in figure 19.3. Notice that the p.d. needs to be greater than about 0.5 V for current to flow in the forward direction: after that, the current increases very rapidly. (For germanium diodes this **turn-on** p.d. is only 0.2 V.) For small p.d.s in the reverse direction, the current is just a few nanoamperes, so effectively the diode is a **valve**, allowing current to pass in only one direction.

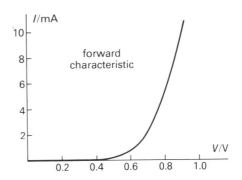

FIGURE 19.3

Rectifying circuits

A diode connected in series with some device to an alternating supply, as in figure 19.4(a), conducts only in those parts of the cycle where the p.d. across it is in the forward direction and greater than about 0.5 V. The current through the device is then in one direction only, but pulsating, as shown in the oscilloscope trace in figure 19.4(b). This is called **half-wave rectification**.

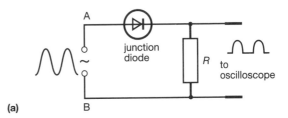

(a)

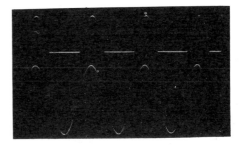

(b)

FIGURE 19.4

Four semiconducting diodes can be joined in a **bridge network** as shown in figure 19.5(a). On one half-cycle current passes through diodes P and S, charging the **storage capacitor** to the peak p.d. of the supply. In the other half-cycle current passes through Q and R instead. The peak p.d. across the storage capacitor and resistor now occurs twice as often as before, so the ripple has a frequency of 100 Hz. But the main advantage of using four diodes is that the power delivered is doubled, since although the output p.d. is much the same, the average current is doubled, because there is a current for twice the length of time. Sets of four diodes are usually encapsulated in a single unit with four terminals, as shown in figure 19.5(b). The result of this is called **full-wave rectification**.

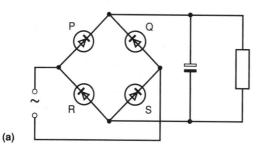

(a)

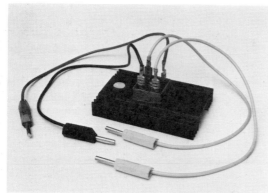

(b)

FIGURE 19.5

EXAMPLE

A full-wave bridge rectifier circuit, as shown in figure 19.5(a), is used to produce a smoothed d.c. output.
(a) Calculate the size of the ripple if the original alternating supply had an r.m.s. p.d. of 12.0 V, the capacitor has a capacitance of 1000 μF and the resistor a resistance of 470 Ω.
(b) What would be the effect of, separately,

making the resistance 4.7 kΩ, and the capacitance 100 μF?

(a) The peak p.d. is $\sqrt{2} \times 12.0\,\text{V} = 17.0\,\text{V}$, and the capacitor is charged up to this p.d. 100 times each second, so we have to ask how much charge the capacitor loses in 0.010 s. Assuming that the p.d. across the capacitor is almost constant, the current I which flows through the resistor is given by

$$I = \frac{V}{R} = \frac{17.0\,\text{V}}{470\,\Omega} = 0.036\,\text{A}.$$

In 0.01 s the charge ΔQ which passes is given by

$$\Delta Q = It = (0.036\,\text{A})(0.010\,\text{s}) = 0.00036\,\text{C}.$$

The fall in p.d. ΔV is given by

$$\Delta V = \frac{\Delta Q}{C} = \frac{(3.6 \times 10^{-4}\,\text{C})}{(1000 \times 10^{-6}\,\text{F})} = 0.36\,\text{V}$$

(b) If the resistance were 10 times greater, the current would be 10 times smaller, and the charge passed would be 10 times smaller. Hence the fall in p.d. would be only 0.036 V. If the capacitance had been 10 times smaller, the same loss of charge would result in a decrease in p.d. which was 10 times greater, i.e. 3.6 V.

Measuring instruments

The direction of the torque on the coil of a moving-coil meter depends on the direction of the current. At low frequencies the coil will therefore turn alternately clockwise and anti-clockwise, but at frequencies as high as 50 Hz the inertia of the coil will prevent it from following the alternation of the supply and it will simply point to the average value of the current, which is zero. Some meters which *can* measure alternating p.d.s and currents are designed on entirely different principles, but nowadays it is more common to use a moving-coil meter or a digital meter fitted with a bridge rectifier. There is a problem if it is necessary to measure *small* alternating p.d.s, since, as we have seen, semiconducting diodes do not conduct even in the forward direction if the p.d. is less than a certain value. But for p.d.s of several volts the error is small, especially if germanium diodes are used.

To measure alternating *currents* the meter must be placed *in* the circuit. We cannot allow there to be a p.d. of more than a fraction of a volt across the meter, or else the meter will exert too great an effect on the current it is measuring; but as we have seen, a small p.d. connected to semiconducting diodes will lead to errors. The solution is to use a **current transformer**. The circuit current passes through the primary coil of the transformer, and the small p.d. across it is stepped up to give a larger p.d. across the bridge rectifier. Figure 19.6(a) shows the circuit and figure 19.6(b) a typical adaptor which can be attached to a moving coil meter to measure alternating current. Alternating current voltmeters and ammeters are calibrated to read the r.m.s. value of the alternating p.d. or current.

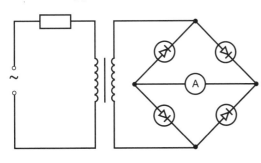

(a)

(b)

FIGURE 19.6

A cathode ray oscilloscope displays the waveform of an alternating p.d. Measurements made from the display give the peak or instantaneous values, and not the r.m.s. values. The example which follows shows how this fact can be used.

EXAMPLE

A light bulb is connected by a two-way switch to either a battery in series with a rheostat, or a sinusoidally varying alternating supply, as shown in the figure. The rheostat is adjusted so that the brightness of the bulb is the same, whichever supply is connected. A cathode ray oscilloscope with the time-base not connected is then adjusted so that there is a small spot at the centre of the screen. The sensitivity is 5.0 V per division. The cathode ray oscilloscope is then connected across the bulb. When the switch connects the battery to the bulb, the spot moves up 3.6 divisions; when the switch connects the alternating supply to the bulb, a vertical line of length 10.4 divisions is seen on the screen. Show that these measurements are consistent with the idea that the peak value of an alternating p.d. is $\sqrt{2}$ times the r.m.s. value.

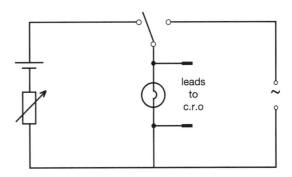

The bulb lights equally brightly in either case, so the steady p.d. must be equal to the r.m.s. value of the alternating p.d. The r.m.s. value is therefore given by the movement of the spot, which represents (3.6 div)(5.0 V/div) = 18 V. The length of the vertical line represents the peak-to-peak value of the alternating p.d., i.e. from the positive peak value to the negative peak value, so the peak value is (5.2 divisions)(5.0 V/div) = 26 V. The ratio (26 V)/(18 V) = 1.44, which is close to the expected value of $\sqrt{2}$ (= 1.41): measurements on a c.r.o. screen cannot be very precise.

19.2 Circuits containing capacitors

A hi-fi system often has loudspeakers of more than one kind: a treble unit to transmit the higher frequencies and a bass unit to transmit the lower frequencies (and perhaps also a third, mid-range, speaker). How are the frequencies directed to the appropriate loudspeaker? To answer this question we need to study what happens when an alternating p.d. is connected to circuits which contain capacitors and inductors.

Suppose a capacitor is connected to an alternating supply, as in figure 19.7(a), and that the circuit contains no resistance. There cannot be a current *between* the plates of the capacitor, but there will be a current in all other parts of the

circuit, since the capacitor will be charged alternately, and for this to happen electric charge must continually be flowing from the supply to the capacitor plates, in alternate directions.

The current will vary in size. How will it vary and what does the size of the current depend on? We can make a start by realising that at any instant the p.d. across the capacitor is equal to the p.d. across the supply. When the supply has its peak p.d., the p.d. across the capacitor will also be at its maximum, and so the capacitor will have its maximum charge. Figure 19.7(b,c) shows graphs of p.d. and charge on the capacitor. The two quantities are in phase. At any time the charge Q on the capacitor is given by $Q = CV$, and the maximum charge Q_0 is given by $Q_0 = CV_0$.

But whenever the charge is a maximum, the current in the circuit must at that moment be zero, so we can plot the two points labelled '1' on the current–time axes shown in figure 19.7(d). And because the current is the rate of flow of charge, the current has its largest size when the charge is changing most quickly, which you can see from the graph occurs when the charge is zero. We can now plot the three points labelled '2' in figure 19.7(d). You might guess that the graph of current against time is going to be sinusoidal, but out of phase with the p.d. and charge, especially if you know enough mathematics to know that the rate of change of a sine function is a cosine function. Just a little calculus proves that this is the case:

If
$$Q = Q_0 \sin(2\pi f t),$$

then
$$I = \frac{dQ}{dt} = 2\pi f Q_0 \cos(2\pi f t)$$

$$= 2\pi f C V_0 \cos(2\pi f t)$$

which shows that the current does indeed vary sinusoidally with time, but not in phase with the p.d.: we say the current *leads* the p.d., as its peak value occurs one quarter-cycle (or $\frac{1}{2}\pi$ radians) before the p.d. reaches its peak. The two graphs are shown in figure 19.8(a).

You can demonstrate this with a double-beam cathode ray oscilloscope, as shown in figure 19.8(b), placing a small resistance in series with the capacitor (it must be small so that the capacitor is still the main factor which controls the current in the circuit). One pair of plates (Y_1 and Earth) is connected across the resistor and

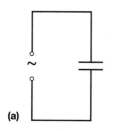

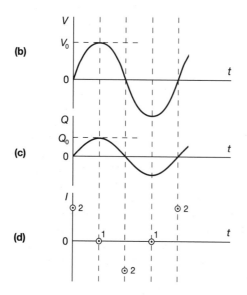

FIGURE 19.7

260

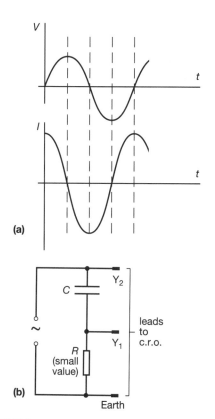

(a)

(b)

Earth

FIGURE 19.8

measures the p.d. across it: because the p.d. across the resistor is in phase with the current in it, this gives a trace which shows how the current varies with time. The other pair (Y_2 and Earth) is connected across the supply terminals and shows how the applied p.d. varies with time.

The equation $I_0 = 2\pi fCV_0\cos(2\pi ft)$ also shows that the peak current I_0 is given by $I_0 = 2\pi fCV_0$, i.e. that it is directly proportional to the frequency f, the capacitance C and the peak p.d. V_0 (but note that the peak current does not occur at the same time as the peak p.d.). This is all to be expected, since the current must be proportional to Q_0, the amount of charge which has to flow in each half-cycle, and $Q_0 = CV_0$, and because current is the *rate of* flow of charge, the current will increase in direct proportion to the frequency.

EXAMPLE

An alternating supply which provides a sinusoidally varying p.d. of frequency 50 Hz is connected to a capacitor of capacitance 0.22 μF.

(a) If the peak value of the supply is 6.0 V, what is the peak value of the current?
(b) What does the peak value of the current become if, separately, (i) the frequency is increased to 100 Hz, and (ii) a second capacitor of capacitance 0.10 μF is connected in parallel with the first capacitor?

(a) Using $I_0 = 2\pi fCV_0$, we have

$$I_0 = (2\pi)(50\,\text{Hz})(0.22\times10^{-6}\,\text{F})(6.0\,\text{V})$$

$$= 4.1\times10^{-4}\,\text{A} = 0.41\,\text{mA}$$

(b)(i) If the frequency is doubled, the current becomes 0.82 mA.
(b)(ii) The second capacitor increases the total capacitance to 0.32 μF, so the current is increased in the ratio (0.32)/(0.22), i.e. from 0.41 mA to 0.60 mA.

In a resistive circuit the power P (given by $P = VI$) is always positive. In a capacitative circuit the power P is again given by $P = VI$ but now, because V and I are a quarter-cycle out of phase, the product VI is sometimes positive and sometimes negative. You can convince yourself that this is true by drawing graphs of V and I against time, and then a graph of VI against time: you will see that the power is negative (power is being returned *to* the supply) as much as it is positive (power is being drawn *from* the supply) so the net power is *zero*, even though a current is flowing.

Reactance

In an alternating current circuit which contains simply a capacitor, it is always possible to work out values of current, as in the last example. In more complicated circuits, however, it is helpful to use the idea of the **reactance** of the capacitor. This is a measure of how much the presence of the capacitor resists the flow of charge in the circuit. It is analogous to resistance, but of course it is *not* a resistance, which is a property of conducting materials. However, reactance, which is given the symbol X, or X_C when it applies to a capacitor, *is* measured in ohms, as you can deduce from the definition:

$$X = \frac{\text{peak p.d. } V_0}{\text{peak current } I_0}$$

Because the peak values are proportional to the r.m.s. values, it is also true that

$$X = \frac{\text{r.m.s. p.d. } V_{\text{r.m.s.}}}{\text{r.m.s. current } I_{\text{r.m.s.}}}$$

For a capacitor, $X_C = \dfrac{V_0}{2\pi f C V_0} = \dfrac{1}{2\pi f C}$

and, since $\omega = 2\pi f$, $X_C = 1/(\omega C)$.

You will notice that, unlike resistance, the *reactance* of a component depends on the frequency of the supply; the larger the frequency, the smaller the reactance, and vice versa. In particular, when f, or ω, is zero, the reactance is infinite, so no current flows. This is what you should expect since when $f = 0$ the current is direct, and you know that no current flows through a capacitor (after the initial charging current). Figure 19.9(a) illustrates how reactance varies with frequency for a $100 \, \mu\text{F}$ capacitor, and figure 19.9(b) shows how the r.m.s. current varies with frequency for the same capacitor in a purely capacitative circuit when the applied p.d. is $10 \, \text{V}$ r.m.s.

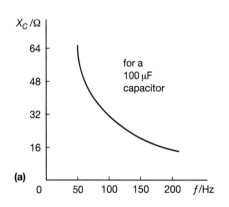

(a)

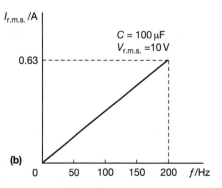

(b)

FIGURE 19.9

EXAMPLE

Using the values given in the last example, calculate the reactance of the capacitor.

(a) $X_C = \dfrac{1}{(2\pi f C)}$

$$= \frac{1}{(2\pi)(50\,\text{Hz})(0.22 \times 10^{-6}\,\text{F})} = 14\,\text{k}\Omega.$$

(b) Doubling the frequency halves the reactance, so now $X_C = 7.2\,\text{k}\Omega$.
(c) Increasing the capacitance reduces the reactance, and $X_C = 9.9\,\text{k}\Omega$.

19.3 Circuits containing inductors

Circuits containing inductors have some of the properties of circuits containing capacitors: the current is not in phase with the supply p.d. and the current depends on the frequency of the supply. The reasons, however, are quite different.

Let us begin by considering a circuit which contains just a power supply and an inductor, as shown in figure 19.10(a): the inductor is assumed to have no resistance. Although in practice it must have some resistance, its effect on the current may well be small compared with that of the inductance. An important thing to remember in this type of circuit is that the e.m.f. induced in the inductor has exactly the same size as the applied p.d. but is opposite in direction: $V = -\mathscr{E}$. This must be true, since the size of the applied p.d. tells us that when the current is I there is a rate of supply of energy to the circuit, given by VI. What is happening to this energy? It is not being converted into internal energy, since we assume that there is no resistance. The only component to which the energy can be delivered is the inductor. If the induced e.m.f. in this is $\mathscr{E}$, then when the current is I, its rate of working is $\mathscr{E}I$. As we shall see later, electrical energy is being converted to **magnetic potential energy** in the inductor. So VI is the same size as $\mathscr{E}I$, and V is the same size as $\mathscr{E}$. In

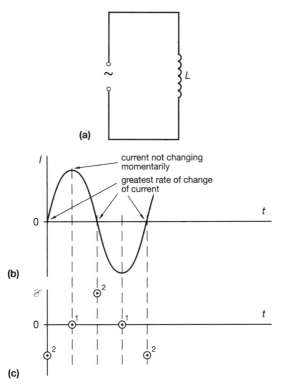

FIGURE 19.10

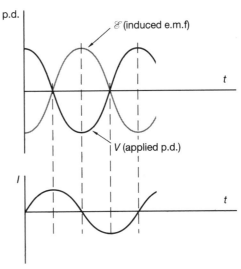

FIGURE 19.11

fact, $V = -\mathscr{E}$, since the e.m.f. acts so as to oppose the changes which are occurring.

We start by drawing a graph of the current in the circuit, as in figure 19.10(b). We know that at the instants when this is not changing, the induced e.m.f. $\mathscr{E}$ in the inductor must be zero, and we can plot two of the points (labelled 1) on the e.m.f. time axes shown in figure 19.10(c). We know too that $\mathscr{E}$ is a maximum when the rate of change of current is greatest (which is when the current is zero). Also, as $\mathscr{E}$ tries to oppose the change, this maximum e.m.f. will be positive when the current is decreasing, and negative when the current is increasing. This enables us to plot three more points (labelled 2) on the graph. You would probably guess that the variation of induced e.m.f. with time is sinusoidal, and this is shown in the graph of $\mathscr{E}$ against time in figure 19.11. Because the applied p.d. V is always equal to $-\mathscr{E}$ we can also draw the graph of applied p.d. V: the graph of I against time is also drawn so that you can more easily see that in a purely inductive circuit the current is out of phase with the p.d. by one quarter-cycle, and it *lags* the p.d.: its maximum

value occurs after the maximum value of the p.d.

More formally, we know that $V = -\mathscr{E}$ and that $\mathscr{E} = -L(dI/dt)$ and that $I = I_0 \sin(2\pi ft)$, we can write

$$V = -\mathscr{E} = L\frac{dI}{dt} = L\frac{d}{dt}\{(I_0\sin(2\pi ft))\}$$

$$= 2\pi fLI_0\cos(2\pi ft).$$

The graphs of V and I which are shown in figure 19.11 are similar to the graphs in figure 19.8 except that this time the current lags the p.d. instead of leading it. The graphs can be displayed on a double-beam oscilloscope using the same circuit as we did for a capacitor, replacing the capacitor with an inductor.

The equation shows that the applied p.d. has a maximum value V_0 of $2\pi fLI_0$, i.e. $V_0 = 2\pi fLI_0$. Rearranging this, we see that

$$I_0 = \frac{1}{2\pi fL}V_0$$

which shows that the peak current I_0 is directly proportional to the peak p.d. V_0, but is inversely proportional to the frequency f and the inductance L. This is what we should expect, since the greater the values of f and L, the greater is the induced e.m.f.

EXAMPLE

An alternating supply which provides a sinusoidally varying p.d. of frequency 50 Hz is connected to an inductor of inductance 13 mH. The peak value of the supply p.d. is 10 V. Assume that the resistance of the inductor is negligible.

(a) What current flows in the circuit?
(b) What does the current become if, separately,
(i) the frequency is increased to 200 Hz, and
(ii) an iron bar is inserted into the inductor and its inductance increases to 1.3 H?

(a) Using

$$I_0 = \frac{1}{2\pi f L} V_0$$

we have

$$I_0 = \frac{1}{(2\pi)(50\,\text{Hz})(0.013\,\text{H})} = 0.24\,\text{A}$$

(b)(i) Increasing the frequency by a factor of 4 reduces the current by a factor of 4, so the peak current is now $0.061\,\text{A} = 61\,\text{mA}$.
(b)(ii) Increasing the inductance by a factor of 100 reduces the current of 0.24 A by this factor, so the current is now only 2.4 mA.

As with a purely capacitive circuit, the power is given by $P = VI$, and the product of V and I is alternately positive and negative. In one quarter-cycle energy is being drawn from the supply to provide magnetic potential energy in the inductor when there is a current in it, and in the next quarter-cycle the current and therefore the magnetic potential energy decreases and energy is returned to the supply. The net power drawn from the supply is zero.

Reactance

As with a capacitor, we can define the reactance X_L of an inductor using

$$X_L = \frac{V_0}{I_0} = 2\pi f L = \omega L.$$

Also

$$X_L = \frac{V_{\text{r.m.s.}}}{I_{\text{r.m.s}}}.$$

The reactance varies with the frequency f. When f is zero, the reactance is zero: this corresponds to the current not changing, i.e. being a direct current, and then the inductor has no effect on the current (after the initial establishing of the magnetic potential energy in the inductor). When f is very large, the reactance is very large, and the current is very small. This is because $V = L(\text{d}I/\text{d}t)$ and since V and L are fixed, the rate of change of current $\text{d}I/\text{d}t$ is fixed. If the frequency is high, the current cannot also be large, if the rate of change is fixed. Figure 19.12(a) shows how reactance X_L varies with frequency for an inductor of inductance 100 mH, and figure 19.12(b) shows how the current varies with frequency in a purely inductive circuit for the same inductor when the applied p.d. is 10 V r.m.s.

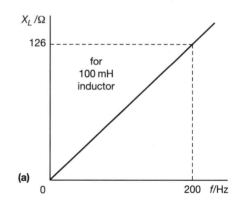

(a)

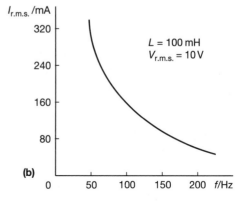

(b)

FIGURE 19.12

EXAMPLE

Using the values given in the last example, calculate the reactance of the inductor.

(a) $X_L = 2\pi fL = (2\pi)(50\,\text{Hz})(0.013\,\text{H}) = 4.1\,\Omega$.
(b)(i) X_L will be 4 times greater, i.e. $16\,\Omega$.
(b)(ii) X_L will be 100 times greater than it was originally, i.e. $0.41\,\text{k}\Omega$.

Filter circuits

The circuit shown in figure 19.13 could be used to connect the output of an amplifier to a loudspeaker system consisting of a treble unit (a **tweeter**) and a bass unit (a **woofer**). These speakers are designed to reproduce high and low frequencies respectively.

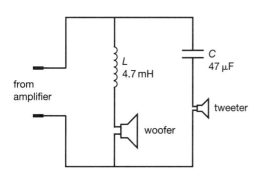

FIGURE 19.13

The audible range for humans is from about 20 Hz to about 18 kHz, so taking three representative values of frequency (30 Hz, 300 Hz and 3.0 kHz), and using the equations $X_C = 1/2\pi fC$ and $X_L = 2\pi fL$ the reactances of the capacitor (capacitance 47 μF) and the inductor (inductance 4.7 mH) can be calculated for these frequencies. You should check that they are as shown in table 19.1.

TABLE 19.1 Calculations for filter circuit

f/Hz	X_C/Ω	X_L/Ω
30	113	0.886
300	11.3	8.86
3000	1.13	88.6

You can see that the low-frequency part of the alternating p.d. from the amplifier will pass mainly through the inductor, and hence the woofer, whereas the high-frequency part will pass mainly through the capacitor, and hence the tweeter. This is known as a **filter circuit**.

19.4 Mixed circuits

Circuits containing capacitance and resistance

You will expect a circuit which contains a resistor as well as a capacitor (an RC circuit) to be more difficult to analyse. However, in such a circuit some familiar rules still hold.

♦ At any instant the current at each point in the circuit is the same.

♦ The current is in phase with the p.d. across the resistor R.

♦ The current leads (by a quarter-cycle) the p.d. across the capacitor C.

♦ The applied p.d. must, at any instant, be equal to the sum of the p.d.s across R and C.

The graph in figure 19.14 shows the result: the current leads the applied p.d. by *less than* one quarter-cycle, but you should draw these graphs for yourself. First draw a graph of the current. Then, on new axes, draw the graph of p.d. V_R across R and then the graph of the p.d. V_C across C. Finally add the two p.d.s together to get the graph of the p.d. $V (= V_R + V_C)$ across the supply.

The size of the current in an RC circuit depends on both R and C, but we cannot simply add the resistance of R to the reactance of C to find the total effect of these two components. This is because the two p.d.s are not in phase. Instead the total effect, which is called the **impedance** Z of the circuit, is given by

$$Z^2 = R^2 + X^2.$$

Just as X is the ratio V_0/I_0 and $V_{\text{r.m.s.}}/I_{\text{r.m.s.}}$ for a reactance X, so

$$Z = \frac{\text{peak p.d. } V_0}{\text{peak current } I_0}$$

and $$Z = \frac{\text{r.m.s. p.d. } V_0}{\text{r.m.s. current } I_0}$$

for an impedance Z.

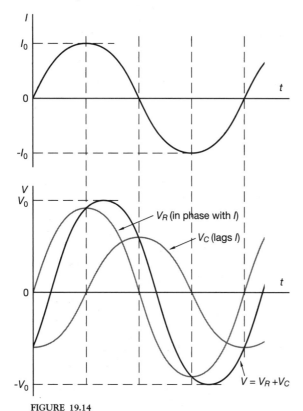

FIGURE 19.14

so the impedance Z of the circuit is given by

$$Z^2 = R^2 + X^2$$
$$= (470\,\Omega)^2 + (677\,\Omega)^2$$
$$\Rightarrow \quad Z = 824\,\Omega$$

so $$I_{\text{r.m.s.}} = V_{\text{r.m.s.}}/Z$$
$$= (8.0\,\text{V})/(824\,\Omega)$$
$$= 9.7\,\text{mA}.$$

If the frequency had been 500 Hz, X_C would have been one tenth the size, i.e. $68\,\Omega$. A similar calculation shows that then $Z = 475\,\Omega$ and $I_{\text{r.m.s.}} = 17\,\text{mA}$.

Circuits containing inductance and resistance

A circuit may consist of a separate inductor and resistor but because inductors are made of many turns of wire, the only resistance in the circuit may be that of the inductor itself. However it is legitimate to think of an inductor with resistance as consisting of an inductor and a resistor in series. In these 'RL' circuits the same principles apply as for 'RC' circuits.

- At any instant the current at each point in the circuit is the same.

- The current is in phase with the p.d. across the resistor R.

- The current lags (by a quarter-cycle) the p.d. across the inductor L.

- The applied p.d. must, at any instant, be equal to the sum of the p.d.s across R and L.

We could use these ideas to show that the current lags the applied p.d. by *less* than a quarter-cycle. This fits into the pattern which we have seen for circuits which carry alternating current, and is included in table 19.2. The impedance Z of an RL circuit is given by $Z^2 = R^2 + X^2$, where X is the reactance of the inductor.

EXAMPLE

A resistor of resistance $470\,\Omega$ is connected in series with a capacitor of capacitance $4.7\,\mu\text{F}$ to a power supply of frequency 50 Hz which provides an r.m.s. p.d. of 8.0 V. What is the current in the circuit, and what would the current be if the frequency of the supply were increased to 500 Hz?

The reactance of the capacitor is given by

$$X_C = \frac{1}{2\pi fC}$$

$$= \frac{1}{(2\pi)(50\,\text{Hz})(4.7 \times 10^{-6}\,\text{F})}$$

$$= 677\,\Omega$$

TABLE 19.2 Phase relationship between I and V in various circuits

Circuit	Phase relationship between I and V
C only	I leads V by a quarter-cycle
RC	I leads V by less than a quarter-cycle
L only	I lags V by a quarter-cycle
RL	I lags V by less than a quarter-cycle

$$2\pi fL = \frac{1}{2\pi fC}$$

i.e.

$$f^2 = \frac{1}{4\pi^2 LC}$$

At this frequency, with the impedance a minimum, the current will be a maximum. The circuit is said to **resonate** at this frequency, since there is a large response (a large current) at this frequency.

EXAMPLE

An inductor with 100 turns of copper wire has resistance $0.70\,\Omega$ and inductance $10\,mH$.
(a) Calculate the ratio of resistance to reactance at a frequency of $50\,Hz$.
(b) If the inductance depends on the square of the number of turns, calculate the new ratio when the number of turns is increased to 1000.

(a) The reactance X_L is given by $X_L = 2\pi fL = (2\pi)(50\,Hz)(0.010\,H) = 3.1\,\Omega$, so the ratio $R/X = (0.70\,\Omega)/(3.1\,\Omega) = 0.22$.

(b) Now the resistance will be 10 times greater, because the wire is 10 times longer, but the inductance will be $10^2 (= 100)$ times greater, so the new ratio $R/X = (7.0\,\Omega)/(314\,\Omega) = 0.022$.

We can see that the larger the number of turns, the more justified it is to treat an inductor as having negligible resistance. This is important in the design of transformers.

LCR series circuits

The rule for calculating the impedance Z when a circuit contains all three types of component is

$$Z^2 = R^2 + (X_L - X_C)^2.$$

When the current leads or lags the applied p.d. depends on whether the inductor or the capacitor has the greater effect. If X_L is greater than X_C, then the current lags the p.d.; if X_C is greater than X_L, the current leads the p.d.

X_L and X_C both depend on the frequency of the supply, and it follows from the equation that the impedance of the circuit will be a minimum when the frequency is such that $X_L = X_C$, since then $Z = R$. If $X_L = X_C$, then

EXAMPLE

A circuit contains a resistor of resistance $22\,\Omega$, a capacitor of capacitance $47\,\mu F$ and an inductor of inductance $0.10\,H$ connected in series to a power supply which provides a peak p.d. of $6.0\,V$ at a variable frequency. Calculate

(a) the frequency which should be used to provide the maximum current,
(b) the size of this maximum current, and
(c) the current which flows when the frequency is $10\,Hz$ less than the frequency calculated in (a).

(a) For maximum current,

$$f^2 = \frac{1}{4\pi^2 LC}$$

$$= \frac{1}{(4\pi^2)(0.10\,H)(47 \times 10^{-6}\,F)}$$

$$\Rightarrow \quad f = 73\,Hz.$$

(b) Since now $X_L = X_C$, the impedance $Z = R$, so $I_0 = V_0/R = (6.0\,V)/(22\,\Omega) = 0.27\,A$.

(c) When $f = 63\,Hz$,
$X_L = (2\pi)(63\,Hz)(0.10\,H) = 39.6\,\Omega$ and

$$X_C = \frac{1}{(2\pi)(63\,Hz)(47 \times 10^{-6}\,F)} = 53.8\,\Omega$$

so $Z^2 = (22\,\Omega)^2 + (53.8\,\Omega - 39.6\,\Omega)^2 \Rightarrow Z = 26\,\Omega$, so the peak current is given by $I_0 = V_0/Z = (6.0\,V)/(26\,\Omega) = 0.23\,A$.

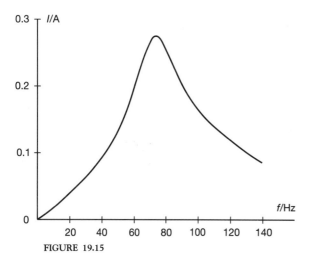

FIGURE 19.15

resistor. But if we connect the capacitor to an inductor (figure 19.17(a)) the result is different: for the time being we shall assume that this circuit has no resistance. The p.d. across the capacitor is given by $V_C = Q/C$ and the p.d. across the inductor is given by $V_L = L(dI/dt)$ and these two p.d.s must be the same, so

$$\frac{Q}{C} = L\frac{dI}{dt}, \quad \text{or} \quad \frac{dI}{dt} = \frac{1}{LC}Q$$

Also, the rate of change of charge on the capacitor (dQ/dt) is equal to $-I$ (the minus sign arises because when I is in the direction shown, the charge is decreasing). So

$$-I = \frac{dQ}{dt} \quad \text{and} \quad -\frac{dI}{dt} = \frac{d^2Q}{dt^2}$$

$$\text{so } \frac{d^2Q}{dt^2} = -\frac{1}{LC}Q$$

The graph in figure 19.15 shows how the current varies with frequency. The sharpness of the peak depends on how much resistance there is in the circuit. With less resistance, the peak is sharper. You should expect the current to be zero if the frequency is zero, since then the capacitor blocks the alternating current. The current is also small at high frequencies, since then the inductor blocks the alternating current.

19.5 Electrical oscillations

If we charge a capacitor and then connect it to a resistor (as in figure 19.16(a)) the charge on it decreases exponentially, as shown in figure 19.16(b). So does the p.d. across it, and the current in the circuit. The electric potential energy of the capacitor is converted into internal energy in the

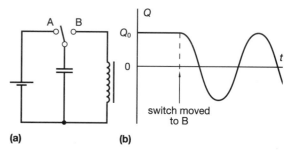

FIGURE 19.17

You should remember that you have seen an equation like this before, when dealing with simple harmonic motion: then

$$\frac{d^2s}{dt^2} = -(\text{constant})s$$

which means that s varies sinusoidally with time. The equation we have here must mean that *the charge Q varies sinusoidally with time*, as shown in figure 19.17(b), so you should expect the p.d. across the capacitor to vary in the same way, since $V_C = Q/C$. The circuit *oscillates*.

This happens because when the charge on the capacitor falls to zero the current is a maximum (compare maximum velocity for zero displacement in a s.h.m.). When the current then starts to

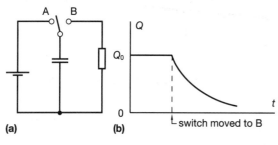

FIGURE 19.16

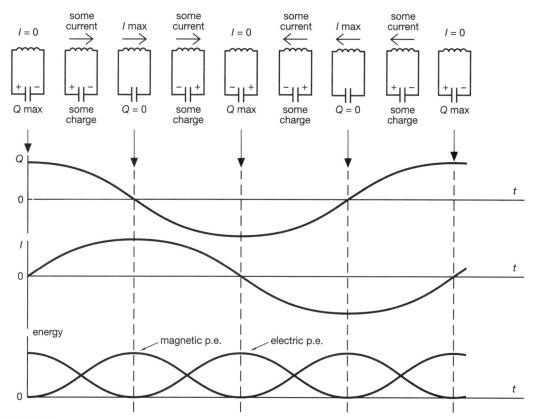

FIGURE 19.18

decrease an e.m.f. is induced in the inductor which tries to maintain the current, so the current continues and the capacitor begins to be charged in the opposite sense. When it is fully charged (with the lower plate now positive) the current is again zero. Then the whole process starts again.

During the oscillation the electric potential energy which was originally stored in the capacitor is being converted into magnetic potential energy in the inductor, and then back to electrical potential energy again, indefinitely: if there is no resistance in the circuit these are the only possible energy conversions. We could write the total energy in the circuit as the sum of the electric p.e. and the magnetic p.e.:

$$\text{total energy} = \text{electric p.e.} + \text{magnetic p.e.}$$

$$= \tfrac{1}{2}Q^2/C + \tfrac{1}{2}LI^2 = \text{constant.}$$

You should now look at the graphs in figure 19.18 which show how Q and I vary with time and relate the circuit diagrams to the various stages of the

graph. There are also graphs which show how the electric and magnetic potential energies vary with time. You should also check that you understand how these values relate to the different stages of the oscillation.

The mass–spring analogy

In the last chapter we used mechanical analogies to help us understand inductance. Analogies are useful again here, especially if we add another property to the table: the compliance (the inverse

TABLE 19.3 Characteristics of the mass-spring analogy for the LC circuit

Mechanical	Electrical
velocity v	current I
acceleration dv/dt	rate of change of current dI/dt
force F	p.d. V
mass m	inductance L
compliance of spring $1/k$	capacitance C

of the stiffness) of a spring $(1/k)$ is analogous to the capacitance C of a capacitor. Let us compare the LC circuit to a mass connected to a spring, as shown in figure 19.19. The mass rests on a horizontal frictionless surface, and the spring can be both stretched and compressed. Then a capacitor being charged is like the mass being given a displacement: the capacitor has electric p.e., and the spring has elastic p.e.

In practice any electrical circuit does have some resistance, and so the oscillations are damped: their amplitude decreases with time, as shown in figure 19.20. The initial electric p.e. is converted into internal energy in the resistance of the circuit. This graph was drawn by a printer connected to a VELA datalogger which measured the p.d. across the inductor at time intervals of 10 ms.

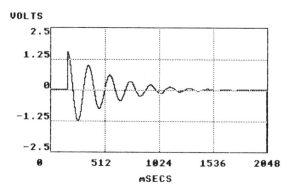

FIGURE 19.20

(a)

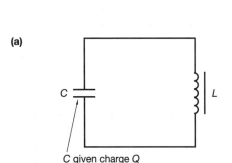

C given charge Q

(b)

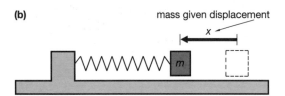

mass given displacement

FIGURE 19.19

When the capacitor is allowed to discharge, this is like the mass moving towards its equilibrium position: just as the inductor gains magnetic p.e., so the mass gains k.e. When the capacitor has no charge, the current is a maximum: when the mass has zero displacement, its velocity is a maximum, and just as the mass continues to move through its equilibrium position, because of its inertia, so does the current continue to flow, even when the capacitor has no charge, because of the inductance of the inductor. Just as there is a continual interchange of electric p.e. and magnetic p.e. in the circuit (if there is no resistance) so for the spring there is a continual interchange of elastic p.e. and k.e., if there is no resistance. If there is resistance, then in both cases the oscillations decrease in amplitude and the electric and mechanical energy is converted into internal energy.

Resonance

This LC loop clearly has a natural frequency of oscillation which can be calculated by realising that in figure 19.19 there must, at any moment, be the same p.d. across both L and C, and there must be the same current in each. So they must have the same reactance. If

$$X_C = X_L, \quad \text{then } \frac{1}{2\pi fC} = 2\pi fL,$$

so

$$f^2 = \frac{1}{4\pi^2 LC} \text{ or } f = \frac{1}{2\pi}\sqrt{\left(\frac{1}{LC}\right)}$$

$$\left(\text{compare } f = \frac{1}{2\pi}\sqrt{\left(\frac{k}{m}\right)} \text{ for a mass on a spring}\right).$$

If it is supplied with an alternating e.m.f. which has this frequency, the oscillations will increase in amplitude until the rate of supply of electrical energy is balanced by the rate of conversion of it into internal energy in the resistance of the circuit.

Radio tuning circuits

One important application of this principle is in radio reception. Many different broadcasting stations are sending out electromagnetic waves of different frequencies at the same time: what is required, of course, is that the radio receiver should select the required station and no other. Figure 19.21 shows the aerial circuit of a radio receiver.

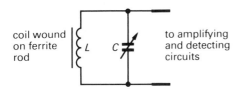

coil wound on ferrite rod — L — C — to amplifying and detecting circuits

FIGURE 19.21

The aerial consists of a ferrite rod with a coil wound round it. The rod is magnetised by the oscillating magnetic field of the electromagnetic waves moving past it: e.m.f.s of many frequencies are therefore induced in the coil L of the resonant LC loop. At most of these frequencies no resonance occurs and the currents in the loop remain small. But if an e.m.f. is induced at the resonant frequency, the amplitude of the current in the loop increases, and an appreciable alternating p.d. is produced across it. The output of the circuit therefore effectively contains only those frequencies very close to the resonant one.

EXAMPLE

The aerial circuit of a radio receiver has a coil of inductance 1.8 mH and a capacitor whose capacitance can be varied. To what value should the capacitor be adjusted to receive waves of frequency 200 kHz?

If the resonant frequency is to be 200 kHz then

$$C = \frac{1}{4\pi^2 f^2 L} = \frac{1}{(4\pi^2)(200 \times 10^3 \,\text{Hz})^2 (1.8 \times 10^{-3}\,\text{H})}$$

$$= 3.5 \times 10^{-10}\,\text{F}$$

$$= 350\,\text{pF}$$

19.6 Transformers

Earlier we saw that if two coils are placed near each other a changing current in one coil produces changing magnetic flux through the other and induces an e.m.f. in it. This principle is used in the **transformer**, which consists of two separate coils of wire which are both wound on a closed iron core, as shown in figure 19.22. The conventional symbol for a transformer is shown also, the straight line indicating the iron core. To one of these coils an alternating supply is connected: this is called the **primary** coil. The other coil is called the **secondary** coil. We shall call the numbers of turns

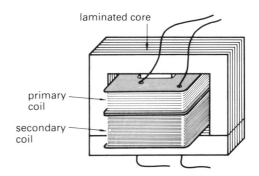

laminated core

primary coil

secondary coil

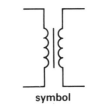

symbol

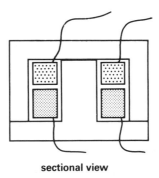

sectional view

FIGURE 19.22

in the two coils n_1 and n_2. The alternating current produces an alternating magnetic flux in the core, and since both coils are wound on the same iron core, the same flux passes through each turn of both coils and therefore induces the same e.m.f. in each of these turns. If we call this e.m.f. (induced in each turn) e, then the e.m.f. $\mathscr{E}_1$ induced in the primary coil is $n_1 e$ and the e.m.f. $\mathscr{E}_2$ induced in the secondary coil is $n_2 e$. So

$$\mathscr{E}_1 = n_1 e \text{ and } \mathscr{E}_2 = n_2 e$$

so $\quad \dfrac{\mathscr{E}_2}{\mathscr{E}_1} = \dfrac{n_2}{n_1} \quad (\dfrac{n_2}{n_1} \text{ is called the } \textbf{turns-ratio}).$

If there is a large number of turns in the primary coil its reactance will be much greater than its resistance: we saw in an example, earlier, that the greater the number of turns, the greater the ratio of reactance to resistance, so this is a reasonable assumption. We shall therefore assume that the resistance is negligible. Then we can say that the p.d. of the alternating supply (call this V_1) is equal to the e.m.f. $\mathscr{E}_1$ induced in the primary coil. The e.m.f. $\mathscr{E}_2$ generated in the secondary is the same as the p.d. V_2 across its terminals when there is no current (as is the case with a cell): if the resistance of the secondary coil is small, as it usually is, we can write $V_2 = \mathscr{E}_2$, and then

$$V_2 = \frac{n_2}{n_1} V_1$$

if the secondary current is not too large.

So the transformer provides a p.d. V_2 which may be larger or smaller than the p.d. V_1 applied to it. For example, if the turns-ratio is 10, V_2 will be equal to $10 V_1$ (it is a **step-up** transformer), and if the turns ratio is $\frac{1}{2}$, then V_2 will be $\frac{1}{2}V_1$ (it is a **step-down** transformer).

You may be wondering if in a step-up transformer we are 'getting something for nothing', since the p.d. which the transformer provides is greater than the applied p.d. The answer, as usual, is 'no'! Suppose we have a step-up transformer with a turns-ratio of 20, and we connect the primary coil to an alternating 12 V supply, as shown in figure 19.23. (We are using r.m.s. values of p.d. and current.) The secondary p.d. will be 240 V, and we might use this to light a 60 W mains bulb. As expected the current is 0.25 A, but when we look at the current which the power supply provides for

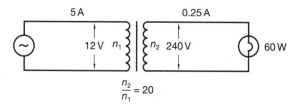

FIGURE 19.23

the transformer we find that it is 5 A, i.e. 20 times 0.25 A, and therefore the power put into the transformer is exactly the same as the power put out by the transformer. There is no miraculous gain of energy. The transformer transforms the p.d. *and* the current, providing us, if we wish, with a higher p.d., but it cannot at the same time provide us with a larger current. It steps *down* the current just as much as it steps *up* the p.d. We can sum this up in the **power equation**

$$V_2 I_2 = V_1 I_1.$$

where I_1 and I_2 are the primary and secondary currents.

Strictly speaking the equations stated so far apply to an 'ideal' transformer, one for which no approximations need be made. In any real transformer, it will not be true that all the primary flux links the secondary coil, or that the resistances of the coils are negligible. So the output p.d. will be a little less than that calculated using $V_2 = (n_2/n_1)V_1$, and the power output $V_2 I_2$ will be a little less than the power input $V_1 I_1$.

There are other reasons for expecting the power output to be less than the power input. There will be internal energy generated

♦ in the coils by the currents in them;

♦ in the core by hysteresis (the reversals of its magnetisation);

♦ in the core by eddy currents induced by the changing flux in the core (though the core is laminated to minimise this effect).

Even so, transformers are in practice very efficient: large ones may have an efficiency of 99%. This implies that the approximations we have made are justified.

EXAMPLE

A laboratory power supply, connected to a 240 V mains supply, contains a transformer which provides a p.d. of 12 V. The front panel states that the maximum current which may be taken from the power supply is 8.5 A. Unfortunately the power supply does not work, and you suspect that a fuse in the primary circuit has failed, but the label indicating the fuse rating has been obscured. What size fuse would you use?

The power output from the transformer, given by $P = VI$, may be as much as $(12\,\text{V})(8.5\,\text{A}) = 102\,\text{W}$. The primary p.d. is 240 V, so the primary current, given by $I = P/V = (102\,\text{W})/(240\,\text{V}) = 0.43\,\text{A}$. Fuses of rating 0.5 A and 1 A are probably available: at first sight 0.5 A looks like the better choice, since it is not much greater than the maximum allowed current, but 1 A is more practicable, since there are often large currents when heaters are first switched on, because their resistance is low when they are cold, so the initial current surge would melt a 0.5 A fuse each time.

The national grid system

One very important use of transformers is to step up the p.d. produced by the alternators in power-stations. The reason for this is that to deliver a particular power we need to supply less current if we use a larger p.d. In this way the current flowing along the power cables in the grid can be minimised. Typically a power-station provides a p.d. of 25 kV, and this is stepped up to 400 kV by a transformer at the power-station. Wires, many kilometres long, then connect the output to a distant town, but the energy loss in these wires is now relatively small since when the p.d. was stepped up, the current was stepped down. The power loss P in the lines is given by $P = VI = V^2/R = I^2R$, where V is the p.d. *between the ends of the cable* and I is the current flowing in it. Since we do not usually know the value of this p.d., we use $P = I^2R$, which makes the point that since $P \propto I^2$ the rate of loss of energy is much less than before (P would become one-hundredth of what it was if I were one-tenth of what it was). It is, of course,

necessary that the output from the power-station should be alternating: transformers do not work with direct current.

Other transformers, in sub-stations, are then used to step down the p.d. for use in factories and houses. This may be done in several stages: the p.d. may be stepped down first to 33 kV or 11 kV, and only when the wires are near the consumer is it stepped down further to the final 240 V.

Although the percentage of energy converted to internal energy in a transformer is small (1% or less) the absolute amount of energy converted is large when the power is large. If a transformer has an input power of 500 MW, as is common in power-stations, internal energy is being generated at a rate of up to 5 MW. A cooling system is therefore needed, and oil is pumped through pipes in close contact with the transformer core, so that the unwanted internal energy can be given to the surroundings. Figure 19.24 shows oil pipes round a small transformer used in the country to step down a p.d. of 11 kV to 240 V to an isolated village. In these pipes the oil flows through natural convection.

FIGURE 19.24

EXAMPLE

A power station has an output power of 2000 MW at a p.d. of 25 kV and a step-up transformer is used to provide a p.d. of 400 kV for the grid system. The wires in the grid system have a diameter of 24 mm and are made of aluminium, which has a resistivity of $2.5 \times 10^{-8} \, \Omega \, m$.

(a) Assuming that the transformer works ideally, at what rate is energy wasted in one kilometre of the wire?

(b) What would have been the power loss if the p.d. had not been stepped up?

(a) The primary current is given by
$$I = P/V = (2000 \times 10^6 \, W)/(25 \times 10^3 \, V)$$
$$= 80 \, kA.$$
The step-up ratio is $(400 \, kV)/(25 \, kV) = 16$, so the secondary current is 16 times smaller than this, i.e. 5.0 kA.

The resistance R of 1 km of the wire is given by

$$R = \frac{\rho l}{A}$$

$$= \frac{(2.5 \times 10^{-8} \, \Omega \, m)(1000 \, m)}{\pi (12 \times 10^{-3} \, m)^2}$$

$$= 5.5 \times 10^{-2} \, \Omega$$

so the power loss $P = I^2 R =$
$(5.0 \times 10^3 \, A)^2 (5.5 \times 10^{-2} \, \Omega) = 1.4 \, MW.$

(b) If the current had not been stepped down, it would still be 80 kA, and so the power loss, which depends on the (current)^2, would be 16^2 ($= 256$) times greater, i.e. 360 MW, which is clearly unacceptable.

Other uses of transformers

Some transformers have more than one secondary coil. For example, a common type of simple oscilloscope has a primary coil connected to the 240 V mains supply, and four secondary coils. One gives a p.d. of 6.3 V to heat the filament, two give

p.d.s of 300 V and 140 V to provide the anodes with the necessary p.d.s, and a fourth gives a p.d. of 12 V to provide the power supply for the operational amplifiers used to provide the amplification. Again, we are not getting something for nothing: when more than one secondary coil is being used, more current, and therefore more power, is drawn from the supply.

An **isolating** transformer is used in situations where there is a risk of electric shock. It usually has a turns-ratio of 1 so its purpose is not to provide a different p.d. but to separate the user of electrical equipment from the mains supply. Figure 19.25 shows how this works. Normally we can get a shock because current from a live wire runs through us to Earth and back to the supply, which is also earthed. With an isolating transformer, touching *either* of the terminals does not make a complete circuit (though, of course, touching *both* would). Isolating transformers are used particularly in wet or damp conditions, where accidental contact with live terminals is more likely: e.g. on building sites, or for shaver sockets in bathrooms.

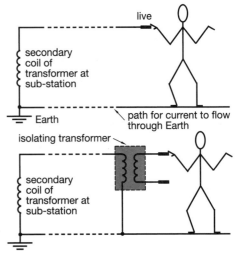

FIGURE 19.25

> **Exercises** on each section of this chapter may be found in the companion textbook, **Practice in Physics**.

SUMMARY

At the end of this chapter you should be able to:

◆ understand that the electricity supply in Great Britain is alternating sinusoidally at a frequency of 50 Hz.

◆ use the equations $V = V_0 \sin 2\pi ft$ and $I = I_0 \sin 2\pi ft$ to calculate values of V and I at different times.

◆ understand that in a resistive circuit the variations of current and p.d. are in phase.

◆ remember that for a sinusoidally varying quantity the root mean square (r.m.s.) value is the fraction $1/\sqrt{2}$ of the peak value.

◆ understand the use of four diodes in a bridge network to produce full-wave rectification.

◆ understand the use of a smoothing capacitor to produce an almost steady p.d. from either a half- or full-wave rectified output.

◆ understand that in a purely capacitative circuit the alternating current leads the p.d. by a quarter-cycle.

◆ understand what is meant by the reactance of a capacitor and how it depends on frequency and capacitance.

◆ understand that in a purely inductive circuit the e.m.f. induced in the inductor is exactly the same size as the applied p.d.

◆ understand that in a purely inductive circuit the alternating current lags the p.d. by a quarter-cycle.

◆ understand what is meant by the reactance of an inductor and how it depends on frequency and capacitance.

◆ understand that an inductor stores magnetic potential energy.

◆ understand that there will be oscillations of p.d. when a capacitor is discharged through an inductor.

◆ understand the analogy between an electrical oscillating circuit and a mass linked to a spring.

◆ understand that a radio receiver contains a circuit which resonates to the frequency of the chosen broadcast.

◆ understand the principle of the transformer.

◆ use the equation $V_2 = (n_2/n_1)V_1$ to calculate the secondary or primary p.d.

◆ understand that the power equation $V_2 I_2 = V_1 I_1$ is a consequence of the principle of conservation of energy.

◆ remember that the transformer equations apply to an 'ideal' transformer, and remember the reasons why real transformers behave slightly differently.

◆ understand the need to minimise the current in the grid system, and how this can be achieved by using high p.d.s.

20 Oscillatory Motion

20.1 Describing oscillations

A swinging pendulum bob, a bouncing rubber ball and a mass on the end of a spring are all examples of objects which vibrate or oscillate. The motions are repetitive: to-and-fro or up-and-down. Both the pendulum and the lump oscillate in such a way that the time taken for a complete cycle T is constant even if the size of the oscillation decreases because the energy of the oscillation is gradually being transformed to internal energy. The bouncing of the rubber ball is quite different; T gets smaller and smaller as the bounce gets smaller. Most oscillating systems are like the pendulum and the lump-on-a-spring; their motion is **isochronous**, i.e. the periodic time is independent of the size of the oscillation.

If the periodic time or **period** of an oscillation (a complete to-and-fro motion) is T and it repeats itself with a frequency f then

$$f = \frac{1}{T}$$

For example, the girl on the swing may take 0.8 s from the highest point at the back to the highest point at the front. So $T = 1.6$ s and $f = (1/1.6)\text{s}^{-1}$ or $0.625\,\text{s}^{-1}$. The unit for frequency, the 'per second' is called a **hertz** (Hz) so here $f = 0.625$ Hz.

Each one of us has, as a baby, had the experience of being rocked to-and-fro in an adult's arms. Even before we were born we were bumped gently up-and-down in our mother's womb as she walked along the road. As children we probably played on swings. When crossing the English Channel by boat we may have felt seasick because a heavy swell moved us and our stomachs up and down in a special way. A helicopter pilot's body is subject to oscillations which can make his eyes wobble in their sockets and produce a blurring of vision. So oscillations can be pleasurable or uncomfortable or dangerous.

Many objects tend to oscillate or vibrate naturally at special frequencies. The sound from a door-bell gong is the result of the gong oscillating when struck. Actually all musical instruments use vibrating strings or vibrating columns of air to produce their characteristic sounds. Tall chimneys oscillate even in steady winds, as do power cables and even suspension bridges. In this chapter you are going to learn to describe how bodies oscillate – and to explain why they do so.

Studying oscillating bodies

Figure 20.1 shows three possible ways of finding out in more detail how an oscillating body actually moves. Method (a) uses an ultrasonic position sensing technique to follow a swinging pendulum. The ultrasonic pulses are reflected from a card attached to the pendulum. The variable output can be fed to a computer and displayed graphically. In (b) a trolley is tethered to two fixed supports by springs so that it can oscillate to-and-fro. By attaching ticker-tape to it a record of half a full oscillation can be produced. A third method (c) uses stroboscopic photography which can capture

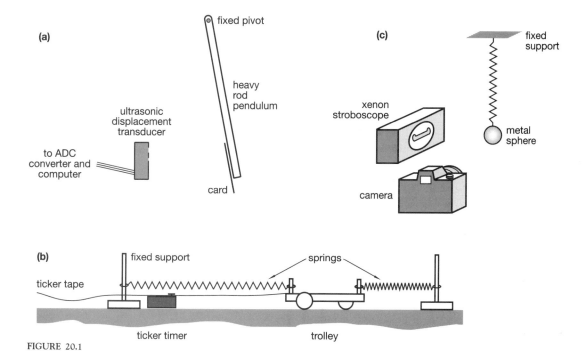

FIGURE 20.1

half an oscillation of the mass on the spring. The techniques are interchangeable: (c) has the advantage of not exerting extra forces on the oscillating body. From each it is possible to produce a displacement–time graph for the motion. Such graphs all have a characteristic *sinusoidal* shape. A particularly neat way of producing an *s-t* curve experimentally is to move a camera steadily in a direction perpendicular to the line of an oscillation. Figure 20.2 shows the result of doing this for the lump oscillating up and down on a spring.

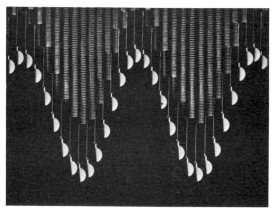

FIGURE 20.2

Simple harmonic motion

From a displacement–time graph you can deduce both the velocity–time and the acceleration–time graph for the oscillator. You draw tangents to the *s-t* curve (as $v = ds/dt$) to find values of v, and tangents to the *v-t* curve (as $a = dv/dt$) to find values of a. Or, of course, you can program the computer to do this for you. You find that the resulting graphs are themselves sinusoidal curves; the positions of their peaks and troughs are related as shown in figure 20.3 which continues the curves for a period or so. All these sine-shaped curves have the same period T (and frequency f). When oscillations are like this, a simple repeating oscillation, they are called 'simple harmonic motions', or just **s.h.m.**

At moments such as X the oscillator is at its maximum (positive) displacement s_0 from the centre and is momentarily at rest, so that $v = 0$. However, this is the point at which v is changing at its maximum rate (from positive to negative values), thus a is at a (negative) maximum. Similarly at moments such as Y, $T/4$ later, the oscillator has zero displacement but its displacement is changing at its maximum rate (from positive to negative values), i.e. v is at a

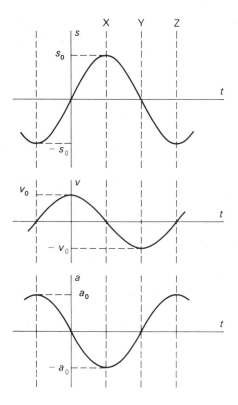

FIGURE 20.3

$$a = -(\text{constant})s$$

at every point, where the constant has the unit s^{-2}. Since $s = s_0 \sin 2\pi ft$ the maximum rate of change of s (which is called v_0) is given by

$$v_0 = \left(\frac{ds}{dt}\right)_{max} = 2\pi fs_0$$

which is often written $v_0 = \omega s_0$ using the symbol ω (Greek omega) where $\omega = 2\pi f$. (This can be shown by differentiation but if you are not studying mathematics it is very useful to remember that, for sinusoidally varying quantities of frequency f, the maximum rate of change is $2\pi f$ times the amplitude. We use it again below, just after the examples.)

A more general equation which gives the velocity, v, at any displacement, s, is

$$v = 2\pi f \sqrt{(s_0^2 - s^2)}$$
$$= \omega \sqrt{(s_0^2 - s^2)}$$

(negative) maximum $-v_0$. At this point v is momentarily not changing; so that $a = 0$. You should continue the argument for moments such as Z or at $t = 0$, and be quite clear how the curves of figure 20.3 are related.

Equations for s.h.m.

The equation for the s-t graph of figure 20.3 is

$$s = s_0 \sin 2\pi ft$$

where f is the frequency. Because the greatest positive and negative values of $\sin 2\pi ft$ are $+1$ and -1, the greatest positive and negative values of s are $\pm s_0$. s_0 is known as the **amplitude** of the oscillation.

With $f = 1/T$ we can see that the term $2\pi ft$ will be 2π rad (or $360°$) when $t = T$, and the curve then repeats itself and does so every period.

Notice that the a-t graph is upside down compared with the s-t graph (i.e. troughs of a coincide with peaks of s). This means that

EXAMPLE

A grandfather clock pendulum of length 0.99 m swings through the vertical every 1.00 s. If the amplitude of the swing is 5.6° calculate the speed of the bottom of the pendulum as it passes through the vertical.

The amplitude

$$5.6° = 5.6 \times \frac{2\pi}{360} \text{ rad} = 0.098 \text{ rad}$$

so the bottom of the pendulum swings from the centre along an arc of length

$$s = r\theta = (0.99 \text{ m})(0.098 \text{ rad}) = 0.097 \text{ m}$$

The frequency of the motion $f = 1/2.00 \text{ s}$
$= 0.50 \text{ Hz}$

so the maximum speed

$$v_0 = 2\pi fs_0$$
$$= 2\pi(0.50 \text{ Hz})(0.097 \text{ m})$$
$$= 0.30 \text{ m s}^{-1}$$

EXAMPLE

The tides at Newquay have a period of 12 hours 40 minutes and oscillate with s.h.m. On one constantly sloping beach the distance from low to high time is 142 m. Find (a) how far below the high tide mark the sea is 1 hour 30 minutes after high tide and (b) the maximum speed at which the tide recedes down the beach.

(a) *It is very helpful to sketch a graph to understand the details of the question.*

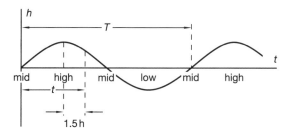

From mid-tide to high tide is $T/4 =$ 3 h 10 min or 190 min. So 1 h 30 min or 90 min after high tide gives t to be 280 min from mid-tide when $h = 0$ and $t = 0$.

Using $h = h_0 \sin 2\pi f t$ and

$$f = \frac{1}{T} = \frac{1}{760 \text{ min}}$$

$$h = (71 \text{ m}) \sin \left(2\pi \times \frac{280 \text{ min}}{760 \text{ min}} \right)$$

$$= (71 \text{ m})(\sin 2.315 \text{ rad})$$

$$= 52.2 \text{ m}$$

i.e. the tide has gone down 18.8 m down the beach in 1.5 h.

Be careful how you use your calculator; $2\pi f t$ will be in radians not degrees.

(b) The maximum speed $v_0 = 2\pi f s_0$

$$\therefore \quad v_0 = 2\pi \frac{71 \text{ m}}{760 \text{ min}} = 0.59 \text{ m min}^{-1}$$

which is over 50 cm per minute, a very noticeable speed.

In figure 20.3 the v-t curve is also sinusoidal (it is a sine curve but displaced a quarter of an oscillation along the axis). Thus the maximum rate of change of v (which is also a_0) is given by

$$a_0 = \left(\frac{dv}{dt} \right)_{\text{max}} = 2\pi f v_0 = \omega v_0$$

So the size of $a_0 = (2\pi f)^2 s_0 = \omega^2 s_0$ and the constant in $a = -(\text{constant})s$ is $(2\pi f)^2$, and we have $a = -(2\pi f)^2 s = -\omega^2 s$.

Any s.h.m. is described by an equation of this form

acceleration $\propto -(\text{displacement})$

This equation is usually accepted as a *definition* of what is meant by simple harmonic motion.

A baby in a 'baby bouncer' (figure 20.4) experiences quite high accelerations – and seems to like it. For example if the amplitude is 20 cm and the frequency is 0.75 Hz the maximum acceleration is

$$a_0 = (2\pi \times 0.75 \text{ Hz})^2 (0.20 \text{ m})$$

$$= 4.4 \text{ m s}^{-2}$$

nearly half of g.

FIGURE 20.4

S.h.m. and circular motion

Figure 20.5 shows a model which can be used to demonstrate the equivalence of s.h.m. and the projection of circular motion along a diameter. Once the frequencies are adjusted to be the same the two shadows are found to move together throughout their motion. This is why ω is often used for $2\pi f$ in s.h.m. for ω is the angular velocity of the turntable if f is the frequency of the mass on the spring.

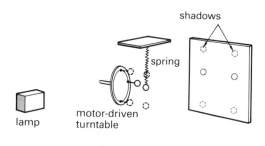

FIGURE 20.5

20.2 Simple harmonic oscillators

In this section we want to discover what it is that makes a body move with an acceleration which is proportional to its displacement from a given point O, i.e. with $a \propto -s$. The minus sign indicates that the acceleration slows the body down as it moves away from O and then speeds it up as it approaches O again.

Clearly any *springy* system will behave a bit like this. Look at the baby in figure 20.4. The further she is from the centre of the bounce the greater the resultant force on her. We can also see that the stiffer the springs the greater the force on her and hence her acceleration at a given displacement, while the bigger the baby the smaller the acceleration.

To analyse the simplest possible oscillator take a mass m moving on a frictionless horizontal surface as in figure 20.6. If the spring obeys Hooke's law and has a stiffness k, then the pull of the spring on the mass is

$$F = -ks$$

measuring s from the position where the spring is unstrained.

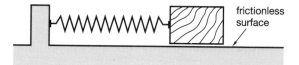

FIGURE 20.6

By Newton's second law, the acceleration a of a body of mass m acted on by a resultant force F is given by

$$a = \frac{F}{m}$$

so that, if the resultant force on the body is equal to the push or pull of the spring on it we can write

$$a = -\frac{k}{m}s$$

i.e.
$$a = -(\text{constant})s$$

and the motion of the body is therefore simple harmonic motion. Comparing this with $a = -(2\pi f)^2 s$, which defines s.h.m., we see that

$$(2\pi f)^2 = \frac{k}{m}$$

and so

$$f = \frac{1}{2\pi}\sqrt{\frac{k}{m}}$$

The period $T\,(= 1/f)$ is thus given by

$$T = 2\pi\sqrt{\frac{m}{k}}$$

This argument deriving T shows that the period is independent of the amplitude of the motion s_0, and hence independent of the maximum velocity v_0 and acceleration a_0. If $k = 3.7\,\text{N m}^{-1}$ and $m = 0.96\,\text{kg}$ we shall have an oscillation for which $T = 3.2s$ for *any* value of s_0 (e.g. for $s_0 = 36\,\text{mm}$ or $s_0 = 17\,\text{mm}$). Of course, other pairs of values of m and k could also produce a period of 3.2s.

The answer to the question 'under what conditions will a body oscillate with s.h.m.?' is thus something like, 'when the resultant force acting on it tends to restore the body to its equilibrium position and is proportional to its displacement from that position'.

Then

$$T = 2\pi\sqrt{\frac{m}{k}}$$

$$= 2\pi\left(\frac{\text{mass of oscillating body}}{\text{net restoring force per unit displacement}}\right)^{1/2}$$

Any mechanically oscillating system must have these properties of **inertia** and of **elasticity** (or an equivalent mechanism for producing a restoring force).

EXAMPLE

To 'weigh' himself an astronaut ties himself into a harness of mass 2.0 kg which is held to the side of his space vehicle by strong springs of stiffness 1.1 kN m^{-1}. He sets himself oscillating and times 10 oscillations as taking 16.4 s. Calculate his mass.

As
$$T = 2\pi\sqrt{\frac{m}{k}} = 1.64\,\text{s}$$

then
$$m = \frac{T^2 k}{4\pi^2}$$

$$= \frac{(1.64\,\text{s})^2(1100\,\text{N m}^{-1})}{4\pi^2}$$

$$= 73\,\text{kg}$$

So the mass of the astronaut is 73 kg.

A vertical mass–spring oscillator

When a mass m oscillates up and down on a spring its weight mg acts along the line of the spring and so is relevant in discussing the mechanics of its oscillation. We expect the motion to be s.h.m. but to prove it consider the mass attached to a light spring of stiffness k. Figure 20.7 shows (a) the spring before the mass is attached; (b) the spring after the mass is attached and a free-body diagram for the mass hanging in equilibrium with the spring extended a distance e; (c) the spring and mass at a moment during an oscillation and a free-body force diagram for the mass then.

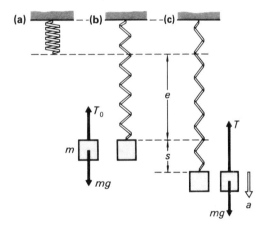

FIGURE 20.7

From (b) we get $T_0 = mg$, and, if the spring obeys Hooke's law $T_0 = ke$, so that

$$mg = ke$$

In (c) we have marked the displacement s from the equilibrium position of the particle and put the acceleration in the *positive s* direction. Using Newton's second law we have

$$ma = mg - T$$

where
$$T = k(e + s)$$

so that
$$ma = mg - k(e + s)$$

$$= mg - ke - ks$$

But $mg = ke$, from above,

so that

$$a = -\frac{k}{m}s$$

The equation represents simple harmonic motion. The period T is therefore given by $T = 2\pi\sqrt{m/k}$, and will be the same no matter what the local value of the gravitational field g. Thus if $k = 420\,\text{N m}^{-1}$ and we measure T to be 0.356s, then m must be 1.35 kg, on Earth or on the Moon. We could use the oscillations to measure m even in an orbiting (free-fall) laboratory as the previous example shows.

Pendulums

Suppose that the period T of a **simple pendulum** (a small bob of mass m on the end of a thin thread of length l) depends only on m, l and the free-fall acceleration g, i.e.

$$T = km^x l^y g^z$$

where k is a dimensionless constant.

As $[g] = LT^{-2}$ we can write $[T] = M^x L^y (LT^{-2})^z$

which gives $x = 0$, $y = \frac{1}{2}$ and $z = -\frac{1}{2}$

Thus a dimensional analysis suggests that

$$T = k\sqrt{\frac{l}{g}}$$

(The same result can be deduced by simply analysing the units of the quantities in the equation.)

A quick experiment will confirm this; T is independent of m and a graph of T against $\sqrt{l}$ turns out to be a straight line through the origin which shows that T is proportional to $\sqrt{l}$.

To derive the expression and to find k consider a free-body diagram for the pendulum at a position displaced an angle θ from its equilibrium position (figure 20.8). Applying Newton's second law at right angles to the string (F has no resolved part in this direction)

$$ma = -mg\sin\theta$$

As $\sin\theta = s/l$ then

$$a = -\frac{g}{l}s$$

This is in the form of the defining equation for simple harmonic motion but it is only s.h.m. if s and a are along a line and this is only

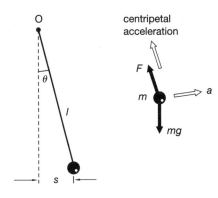

FIGURE 20.8

approximately true here for small values of s/l, i.e. for *small amplitude* oscillations. With this restriction then,

$$T = 2\pi\sqrt{\frac{l}{g}}$$

Values of T calculated from this equation are within $\pm1\%$ of the experimental values if the amplitude θ_0 of the motion is below about 15°.

The acceleration of free fall, g, can be found experimentally using this expression – see page 17 for details.

EXAMPLE

How long is a 'seconds pendulum', that is, one which passes through its lowest point once every second?

The period of this pendulum is 2.00s, so that in a laboratory where $g = 9.80\,\text{m s}^{-2}$ the length l of the pendulum is given by

$$l = \frac{T^2 g}{4\pi^2} = \frac{(2.00\,\text{s})^2 (9.80\,\text{m s}^{-2})}{4\pi^2}$$

$$= 0.993\,\text{m}$$

i.e. the pendulum is just under 1 m long.

20.3 Oscillators and energy

Figure 20.9(a) shows a possible way of describing energy flow for a mechanical oscillator in which the energy is transferred from kinetic energy to potential energy and back again continuously, e.g.

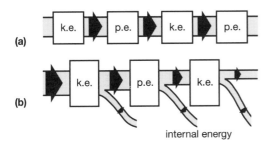

FIGURE 20.9

the girl on the swing or the astronaut in his weighing machine. The motion is said to be **undamped**. At any time the sum of the kinetic and potential energies is constant.

Of course in practice oscillations always die away. They are said to be **damped**. Figure 20.10 shows a 'time trace', i.e. displacement–time graph for a damped oscillator. (Notice that the period remains constant, even though there is heavy damping.)

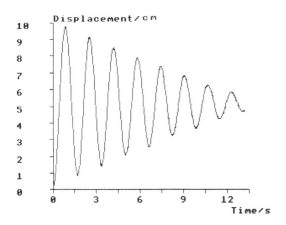

Oscillations of a metre-rule

FIGURE 20.10

The mechanical energy of the oscillator is converted to internal energy of the surroundings. Damping forces are frictional forces, either viscous forces in air or in liquids or solid frictional forces as in the wiper of the potentiometer for the heavy bar. Sometimes the damping of a mechanical system is a nuisance, sometimes it is desirable. The girl is disappointed by the damping on the swing but if a car continued to oscillate after hitting a

to oscillate after hitting a bump the effect would be most uncomfortable.

A system is said to be lightly damped if it undergoes a very large number of oscillations before coming to rest; for example a guitar string. Many mechanical systems such as motor-cycle springs exhibit heavy damping and, once displaced, approach their equilibrium position without oscillating much. Some are fitted with shock absorbers, which often consist of a piston moving in a cylinder full of very viscous oil (figure 20.11).

FIGURE 20.11

Energy graphs for oscillators

The total energy, E, of an undamped oscillator is constant, so a graph of E against displacement, s, or time, t, is just a horizontal line a fixed distance above the s or t axis. The following example shows how the elastic potential energy, E_p, varies with s for a mass–spring oscillator.

EXAMPLE

An air-track glider of mass 0.22 kg is attached to two springs the other ends of which are fixed to posts at the ends of the air track. The graph shows the elastic potential energy of the glider as it oscillates between two points 320 mm apart. Calculate (i) the speed of the glider when it is 100 mm from its equilibrium position, (ii) the restoring force (the pull of the springs) on the glider when it is 100 mm from its equilibrium position, and (iii) the acceleration of the glider as it passes through this position.

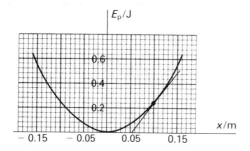

(i) At $x = 0.16\,\text{m}$ the e.p.e. of the spring is $0.64\,\text{J}$, and since there the k.e. of the glider is zero, the total mechanical energy of the system is also $0.64\,\text{J}$. The graph shows that when $x = 0.10\,\text{m}$, the e.p.e. of the spring is $0.25\,\text{J}$, so there the k.e. of the glider is $(0.64\,\text{J} - 0.25\,\text{J}) = 0.39\,\text{J}$.

If the speed of the glider at this moment is v, then

$$\tfrac{1}{2}mv^2 = 0.39\,\text{J}$$

$$\therefore \quad v^2 = \frac{2 \times 0.39\,\text{J}}{0.22\,\text{kg}} = 3.6\,\text{J}\,\text{kg}^{-1}$$

and so $\quad v = 1.9\,\text{m}\,\text{s}^{-1}$

(ii) As $W = Fs$ or here $\Delta W / \Delta x = F$, a tangent drawn at $x = 0.10\,\text{m}$ has a gradient equal to the restoring force F at this point, i.e.

$$F = \frac{(0.48 - 0)\,\text{J}}{(0.150 - 0.055)\,\text{m}}$$

$$= \frac{0.48\,\text{J}}{0.095\,\text{m}} = 5.1\,\text{N}$$

(iii) The accleration of the trolley at this position is thus, using Newton's second law,

$$a = \frac{5.1\,\text{N}}{0.22\,\text{kg}} = 23\,\text{m}\,\text{s}^{-2}$$

You should be able to draw the graph of k.e. $\tfrac{1}{2}mv^2$ against displacement for the glider in the above example. It is, of course, an upside-down parabola drawn so that k.e. + e.p.e. is constant.

A graph of k.e. or of e.p.e. against *time* for a body moving with s.h.m. is a sinusoidal curve of twice the frequency of the motion. It is always positive, i.e. above the axis.

20.4 Resonance

When a washing machine spin-drier is started the casing of the machine sometimes vibrates alarmingly at certain stages during the speeding-up process. The vibrations can be so violent that the motor fails to accelerate the tub through one of these stages. If we stop the machine and redistribute the clothes evenly the problem is often overcome. The system is then said to be 'balanced'. The problem is one of resonance; at a certain frequency or frequencies the spin-drier motor is making the tub vibrate at its own natural frequency. This will depend on the way in which the mass of the clothes is distributed. At these resonant frequencies a lot of energy is transferred from the driving force, the motor, to the oscillating system, the tub, instead of increasing the k.e. of the tub.

Large structures such as bridges and buildings (and even human beings) have resonant frequencies at which a small driving force may set up large-amplitude oscillations. The most spectacular example was the Tacoma Narrows Bridge collapse of 1940 – figure 20.12. The bridge, a large suspension bridge, obtained enough energy from a steady wind of $19\,\text{m}\,\text{s}^{-1}$ (eddies providing the

FIGURE 20.12

driving force) to reach an amplitude of oscillation beyond that which the structure could support. Factory chimneys, power transmission cables and cooling towers can also be set oscillating by steady winds. The resonant properties of the human stomach (in its cradle of supporting muscles) are at the roots of travel sickness.

Forced oscillations

If a periodic force of frequency f is applied to a pendulum or a mass–spring oscillator with a natural frequency f_0 the oscillator oscillates at the frequency of the driving force (though it may take a little time to settle down to a steady oscillation). The amplitude, s_0, at which it settles depends on how close f is to f_0. It also depends a little on the degree of damping. The graphs of figure 20.13 summarise the results of such experiments where f_0 is about 1.35 Hz. The peak of the graph is called a resonance peak and occurs when $f = f_0$.

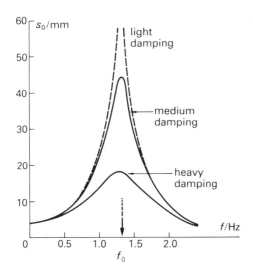

FIGURE 20.13

For the washing machine mentioned at the start of this section there is more than one natural frequency, as is nearly always the case with any but the simplest system. At resonance there is a maximum transfer of energy from the driving mechanism to the driven oscillator.

Barton's pendulums

Instead of studying one driven oscillatory body as the frequency of the driving force is varied we can, with a very simple system of pendulums, study the effect of a single driving frequency on a number of oscillators of different natural frequencies. Figure 20.14(a) shows such a system. P is a massive driver pendulum which forces the cord ABC to oscillate at a fixed frequency f_d (which depends on the length OP). It is set in motion and given an occasional push if necessary. Attached to AB, and thus forced to oscillate at f_d, are a number of very light pendulums of varying length. After some initial confusion their motions are like those shown in the photograph on the left of figure 20.14(b). The air damping is quite heavy on the light (e.g. polystyrene) spheres but with two metal washers attached to each sphere the damping is (relatively) less and the photograph on the right is obtained. Each photograph has an instantaneous

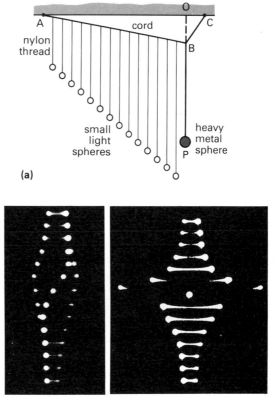

(b)

FIGURE 20.14

flash photo superimposed on a time exposure of the pendulum bobs.

The demonstration shows that all the driven pendulums have the same frequency f_d but that their amplitudes vary. The sphere which oscillates with the greatest amplitude is found to be the one which would oscillate at f_d if it were given a push and set swinging freely, i.e. the one whose natural frequency f_0 is equal to f_d. The photographs also show that the driven pendulums are not in phase. By looking along the line of the pendulums we see that they all lag behind the forcing pendulum and at resonance the phase lag is exactly 90°, a quarter of an oscillation. The shortest pendulum is almost **in phase** with the driver P and the longest is almost **in antiphase** with it.

Exercises on each section of this chapter may be found in the companion textbook, **Practice in Physics**.

SUMMARY

At the end of this chapter you should be able to:

◆ describe how to produce displacement–time graphs for oscillating objects.

◆ remember that the time for one complete oscillation is called the period T and the number of oscillations per unit time is called the frequency f; $f = 1/T$.

◆ understand that for simple harmonic oscillators the acceleration of the oscillating body is proportional to its displacement from a fixed point and that the acceleration is always directed towards this point.

◆ explain what is meant by simple harmonic motion.

◆ use the relationships:

$a \propto -(\text{constant})s$ and $a = -(2\pi f)^2 = -\omega^2 s$

◆ understand that, for a body moving with s.h.m.:
  – the period is independent of the maximum displacement or amplitude s_0;
  – the displacement varies sinusoidally with time;
  – the maximum velocity v_0 occurs when the displacement is zero.

◆ use the equations: $s = s_0 \sin 2\pi f t$ and $v_0 = 2\pi f s_0 = \omega s_0$

◆ explain why, when the resultant restoring force acting on a body is proportional to its displacement, the body will move with s.h.m.

◆ use the equations $T = 2\pi\sqrt{k/m}$ and $T = 2\pi\sqrt{l/g}$ for the periods of a mass on a spring and a simple pendulum respectively.

◆ explain the nature of the energy transfers in an undamped oscillator for which the total energy is constant and describe the nature of damped harmonic oscillations.

◆ describe an experiment to demonstrate the resonant oscillation of a mechanical system, e.g. a pendulum.

21 Mechanical Waves

21.1 Describing waves

Waves are the means by which we communicate; sound and light are both wave motions as is radio communication.

Waves transmit energy from place to place.

Mechanical waves are waves which transmit mechanical energy through solids, liquids and gases by vibrations or disturbances in the medium. Table 21.1 on the next page shows some examples.

In studying waves it is easiest to start by considering waves on ropes and ripples on water where the properties and behaviour of the waves are part of our everyday experience. These wave motions both leave the medium through which they pass *undisturbed*, i.e. after a pulse on a rope or a ripple on a pond has passed, the rope and the water return to their original condition. This is true for all wave motions.

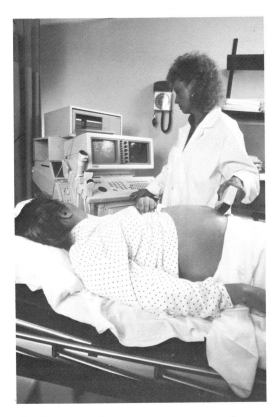

The photograph shows a pregnant woman having an ultrasonic scan to confirm that her baby is developing properly in the womb. Such a technique is relatively new: the use of ultrasonic waves was developed only in the 1950s, and then only as a research tool. They can now be used for such different applications as drilling square holes in glass, searching for shoals of fish and measuring the heartbeat of an unborn child.

You probably know that bats use ultrasound echoes to navigate. Have you ever wondered what a bat 'sees' as it flies along? Ultrasound, like sound, is a mechanical wave motion. Seismic waves in the Earth's crust and water waves, both large ocean swells and ripples in the bath, are also examples of mechanical waves. They travel at a wide variety of speeds in solids, liquids and gases and they all transfer mechanical energy from place to place.

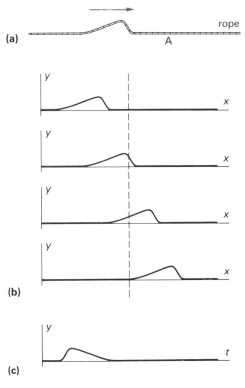

FIGURE 21.1

TABLE 21.1 Examples of mechanical waves

Type of wave	Cause	Use (brackets indicate nuisance)
sound waves in air	vibrating solid bodies, vibrating air columns, the human larynx	(undesirable noise), desired communication, wind instruments
ultrasonic waves in fluids	oscillating quartz crystals, bats	drilling, sonar, echo location, medical diagnosis
seismic waves in Earth's crust	explosions, earthquakes	geophysical prospecting, (earthquakes), nuclear arms control
waves on ropes, etc.	plucked or bowed strings, winds on cables	stringed instruments, (galloping suspension bridges)
water surface waves	winds at sea, tidal effects, vibrating surface objects	possible energy source, (bores), ripple tanks

We describe waves by drawing graphs either of displacement against position (figure 21.1(a) – three are shown below, at regularly spaced times) or displacement against time – figure 21.1(b); notice that the shape is reversed – you should make sure you understand how this comes about.

Transverse and longitudinal waves

Waves on ropes and strings are called **transverse** waves because their displacements are *perpendicular* to the direction in which the wave energy is travelling. Displacements can, however, be *parallel* to the wave direction as on a slinky spring; these waves are called **longitudinal** waves.

Transverse waves can involve particle displacements in more than one plane; you can vibrate the end of a rope up-and-down or side-to-side. A transverse wave which only has oscillations in one plane is said to be **polarised**. Longitudinal waves cannot be polarised, and sound is a longitudinal wave motion.

Sound waves

Sound can be produced by any vibrating body which transfers its energy of oscillation to the air around it, e.g. a loudspeaker, a guitar or the larynx. The wave energy propagates as a series of **compressions** and **rarefactions**, i.e. as a series of small pressure variations of amplitude p_0 above and below atmospheric pressure P. The size of p_0 at the

ear when listening to normal conversation may be 10^{-2} Pa, a *very* small fraction of P (which is about 10^5 Pa). The displacements of air molecules in a sound wave are extremely small, but the ear is able to detect waves in which the movements are much less than the diameter of an atom ($\approx 10^{-10}$ m).

To get a mental picture of how sound waves propagate look at figure 21.2.

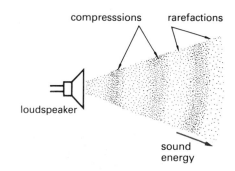

FIGURE 21.2

When the air is pushed to the right by the loudspeaker it is compressed a little. The molecules now make more collisions per second in this denser air and there is a net movement of the molecules away from the region of compression. As the mean speed of the molecules is about $500 \, \text{m s}^{-1}$ we might expect sound to travel at this speed in air. In practice the speed of propagation in air is found to be about $340 \, \text{m s}^{-1}$ which is of the same order of magnitude.

There is no net movement of the molecules away from the source, because when the loudspeaker moves to the left they move back into the region of rarefaction and so on. A sound wave in a gas can thus be thought of as a series of tiny variations in the pressure (and density) of the gas superimposed on the random motion of the gas molecules.

Wave energy

The energy transfer in a mechanical wave occurs continuously from one part of the medium carrying the wave to the next part. When the energy spreads out in three dimensions it follows an inverse square law, i.e. the energy per unit time P crossing the surface of any sphere drawn around the source is the same. The **intensity** of the wave energy, its power per unit area I, is given by

$$I = \frac{P}{4\pi r^2}$$

i.e.

$$I \propto \frac{1}{r^2}$$

This law is really a deduction from the principle of conservation of energy. When the wave energy is gradually converted to internal energy in the medium or the surroundings, the intensity of the wave is further reduced (or **attenuated**); but the transfer from the source to the receiver is usually very efficient.

EXAMPLE

A loudspeaker at a concert is emitting sound energy at the rate of 300 W. What is the intensity of the sound (a) 15 m and (b) 90 m from the source? What assumptions are involved in your calculations?

Assumptions:

- the loudspeaker acts as a point source;
- there is no attenuation of sound energy;
- reflections of the sound are ignored.

It is often simplest to give a list like this unless you are asked, for example in an examination question, to give the assumptions at particular places.

(a) As $P = 300$ W, then at 15 m the intensity

$$I = \frac{300\,\text{W}}{4\pi(15\,\text{m})^2} = 0.106\,\text{W}\,\text{m}^{-2}$$

(b) At 90 m, which is 6×15 m, the intensity will be

$$\frac{1}{6^2} \times 0.106\,\frac{\text{W}}{\text{m}^2} = 0.0029\,\text{W}\,\text{m}^{-2}$$

The decibel scale

In the above example a fall of intensity by a factor of 36, i.e. $I_2/I_1 = 1/36$ could be described as a *fall of 15.6* **decibels** (dB). This is because of the definition of the decibel scale:

$$\text{intensity change/dB} = 10\,\lg\,(I_2/I_1)$$

i.e. $10\,\lg\,(I_2/I_1)$ gives the change in sound intensity in decibels. So here intensity change $= 10\,\lg\,(1/36) = -15.6\,\text{dB}$. Similarly a rise of intensity of 3 dB would mean an increase in intensity of I_1 to I_2 where $10\,\lg\,(I_2/I_1) = 3$, i.e. $I_2/I_1 = 2.0$ which is a doubling of intensity.

21.2 Wave speeds

The speed of a mechanical wave depends on the elasticity and the inertia of the medium through which it is travelling. Think of a pair of climbing ropes lying on a smooth slab; if one has three times the mass per unit length of the other then a wave pulse (a flick), sent along them both will travel along the more massive one more slowly. Likewise if two similar ropes are hanging over a cliff, one with a free end and the other supporting a climber, then a wave pulse will travel down the taut one much more quickly than the other.

FIGURE 21.3

Transverse pulses

Suppose we guess that the speed c of a transverse wave pulse along a rope depends only on the tension T in the rope and its mass per unit length μ i.e.

$c = kT^x\mu^y$ where k is a constant with no units.

T will change slightly as the wave pulse passes as the length of the rope must momentarily increase; let us assume that such changes are very small so that we can talk about *the* tension. Our guess then means that as the units of c are $m\,s^{-1}$, of T are N or $kg\,m\,s^{-2}$ and of μ are $kg\,m^{-1}$, we can write

$$m\,s^{-1} = (kg\,m\,s^{-2})^x (kg\,m^{-1})^y$$

which gives $x = \frac{1}{2}$, $y = -\frac{1}{2}$, so we have $c = k\sqrt{T/\mu}$. We can show experimentally that k = 1,

i.e.
$$c = \sqrt{\frac{T}{\mu}}$$

and that c does *not* dpend on the shape of the pulse – i.e. on how you flick the rope.

EXAMPLE

A transverse wave is seen to travel along a power cable at about $50\,m\,s^{-1}$. If 100 m of the cable has a mass of 60 kg, what is the tension in it?

$$\mu = \frac{60\,kg}{100\,m} = 0.60\,kg\,m^{-1}$$

$$\therefore \qquad c = \sqrt{\frac{T}{\mu}} \quad \Rightarrow \quad T = \mu c^2$$

and so $T = (0.60\,kg\,m^{-1})(50\,m\,s^{-1})^2$

$$= 1500\,kg\,m\,s^{-2} \text{ or } 1500\,N$$

As the tension in such a cable varies from point to point the tension we have calculated is only an average value.

The speed of sound

The speed of sound c in air at 288 K (15°C) is $340\,m\,s^{-1}$. An estimate of this speed can be made by a simple clapping experiment using sound echoes, i.e. measuring directly the time taken for a sharp clap to return from a cliff or large building a known distance away. A laboratory experiment with a timer which can measure to $10\,\mu s$ involves arranging to switch the timer on when a hammer hits a metal block and switching it off when the pulse is received by a microphone a few metres away.

A more precise method measures the wavelength, λ, of a sound of known frequency, f, and uses $c = f\lambda$ (see page 293) to get a value for the speed. Figure 21.4 shows one possible arrangement. The signal generator produces a varying potential difference at a small speaker L. This p.d. is also fed to the X-plates of an oscilloscope, the time-base of which is switched off. A microphone M receives the sound wave and

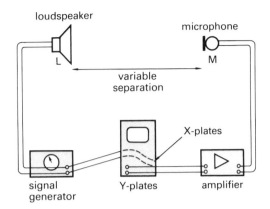

FIGURE 21.4

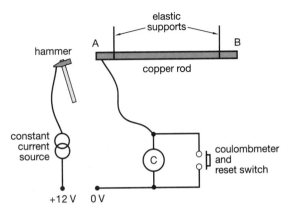

FIGURE 21.5

after amplification its output potential difference is fed to the Y-plates of the c.r.o. You arrange for a diagonal line to appear on the screen and then move M or L until the same line appears again. The distance moved is one wavelength.

For $\lambda \approx 0.5\,\text{m}$ a movement of M of $\approx 2\,\text{mm}$ is enough to turn the line into a thin slanting ellipse. For $\lambda \approx 0.1\,\text{m}$ it is necessary to move L through several wavelengths to achieve similar precision. The frequency f must be read from the signal generator which is unreliable and should be checked against a frequency meter. Any error here affects the value of c, which is equal to the product $f\lambda$. With this arrangement we can quickly measure c for a variety of wavelengths. In the audible range the product $f\lambda$, i.e. the speed of sound is found to be independent of λ (and thus also of f).

Longitudinal waves in solids

We can measure the speed of longitudinal mechanical waves along a solid metal rod using the apparatus shown in figure 21.5.

When the hammer first makes contact with the rod it starts a compression pulse at A and as this wave pulse moves towards B the hammer remains in contact with the rod. The wave pulse is reflected at B as a rarefaction pulse which when it reaches A breaks the contact between the rod and the hammer. If the length of the rod AB is d, then $c = 2d/t$.

To measure the time of contact, t, a constant current supply and coulombmeter are used. If the

current is I and the charge Q then $Q = It$ and so $t = Q/I$. If I and Q are measured t can be found. The switch enables the coulombmeter to be discharged.

For a solid material of density ρ and with a Young modulus E, it can be shown that (check the units for yourself)

$$c = \sqrt{\frac{E}{\rho}}$$

For copper $E = 1.30 \times 10^{11}\,\text{N}\,\text{m}^{-2}$ and $\rho = 8.93 \times 10^{3}\,\text{kg}\,\text{m}^{-3}$, so that for copper

$$c = \left(\frac{1.30 \times 10^{11}\,\text{N}\,\text{m}^{-2}}{8.93 \times 10^{3}\,\text{kg}\,\text{m}^{-3}}\right)^{1/2}$$

$$= 3.8 \times 10^{3}\,\text{m}\,\text{s}^{-1}$$

and this prediction can be tested with the apparatus in figure 21.5.

Seismic waves

Seismic, or earthquake, waves result from a fracture or sudden deformation of the Earth's crust. The energy released is transmitted from the source by both longitudinal and transverse waves. The longitudinal waves, called P (or primary) waves, travel faster than the transverse waves, called the S (or secondary) waves. (It helps you to remember which is which if you think of P as push and S as shake.) The speeds of the P and S waves, near the Earth's surface, are about $7500\,\text{m}\,\text{s}^{-1}$ and $4000\,\text{m}\,\text{s}^{-1}$ respectively. Both speeds increase with depth for about the first 3000 km but beyond this depth the

Earth's core behaves like a liquid in which the P waves can, but the S waves cannot, travel.

EXAMPLE

A recording station observes that there is a time interval of 600 s between the arrival of the P and S waves from an earthquake as shown. At what distance from the station did the earthquake occur? Use the speeds given above and assume that the P and S waves have travelled along the same path to the station.

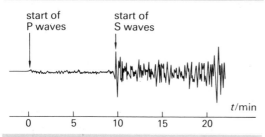

Suppose the P wave takes a time t to travel from the earthquake to the station; then the S waves will have taken $t + 600$ s. They travel the same distance so that

$$(7500 \text{ m s}^{-1})t = (4000 \text{ m s}^{-1})(t + 600 \text{ s})$$

$$\Rightarrow \qquad t = 686 \text{ s}$$

$$\therefore \qquad s = (7500 \text{ m s}^{-1})(686 \text{ s})$$

$$= 5150 \times 10^3 \text{ m or } 5150 \text{ km}$$

Man-made seismic waves are used in prospecting for oil, gas and other useful deposits in the Earth's surface.

Water waves

Water waves are neither wholly transverse nor longitudinal: the motion of a particle of water in the path of a wave travelling on deep water is roughly circular. You can feel this if you bathe from a boat in deep water. As you float the waves clearly carry you round in a circle.

Types of water wave

- Small-wavelength water waves ($\lambda \approx 10$ mm) are called **ripples**. They depend on the surface tension of the water surface for their

elastic property and are familiar in the laboratory in ripple tanks. Typical speeds in ripple tanks with water of different depths are 0.2 m s^{-1} to 0.3 m s^{-1}.

- Ocean waves or **gravity waves** rely on the Earth's gravitational field g acting on the displaced water for restoring force. Their speed c in deep water is given by

$$c = \sqrt{\frac{g\lambda}{2\pi}}$$

where λ is their wavelength.

For example, λ might be 2 m for choppy water in the English Channel. This gives $c = 1.8 \text{ m s}^{-1}$; but for a roller in the Southern Pacific, λ could be as high as 1 km which gives $c = 40 \text{ m s}^{-1}$.

In shallow water (e.g. a river bed or water trough) their speed is equal to $\sqrt{gh}$ and so depends on the depth of the water h. The Severn bore (figure 21.6) travels at different speeds as it moves up the river.

FIGURE 21.6

21.3 The principle of superposition

You can hear a person talking even though other sound waves are crossing the line between your ear and the speaker; you can see across someone else's line of sight; you can receive a radio programme even though the path from your aerial to the transmitting station is crossed by many other radio and TV carrying waves. *Waves do not seem to bump into one another.* They pass through one another and emerge the other side unaffected.

Figure 21.7 shows two sets of photographs of wave pulses on (a) a spring and (b) a string as they 'cross'. As the wave pulses cross they are said to **superpose** and

the displacement of any point on the string or spring is the sum of the displacements caused by each disturbance at that instant.

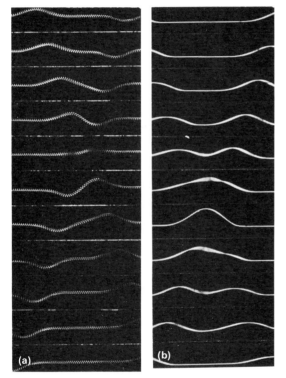

FIGURE 21.7

This is called the **principle of superposition** and is the key to understanding many wave phenomena. It applies to all types of wave motion. To superpose two waves is to add the displacements they each produce. Thus the photographs are a set of y-x graphs – you should try drawing a series for two simple wave pulses, e.g. square pulses, which cross.

Sinusoidal waves

The most important wave shape is the sinusoidal profile. The distance between corresponding points

on neighbouring waves is called the **wavelength** of the wave and is denoted by λ. The **period** T of the wave is the period of oscillation of each point on the wave. Its frequency f is the number of oscillations each point undergoes per second.

$$f = \frac{1}{T}$$

The wave pattern moves forward one complete wavelength in the time T which any point takes to complete one oscillation. Thus the wave speed c is given by

$$c = \frac{\lambda}{T}$$

or $$c = f\lambda$$

which is a general relation for all **periodic waves**.

Figure 21.8 shows (a) displacement–position and (b) displacement–time graphs for sinusoidal waves. The amplitude is y_0 in each case.

As the curve (a) repeats itself in a distance λ and is a sine curve, we can write its equation as

$$y = y_0 \sin 2\pi \frac{x}{\lambda}$$

while the curve (b) repeats itself in time T and so

$$y = y_0 \sin 2\pi \frac{t}{T} = y_0 \sin 2\pi f t$$

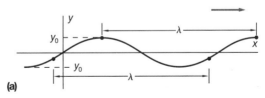

(a)

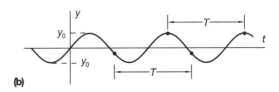

(b)

FIGURE 21.8

Phase difference

When you oscillate one end of a slinky spring from side-to-side you send a progressive wave along the slinky. Each point on the slinky oscillates from side to side but seen in slow motion you would notice that different points are not oscillating in time with one another. It is as if each part of the slinky pulls the part in front of it up after it – in fact this is just what does happen. We say that neighbouring parts of the slinky are oscillating out of phase with each other. But further on you will see some points which move exactly together, move exactly in step; we say they are oscillating **in phase** with one another and you would notice that they are λ, 2λ or 3λ etc. apart.

For points in phase the difference in angle in the equation $y = y_0 \sin 2\pi ft$ is 2π rad, 4π rad, 6π rad etc. (or 360°, 720°, etc.) and this is called the **phase difference** between their motions.

When the two points are $\lambda/2$, $3\lambda/2$, $5\lambda/2$, etc. apart they are said to be **in antiphase** (or wholly out of phase). They are oscillating out of step with one another and their phase differences is π rad, 3π rad, etc. or 180°, 540°, etc.

When two different waves arrive at the same point we talk about the **phase difference** ϵ between the two waves at that point. When the frequencies are the same the result of superposing the waves depends on this phase difference.

If $\epsilon = 0$ or 2π etc., i.e. the two oscillators are in phase, the result of superposition is an oscillation of large amplitude: the total is the *sum* of the wave amplitudes – a constructive effect. But if $\epsilon = \pi$ or 3π etc., i.e. the two oscillations are in antiphase, the result of superposition is an oscillation of small amplitude: the total is the *difference* of the wave amplitudes – a destructive effect.

Summarising: for two sinusoidal waves of the same frequency f and the same amplitude y_0, each transporting energy at a rate P, at a given point we have

- if $\epsilon = 0$:   oscillations in phase; resultant oscillation of frequency f and amplitude $2y_0$; energy $4P$, n.b. *not* $2P$. (Just as for a simple harmonic oscillator the energy carried by the wave is proportional to the square of the amplitude.)

- if $\epsilon = 180°$:   oscillations in antiphase; no resultant oscillation; energy zero.

Finally we can talk about the phase difference between two sources of waves S_1 and S_2. If they have the same frequency then they may be in phase or in antiphase (or may be something in between). You might try to see what happens when you reverse one of the speaker connections of a stereo system.

The ripple tank: two-dimensional superposition

When two sources S_1 and S_2 of about the same amplitude are placed a few wavelengths apart a pattern of superposition is produced. Such a pattern is usually referred to as an **interference pattern**. The simplest two-dimensional

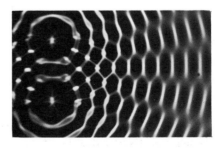

FIGURE 21.9

demonstration is with ripples as shown in the photograph of figure 21.9.

Consider the point P in figure 21.10. If the sources are in phase then the phase difference between the two waves arriving at P from S_1 and S_2 depends on the difference in lengths of the paths S_1P and S_2P, i.e. on how many wavelengths there are in S_1P and in S_2P. For example, if as shown $S_1P = 4\lambda$ and $S_2P = 5\lambda$ then at P the two waves will be in phase. The result of superposition at P is thus constructive; i.e. a maximum displacement amplitude.

At Q, on the other hand, $S_1Q = 7\lambda$ but $S_2Q = 5\frac{1}{2}\lambda$ the path difference is $1\frac{1}{2}\lambda$, so at Q the two waves will be in antiphase and the result of the superposition will be destructive, a zero or minimum displacement amplitude. Choose some other positions and make sure you can predict which are maxima and which are minima.

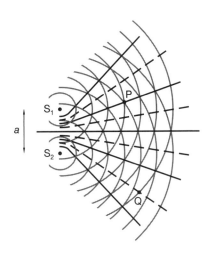

FIGURE 21.10

In general:

♦ for *maximum* displacement $S_2P - S_1P = n\lambda$.

♦ for *minimum* displacement $S_2P - S_1P = n\lambda + \lambda/2$ where $n = 0, 1, 2$ etc.

Places which have minimum displacement lie on lines – dotted in figure 21.10 – which follow the fuzzy patches of flat water in the photograph of figure 21.9. The lines of maximum displacement are the full lines. (Notice that the lines are not straight – they are hyperbolic.)

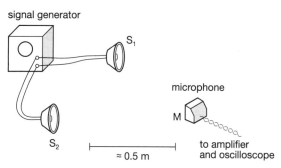

FIGURE 21.11

Sound: superposition in three dimensions

In figure 21.11 S_1 and S_2 are identical small speakers which are in phase and M is a sensitive microphone. Suppose M is placed at a point on the line bisecting S_1S_2, then it is at a maximum as $S_2M = S_1M$, the path difference is zero. A whole set of positions in three dimensions can be found where $S_2M - S_1M = \lambda$ and so on. In practice this sort of experiment is hard to do in a laboratory as the microphone will also pick up reflected sound which makes the positions of minima unclear.

Antisound

The principle of superposition is increasingly being used to reduce **noise pollution**. A microphone picks up the unwanted sound which is then phase inverted (antisound), amplified and played back via a speaker. One particular application is shown in figure 21.12 where the driver can be fed a mixture of sound and antisound so as to guard him from the noisy environment.

FIGURE 21.12

Superposition and energy

Suppose, in a sound experiment like that shown in figure 21.11, S_2 is switched off by disconnecting it from the signal generator, and we find that the height of the trace on the oscilloscope ($\propto$ wave amplitude) is h. If S_2 is now switched on the length of the trace increases to $2h$. This means that the amplitude of the wave detected by M has doubled when the waves arriving at M are in phase. As the intensity of the wave $\propto$ (amplitude)^2, then the intensity with both S_1 and S_2 connected is four times that with S_1 alone. At places where the waves from S_1 and S_2 arrive in antiphase very little sound energy is detected by the microphone.

The result of superposing the waves is thus to *redistribute* the energy in the interference pattern. There is four times the energy from one source at a maximum and zero energy at a minimum. The average energy taken over all positions for M is just twice the energy taken from one of the sources; as we would expect. When the microphone is moved about in front of S_1 and S_2 the trace oscillates in size. The extreme values of $2h$ and zero occur only near the centre line as at other points M is closer to one of the sources than the other and the amplitudes of the two incident waves are then not equal.

EXAMPLE

Two small speakers at S_1 and S_2 emit sound waves of the same frequency 1100 Hz. Assume that the diagram is to scale but reduced by a factor of 200. A maximum is detected at P and as the microphone is moved to Q, which is also a maximum, one other maximum is found between them. Use this information to find the speed of sound.

$$S_2P - S_1P = 48\,\text{mm} - 45\,\text{mm} = 3\,\text{mm}$$
which is 600 mm or 0.60 m in reality.

$$S_2Q - S_1Q = 69\,\text{mm} - 69\,\text{mm} = 0$$

So $\qquad$ $S_2P - S_1P$ must be 2λ i.e. $\lambda = 0.30\,\text{m}$

and using $\quad c = f\lambda$

$$c = 1100\,\text{Hz} \times 0.40\,\text{m} = 330\,\text{m\,s}^{-1}$$

21.4 Hearing

The ear–brain system is a most remarkable device; you can listen to a record (the sound from which is transmitted by a pair of speakers) and yet hear the individual players, singers and background noises separately. In this sense the ear acts as a sort of frequency analyser.

However, the ear is not good at distinguishing the direction from which the sound comes. To judge direction we use both ears. At high frequencies we note the different intensities of the signals picked up by each ear; at low frequencies we note the different times of arrival of the signal at each ear. The brain can detect time differences of as little as 30 µs.

The Doppler effect

When an ambulance passes close to you the pitch of its siren notes falls as it passes. This is called the **Doppler effect** in sound and an analogous effect can be produced with all types of waves.

P
•

S_1 •

S_2 •

•
Q

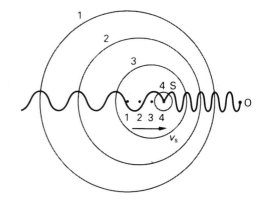

FIGURE 21.13

Suppose you are at O in figure 21.13 and the ambulance S is moving towards you at a speed v. Figure 21.13 shows that the waves produced are compressed and so you receive a frequency f' which is not equal to the frequency of the siren. In fact

$$f' = f\left(\frac{1}{1 - v/c}\right)$$

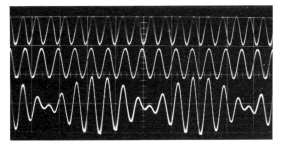

FIGURE 21.14

where c is the speed of sound in air. If the ambulance is moving away from you the minus sign becomes a plus sign. For $f = 2000\,\text{Hz}$ and $v = 30\,\text{m s}^{-1}$ you will find the two values of f' are, taking c to be $340\,\text{m s}^{-1}$, 2196 Hz, i.e. up 196 Hz and 1838 Hz, i.e. down 162 Hz. When $v \ll c$ you can show that, approximately, $\Delta f = vf/c$.

Clearly something odd happens when $v = c$, as this produces a zero in the denominator. An example of this is an aeroplane whose speed reaches the speed of sound. At the sound barrier pressure waves are produced by the wings and tail of the plane and you hear a loud bang when such a wave reaches you.

In medical science the Doppler effect enables the speed of blood flow to be measured. It can also detect the movement of pulsating organs, (e.g. the heart of a growing foetus) and hence check that they are functioning properly.

Beats

If you connect two different signal generators to the two loudspeakers and set their scales at the same frequency, it is unlikely that their calibrations are exactly the same. Suppose they differ by 0.1% at 2000 Hz, i.e. $f_1 - f_2 = 0.1\%$ of $2000\,\text{Hz} = 2\,\text{Hz}$.

Assuming that the speakers are equally loud you will hear a sound of frequency about 2000 Hz which rises and falls (pulses) twice every second, i.e. at 2 Hz. This is called a **beat phenomenon**. It is easy to adjust one of the signal generators so as to 'speed up' the beats or to slow them down to zero. In the latter case $f_1 = f_2$ to within very precise limits as the ear can detect a slow beat which rises and falls only once every 10 seconds.

Figure 21.14 shows oscilloscope traces (the time-base is the same in all three cases) for a sinusoidal wave at 700 Hz, at 600 Hz and then for the two waves superposed at the microphone. The beat phenomenon is clear in the bottom trace and, in this case, has a pulsing or beat frequency of $(700 - 600)\,\text{Hz} = 100\,\text{Hz}$, that is the bottom trace bulges 100 times per second. In general

$$\text{beat frequency} = f_1 - f_2$$

where f_1 is the larger of the two frequencies.

A piano tuner listens for and counts beats when she is tuning a piano, starting only from the sound of one standard tuning fork.

Exercises on each section of this chapter may be found in the companion textbook, *Practice in Physics*.

SUMMARY

At the end of this chapter you should be able to:

♦ understand that waves transmit energy from place to place.

♦ remember that there are lots of different types of mechanical waves, including sound waves in air and waves on the surface of water.

♦ explain the nature of the particle displacements in transverse and longitudinal mechanical waves.

♦ understand the inverse square law for the intensity of a wave (its power per unit area) and use the relationship

$$I \propto 1/r^2$$

♦ remember that the speed of sound in air is about $340\,\mathrm{m\,s^{-1}}$ under normal atmospheric conditions.

♦ use the equation $c = \sqrt{T/\mu}$ for the speed of a transverse mechanical wave on a rope.

♦ explain that wave speed is related to wavelength and frequency by the equation $c = f\lambda$.

♦ use the principle of superposition which states that the displacement of any point in the path of a wave is the sum of the displacements affecting the point at that instant.

♦ draw the result of superposition as a sequence of displacement–time graphs.

♦ understand that two waves arriving at one point may be in phase or in antiphase, or there may be a phase difference ϵ, measured in radians or degrees, between them.

♦ understand that two wave sources may be in phase or in antiphase.

♦ explain that the result of superposition is called an interference pattern and that it can be two-dimensional, as on the surface of water, or three-dimensional, as with sound.

♦ remember that for two sources which are in phase the result of superposition at a point P is given by:

$S_1P - S_2P = n\lambda$ maximum or constructive superposition,

$S_1P - S_2P = n\lambda + \lambda/2$ minimum or destructive superposition.

♦ understand that energy is not lost in an interference pattern, it is simply redistributed.

♦ remember that the intensity of a wave is proportional to its amplitude squared.

♦ explain the change in pitch or frequency heard when an observer and a source of sound are moving relative to one another.

♦ remember that two sound sources of frequencies f_1 and f_2 can produce beats of frequency $f_1 - f_2$.

22 Electromagnetic Waves

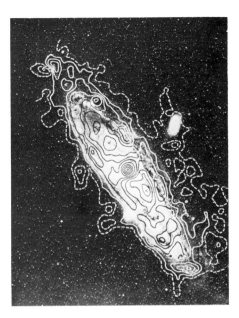

22.1 The electromagnetic spectrum

In Chapter 21 the behaviour of mechanical waves is described. Electromagnetic (e-m) waves, from radio waves to γ-rays, share many of the general wave properties of mechanical waves; for instance:

- they transfer energy from place to place;
- they reflect (one example is the dish-shaped satellite TV aerial) and refract;
- they exhibit diffraction effects;
- they obey the principle of superposition.

How do we know about our environment, about the solar system or about the more distant universe? How do we investigate the structure of atoms, the bonds which hold atoms together to form molecules or the arrangement of atoms in giant molecular structures? The answer to these and to many other 'big' questions is 'by collecting and analysing emitted, absorbed, scattered or reflected electromagnetic waves'. The photograph shows the nearby Andromeda galaxy with the relative strengths of its radio emissions superimposed on the optical picture.

The element helium was first discovered as a component of the sun's outer atmosphere by looking in great detail at the spectrum of sunlight. Certain wavelengths were found to be absorbed which did not coincide with those of any known element. Also, radioastronomers have detected a universal radiation arriving equally from all directions which is believed to be evidence for the 'hot big bang' theory of the beginning of the universe about fifteen thousand million years ago. This background radiation is most intense at a microwave wavelength of just under one millimetre.

All **electromagnetic waves** also share the following properties:

- they are generated by accelerating charged particles;
- they travel at a speed of $3.00 \times 10^8 \, \text{m s}^{-1}$ in free space.

Table 22.1 summarises some properties of electromagnetic waves of different frequencies. You will notice that the frequency ranges overlap. This is because we give different names to the same waves if they are produced by different means. And sometimes we remember the wavelengths rather than the frequencies of electromagnetic waves, at other times vice versa. Using the table and $c = 3.00 \times 10^8 \, \text{m s}^{-1}$ you can check that radio waves include waves of $\lambda = 1500 \, \text{m}$ and $\lambda = 200 \, \text{mm}$, while infra-red waves include those for which λ goes from a few μm up to a few mm.

The table also gives some selected information about the electromagnetic spectrum. Electromagnetic waves of all wavelengths can be found beyond the Earth's atmosphere and are the result of cosmic, galactic or stellar interactions or explosions. The 'big bang' theory of the origin of our universe is supported by the existence of the

TABLE 22.1 The electromagnetic spectrum

Name and frequency range	Source(s)	Selected applications
radio waves 10^4–10^9 Hz	electrical or electronic equipment	radio communications, worldwide location systems (OMEGA), radioastronomy
microwaves 10^8–10^{12} Hz	klystrons, magnetrons, Gunn effect circuits, masers	radar (including speed traps), television communication, microwave cooking, analysis of molecular structure
infra-red 10^{10}–10^{14} Hz	warm and hot bodies (including human beings), lasers, l.e.d.s	reconnaissance and tracking, domestic heating, beams for switching circuits, organic analysis, fibre communication
visible light close to 10^{15} Hz	the Sun, fluorescent and hot filament lamps, lasers, l.e.d.s	vision, photosynthesis, all optical systems, identifying gaseous elements, surveying, digital displays
ultra-violet 10^{15}–10^{17} Hz	the Sun, mercury vapour lamps, arcs and sparks	lamps used in medicine, sterilising food and utensils, detecting forgeries, fluorescence in washing powders, microscopy
X-rays 10^{15}–10^{25} Hz	X-ray tubes, pulsars	radiography, radiotherapy, astronomy, X-ray crystallography
γ-rays 10^{17}–10^{32} Hz	nuclear reactions, cosmic radiation	radioactive tracers (in medicine and elsewhere), cancer treatment, non-destructive testing, induced mutations

3 K background radiation, i.e. radiation which would be emitted by a body at a temperature of 3 K ($-270°C$), just as sunlight is characteristic of a body of about 6000 K. In the second half of the twentieth century astronomy has gradually developed to include observation at all wavelengths.

EXAMPLE

The spectral output of an infra-red laser used for optical fibre communication is described as being 1.550 μm with a spread of only 2 nm. What frequency range does this represent? Take $c = 2.998 \times 10^8 \, \text{m s}^{-1}$.

As 1.550 μm is 1550 nm then the wavelength range is from 1549 nm to 1551 nm. Using $c = f\lambda$, the frequency range is from

$$f = \frac{2.998 \times 10^8 \, \text{m s}^{-1}}{1549 \times 10^{-9} \, \text{m}}$$

$$= 1.935 \times 10^{14} \, \text{Hz}$$

to

$$f = \frac{2.998 \times 10^8 \, \text{m s}^{-1}}{1551 \times 10^{-9} \, \text{m}}$$

$$= 1.933 \times 10^{14} \, \text{Hz}$$

i.e. a spread of 0.002×10^{14} Hz or 0.2 THz.

This is quite a large frequency spread! Note that you cannot calculate the change in f by just dividing the 2 nm into c.

The speed of light

The speed c at which light travels in a vacuum has, since the early 1980s, been *defined exactly* by $c = 2.99792458 \times 10^8 \, \text{m s}^{-1}$.

It is one of the fundamental constants. *All* electromagnetic waves travel in a vacuum at this speed. The simultaneous reception of waves from events which occur far away in space by both optical and radio telescopes provides us with very convincing evidence of this. We used to measure c using whichever were the most convenient electromagnetic waves. A microwave of frequency 10 GHz has a wavelength λ of 0.030 m or 30 mm. A stationary wave experiment using $f = 10$ GHz electromagnetic waves (see page 315) could lead to a measured value of their wavelength and hence to a value for c ($= f\lambda$).

Direct measurements of the speed of light in glass involve sending a very short pulse down an optical fibre which is tens of metres long and measuring the time it takes using an oscilloscope with a fast time base. Figure 22.1 shows the outgoing and the returning pulse. If the time base is set to 0.05 µs div^{-1} then the time for the pulse to go down and back is 0.055 µs. So for this fibre which was 6.0 m long,

$$c = \frac{2 \times 6.0\,\text{m}}{5.5 \times 10^{-8}\,\text{s}} = 2.2 \times 10^8\,\text{m s}^{-1}$$

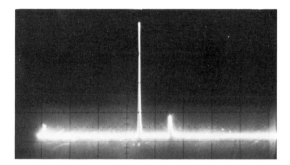

FIGURE 22.1

EXAMPLE

At present the BBC broadcasts in one region using 247 m, 1214 Hz and 1500 m, 200 Hz. What do these numbers tell you about electromagnetic waves?

Using $c = f\lambda$ for each in turn yields firstly:

$$c = (1214 \times 10^3\,\text{Hz})(247\,\text{m})$$
$$= 3.00 \times 10^8\,\text{m s}^{-1}$$

and secondly:

$$c = (200 \times 10^3\,\text{Hz})(1500\,\text{m})$$
$$= 3.00 \times 10^8\,\text{m s}^{-1}$$

so that this information tells us that two electromagnetic waves different in their wavelengths by a factor of 6 travel at the same speed.

Radar

The name comes from RAdio Detection And Ranging. Radar enables distances to be measured if c is known; it is an echo technique. Modern fighter aircraft and Cruise missiles use radar to continuously measure their height above the ground and check this against a computer-stored record of the landscape. This enables them to navigate very precisely and to follow pre-planned paths. Very large distances can also be measured, e.g. a short radar pulse transmitted from the Earth and reflected at the Moon's surface might be timed as taking 2.498 s for the round trip, thus determining the Earth–Moon surface-to-surface distance d at that moment as

$$2d = (2.998 \times 10^8\,\text{m s}^{-1})(2.498\,\text{s})$$
$$d = 3.745 \times 10^8\,\text{m} \quad \text{or} \quad 374\,500\,\text{km}$$

Doppler shift

When a source of electromagnetic waves S and an observer O are in relative motion the frequency of the waves as measured by O is different from the frequency f at which they are emitted by S. The same is true for sound waves (page 296). If O and S are moving at speed v either directly towards or away from one another then the change of frequency, Δf, is given by

$$\frac{\Delta f}{f} \approx \frac{v}{c}$$

provided $v \ll c$; Δf is called the electromagnetic Doppler shift. If v is a relative speed of *approach* then Δf is an *increase* in frequency, and there would be a consequent *decrease* in wavelength. The reverse is true if O and S are receding from one another.

A significant consequence of the Doppler effect is the **red shift** which enables astronomers to measure how fast distant galaxies are moving away from us and thus to measure the rate of expansion of the Universe.

When e-m waves are reflected from a moving object, the waves appear to come from the image of the source. This is moving at $2v$, where v is the speed of the reflecting surface, and so in this case

$$\frac{\Delta f}{f} \approx \frac{2v}{c}$$

Suppose a car is approaching a source of radar waves of frequency 1.0×10^{10} Hz at $15\,\text{m s}^{-1}$. As $c = 3.0 \times 10^{8}\,\text{m s}^{-1}$ we have $v/c = 5 \times 10^{-8}$ and so $\Delta f = 1000$ Hz.

The waiting police – figure 22.2 – notice an increase of 1000 Hz because the approach of the car in no way affects the speed of the reflected waves, only their frequency. If now the reflected waves are compared with the transmitted waves a *beat* frequency equal to the difference between the two will be detected, i.e. a beat frequency of 1000 Hz. This can be readily turned into a digital reading giving the speed of the approaching car providing that it is moving directly towards the detector.

FIGURE 22.2

Relativity and c

A knowledge of c is needed for the mass–energy relation $\Delta E = c^2 \Delta m$. The speed of light turns out to be the *maximum speed* at which energy can be transmitted in our universe. The speed at which light and other electromagnetic waves propagate in a vacuum c is also found to be independent of the speed of their source. This is extremely surprising; it means that if a star is moving away from the Earth at a speed $v = 2 \times 10^{8}\,\text{m s}^{-1}$ (so that $v \approx 2c/3$) the speed of the light from the star measured on Earth is $3 \times 10^{8}\,\text{m s}^{-1}$ and not, as one might at first guess, $(3-2) \times 10^{8}\,\text{m s}^{-1}$. This result has been rigorously tested in a number of ways. For example, we can measure the speed of γ-rays which come from sub-atomic particles (pions) moving through a laboratory at more than 99.9% of the speed of light. When such a pion emits a γ-ray in the forward direction the speed of the γ-ray is still c! This constancy of the speed of light is a cornerstone of the **theory of special relativity** from which is derived the mass–energy relationship mentioned above.

22.2 The nature of electromagnetic waves

All the methods of generation, especially those producing long wavelength e-m waves, can be seen to involve *accelerating charged particles* – usually electrons.

Travelling fields

Suppose a positively charged particle held at P is moved very quickly to Q. During this process it would of course speed up and slow down. At P there would be a radial electric field associated with the charge and zero magnetic field (it is at rest) and similarly at Q. Of course the electric field lines extend in all directions from the point charge but the figure shows only a group of lines in two dimensions to the right. If we assume that during the motion of the charge from P to Q the field lines remain unbroken then they will bend in the way indicated unless the effect travels infinitely quickly.

While the charge is moving it also produced a magnetic field which, to the right of PQ, will be into the page (you might work out what happens in other directions). The 'kink' in the electric field and its accompanying patch of magnetic field is an **electromagnetic wave pulse** – figure 22.3. It is this kink that travels away from PQ at a speed which

TABLE 22.2 The nature of e-m waves

E-M waves	Method(s) of generation	Detected by
radio waves	oscillating electrons in conductors and dipole antennae	tuned oscillatory electric circuits
micro waves	oscillating electrons in resonant cavities	resonant or tuned cavities
infra-red	random energy changes in molecules of liquids and gases, oscillating molecules in hot solids	special photographic plates, thermopile, photo-conductive cell
visible light	energy transitions of outer orbital electrons in atoms and molecules	eyes, photographic plate, photocell
ultra-violet	energy transitions of orbital electrons, randomly accelerating ionised particles	fluorescence, photoelectric cell or photographic plates
X-rays	rapid deceleration of high energy electrons, energy transitions of innermost orbital electrons	ionisation chamber, photographic plates
γ-rays	changes of energy levels of nucleons, mass–energy conservation processes	GM tubes, phosphorescence, scintillation detectors

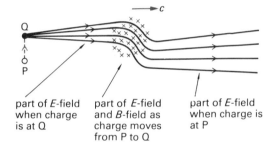

part of *E*-field when charge is at Q

part of *E*-field and *B*-field as charge moves from P to Q

part of *E*-field when charge is at P

FIGURE 22.3

we can measure to be $3.00 \times 10^8 \, \text{m s}^{-1}$. Of course, if the charge is made to pulse up and down at a frequency f, a continuous wave of the same frequency is generated. A long way from its source, this sinusoidal electromagnetic wave can be represented by varying E- and B-fields which are at right angles – figure 22.4.

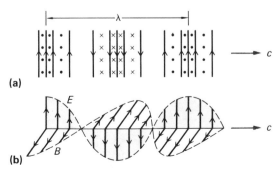

(a)

(b)

FIGURE 22.4

James Clerk Maxwell used ideas of this kind to predict the existence of electromagnetic waves in 1867, but it was not until 1888 that Heinrich Hertz successfully transmitted and received what we would now call radio waves, and a further 30 years before ship-to-shore radio communication was commonplace.

Maxwell's theory showed that the waves will travel in a vacuum at a speed c given by

$$c^2 = \frac{1}{\epsilon_0 \mu_0}$$

The value of $\mu_0 = 4\pi \times 10^{-7} \, \text{N A}^{-2}$ is fixed by the way we define the ampere and as, nowadays, the value of c is not measured but given an agreed value, it is this equation which determines that $\epsilon_0 = 8.854 \times 10^{-12} \, \text{F m}^{-1}$. You should check that the units are correct.

Superposition effects

Now that you know what electromagnetic waves are, it is easy to see what happens when two waves of the same frequency (and wavelength) superpose. The *E*-fields add as vectors and so you get constructive and destructive superposition giving rise to patterns of maxima and minima like those for ripples and sound waves described in the previous chapter (pages 294/295).

Because the frequencies of e-m waves are so high it is not possible to tune two sources to exactly the same frequency. However, if we reflect the waves from one transmitter T, the waves from T and from its image form two coherent sources S_1 and S_2. Figure 22.5 shows an experiment to investigate the superposition of microwaves of wavelength 30 mm.

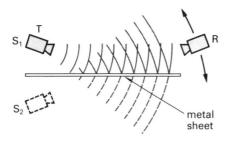

FIGURE 22.5

In (i)
$$(TP + PR)^2 = (50\,cm)^2 + (2 \times 20\,cm)^2$$
$$TP + PR = 64.0\,cm$$

In (ii)
$$(TQ + QR)^2 = (50\,cm)^2 + (2 \times 22.5\,cm)^2$$
$$TQ + QR = 67.3\,cm$$

As the reflected wave in (ii) must have travelled λ further than the reflected wave in (i) then
$$\lambda = (TQ + QR) - (TP + PR)$$
$$= 67.3\,cm - 64.0\,cm$$
$$= 3.3\,cm$$

Beware of looking for a formula to solve questions like this; superposition depends on path difference.

EXAMPLE

Using a microwave transmitter and receiver like those shown in figure 22.5, T registered a maximum when the distance from the metal sheet to the line joining T and R was 20 cm. The metal sheet was then slowly moved away from T and R and the *next* maximum was found after it had moved a distance 2.5 cm. If the distance from T to R was 50 cm, calculate the wavelength of the microwaves.

It is vital to draw a sketch of each of the situations and to mark all the relevant distances on them.

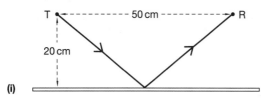

(i)

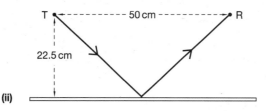

(ii)

Phase change of reflection

When R is placed right up to the reflecting surface in figure 22.5 (the metal sheet) it is found that the result of superposition is to produce a *minimum*, effectively a signal of zero amplitude. The path difference is, however, zero at that point. We can only conclude that the reflected wave has undergone a **phase change** of 180° at the metal sheet and so the waves reaching R are in antiphase. This phase reversal is equivalent to an extra path of $\lambda/2$. (You will see a similar phase change on reflection for mechanical waves when a pulse on a rope tied to a post reflects 'upside down'.) Superposition experiments with light are discussed fully in the next chapter, as is the diffraction of light and other e-m waves.

22.3 Plane polarised waves

Radio waves are plane polarised waves. By this we mean that the variations in the electric field occur only in one plane as the wave propagates. For instance in figure 22.4 this plane, called the plane of polarisation, is vertical. Of course if the variations of E lie in one plane then the variations

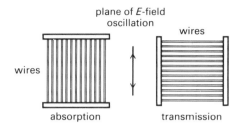

plane of *E*-field
oscillation

wires

wires

absorption

transmission

FIGURE 22.6

of *B* lie in a plane perpendicular to the variations of *E*, and so are perpendicular to the plane of polarisation. All radio waves and microwaves from dipole transmitting aerials are plane polarised and the variations of *E* occur in a plane formed by the dipole itself and the direction of propagation.

The 30 mm microwave receiver used in the laboratory has a receiving dipole aerial which must be vertical, so as to pick up the plane polarised waves. If you rotate the microwave receiver gradually about a horizontal axis, i.e. turn it on its side, the signal falls until at a 90° rotation no signal is received. On continuing the rotation until the receiver has turned through 180°, you will find that the signal rises again to its initial value.

A dipole receiving aerial absorbs the energy of the incident wave. Figure 22.6 describes the interaction of plane polarised microwaves with a grid of wires (only the electric field is shown). Absorption takes place when the incident *E*-field sets the free electrons in the wire oscillating *along* the length of the wire and thus absorbs the wave energy.

Polarised light

Infra-red waves, visible light, and all waves at the short-wavelength end of the electromagnetic spectrum are generated by *random* processes and consist of a collection of tiny waves each starting from different atoms or molecules at different times. Thus although the wavetrain from each excited molecule or atom is plane polarised the resulting radiation consists of waves with *E*-fields oscillating in all possible planes. Such a wave is said to be unpolarised. There are certain sheets of crystals (e.g. **Polaroid**, which consists of long molecules whose axes have been aligned during manufacture) which interact with light in the same way as the grid of wires with microwaves. After

passing through such a sheet the light is plane polarised with only the resolved part of the *E*-field in one plane emerging. Rotating a sheet of Polaroid through which light is passing can thus quickly establish its state of polarisation. If the transmitted light:

- shows no change in intensity – the incident light was unpolarised.
- varies slightly in intensity twice per revolution – the incident light was partly plane polarised.
- is cut off completely twice per revolution – the incident light was plane polarised.

Light reflected from the surface of water or glass is partially plane polarised. This is the key to understanding the use of Polaroid to remove glare. If the glare is the result of reflection from a horizontal surface, e.g. a wet road or the sea, Polaroid spectacles can cut off some (and at certain angles most) of the reflected light if the Polaroid is orientated to absorb horizontal variations of *E*. You can easily test this by looking at the glare through one Polaroid 'lens' while rotating the spectacles.

Photoelastic stress analysis

A major industrial and research application for the properties of polarised light is the analysis of stresses in structures of all sorts. The structure, e.g. the beam gripped at two points shown in figure 22.7, is first reproduced on a small scale using perspex. The model is then viewed between crossed sheets of Polaroid, i.e. light is first plane polarised by one sheet, passes through the perspex model and then is viewed or projected onto a

FIGURE 22.7

screen after passing through a second sheet which is called the analyser.

The effect is that the analysed light exhibits patterns of superposition (interference patterns) which reveal the lines of equal stress in the sample. Using monochromatic light and perspex of known thickness the number of the lines, starting from an obviously unstressed position, can be used to measure the stress at desired points in the structure. You can demonstrate this phenomenon in the laboratory by pulling a strip of polythene, perhaps with a nick cut in it, between two sheets of Polaroid placed in the beam from a projector.

22.4 Radio and television

Radio waves cover a wide range of frequencies (and wavelengths). They are commonly classified as shown in table 22.3.

Aerials

A radio designed to pick up both FM ($f \approx 90\,\text{MHz}$) and AM ($f \approx 1\,\text{MHz}$) broadcasts will use (i) a conducting aerial for VHF – the best position for the aerial is parallel to the E-field of the transmitted signal; and (ii) a coil wrapped on a ferrite rod (to increase the B-field) for the medium wave – the best position for the rod is parallel to the B-field of the transmitted signal. Have a look inside the radio to see the ferrite rod and copper coil.

For TV reception a dish aerial such as that on page 327 is used for picking up satellite TV. The rooftop aerials which pick up UHF signals, for example BBC TV, are made up of two rods (a

dipole receiver) each $\lambda/4$ long. A reflector is placed $\lambda/4$ behind the dipole and a series of directional rods in front of it. The whole is mounted on and insulated from a framework which can be attached high up on a building. In figure 22.8 the supporting framework is not shown.

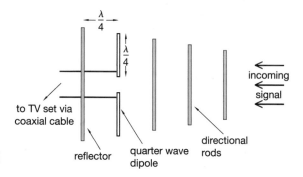

FIGURE 22.8

EXAMPLE

By looking at a TV aerial (a) estimate the frequency of the incoming e-m wave, and (b) explain the function of the reflector.

(a) The dipoles are each about 0.1 m long.

$$\frac{\lambda}{4} = 0.1\,\text{m} \quad \Rightarrow \quad \lambda = 0.4\,\text{m}$$

Using $\quad c = f\lambda$

$$f = \frac{c}{\lambda} = \frac{3.0 \times 10^8\,\text{m s}^{-1}}{0.4\,\text{m}}$$

$$= 7.5 \times 10^8\,\text{Hz or } 750\,\text{MHz}.$$

TABLE 22.3 Classification of radio waves

	Classification of carrier wave	Frequency range	Wavelength range
VLF	very low frequency	3–30 kHz	100–10 km
LF	low frequency	30–300 kHz	10–1 km
MF	medium frequency	300–3000 kHz	1000–100 m
HF	high frequency	3–30 MHz	100–10 m
VHF	very high frequency	30–300 MHz	10–1 m
UHF	ultra high frequency	300–3000 MHz	100–10 cm
SHF	super high frequency	3–30 GHz	10–1 cm
EHF	extra high frequency	30–300 GHz	1–0.1 cm

(b) The incoming wave travels an extra 0.2 m ($2 \times \lambda/4$) if it goes past the dipole and is reflected back to it. But, on reflection, it undergoes a phase reversal (a phase change of π rad) which is equivalent to a path difference of $\lambda/2$. So the total effective path difference between the direct and the reflected wave is λ.

They therefore arrive in phase and a stronger signal is received when they superpose.

Radio reception

Radio waves can reach a receiving aerial from a transmitter by two different paths.

Some of the wave travels directly over the surface of the ground; with long-wave (LF) signals diffraction effects enable the waves to follow the curvature of the Earth and to be relatively unaffected by obstacles such as hills and buildings – figure 22.9). But VHF signals, with a shorter wavelength, do not diffract as much, so the transmitter and receiver need to be fairly close to each other.

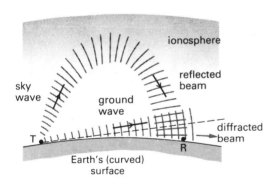

FIGURE 22.9

Another part of the wave travels up into the air and is reflected from the **ionosphere** (layers of ionised molecules between 80 km and 500 km from the Earth's surface). Close to the transmitter the **ground wave** predominates but at distances of 100 km or more the **sky wave** may be of about the same amplitude; when the waves superpose the minima can then be very pronounced.

Conditions in the ionosphere are changing continually. At one moment the receiving aerial

may be at a maximum in the interference pattern and a strong resultant signal is received. At another the resultant signal may drop off almost to nothing. This is the fading in the medium wave band which is often noticeable at night (when the sky wave is more effectively reflected). The reflection is an example of total internal reflection; the wave is refracted in a curve as shown. For $\lambda \approx 30$ m (i.e. TV waves) the sky wave is not reflected, but we can use a satellite to pick up the signal from the transmitter, amplify it and re-emit it using a different carrier frequency. Such **communications satellites** are parked above the equator in geostationary orbits and their signals are received by parabolic dish aerials – see figure 24.2 (page 327).

E-M wave energy

The energy of e-m waves propagating from a source is conserved. This means that the intensity of the wave falls off according to an inverse square law:

$$I \propto \frac{1}{r^2}$$

You can see that the ratio of the intensities at two different distances r_1 and r_2 from the source is given by

$$\frac{I_1}{I_2} = \left(\frac{r_2}{r_1}\right)^2$$

Wave intensity is measured in W m^{-2} and if, for example, the intensity (or strength) of a radio signal was $2 \times 10^{-15} \, \text{W m}^{-2}$ at a distance of 16 km from a transmitter, the signal strength at 32 km (twice the distance) would have fallen to 1/4, i.e. to $0.5 \times 10^{-15} \, \text{W m}^{-2}$.

This reduction in intensity is often expressed in decibels, dB, and is calculated as follows:

$$\text{intensity loss} = 10 \lg(P_2/P_1)$$
$$= 10 \lg(0.25) = -6 \, \text{dB}$$

You can see that, similarly, a drop in intensity by a factor of two is described as a loss of -3 dB.

When an e-m wave is travelling in a medium which absorbs some of the energy, e.g. light travelling through water or X-rays through human tissue, we say the wave is **attenuated**, and the simple inverse square law no longer holds.

22.5 Electromagnetic wave spectra

There are essentially two ways of investigating a beam of waves so as to discover the range of wavelengths (or frequencies) it contains. One, used at wavelengths near the optical part of the electromagnetic spectrum is to disperse the waves using a prism or a diffraction grating. The other, used at radio and microwave wavelengths, is to use a tuned circuit. As most of the information we have for the structure of atoms and molecules and for the nature of the universe beyond the Earth comes from the waves which atoms and stars emit, the detailed analysis of what wavelengths are present in a beam of waves is of vital interest.

Optical line spectra

The details of how you produce a spectrum using a diffraction grating are given in Section 23.4. When the source of the waves is a gas whose molecules have been excited in some way the spectrum is always a **line spectrum** *unique* to that element. Figure 22.10 shows diagrammatically the line spectrum of neon and indicates the relative intensities of the lines in two ways; as expected (think of a neon advertising sign) many of them are in the red region of the spectrum.

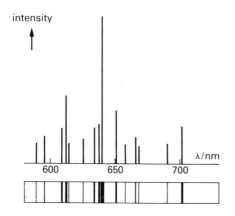

FIGURE 22.10

One of the commonest laboratory sources is a low-pressure sodium vapour lamp. In the sodium spectra there are two prominent and very close lines in the yellow region with wavelengths of 589.0 nm and 589.6 nm – the source can be treated as a monochromatic (one-colour) source for many purposes.

The fact that each element produces a unique emission spectrum can be used to provide a chemical analysis of, for example, a sample of moon dust or a metal alloy. The sample must be vaporised with a high temperature arc or spark, e.g. an electric discharge across the gap between two carbon rods. Individual atoms are separated and are given energy. They are said to be **excited**; they emit their characteristic light in returning to their ground state.

Line spectra are thus evidence for the structure of atoms. Their analysis and the expression for photon energy, $E = hf$, enables energy level diagrams for atoms to be deduced. For more on the quantum theory and the 'particle-like' nature of e-m waves you need to refer to Chapter 26 of this book.

Gas atoms which emit a series of discrete wavelengths when excited, also absorb exactly those wavelengths when e-m waves containing them pass through the gas. This is a resonance phenomenon and is called **absorption**.

Figure 22.11 shows how to demonstrate absorption in the laboratory. The sodium flame, a bunsen flame burning under a sodium flame pencil, casts a shadow on the screen as energy at wavelengths characteristic of sodium atoms is absorbed from the incident light and then re-radiated uniformly in all directions. The shadow disappears almost completely if the pencil is removed but the bunsen left in position.

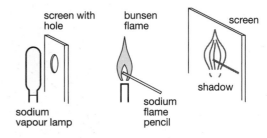

FIGURE 22.11

The Sun's spectrum is found to be crossed by many narrow dark lines. These absorption lines correspond to elements in the Sun's chromosphere, the cooler outer gaseous atmosphere of the sun, and are called **Fraunhofer lines**. The element helium was first discovered by a careful analysis of the Sun's absorption spectrum.

Infra-red spectra

Thermal radiation is discussed in Chapter 11 (page 147). The absorption spectra produced when thermal radiation from a source at about 800 K passes through solids, liquids or gases and is then dispersed are called infra-red absorption spectra. They can be used to identify particular gaseous molecules such as CO_2 or SO_2 but are more important than this in that they can give us information about the bonding in complex organic molecules. Infra-red spectra are examples of *resonant* absorption; the sample absorbs at the resonant frequencies of its bonds (acting a bit like two masses connected by a spring) and we study the radiation which is left over.

Figure 22.12 shows the infra-red absorption spectrum of trichloromethane (chloroform $CHCl_3$). The main absorption lines are labelled with the bonds which produce them.

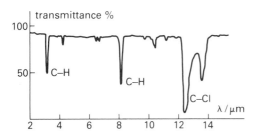

FIGURE 22.12

X-ray and γ-ray spectra

The X-ray source emits a continuous spectrum superimposed upon which there is a line spectrum

consisting of two prominent lines. Like the optical spectra of gases and vapours the line spectrum from an X-ray tube is characteristic of an element. The details are discussed further in Section 26.4 which starts on page 366.

All γ-rays originate in the nucleus of an atom. They are radioactive radiations but are identical with short-wavelength X-rays. For a given γ-ray source only certain sharply defined wavelengths are observed, which are characteristic of the type of nucleus concerned, i.e. γ-ray spectra are line spectra. At these very short wavelengths it is usual to refer not to the wavelength, but to the γ-ray photon energy, and to express the energies in electron-volts.

e.g. $^{212}_{83}Bi$ decays to $^{208}_{81}Tl$ by α-decay

but the resulting nucleus may be left in a number of **excited states**. It reverts to its ground state by the emission of γ-rays with energies of 0.04 MeV, 0.33 MeV, 0.47 MeV, 0.49 MeV and 0.63 MeV, these forming a γ-ray line spectrum.

Radio and microwave spectra

Radiation from space is detected using parabolic dish aerials and tuned circuits. The spectrum of the radiation is thus plotted by scanning across the range of resonant frequencies of a tuned circuit. Clouds of atomic hydrogen emit a characteristic line at 0.21 m and complex molecules have been discovered in interstellar dust by their emission and absorption spectra. In the last 20 years methanal (formaldehyde, HCHO), carbonyl sulphide (OCS) and isocyanic acid (HNCO) have been found by emission lines at wavelengths of 62 mm, 2.7 mm and 14 mm respectively. There are many more. Thus molecules containing H, C, O, N and S are all known to exist in forms from which more complex reproducing organisms, life itself, may well have evolved.

Exercises on each section of this chapter may be found in the companion textbook, *Practice in Physics*.

SUMMARY

At the end of this chapter you should be able to:

♦ understand that e-m waves carry energy from place to place and those from a point source obey the inverse square law $I \propto 1/r^2$.

♦ remember that all e-m waves travel at $3.00 \times 10^8\,\mathrm{m\,s^{-1}}$ in a vacuum and that this speed is independent of the speed of the source or the observer.

♦ explain that the electromagnetic spectrum has different sections within each of which the waves are produced in a particular way.

♦ remember the order of magnitude of the wavelength or frequency of different parts of the electromagnetic spectrum.

♦ remember that there is a change in the received frequency (and wavelength) of e-m waves from a moving source.

♦ understand that e-m waves consist of oscillating electric and magnetic fields and that their speed in a vacuum is given by $c = 1/\sqrt{\epsilon_0 \mu_0}$.

♦ understand that all e-m waves are transverse waves which can be polarised and that the plane of polarisation is that of the electric field oscillations.

♦ remember that light is not usually polarised and that reflected light is partially plane polarised.

♦ understand that e-m waves obey the principle of superposition and explain how to achieve coherent sources of e-m waves with microwaves.

♦ remember that when e-m waves are reflected they sometimes undergo a phase change of 180° which is equivalent to an extra path length of $\lambda/2$.

♦ explain that radio waves cover a huge frequency range: from about 10 kHz (VLF) to about 100 GHz (EHF).

♦ understand that the line spectra resulting from the emission and absorption of e-m waves is the source of a great deal of our knowledge of the universe and of the structure of atoms and molecules.

23 Interference Patterns

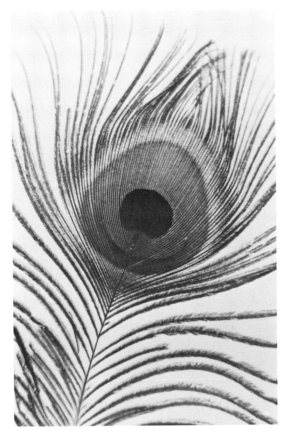

When you see colours in a soap bubble you are seeing an interference pattern: the photo on the cover of this book is of soap bubbles. Try to make a flat soap film between the thumb and first finger of your hand and look at the light reflected from it as it gets thinner. The colours you see in puddles on the road are also the result of the superposition of light; they occur when a thin layer of oil lies on top of the water surface. Nature has used a similar trick to produce the colours we see in the thin transparent wings of dragonflies and other insects; the colours change as the angle at which you view the wing changes. Perhaps the most spectacular example of superposition producing colours in nature is the pattern on the feathers in a

peacock's tail – though the exact mechanism here is more complex than for soap bubble colours.

The simplest quick demonstration that light is a wave motion is to look through an umbrella at a distance sodium street lamp. The light refracts through the tiny gaps between the fibres and an interference pattern is formed. To convince someone that different colours have different wavelengths get them to look at the reflection of a small light source in the surface of a compact disc.

23.1 Stationary waves

Sinusoidal waves on strings and springs can be produced in the laboratory using the sinusoidal output from a signal generator to drive a vibrator – a sort of robust moving-coil loudspeaker.

Transverse waves

Figure 23.1 shows (a) how a transverse wave can be produced on a thin rubber cord.

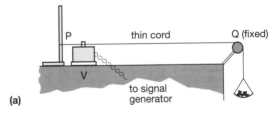

(a)

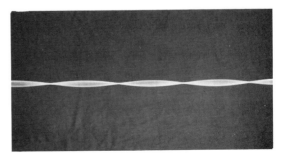

FIGURE 23.1

311

The tension in the cord can be adjusted by placing small masses in the hanging tray. For a fixed tension, the frequency f of the vibrator V is slowly raised from its lowest value. The transverse waves from the source V travel in both directions along the cord and are reflected at the fixed ends P and Q. We thus have a system in which waves travelling along the cord in one direction are superposed on waves travelling in the opposite direction. There are some values of f which produce a fixed pattern of superposition in which some places show zero resultant oscillation (where the waves are in antiphase all the time) and some places which show large amplitudes of oscillation (where the waves are always in phase). The observed pattern shown in figure 23.1(b) is called a **stationary wave pattern** or, more simply, stationary waves.

Nodes and antinodes

Figure 23.2 is a series of displacement–position graphs showing the resultant wave profile at successive intervals of time for a stationary wave. The graphs are drawn every eighth of a cycle, i.e. at time intervals of $T/8$. Points like N have zero displacement at all times; they are called nodal points or **nodes**. Neighbouring nodes can be seen to be *half a wavelength* $\lambda/2$ apart. Similarly there are points like A with the largest amplitude of oscillation, called antinodal points or **antinodes**, which are also $\lambda/2$ apart.

The cord in the demonstration of stationary waves has two fixed ends which must both be nodes. There are therefore only certain wavelengths (and frequencies f) for which a stationary wave pattern fits onto PQ. Energy is transferred efficiently from V to the cord and large-amplitude oscillations are built up when the driving frequency f_d has these values. In this sense the experiment is a *resonance* experiment, and the frequencies at which the stationary wave patterns appear are the **natural frequencies** of the cord.

Using a flashing stroboscope to illuminate the rope you can see that all bits of the rope between adjacent nodes move up together and then down together, i.e. they are in phase. They are, however, in antiphase with the next half-wavelength loop so the rope moves with a sort of flip–flap motion; it is vibrating.

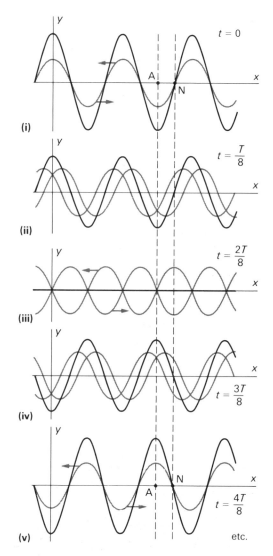

FIGURE 23.2

Vibrating strings

The violin, cello, guitar (figure 23.3) and piano are all musical instruments which depend on the resonant vibration of stretched strings. In the piano a complete set of strings of fixed length and tension are available. But in the others the player makes do with a few strings by altering their effective length with his or her fingers.

Harmonics

The length of a resonating string of length l must be a whole number n of half-wavelengths if we are to get a stationary wave pattern.

i.e.
$$l = n\frac{\lambda}{2}$$

and as $c = f\lambda$, there is a definite value of f associated with a given wavelength

$$f = \frac{c}{\lambda} = \frac{c}{2l/n} = \frac{nc}{2l}$$

For a stationary wave pattern we must have $f = c/2l, 2c/2l, 3c/3l$ etc. only, and the experiment in figure 23.1 clearly shows this.

The speed of a transverse wave on a cord or string is given by $c = (T/\mu)^{1/2}$ so that under given conditions of tension T and mass per unit length μ, the frequencies which produce a stationary wave pattern are

$$f = \frac{n}{2l}\sqrt{\frac{T}{\mu}}$$

FIGURE 23.3

Each value of n (1, 2, 3 etc.), corresponds to a certain mode of vibration. The **fundamental** frequency (or first harmonic) occurs when the cord is half a wavelength long. The other resonant frequencies, called **overtones** or **harmonics** (the first overtone is the same as the second harmonic, etc.), are illustrated above in figure 23.4.

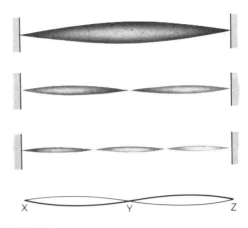

FIGURE 23.4

EXAMPLE

Middle C on a piano has a frequency of 261 Hz. The wire which produces this note is 0.60 m long and has a total mass of 8.0 g. Calculate the tension in the string. If the tension increased by 2% what effect would this have on the frequency of the note?

$$c = \sqrt{\frac{T}{\mu}}$$

and here
$$\mu = (8.0 \times 10^{-3}\,\text{kg}) \div (0.60\,\text{m})$$
$$= 13.3 \times 10^{-3}\,\text{kg}\,\text{m}^{-1}$$

But
$$\frac{\lambda}{2} = 0.60\,\text{m} \quad \text{(node to node)}$$

$\therefore$
$$c = f\lambda = (261\,\text{Hz})(1.2\,\text{m})$$
$$= 313\,\text{m}\,\text{s}^{-1}$$

and hence $T = c^2\mu$
$$= (313\,\text{m}\,\text{s}^{-1})^2(13.3 \times 10^{-3}\,\text{kg}\,\text{m}^{-1})$$
$$= 1300\,\text{N}$$

Beware of trying to find a single formula which gives the answer. You can be confident that enough data has been given to solve the problem so if you use relevant physics a solution will emerge.

Increasing the tension by 2% will not noticeably affect the length of the string, i.e. neither l nor μ change. The speed of the wave will increase by about 1% as $c = (T/\mu)^{1/2}$ and so the frequency will increase by 1% as $f = c/\lambda$.

Longitudinal waves

Figure 23.5 shows an experimental arrangement for demonstrating longitudinal stationary waves on a spring.

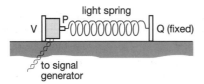

light spring

V P Q (fixed)

to signal
generator

FIGURE 23.5

The frequency of the vibrator V is again raised slowly from its lowest value. The stationary wave patterns or resonances are obvious to the eye as a blurring of the spring at antinodes for certain values of f. The distance between adjacent nodes or antinodes is again $\lambda/2$. In this case Q, the fixed end of the spring, will always be a node but if it is free, you can sometimes get stationary waves where Q is an antinode.

Vibrating air columns

Sound is a longitudinal wave so the above experiment helps to explain the formation of stationary sound waves in a tube. Musical instruments like clarinets or trumpets have resonances at which their lengths are related to the wavelength of the note sounded. In a flute or a recorder, for example (figure 23.6), both ends of the vibrating air column are antinodes; they are like the end of the spring above when it is free.

Figure 23.7 shows one way of representing these stationary waves. (The shaded regions are really graphs of longitudinal displacement against position.) The A and N – for antinode and node – are for the *displacement* of the air. Where there is maximum *displacement* (a displacement *antinode*) there is zero *pressure change* (a pressure *node*). You can imagine that if one end of the tube is closed a different set of resonant frequencies will occur.

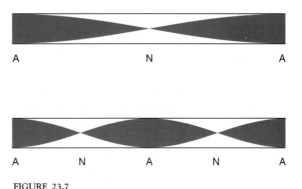

A   N   A

A   N   A   N   A

FIGURE 23.7

FIGURE 23.6

$(635 - 203)\,\text{mm} = 432\,\text{mm}$, the wavelength of the sound in the tube is given by

$$\frac{\lambda}{2} = 432\,\text{mm}$$

so that $\qquad \lambda = 864\,\text{mm}$

Thus $\qquad c = f\lambda = (384\,\text{Hz})(0.864\,\text{m})$

$$= 332\,\text{m s}^{-1}$$

Electromagnetic waves

It is easy to set up stationary waves using microwaves of $\lambda \approx 3\,\text{cm}$ but not so easy to demonstrate the effect with light for which $\lambda \approx 10^{-6}\,\text{m}$.

Figure 23.8 shows a microwave transmitter T facing a metal plate. Instead of the usual receiver with its directional horn we need a small probe P which receives equally well from all sides. As P is moved along the perpendicular from T to the reflecting surface, maxima and minima can be detected at regular intervals. The distance between successive minima (nodes) is $\lambda/2$ and by measuring the distance across, for example, 10 nodes the wavelength of the microwaves from the transmitter can be found. The uncertainty is only $\pm 1\,\text{mm}$ in $300\,\text{mm}$ or about $\pm 0.3\%$. At the nodes the received signal is not zero as the amplitude of the reflected wave is less than that of the wave direct from T. If the frequency f of the microwaves is known then measuring in this way enables the speed c $(= f\lambda)$ of the microwaves to be found. In this experiment there is no need to position T at a whole number of half wavelengths from the reflecting sheet as T does not have to be at a node (or an antinode) in the stationary wave pattern.

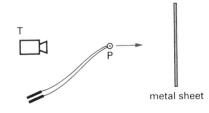

metal sheet

FIGURE 23.8

Thin-film interference

To demonstrate stationary waves with light in the laboratory you could place two clean microscope slides or glass blocks on top of one another and look at the reflection of a white fluorescent light in the blocks (a sodium lamp is even better). After a moment or two the eye accidentally focuses on the plane of the gap between the blocks and a yellowy-orange pattern (if the sodium lamp is being used) rather like a contour map – figure 23.9 – is seen which changes if the blocks are pressed together or twisted while in contact. Once found it is easy to study the pattern in detail.

FIGURE 23.9

These patterns are not, strictly, stationary wave patterns but are the result of the superposition of light being reflected from the glass surface on either side of the thin air film between the blocks. They are like the colours seen on puddles which have a film of oil on their surface or the colours which are seen in soap bubbles – the two commonest everyday examples of the fact that light, because it superposes, must be wavelike.

23.2 Diffraction

Waves can bend round corners. Medium-wave radio signals can be detected over the horizon or behind a hill, sound waves carry conversations over walls or round open doors. The common feature in these changes in the direction of energy propagation is that where part of a wavefront is blocked or limited the remaining part spreads into the 'shadow' area. The phenomenon is called the diffraction of waves.

The photographs in figure 23.10 show the diffraction of water waves in a ripple tank. The

FIGURE 23.10

diagram on the right is a **polar diagram** which represents the distribution of energy beyond the gap in the third photograph: the straight dashed line coincides with the calm water (no energy).

The detail of the diffraction in the figure is seen to depend on the size of the gap – or, more correctly, on the ratio of the wavelength λ to the width d of the gap. We can see that a greater proportion of the wave energy spreads to the shadow area when λ/d is large than when it is small. A close look reveals that the region beyond the barrier contains a complex wave pattern. Obviously water waves could not propagate beyond the gap *without* spreading sideways at all for this would involve a 'wall' of water with vertical edges. The water thus falls outwards and in so doing the energy of the wave is diffracted.

EXAMPLE

A loudspeaker system in a large hall consists of a number of rectangular speakers, each measuring 1.0 m by 0.1 m and set about 3 m above the ground. Explain why they are this shape and how they should be fixed, i.e. should the long side be vertical or horizontal? The speed of sound in air is about $340\,\mathrm{m\,s^{-1}}$.

The sound from the rectangular speaker acts like sound coming through a door, the amount of spreading depends on the dimensions of the speaker and the wavelength of the sound. For a

sound wave of frequency 1500 Hz, $c = f\lambda$ gives a wavelength of about 0.20 m.

In a large hall you would not want to lose too much sound (energy) upwards nor would you wish to send sound from the loudspeaker only to those listeners sitting directly in front of it.

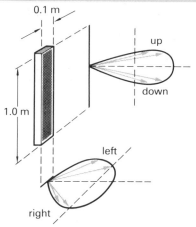

Therefore the diagram shows the speaker correctly mounted with its long side vertical. The width of the speaker is now less than the wavelength of the sound, so the sound spreads out horizontally, but the height of the speaker is greater than the wavelength of the sound, so not much sound spreads upwards or downwards. At 300 Hz (for which $\lambda \approx 1$ m) there will be more vertical and horizontal spreading of the sound.

The diffraction of light

Light bends round corners too. But it is hard to convince people of this. Shadows have sharp edges – don't they? The demonstration shown in figure 23.11(a) using a laser and a very narrow slit shows that shadows are not what you think – light does diffract.

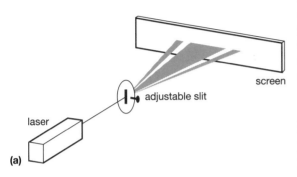

(a)

(b)

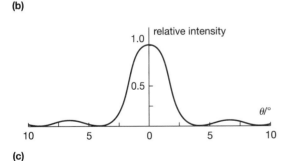

(c)

FIGURE 23.11

As the slit must be very small – to produce the pattern in the diagram (b) it was $10\,\mu m$ – it is clear that the wavelength of the laser light is *very* small. In fact, for a helium–neon laser, $\lambda = 633\,nm$ or $0.633\,\mu m$.

Explaining the pattern

In order to predict where the wave energy transmitted through the slit will go we need to imagine each point on the wavefront in the gap as a secondary source sending waves to the right of the gap. By superposing these secondary waves we can, in principle, find the resulting wave amplitude at any point. In practice this is very difficult. The only case where it is possible to make any simple prediction is when we consider the contributions from each source in a given *direction* θ from the straight through direction and try to find the angle at which the intensity first falls to a minimum.

Suppose we consider ten sources p, q, r, s, t and i, j, k, l, m as shown in figure 23.12. The waves from p and i, along pP and iI respectively, will have a phase difference which depends on their path difference iN. If iN $= \lambda$ these two secondary waves will be in phase. But if iN $= \lambda/2$ they will be in antiphase (180° out of phase) and the contributions from p and i will cancel (the principle of superposition). If in this case $\theta = \theta_m$, then

$$\frac{\lambda}{2} = iN = \frac{d}{2}\sin\theta_m$$

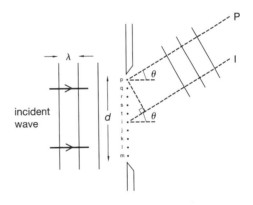

FIGURE 23.12

All other pairs of sources such as q and j, etc. will send waves in the direction θ_m which are in antiphase, i.e. they will cancel and the total amplitude of the wave in the direction θ_m is zero. For zero intensity therefore

$$\sin\theta_m = \frac{\lambda}{d}$$

You can check that for $\lambda = 633\,\text{nm}$ the slit width for the graph in figure 23.11(c) is just over $9\,\mu\text{m}$.

To summarise what $\sin\theta_m = \lambda/d$ tells us:

♦ when $d \gg \lambda$ (which makes $\sin\theta_m$ and θ_m very small) very little diffraction is noticeable: the edge of the central diffraction peak more or less coincides with the edge of the geometrical shadow of the slit.

♦ when $d \approx \lambda$ (which makes $\sin\theta_m \approx 1$, and $\theta_m \approx 90°$) the central maximum spreads over most of the region in which we would expect a shadow.

♦ at intermediate values there is a central diffraction peak of width $2\theta_m$ (i.e. θ_m on either side of the centre) where $\theta_m = \arcsin \lambda/d$.

You can easily see diffraction patterns without a laser by mounting two razor blades so as to make a narrow adjustable slit. Hold this up to the eye and view a line filament bulb a metre or more away with the slit parallel to the filament. You can see the effect of using different wavelengths by placing filters in front of the slit and the width of the slit can be varied by pushing the razor blades closer together.

When considering diffraction in light you can often approximate $\sin\theta$ to θ/rad, since θ is usually smaller than 5° except when using diffraction gratings. With sound or with microwaves (see the example below) you cannot make this approximation as the angle θ shown is about 30°.

EXAMPLE

A microwave source of wavelength 30 mm is used to demonstrate diffraction at a slit as shown. The lens L_2 and the receiver can be swung together on an arm pivoted at the centre of the slit. If the width of the slit is 100 mm, at what angles will the received intensity first fall to a minimum?

The first minimum occurs at θ_m

where
$$\sin\theta_m = \frac{\lambda}{d} = \frac{30\,\text{mm}}{100\,\text{mm}}$$

so that
$$\theta_m = 17.5°$$

In tackling problems like this be careful about the use of the deg/rad function on your calculator.

A similar result is found to hold for the case of circular apertures such as the pupil of the eye, a pinhole or the outline of a lens, but there is a factor of 1.22 in the result. We will not bother to include this factor in what follows.

Diffraction and the eye

As you travel along a road the letters on a signpost only becomes decipherable when you are within a certain distance of the sign. We say that you can then **resolve** the letters. A test is to draw two black lines 2.0 mm apart on white paper. They can be separately distinguished by the eye from distances up to about 6 m; if the lines are only 1.0 mm apart the distance is only 3 m and so on. The angle subtended at the eye

$$= \frac{2 \times 10^{-3}\,\text{m}}{6\,\text{m}} \approx 3 \times 10^{-4}\ \text{rad or 1/60 of a degree.}$$

It is this angle which determines whether or not the two lines can be resolved. (In a road sign the two lines might be the sides of a letter H or a letter U.)

If you think of the images of the two lines on the retina you will realise that they are two diffraction patterns each like that shown in figure 23.11(b). The smaller the aperture of the eye (the pupil) the wider each diffraction pattern and the more likely it is that the central peaks overlap.

wax lens

Telescopes and diffraction

Figure 23.13 shows the image of three sources formed on a photographic film placed in the focal plane of a converging lens. The objects might be stars and the lens the objective of a telescope, or illuminated specks of dust and the lens the objective of a microscope. The two photographs are taken with different circular holes of different diameter d and placed in front of the lens. In (a) d is large and the images clearly resolved while in (b) d is small and the left images can only just be detected as two separate sources.

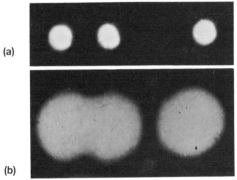

(a)

(b)

FIGURE 23.13

The same principle applies in a *reflecting* telescope where the objective is a concave mirror of diameter d. The largest telescopes use mirrors instead of lenses (see section 24.7).

In designing optical telescopes for use at the limits of resolution of the lens d the diameter must be made as large as possible. Thus in a 2 m or a 5 m astronomical telescope the number quoted is the *aperture* d; this is the telescope's most important and expensive feature. To manufacture, support and steer a concave optical mirror of diameter up to 5 m is a remarkable engineering achievement. In improving the ability of a telescope to resolve we are, of course, automatically collecting more energy and can thus detect fainter objects, for instance, more distant galaxies. For radio telescopes where the wavelength λ is of the order of centimetres, the reflecting dish must be very large so that the ratio λ/d is small.

Radioastronomy

Radio signals from space were first identified by Karl Jansky in 1932. The problems of radioastronomy are similar to those in optical astronomy, namely the design of telescopes (i) which can collect large quantities of energy, and (ii) which can resolve one radio source from another. The following example illustrates both problems for the Jodrell Bank telescope which is illustrated in figure 23.14.

FIGURE 23.14

EXAMPLE

An 80 m diameter dish collects energy from a radio source known (from its Doppler shift) to be a hundred million light years away from the solar system. If the input power which the telescope measures is 10^{-18} W, estimate the power output of that source.

As there are about 3×10^7 s in one year, then 100×10^6 light-years is a distance r, where

$$r = (100 \times 10^6)(3 \times 10^8 \, \mathrm{m\,s^{-1}})(3 \times 10^7 \, \mathrm{s})$$

$$\approx 10^{24} \, \mathrm{m}$$

If the power of the source is P and assuming an inverse square law for the energy spreading from it, the intensity I at a distance r from it is given by $I = P/4\pi r^2$.

But the intensity at the telescope is the detected input power divided by the area of the dish i.e.

$$I = \frac{1 \times 10^{-18}\,\text{W}}{\pi (40\,\text{m})^2} = 2.0 \times 10^{-22}\,\text{W}\,\text{m}^{-2}$$

$$\therefore \qquad \frac{P}{4\pi (10^{24}\,\text{m})^2} = 2 \times 10^{-22}\,\text{W}\,\text{m}^{-2}$$

$$\Rightarrow \qquad P = 2.5 \times 10^{27}\,\text{W}$$

Young's fringes: by measuring the distance between adjacent bright fringes (or adjacent dark fringes) you can find the wavelength of the light used. An eyepiece which can be moved along a scale – like a travelling microscope – is ideal and some eyepieces have a built-in scale which you see when you look through it. The graph (figure 23.15(c)) shows how the intensity varies across the bottom photograph.

23.3 Young's slits

It is not possible with light to produce the equivalent of two loudspeakers which are in phase. A demonstration of the principle of superposition with light can only be achieved by taking a wavefront from a single lamp and using it twice; you cannot use two separate lamps to produce superposition effects. The light from the two slits is then said to be **coherent**, i.e. there is a constant phase difference between the light from the two sources. If the light from one lamp passes through two parallel slits which are close together, the light will diffract through them and overlap.

A laser experiment just like that in figure 23.11 on page 317 but with a double slit works beautifully, and an interference pattern is immediately seen on the screen where the two diffracted beams overlap and superposition occurs.

If you have no laser the experiment takes much longer, and needs to be much more carefully set up, mainly because there is not so much light available. One alternative to a laser would be a white-light source with a coloured filter in front of it but although there will be only one 'colour' this will not be truly monochromatic light, since the filter lets through a range of wavelengths. Instead we shall describe the use of a sodium lamp (which is essentially monochromatic).

The experiment needs to be performed in a darkened room and the distances between the single and double slits and the double slits and the screen are indicated by the scale in figure 23.15(a). Each slit of the double slit needs to be very narrow, less than 0.1 mm, to produce a lot of diffraction. The photographs (figure 23.15(b)), which are life-size, show the fringes obtained for slit separations from 1.4 mm (at the top) to 0.2 mm (at the bottom). The pattern of superposition is called

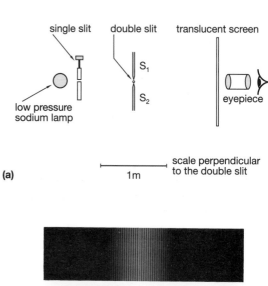

(a)

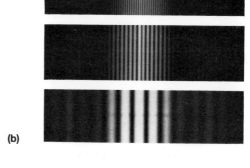

(b)

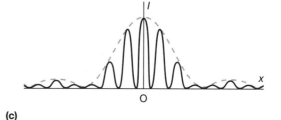

(c)

FIGURE 23.15

Measuring the wavelength of light

Consider figure 23.16. For a maximum at P we need

$$S_2P - S_1P = n\lambda$$

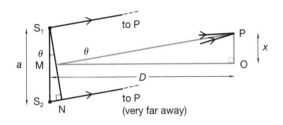

FIGURE 23.16

If P is a very long way from the slits (i.e. $D \gg a$), then S_1P and S_2P are almost parallel

and

$$S_2P - S_1P = S_2N = n\lambda$$

where N is the base of the perpendicular from S_1 to S_2P. (See page 295 for details of path difference and superposition.) Looking at triangles S_1S_2N and MPO which are similar

$$\frac{S_2N}{a} = \frac{OP}{NP} = \frac{OP}{D}$$

so that

$$S_2N \approx \frac{ax}{D}$$

Thus the condition for a maximum at P is that

$$\frac{ax}{D} = n\lambda$$

If we now move the eyepiece (e.g. with a micrometer screw) from one maximum, say $n = 3$, to the next maximum, $n = 4$, it moves through a distance called the **fringe separation**. Then

$$\lambda = \frac{a}{D}(\text{fringe separation}) = \frac{a\Delta x}{D}$$

The proof assumes that the slit-to-screen distance D is so large compared with the slit separation a that we can treat the lines S_1P and S_2P as being parallel.

This is a straightforward but not very precise method for finding the wavelength of light. We cannot measure a and the fringe separation

precisely and so we must expect an uncertainty of at least ±5%, even if the experiment is performed with care. What *can* be quickly achieved with Young's fringes is a quick check on the relative wavelengths of two colours. For example, it is easy to show that $\lambda_{\text{red}} > \lambda_{\text{green}}$ by using a white-light source and placing first a red and then a green filter in front of the eyepiece.

With a white-light source, there is a set of overlapping fringe systems. As the eye is very much more sensitive to wavelengths in the range 500 nm to 600 nm than in the rest of the visible spectrum, a set of Young's fringes in white light *is* observable for values of n up to about 3 or 4 before the overlapping of different colours obscures the pattern. The inner fringes have coloured edges.

EXAMPLE

In a Young's slit experiment with sodium light, the eyepiece was moved 3.2 mm in counting from 0 to 4 fringes. If the distance from the double slit to the eyepiece was 0.96 m calculate the slit separation. Take the wavelength of the light to be 590 nm.

The fringe separation = 3.2 mm/4 = 0.80 mm.

$$\lambda = \frac{a\Delta x}{D}$$

$$\Rightarrow \quad a = \frac{\lambda D}{\Delta x} = \frac{(590 \times 10^{-9}\,\text{m})(0.96\,\text{m})}{0.80 \times 10^{-3}\,\text{m}}$$

$$= 0.71\,\text{mm}$$

Superposition with microwaves

A large-scale version of the Young's slit experiment can be performed to measure the wavelength of microwaves. The transmitter is placed about 0.5 m from a double slit consisting of two metal sheets with a narrow strip of metal (about 50 mm wide) between them – figure 23.17. The receiver detects the characteristic pattern of superposition beyond the double slit and, by covering and uncovering one of the slits, both the diffraction effect and the redistribution of energy in the interference pattern can be studied. As D is *not* very much larger than S_1S_2 in this demonstration we cannot make the approximation which led to $\lambda = a\Delta x/D$. To get a rough value of λ you simply measure S_1P and S_2P with a ruler – compare the two-source experiment with sound on page 295.

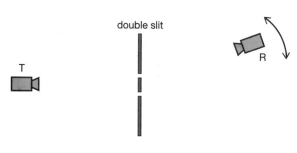

double slit

FIGURE 23.17

23.4 The diffraction grating

A diffraction grating consists of a lot of very narrow parallel slits. A beam of light striking it diffracts through the slits and the diffracted wavefronts then superpose. If you hold a diffraction grating close to your eye and look at a narrow gas-discharge tube, e.g. of neon, you will see sharp images of the source in each of the colours (wavelengths) present in neon light. The grating is thus ideal for studying **spectra**; i.e. for analysing the wavelengths emitted by a source of light. A diffraction grating **disperses** the light into its component colours as does a refracting prism.

A typical grating will have, in each millimetre of its width, about 500 narrow parallel slits through which the light is diffracted. It might be 30 mm across and will thus contain a total of 15 000 diffracting slits spaced evenly with their centres 2×10^{-3} mm (2 μm) apart. With this separation s the width d of each slit must be less than 2 μm; in practice it is about 1 μm, so light of wavelength 0.6 μm diffracts into a wide arc on passing through each narrow slit. In fact, using these numbers, the first minimum does not appear until the angle of diffraction is nearly 40°. Needless to say they are not easy to make: the gratings you use are plastic replicas of master gratings.

The theory of the diffraction grating

A set of plane wavefronts is shown arriving at a row of slits (which is part of a diffraction grating) in figure 23.18(a). If this had been drawn to scale then for wavelength λ, the distance between the slits is about 3λ and each slit is less than $\lambda/2$ wide. With such a narrow slit the waves diffract as shown and, by holding the page up and looking along the thick lines called orders, you will see the way the semicircular waves 'join up' in certain directions. (There are similar orders which deviate below the straight through direction, i.e. tilting down the page.)

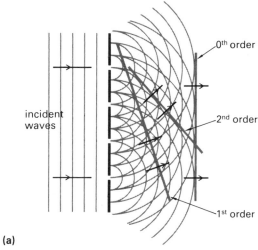

(a)

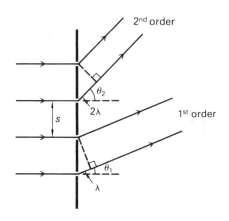

(b)

FIGURE 23.18

322

The ray diagram in (b) enables us to see that at certain angles the path difference between the beams diffracted from adjacent slits will be λ, 2λ etc. Consider the case where the path difference between adjacent diffracted beams is λ. The next slit will contribute a diffracted beam in the direction θ_1 with a path difference of 2λ, the next with a path difference of 3λ, the next 4λ etc., so that all these beams diffracted at θ_1 are in phase (you should study the arcs from the slits to the first-order line in figure 23.18). The condition for this first maximum at θ_1 is therefore

$$n\lambda = s \sin \theta_1$$

The next maximum, at θ_2, needs a path difference of 2λ between diffracted beams from adjacent slits, and so on. In general where there are maxima at angles on each side of the centre

$$n\lambda = s \sin \theta_n$$

For $s = 2.0 \times 10^{-6}$ m and $\lambda = 6 \times 10^{-7}$ m this gives

$n = 0$: $\sin \theta_0 = 0$, $\theta_0 = 0°$, zero order, brightest

$n = 1$: $\sin \theta_1 = 0.3$, $\theta_1 = 17°$, first order

$n = 2$: $\sin \theta_2 = 0.6$, $\theta_2 = 37°$, second order

$n = 3$: $\sin \theta_3 = 0.9$, $\theta_3 = 64°$, third order, faint

$n = 4$: $\sin \theta_4 = 1.2$, not possible, so no fourth order.

For $n \geqslant 4$, $n\lambda > s$ so that $\sin \theta_n > 1$ and no higher orders exist for the grating used with this light. Other combinations of grating and light will give different numbers of orders.

Spectra

When white light is incident on a diffraction grating each colour (wavelength) produces a maximum in each order. This sounds very complicated but what you see is the complete spectrum from red to violet that you see in a rainbow. One of these spectra is produced in *each* order on *both* sides of the central maximum which is white. (For $n = 0$, $\theta_0 = 0$ for all λ.) The grating is dispersing the light in each order.

Figure 23.19 shows in principle what is happening by showing what would happen when a diffraction grating is illuminated with light from

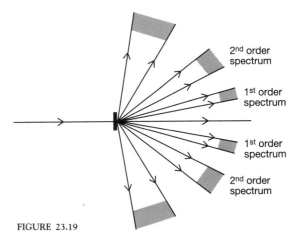

FIGURE 23.19

orange to green, i.e. over only a limited range of λ. In practice the different spectra overlap in the second and third order when white light is used, though you do not notice this much as the eye is not very sensitive to indigo or violet.

A 'good' diffraction grating has a very small s and produces a wide first-order spectrum. The following examples illustrate some of these points.

EXAMPLE

A physicist studying the spectrum of white light from a star wants to produce an angle of 15° between the blue ($\lambda = 500$ nm) and the red ($\lambda = 750$ nm) end of the spectrum. About how many lines per grating per millimetre are needed in the grating?

Suppose the first-order maximum for blue light is at θ: then, for blue

$$500 \text{ nm} = s \sin \theta$$

and for red

$$750 \text{ nm} = s \sin (\theta + 15°)$$

so we need

$$\sin (\theta + 15°) = \left(\frac{750}{500}\right) \sin \theta$$

There is a way of finding an exact solution (which is $\theta = 25.86°$) but here you are asked only for an approximate solution.

Suppose $\theta = 15°$ (this is a first guess). Using a calculator – it doesn't work. More guesses quickly lead to $\theta \approx 25°$ so

$$500\,\text{nm} = s \sin 25°$$

and the grating needs to have

$$\frac{1}{s} = 850\,000 \text{ slits per metre}$$

or about 850 slits per mm.

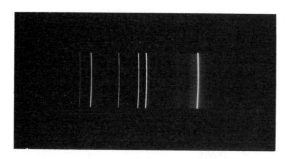

EXAMPLE

A helium discharge tube is found to emit yellow light of wavelength 590 nm which, in its second order, on being dispersed by a diffraction grating, coincides with the third order of a deep violet line from the same discharge tube. What is the wavelength of the violet light?

The path difference in the *second* order for the yellow light is

$$2\lambda_y = 2 \times 590\,\text{nm} = 1180\,\text{nm}$$

and this must be the path difference also for the violet light in the *third* order.

so that $1180\,\text{nm} = 3\lambda_v$

$$\Rightarrow \qquad \lambda_v = \frac{1180\,\text{nm}}{3} = 390\,\text{nm}$$

Notice that in this calculation we did not need to know the grating spacing nor the angle at which these maxima occurred. This same overlap occurs with every grating.

Reflection gratings

We can also make *reflection* diffraction gratings by scratching parallel grooves on a metal surface. The theory is just the same as for a transmission grating, each narrow strip of the reflector diffracting the light into a wide angle and the consequent superposition of the diffracted beams at a given angle giving rise to sharp maxima. A compact disc acts as a simple reflection grating if it is held so that the light strikes the surface obliquely – this is probably the simplest demonstration of the wave nature of light.

Crystal diffraction

The spacing of the atoms in a crystal is of the order of 10^{-10} m. Electromagnetic waves with λ a tenth or a fifth of this are X-rays. When a beam of X-rays is incident on a single layer of atoms in a crystal each atom scatters a minute proportion of the beam, i.e. acts as a source of X-ray wavelets. In a crystal there are many such layers, parallel and regularly spaced (about 10^5 in a crystal 0.01 mm thick) and each layer produces a weak reflected beam. These beams from successive layers are usually out of phase, but when the beam strikes the crystal at certain angles the path difference between beams from adjacent layers can be λ, or 2λ, etc.

If t is the separation of the layers of atoms there is strong reflection at certain angles and by detecting them t can be measured.

Nowadays such **X-ray crystallography** is used to study the structure of all sorts of crystals. It was information of this kind which helped Watson and Crick to predict the structure of DNA in 1953.

Exercises on each section of this chapter may be found in the companion textbook, *Practice in Physics*.

SUMMARY

At the end of this chapter you should be able to:

◆ explain that two waves travelling in opposite directions produce a stationary wave.

◆ describe how to demonstrate stationary waves on stretched strings and with microwaves.

◆ remember that the distance between adjacent nodes in a stationary wave interference pattern is $\lambda/2$.

◆ understand that a cord or string fixed at both ends can only vibrate in such a way that its length is a whole number of half-wavelengths.

◆ understand that all musical instruments depend on the resonant vibration of stretched strings or air columns.

◆ explain that waves diffract; that is, their energy spreads into the shadow area when part of a wavefront is blocked.

◆ explain the way in which diffraction and interference patterns are both the result of the superposition of waves.

◆ draw graphs of intensity against position for diffraction at a single slit and for two-slit interference with light.

◆ use the equation for the first minimum in a single slit diffraction pattern:

$$\sin \theta_m = \frac{\lambda}{d}$$

◆ understand how diffraction effects limit the ability of optical instruments to resolve detail.

◆ understand how to produce two coherent optical sources.

◆ use the relationship

$$\text{wavelength} = \frac{(\text{slit separation})(\text{fringe width})}{\text{slit to screen distance}}$$

◆ understand that a diffraction grating produces a series of narrow maxima at angles θ_n given by the formula

$$n\lambda = s \sin \theta_n$$

where s is the slit separation in the grating.

◆ explain how a diffraction grating can produce spectra.

24 Optics

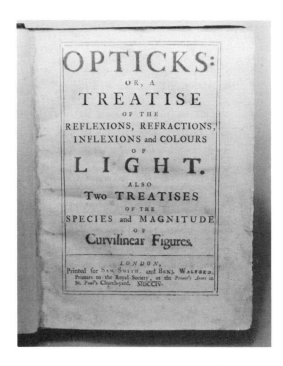

The study of light, optics, is one of the oldest branches of physics. The Romans used spectacles and it is said that at the battle of Syracuse huge mirrors, designed by Archimedes, were used to reflect sunlight so as to damage the Roman ships. The first physics book written in English, as opposed to Latin, was entitled Opticks and was printed just under 300 years ago; the author was Isaac Newton. Nowadays one in eight of us wears contact lenses or spectacles, almost all families possess a camera and many a projector, and school or college teachers cannot imagine life without a photocopier.

The fact that light, left to itself, travels in straight lines is used, for example, by surveyors who employ laser beams as weightless rules. Bending beams of light round corners, channelled inside optical fibres, enables surgeons to illuminate places such as the inside of the stomach and allows all the instruments on the dashboard of a car to be lit up from a single lamp.

24.1 Reflection

We can describe what happens to the energy from wave sources in two ways: either by drawing **wavefronts** or by drawing **rays**. In the study of optical fibres or of satellite dishes we use ray diagrams. Rays are the (imaginary) lines which we draw to show the paths taken by the light energy. You see them as the streaks of light such as these shown in figure 24.1. They show you where the rays go when reflected at a plane mirror. You should know that, in this case, the reflected rays appear to come from behind the mirror – the dotted lines show this. The exact point from which they come, the image, is exactly as far behind the mirror as the filament of the lamp is in front of it and the line joining the image and the filament is perpendicular to the mirror.

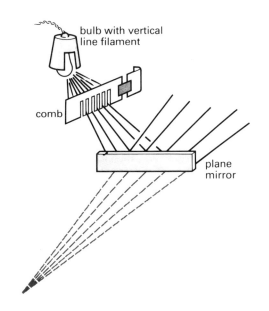

FIGURE 24.1

Each ray reflects so that the **angle of incidence** is equal to the **angle of reflection**; we measure both angles from the perpendicular to the mirror at the point where the ray is reflected.

You see wavefronts when you see ripples on a pond or in a ripple tank. Rays and wavefronts are always perpendicular to each other so it is only necessary to draw one or the other but when you draw wavefronts it is often useful to add a little arrow to show in which direction the energy is going.

Dish reflectors

When a parallel beam of rays strikes a circular reflector as in figure 24.2(a), the law of reflection predicts that they do *not* meet at a point. You can see that to produce a single focus for the reflected rays, and hence the energy, the edges of the dish need to be less curved.

Figure 24.2(b) shows this and is the shape you will find for satellite dishes; it is a **parabolic** shape. If instead of a microwave receiver, the dish was a transmitter, then a source at the focus D of the parabola will produce an outgoing beam which is a set of parallel rays. The same ideas are used in the design of searchlights or narrow-beam torches.

FIGURE 24.3

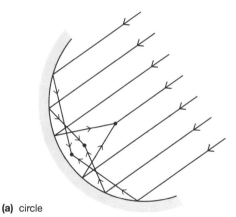

(a) circle

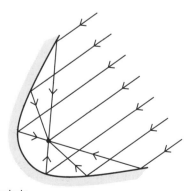

(b) parabola

FIGURE 24.2

24.2 Refraction

Refraction occurs when wavefronts or rays pass from one medium to another. At the boundary there is always some reflection but we will ignore this unless it is of particular importance.

Snell's law

Experiments with ray streaks passing from air to glass and from glass to air show that as θ_1 (figure 24.4) varies so does θ_2 and that

$$\frac{\sin \theta_1}{\sin \theta_2} = \text{constant for two media}$$

This result is called **Snell's law**. If the direction of travel is reversed we still label the angle measured to the perpendicular in medium 1 as θ_1 and in medium 2 as θ_2 (figure 24.4).

If medium 1 is a vacuum we call the constant the **refractive index**, n, of medium 2

$$\frac{\sin \theta_{\text{vac}}}{\sin \theta_{\text{med}}} = n$$

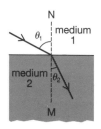

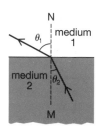

FIGURE 24.4

The refractive index of air under normal laboratory conditions is 1.0003 so we usually take it to be 1 and measure n for other substances (such as glass) by studying refraction from air to glass or from glass to air using semicircular (see figure 24.8, page 330) or rectangular glass blocks.

When light goes from medium 1 with refractive index n_1 to medium 2 with refractive index n_2 the equation becomes

$$\frac{\sin \theta_1}{\sin \theta_2} = \frac{n_2}{n_1}$$

or $\qquad n_1 \sin \theta_1 = n_2 \sin \theta_2$

and this is often the most useful way to remember Snell's law even for cases where $n_1 = 1$, i.e. one medium is air.

EXAMPLE

A narrow beam of light strikes a layer of oil on the surface of a tank of water at an angle of 58° as shown. If the refractive index of oil is 1.28 and that of water 1.34 calculate:

(a)   the angle of refraction in the oil
(b)   the angle of refraction in the water
(c)   the angle of refraction in the water if the layer of oil was removed.

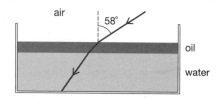

(a) $\theta_{air} = 58°$, so for the first boundary

$$\frac{\sin \theta_{air}}{\sin \theta_{oil}} = 1.28 \text{ (Snell's law)}$$

$= 0.6625$ and so

$\Rightarrow \qquad \sin \theta_{oil}$

$$\theta_{oil} = 41.5°$$

(b) Using $n_1 \sin \theta_1 = n_2 \sin \theta_2$ for the second boundary

$1.28 \sin 41.5° \qquad = 1.33 \sin \theta_{wat}$

which gives $\theta_{wat} \qquad = 39.3°$

(c) This time there is only one boundary, $\theta_{air} = 58°$

so $\qquad \dfrac{\sin 58°}{\sin \theta_{wat}} = 1.34$

and the new $\theta_{wat} \qquad = \arcsin 0.6329$

$$= 39.3°$$

Notice that the answers to (b) and (c) are the same. A parallel-sided layer like the oil will make no difference to the angle at which the ray emerges from the water. And the angles would be the same if the direction of travel of the light was reversed.

Dispersion

When a narrow beam of white light, e.g. sunlight, is refracted at the two surfaces of a prism the emerging light is tinged with colour. Figure 24.5 gives the language we use to describe what is happening: A is called the **angle of the prism**, D is the average **angle of deviation** of the light and δ is called the **angle of dispersion**. You should realise that the glass of the prism must have different refractive indexes for different colours, e.g. for red light n might be 1.62, but for blue light n might be 1.64 for this particular type of glass. You often see

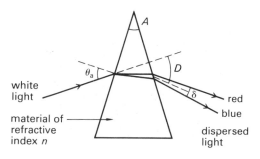

FIGURE 24.5

colour effects produced by this dispersion of light, for example when you look at a line on the bottom of a swimming pool or when you see a rainbow. In the second case, however, a detailed explanation is quite complicated. A serious consequence of dispersion occurs in the design of telescopes where glass lenses produce colour effects which are a great nuisance.

Refraction and wave speeds

The *reason* for refraction is that the speed of the wave changes. The ratio of the speeds (in the two media) is equal to the refractive index n, i.e. Snell's law can be rewritten as

$$\frac{\sin \theta_1}{\sin \theta_2} = n = \frac{c_1}{c_2}$$

where c_1 and c_2 are the wave speeds in medium 1 and medium 2 respectively.

Red and blue light travel at different speeds in glass so they refract by different amounts and disperse. They travel at the same speed $c = 3.00 \times 10^8 \, \text{m s}^{-1}$ in a vacuum (and effectively so in air).

To demonstrate this change of speed effect you can use a ripple tank where the wave speed depends on the depth. Figure 24.6(b) shows the resulting change of direction and of wavelength.

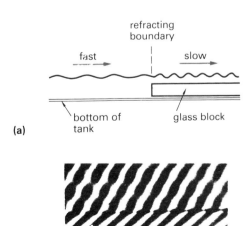

(a)

(b)

FIGURE 24.6

The frequency remains the same, so $c = f\lambda$ tells us that if c changes, λ must also change (in proportion).

EXAMPLE

A thin plane membrane is used to separate some air at 288 K from some warm air at 392 K. A plane sound wave travels in the cooler air and is incident on the membrane at an angle of 20°. Find the angle of deviation given that the speed of sound in air $c = kT^{1/2}$ where k is a constant. For a particular gas at two temperatures T_1 and T_2

$$\frac{c_1}{c_2} = \frac{\sin \theta_1}{\sin \theta_2} = \frac{kT_1^{1/2}}{kT_2^{1/2}}$$

$$= \sqrt{\frac{T_1}{T_2}}$$

$$\therefore \qquad \sin \theta_2 = \sin \theta_1 \times \sqrt{\frac{T_2}{T_1}}$$

$$= \sin 20° \times \sqrt{\frac{392 \, \text{K}}{288 \, \text{K}}}$$

$$= 0.40$$

Thus $\theta_2 = 23.5°$, and the wavefront is deviated through only 3.5°.

On a hot day, if you look at something through rising currents of hot air, it seems to shimmer because light from it is refracted randomly by the air.

Measuring n

Methods which depend on applying Snell's law directly to two angles measured with a protractor in a **ray streak** experiment are very imprecise and any one determination of n for glass by this method may have an uncertainty of as much as $\pm 10\%$. To improve this a number of readings are taken, not of θ_a and θ_g but of their sines found directly from measuring the sides of right-angled triangles, so that it is lengths and not angles which we measure. Further, a graph of $\sin \theta_a$ against $\sin \theta_g$ will average out uncertainties in individual readings and enable n to be found from its slope. This is a case where a proper approach to a simple experiment can reduce the uncertainty considerably to (say) ±2% giving, e.g. $n = 1.52 \pm 0.03$.

When you look down into a swimming pool, the depth markings, 1 – 2 – 3 metres, do not seem to be 1 m apart. Measuring **real and apparent depth** can lead to a value for n (see also page 329). It can be used for both liquids and solids and, with a travelling microscope, a narrow depth of field gives a quick and precise result. To achieve an uncertainty of about ±1% the distance measured should be more than 100 mm and it is very important that the focusing of the microscope is carefully and repeatedly done.

Applications

But why bother to measure n? Is it important to know its value? Yes it is: diamond has a refractive index of 2.417 and measuring this proves it is a *real* diamond. Geologists measure refractive index to analyse specimens of rock crystals. And in the food industry the strength or condition of many liquids can be more easily and quickly tested by measuring n than by measuring, for example, their density or sweetness. A detailed knowledge of n is also required in designing lens systems for optical instruments or when making spectacles and contact lenses.

Figure 24.7 shows in principle how you could measure n for pineapple juice which will tell you the ripeness of the fruit – you will need to try this experiment to appreciate how it works. You look into the block and move your eye (or rotate the block) until the smear of pineapple juice *just* disappears. Here $n = OL/MB$ and if you are doing this every day you will soon have a technique which only takes a few seconds and is thus a quick quality control check.

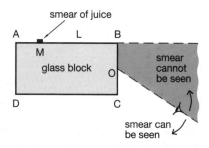

FIGURE 24.7

24.3 Total internal reflection

Figure 24.8 shows that, for low angles of incidence, some light is reflected and most is refracted when going from glass to air. For a large angle of incidence, though, there is *no* refraction. There is total reflection. It is called **total internal reflection (t.i.r.)**; the light must be trying to go from high to low refractive index (the wave must be moving in a slow-speed region and meeting the boundary of a high-speed region) for it to occur.

(a)

(b)

FIGURE 24.8

In between the two conditions shown in the photo there will be one where the refracted beam is parallel to the surface of the glass. As an example of what happens look at figure 24.9 which shows three rays from an underwater source of light S. We have to consider light of only one colour as the different colours reach the 'cut-off' at slightly different angles of incidence θ_a when white light is used. Applying Snell's law at Q we have

$$n_w \sin \theta_w = n_a \sin \theta_a$$

where n_w is the refractive index of water and n_a (=1) that of air. Therefore when

$$\theta_w = \theta_c \quad \text{and} \quad \theta_a = 90°$$

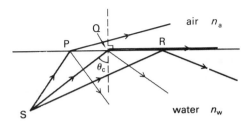

FIGURE 24.9

$$n_w = \frac{1}{\sin\ \theta_c}$$

as $\sin 90° = 1$. θ_w is called the **critical angle** for water and will be 49° if $n_w = 1.33$.

If you know the speed of light in air you can find the speed of light in water from this value of n as

$$\sin\ \theta_c = \frac{c}{c_w}$$

For example if $c = 3.00 \times 10^8\,\mathrm{m\,s^{-1}}$ then $c_w = 2.25 \times 10^8\,\mathrm{m\,s^{-1}}$.

Optical fibres

Optical fibres, which make use of total internal reflection, are one of the most rapidly growing areas of the application of optics. They are used in medicine and in communications where they are replacing metal wires for carrying telephone and other signals. See Section 28.3. The optical fibre

FIGURE 24.10

shown in figure 24.10 can carry as many telephone messages as the metal cable shown in the background.

The fact that when t.i.r. occurs all the energy is reflected means that we can bounce a light beam off the inside of a glass tube many thousands of times without loss. Figure 24.11(a) shows the principle.

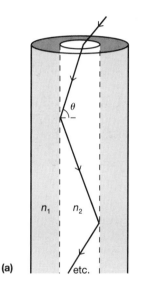

(a)

(b)

FIGURE 24.11

The fibre consists of a glass **core** of high refractive index n_2 and a glass **cladding** of lower refractive index n_1. The whole, which is typically only 125 μm (1/8 mm) in diameter is usually one of a bundle of fibres which surround a strong metal cable and are sheathed in plastic. The whole bundle is, however, flexible, and figure 24.11(b) shows how the light can be turned round corners without leaking out provided always that

$$n_2 \sin\ \theta > n_1$$

The core can be about 50 μm in diameter in a step-index **multimode fibre** or only 2 μm in a **monomode fibre**. The former are used to transmit light to inaccessible places, e.g. inside the body, and to bring back a series of tiny beams which can make up a picture. Here the relative positions of the fibres at the ends of the endoscope have to be preserved. The latter are used for communicating a rapid series of pulses which can carry many thousands of telephone calls simultaneously. In practice, the light energy is absorbed, or attenuated, as it travels along the communications fibre and the signals it is carrying have to be regenerated about every 30 km.

EXAMPLE

An optical fibre core is made of glass of refractive index 1.472 and is surrounded by a cladding of refractive index 1.455. Calculate the critical angle for the fibre. If the core is 125 μm in diameter calculate the time delay between the fastest (straight along the centre, or axial) and the slowest (bouncing from side to side at the critical angle) modes of propagation of light pulses down a fibre which is 200 m long. Take $c = 3.000 \times 10^8 \, \text{m s}^{-1}$.

If the critical angle is θ_c, then

$$1.472 \sin \theta_c = 1.455 \sin 90°$$

$$\therefore \quad \theta_c = 81.3°$$

The speed of light in the fibre for both modes of propagation can be found from the refractive index.

$$n = c/c_g$$

$$\therefore \quad c_g = \frac{c}{n} = \frac{3.000 \times 10^8 \, \text{m s}^{-1}}{1.472}$$

$$= 2.038 \times 10^8 \, \text{m s}^{-1}$$

When the fastest axial ray travels a distance of $2l$ down the fibre, the slowest ray travels a distance $2x$ where

$$\frac{l}{x} = \sin 81.3°$$

$$\Rightarrow \quad x = 1.0117 \, l$$

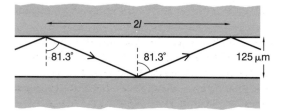

A sketch diagram is essential here to understand what is going on. Notice that the diameter of the fibre is not needed in the calculation. You need to keep more than 3 significant figures at this stage.

The ray which arrives first takes

$$t_f = \frac{2000 \, \text{m}}{2.038 \times 10^8 \, \text{m s}^{-1}}$$

$$= 9.418 \, \mu\text{s}$$

And the ray which arrives last takes 1.0117 times this

i.e. $$t_s = 9.929 \, \mu\text{s}$$

So the time difference is $(9.929 - 9.418) \, \mu\text{s}$

$$= 0.511 \, \mu\text{s}$$

The calculation in the previous example means that a pulse spreads out as it travels along the fibre, as it will propagate along all possible paths. A series of sharp rectangular transmitted pulses (the 1s and 0s of a binary code) therefore arrives at the end of the fibre spread out. There is thus a danger of 'losing' the digital information in the signal if the time delay becomes too large.

24.4 Images

'What are images?' can best be answered by asking 'How do we see images?'. Figure 24.12 shows both wavefronts and rays; the rays diverge from O, some of them are intercepted by an **optical system**, e.g. a mirror or a lens, and after passing through the system they converge on a point I which is where the eye sees a real image of O. The wavefronts – the grey curves – are put in on this diagram but are

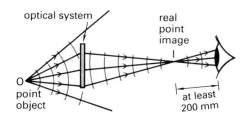

FIGURE 24.12

not usually drawn. For a **real image** the energy is concentrated at I as the waves pass through the image position. This means that, for light, real images can be caught on a screen or recorded on a photographic plate.

There is no energy at I for a **virtual image**. The rays only appear to diverge from I – the wavefronts and rays are drawn in figure 24.13. The eye here *thinks* that the rays are coming from I but no light energy passes through this image position and it cannot be caught on a screen.

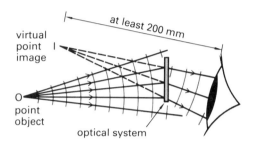

FIGURE 24.13

The conditions for producing a perfect image, with *all* the rays passing through a point or appearing to come from a single point, are rarely met. Perfection *is* achieved in the image produced by a truly flat, plane mirror. Figure 24.1 on page 326 shows light streaks illustrating how this image is formed. The optical image is three-dimensional with each point of the object producing a virtual image as far behind the mirror as the point is in front. (The reflection of a circular wave at a straight barrier in a ripple tank can be used to help you understand what is happening in terms of wavefronts.) When looking into a plane mirror the image is seen to be the same size as the object. On the other hand the real image formed by the satellite dish on page 327 is not a perfect image if a circular dish is used but can be improved using a parabolic dish.

Real and apparent depth

Figure 24.14 will help you to understand why, when you look down through a glass block, it does not seem as deep as you know it is.

Applying Snell's law to the refraction at P, we get

$$n_a \sin \theta_a = n_g \sin \theta_w$$

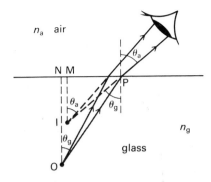

FIGURE 24.14

But $\sin \theta_a = \text{MP/IP}$ and $\sin \theta_w = \text{NP/OP}$, so that putting $n_a = 1$,

$$n_g = \left(\frac{\text{OP}}{\text{NP}}\right)\left(\frac{\text{MP}}{\text{IP}}\right)$$

Obviously NP and MP are equal if you are looking straight down, and then OP = ON and IP = IM so that

$$\frac{\text{real depth}}{\text{apparent depth}} = \frac{\text{ON}}{\text{IM}} = n_g$$

This refracted image is not perfect because of the dispersion of white light. This dispersive imperfection is called **chromatic aberration** and the sort of imperfection produced in a circular dish image, such as that shown in figure 24.2, is called **spherical aberration**. The words themselves are not important but you should realise that both colour effects and shape effects can make images less than perfect. In cameras the aberration effects are reduced by using several lenses close together – a multiple lens system.

Locating images

Real images can be located by judging where they are best focused on a screen; but this is not a very precise process if you want to find out exactly where the image is. For virtual images it doesn't work at all. To locate the position of a virtual image think of a glass-fronted cupboard with an unlit candle standing on a shelf. If you held a lit candle in front of the glass and looked at its reflection, you could arrange for the reflected image to coincide with the unlit candle (and hence to give the impression that the one in the cupboard is lit). To judge that the image I and the unlit candle C are in exactly the same place you move your head from side to side and see if I and C stay together. Moving C until they do now allows you to locate the position of I. Figure 24.15 shows the situation when they do not quite coincide.

head is moved to right in each case

image moves with head: I beyond C

image moves against head: I nearer than C

FIGURE 24.15

This idea, called the **no-parallax** method of locating images, works for both the real and virtual cases and can be made quite precise if the image is so small, e.g. a fine pin, that tiny movements can be detected. In this way you can locate the images produced by all sorts of lenses and curved mirrors.

EXAMPLE

A diver, seated on the end of the 3-metre board, notices a faint image of the board formed by reflection in the water surface. He also sees the image, in the water, of a line of tiles on the bottom of the pool. He moves his head from side to side and finds that the two images coincide. If the refractive index of water is 1.33, how deep is the pool?

Draw a sketch diagram to help you to understand the question.

If the pool depth is h, then the apparent depth, h' is given by

$$\frac{h}{h'} = 1.33$$

The faint image reflected by the flat water surface is as far below it as the board is above it.

$$h' = 3.0\,\mathrm{m}$$

and so

$$h = 1.33 \times 3.0\,\mathrm{m}$$
$$= 4.0\,\mathrm{m}$$

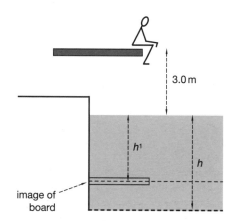

image of board

24.5 Lenses

The **focal length** of a converging lens is the distance along the axis of the lens from its centre P to the **principal focus** F, i.e. to the place where a set of rays which are parallel to the axis are focused – figure 24.16.

Other parallel beams of light from very distant objects which enter a lens at small angles to the axis of the lens are all brought to a focus in a plane which contains F. This is called the focal plane of the lens. You can use ray streaks to demonstrate these patterns.

Similar diagrams can be drawn for a diverging lens.

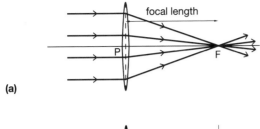

(a)

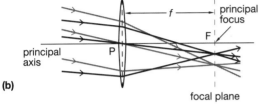

(b)

FIGURE 24.16

Magnification

For an extended object the linear magnification m is measured perpendicular to the axis of the lens and is given by

$$m = \frac{\text{height of image}}{\text{height of object}} = \frac{v}{u}$$

where v is the distance from the image to the lens and u is the distance from the object to the lens. Figure 24.18 shows the path of some rays from a point on the object to a point on the image. Notice that the ray through the centre of the lens is not deviated.

The power of the lens

A lens is powerful if PF, its focal length, is small. We define the power of the lens as F, where $F = 1/f$. The unit of F is m^{-1}, which in this situation is called the **dioptre** (D). A typical pair of spectacle lenses for a short-sighted person might be: left eye $-2.0\,D$, right $-1.5\,D$, the negative signs indicating that they are both diverging lenses. A long-sighted person's spectacles contain converging lenses. In figure 24.17 you can see diverging spectacle lenses which produce a diminished image.

When lenses are placed in contact the effective power is the sum of the individual powers; adding a $+3.0\,D$ lens to a $-1.0\,D$ lens produces a lens combination of power $+2.0\,D$.

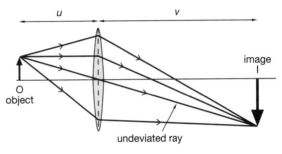

FIGURE 24.18

A simple lens is not very good for values of $m > 5$ as the image then becomes distorted.

A single converging lens can produce a whole range of real images depending on the size of u. Figure 24.19 shows the relative size of u (object–lens distance) and v (lens–image distance) for a 20 cm focal length converging lens.

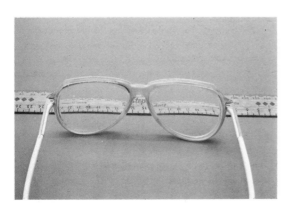

FIGURE 24.17

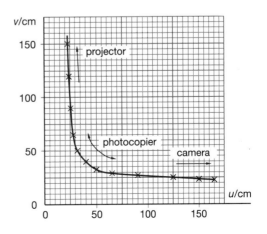

FIGURE 24.19

No matter how big u becomes, v is never less than 20 cm. In fact as $u \rightarrow \infty$, $v \rightarrow f$. As u becomes smaller, v increases until at $u = f$, $v = \infty$ and for smaller values of u the image is virtual and the converging lens becomes a magnifying glass – see figure 24.20 and also figure 24.24.

FIGURE 24.20

You can use the graph to find object–image distances for this 20 cm lens. You will find that the *minimum* value of $u + v = 80$ cm, i.e. $4f$. You cannot produce a real image with a converging lens of focal length f if the object and screen are less than $4f$ apart.

The lens formula

You can predict the position of an image if you know the position of the object and the focal length of the lens. The relationship is

$$\frac{1}{f} = \frac{1}{u} + \frac{1}{v}$$

where u and v have their usual meanings.

In order to apply this equation we measure u, v and f in the following ways (from the optical centre P of the lens): distances to *real* objects and images are taken to be *positive*; to *virtual* images they are taken to be *negative*. The focal lengths of converging and diverging lenses are taken to be positive and negative respectively. You can see that for converging lenses the image is virtual (v negative) when u is less than f and that for diverging lenses the image is always virtual, i.e. v is always negative.

EXAMPLE

A converging lens of focal length 20 cm is placed 15 cm from an insect which is 3.0 mm long. Where is the image of the insect and how large is it?

Using the lens formula

$$\frac{1}{20\,\text{cm}} = \frac{1}{15\,\text{cm}} + \frac{1}{v}$$

$$\Rightarrow \qquad v = -60\,\text{cm}$$

so the image is virtual.

In this example the converging lens is used as a magnifying glass. The image is on the same side of the lens as the object – the insect.

The magnification

$$m = \frac{v}{u} = \frac{60\,\text{cm}}{20\,\text{cm}} = 3\times$$

so the image is 3×3.0 mm, i.e. the image of the insect is 9.0 mm long.

Measuring focal lengths

You should always try to get a rough idea of the focal length of a lens before trying to measure it precisely. You can let light from a bright distant object fall on a converging lens and measure the image distance. As here $u \approx \infty$, then $v \approx f$. For a diverging lens you could find the weakest converging lens which, when held against the diverging lens, produces a converging combination: then the sizes of the two focal lengths will be nearly equal though this provides only a very rough guide.

The plane mirror method

A point object placed at the principal focus of a converging lens will coincide with its image if a plane mirror is placed behind the lens and perpendicular to the axis (figure 24.21). Using a brightly-lit object pin and the no-parallax technique to see when object and image coincide, the focal length of a lens for which $f \approx 200$ mm can be measured with an uncertainty of about 1% – quite good enough for most purposes.

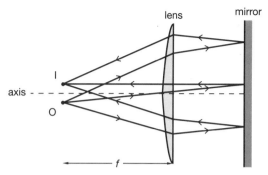

FIGURE 24.21

24.6 The eye

The Moon looks about the same size as the Sun. But so does a disc 20 mm in diameter (a one pence piece) held just over 2 m away from the eye. All three form images on the retina which are the same size; each subtends the same angle at the eye – about 10^{-2} rad or 0.5°; figure 24.22 illustrates this idea.

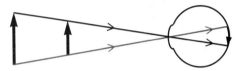

FIGURE 24.22

EXAMPLE

A converging lens of focal length 0.18 m is used to produce an image of an illuminated scale on a screen. If a magnification of 9.0 is required, determine the position of the lens in relation to the object and the screen.

As $\quad m = 9, \quad$ then $\quad \dfrac{v}{u} = 9$

i.e. $\quad v = 9u$

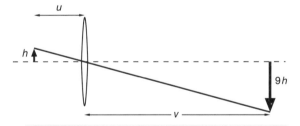

Substituting in the lens formula

$$\frac{1}{0.18\,\text{m}} = \frac{1}{u} + \frac{1}{9u}$$

∴ $$\frac{1}{0.18\,\text{m}} = \frac{10}{9u}$$

⇒ $\quad u = 0.20\,\text{m} \quad$ and $\quad v = 1.8\,\text{m}$

This and many other lens problems can be solved using a scale drawing. Here a ray from the top of the object and parallel to the axis is refracted to cross the axis at the principal focus which is 0.18 m from the lens. Adding this ray to the above diagram creates a scale diagram from which the sizes of u and v can be read.

Visual angle

The angles subtended by an object or by its image when viewed through an optical instrument – e.g. a magnifying glass, microscope or telescope – are called the **visual angles** of the object and the image respectively. The visual angle represents the apparent size of the object or image and hence determines the detail which can be seen. Clearly an object (or image) of given size has a larger visual angle if it is close to the eye, so the one pence piece mentioned above is seen in greater detail if it is placed 0.5 m rather than 2 m away.

Figure 24.23(a) shows how the eye forms an image on the retina; (b) shows how we normally only draw rays from the top of the object and how the eye is usually represented symbolically without showing the formation of the retinal image.

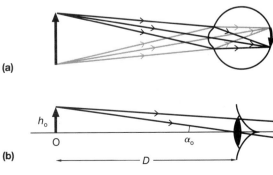

FIGURE 24.23

Suppose in figure 24.23(b) the object O is placed as near to the eye as it can comfortably focus in order to study O. This position is called the **near point** and we take it to be 0.25 m from the eye. In practice this distance $D (= 0.25\,\text{m})$, called the **least distance of distinct vision**, varies from person to person and with age for any one person; 0.25 m is an agreed value for what we call a 'normal' eye but for babies it is much smaller, less than 10 cm. The visual angle of the object α_o is given in radians (for small angles) by

$$\alpha_o = \frac{h_o}{D}$$

When the object O is viewed through a lens used as a simple magnifying glass (figure 24.24), O can be placed much closer to the eye than it could be with the unaided eye – figure 24.23(b). The virtual image I is usually arranged to be at the near point to get the clearest view. I subtends an equal angle α_i at the eye, where

$$\alpha_i = \frac{h_i}{D}$$

where h_i is the size of the virtual image

so
$$\frac{\alpha_i}{\alpha_o} = \frac{h_i}{D} \div \frac{h_o}{D} = \frac{h_i}{h_o}$$

as expected. Using this simple magnifying glass enables more detail to be studied than with an unaided eye *because* α_i is bigger than α_o.

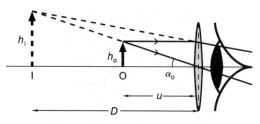

FIGURE 24.24

Diagrams of this type are usually drawn when studying optical instruments. The eye is taken to be very close to the optical centre of the lens through which it is looking, so that α_i (the visual angle of the image of the eye) is also the angle subtended by the image at the lens. In figures

24.23(b) and 24.24 only the top half of the ray diagram is drawn. This again is common practice and the visual angles α_o and α_i are, as here, taken to be half the angles subtended by the whole object or image at the eye. The usefulness of expressing magnifications in terms of visual angles becomes obvious when the object is something like the Moon for which the value of u is not known or when the final image is arranged to be at infinity.

EXAMPLE

An eye sees a tree which is 14 m high and 90 m away. If the eyeball measures 25 mm from the lens to the retina calculate (a) the size of the image and (b) the focal length of the eye lens.

(a) If h is the size of the image, then

$$\frac{h}{14\,\text{m}} = \frac{0.025\,\text{m}}{90\,\text{m}}$$

$$\Rightarrow \qquad h = 3.9 \times 10^{-3}\,\text{m or } 3.9\,\text{mm}$$

(b) Using the lens formula

$$\frac{1}{f} = \frac{1}{90\,\text{m}} + \frac{1}{0.025\,\text{m}}$$

$$\Rightarrow \qquad f = 0.025\,\text{m}$$

You see that 90 m is effectively at infinity for a human eye ($f = 0.02499$ m to 4 sig. fig.)

Accommodation

The power of a normal eye is $\approx +40\,\text{D}$ when it is relaxed. The shape of the eye's lens can be altered so that images of objects at different distances from the eye can be formed on the retina. This adjustment of shape, which is achieved muscularly, is called **accommodation**. The power must increase to about $+44\,\text{D}$ to see clearly an object at 0.25 m – the near point. The *range* over which we can accommodate decreases with age. It falls from about 12 D in a baby to 4 D for the normal eye and to perhaps 1 D in old age.

A **short-sighted** person's eyes are too powerful. He will be able to see clearly objects at less than 0.25 m, perhaps only 0.10 m away ($+50\,\text{D}$), but will be unable to see distant objects clearly. His far point may be only 0.30 m from the eye and to see distant objects he wears diverging spectacles or contact lenses. On the other hand the eyes of a **long-sighted** person are not powerful enough. He can usually see distant objects clearly but his near point may be 1.0 m away (+41 D) or sometimes so far away that all objects are difficult to see clearly without aid. This defect is corrected by using converging lenses.

24.7 Optical instruments

The graph of figure 24.18 mentions three single-lens optical instruments. You may also have used microscopes and binoculars. In modern optical instruments a single lens is usually replaced by a system of several lenses made of different types of glass, and most instruments also involve mirrors or prisms which add to the complexity of the optical design (figure 24.25). We will briefly discuss only the principle of a simple camera and a refracting telescope.

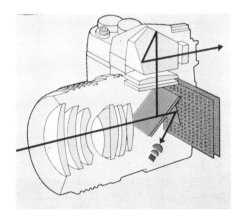

FIGURE 24.25

The camera

Figure 24.26 shows the essential parts of a simple camera. The converging lens forms a real image on the film. The diameter of the **aperture stop**, which is one way of controlling the amount of light reaching the films, is specified as a fraction of the focal length f of the lens. It can be set at $f/1.4$, $f/2$, $f/2.8$, $f/4$, $f/5.6$, $f/8$, $f/11$, $f/16$, $f/22$ or $f/32$. The numbers are chosen so that moving between adjacent **f-numbers** changes the area of the stop by a factor of 2; the larger the f-number the smaller the aperture.

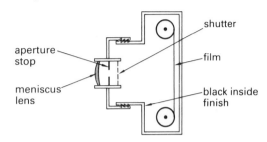

FIGURE 24.26

The shutter opens for a small time interval t when a picture is taken. The correct combination of **exposure time** t and f-number depends on the brightness of the object and the type of film used.

Depth of field

When you want to see clearly objects at different distances, you change the shape of your eye lens: in a camera, the distance between the lens and the film, the image distance v, has to be changed. The lens is mounted on a screw thread, so that rotating the lens changes its distance from the film.

FIGURE 24.27

Looking back at the graph of figure 24.19 you can see that, for small object distances u, very small variations of u produce quite large changes in v, the image distance. We say that there is a small **depth of field**, a small range of the photo which is properly focussed, see figure 24.27. Using a small aperture has the effect of increasing this depth of field. For landscape photography the problem does not exist as you can see by looking at large values of u on the graph. So the size of the aperture stop, the f-number, not only controls the light entering the camera but also the depth of field.

The telescope

With microscopes the object can be illuminated strongly and placed wherever we wish but the purpose of an **astronomical telescope** is to study inaccessible objects. They are so far from the telescope that we can draw a beam of rays diverging from a point on the object (e.g. a lunar crater) as a parallel beam when it enters the telescope.

In a refracting telescope a weak converging **objective** lens produces a real image of the object in its focal plane – figure 24.28; the focal length is typically a few metres in large instruments. The size of this image is $h = f_o \alpha_o$ and so the *weaker* the lens the *larger* the image. It is viewed by a magnifying glass (called the **eyepiece**). As telescopes are generally used for long periods the final image is arranged to be at infinity so that the eye muscles can be relaxed.

The figure has these two optical arrangements put together; it shows that the lens-to-lens distance, the length of the telescope, is $f_o + f_e$. The eye is placed as shown, since this is the place where it can collect the light passing through all parts of the objective lens.

Angular magnification

The angular magnification or magnifying power of an optical system M is defined by the equation

$$M = \frac{\alpha_i}{\alpha_o}$$

The eye then judges the image to be M times larger than the object. If h is the size of the intermediate real image then the angular magnification M is given by

$$M = \frac{\alpha_i}{\alpha_o} = \frac{h}{f_e} \div \frac{h}{f_o} = \frac{f_o}{f_e}$$

The most powerful refracting telescope in the world has $f_o \approx 20\,\mathrm{m}$ and $f_e \approx 6.5\,\mathrm{mm}$, so that the greatest available value of M is about 3000. Its objective lens is about 1 m in diameter.

When using a telescope to photograph a distant object the film or plate is placed to record the real image formed by the objective lens and the eyepiece lens is dispensed with. The telescope is then nothing more than a camera with a very long focal length. The photograph can then be examined with a magnifying glass or a low-powered microscope.

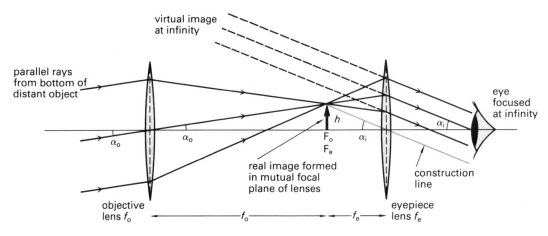

FIGURE 24.28

Reflecting telescopes

The largest optical telescopes (figure 24.29) all use a concave objective mirror rather than a converging objective lens. The reasons are that:

- there is then no chromatic aberration.

- spherical aberration can be eliminated for objects on the axis of the mirror by making the reflecting surface paraboloidal.

- the mirror can be supported from beneath over all its area (a lens can be held only around its edge).

- it is difficult to produce a large lens free from internal variations in refractive index whereas for a mirror only its surface needs to be perfect.

The large apertures possible with reflecting telescopes – up to 200 inches, about 5 m – both increase their ability to resolve and also enable fainter stars to be seen as they gather more light from a given source.

FIGURE 24.29

Exercises on each section of this chapter may be found in the companion textbook, *Practice in Physics*.

SUMMARY

At the end of this chapter you should be able to:

- draw ray diagrams showing the behaviour of light as it reflects and refracts.

- use the law of reflection: that for a reflected ray the angle of incidence is equal to the angle of reflection.

- use the law of refraction: that for a ray refracted from medium 1 to medium 2

$$n_1 \sin \theta_1 = n_2 \sin \theta_2$$

- understand that the ratio of the speeds of light in two media, c_2/c_1, is equal to the ratio of the refractive indexes of the media, n_1/n_2.

- describe an experiment to measure the refractive index of a solid material and of a liquid.

♦ remember that the refractive index of a given material varies with the wavelength or colour of the light and that this is called dispersion.

♦ explain that total internal reflection can only occur when light meets the boundary of a medium with a lower refractive index than the medium in which it is travelling.

♦ describe the construction of an optical fibre and explain the principles behind its use.

♦ remember that the focal length f of a lens is the distance from the centre of the lens to its principal focus.

♦ understand how images are formed and explain why most images are imperfect.

♦ draw diagrams to show the formation of images by converging lenses.

♦ remember what is meant by the power of a lens.

♦ describe how to measure the focal length of a converging lens.

♦ use the lens formula relating the object distance u to the image distance v

$$\frac{1}{f} = \frac{1}{u} + \frac{1}{v}$$

♦ explain how to correct for short-sighted and long-sighted eyes.

♦ understand the meaning of visual angle and remember that the angular magnification of an optical instrument is equal to the visual angle subtended by the image divided by the visual angle subtended by the object.

25 Probing the Nucleus

Our bodies are also bombarded by neutrinos from the Sun; several million of them have passed through each of your eyes since you started to read this paragraph. These neutrinos come from nuclear reactions which occur deep within the Sun, our local star. These reactions release huge quantities of energy some of which reaches us as sunlight. All life stems from this flow of energy and so you could say that an understanding of nuclear reactions is answering that most fundamental of questions: where do we come from?

25.1 Radioactivity

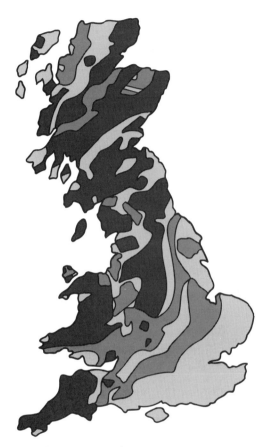

Chemical reactions such as respiration, combustion or rusting are going on all around us all the time. So also are nuclear reactions. The air we breathe contains radioactive carbon, the radioactive gas radon oozes out of the ground on which we walk and the walls of the houses in which we live. The geographical variation of this background dose is shown in the adjoining map. Are we in any danger from this natural radioactivity which causes ionisation in the cells of our bodies? During our evolution it is the mutations caused by ionising radiations that have allowed natural selection to proceed. So, far from being a danger, the presence of natural radioactivity has been central to our development as a species.

FIGURE 25.1

In 1896 the French scientist Henri Becquerel discovered a hitherto unknown type of radiation which was emitted spontaneously by any sample of the element uranium or its compounds; the radiation was first detected by its ability to blacken a photographic plate. The phenomenon was called **radioactivity**. Becquerel's discovery led to an intense search for other radioactive materials. Within a few years a long list of radioactive elements has been discovered; notable among these discoveries was that of radium by Madame Curie (figure 25.1) and her husband. This element is a million more times more radioactive per unit mass than uranium.

You have three different *kinds* of radioactive source in your laboratory. A **Geiger–Müller** or **GM tube** shows that they are different because the radiations from them have different **penetrating abilities**.

The arrangement shown in figure 25.2 enables the three radiations to be quickly distinguished. These are:

- ◆ α-sources.

- ◆ β-sources.

- ◆ γ-sources.

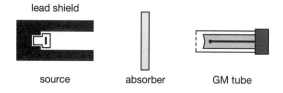

lead shield

source   absorber   GM tube

FIGURE 25.2

α-sources

α-sources, for example, americium, emit alpha *particles*. These are identical with helium nuclei. They travel only a few centimetres in air and are stopped by a single sheet of paper. To detect them using a GM tube the end window is made of a very thin ($\approx 10^{-2}$ mm) sheet of mica, which bows inwards as the gas pressure inside is about one-tenth of atmospheric pressure. The particles from a given α-source all have the same energy (or

sometimes two particular) energies: for americium-241 they are 2.44 MeV and 2.49 MeV. Different sources produce α-particles of other energies and many α-sources, including americium-241, also emit γ-rays.

EXAMPLE

The graph shows the mean range in air of α-particles of different initial energies.

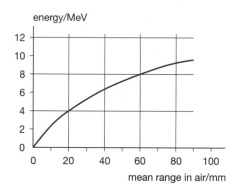

(a) Calculate the initial speed of an α-particle which has enough energy to travel 50 mm in air. Take the electronic charge to be 1.6×10^{-19} C and the mass of an α-particle to be 3.3×10^{-27} kg.

(b) Explain why the range measured using apparatus like that in figure 25.2 will be less than 50 mm and why there will *not* be a sharp cut-off at the detected range.

(a) A range of 50 mm ⇒ energy of 7.2 MeV

$$\therefore \text{ energy} = 7.2 \times 10^6 \times 1.6 \times 10^{-19} \text{J}$$

$$= 11.5 \times 10^{-13} \text{J}$$

Initially this energy is all k.e.

i.e. $\frac{1}{2}mv^2 = 11.5 \times 10^{-13}$ J

$\therefore$ $v^2 = \dfrac{2 \times 11.5 \times 10^{-13} \text{J}}{3.3 \times 10^{-27} \text{kg}}$

$\Rightarrow$ $v = 2.6 \times 10^7$ m s^{-1}

(b) The measured range will be less than 50 mm because the α-particles have to penetrate the end window of the GM tube before they can be detected. Some of the α-particles lose energy before they emerge from the source. The ranges in air of these α-particles will not all be 50 mm and so the cut-off is not sharp.

β-sources

β-sources, for example strontium, emit *beta particles*. These are identical with electrons. They are stopped by about 0.5 mm of lead or several mm of aluminium. The particles from a given source have a range of energies with a well-defined maximum: for strontium-90 the maximum is 2.3 MeV. These β-particles travel up to 9.8 m in air or 4.4 mm in aluminium. A useful, though approximate, relationship is that the product (maximum range in a material) × (density of material) is constant for β-particles of a given maximum energy. Energetic β-particles will sometimes produce γ-rays when they interact with atoms.

γ-sources

γ-sources, for example cobalt-60, emit gamma photons. These are short bursts of *electromagnetic radiation* just like X-rays. γ-photons from a given source have a given energy or often two or more particular energies: for cobalt-60 the energy is 1.3 MeV. As noted above, γ-rays are sometimes detected when an α-source or a β-source is being studied, and in such cases the γ-rays will be of definite wavelengths corresponding to different energies.

To cut down γ-radiation to even half its original intensity requires about 10 mm of lead or about 30 times as much as needed to produce the same reduction for β-particles. But doubling the thickness, to 20 mm, does not cut off the γ-rays completely, it produces only a further reduction of one-half. For these γ-rays we call 10 mm of lead the **half-value thickness**, $x_{\frac{1}{2}}$ – figure 25.3.

The half-thickness varies with the energy of the γ-photons; it is 4 mm of lead at 0.5 MeV and 12 mm of lead at 1.0 MeV.

Some radioactive sources, such as radium, emit all three kinds of radiation. This is because the source does not contain only radium but also other radioactive elements the atoms of which are formed by the decay of the radium. These *decay products* are numerous in the radium in your laboratory and enables α-, β- and γ-radiation to be detected coming from one source.

Magnetic deflection

β-particles are readily deflected in a magnetic field. The direction of the deflection shows that the β-particles are negatively charged. In figure 25.4, if the magnetic field is down into the paper, the beam is deflected towards the lower edge of the diagram, as shown. The left-hand motor rule then shows that the beam is behaving like an electric current directed *towards* the source; the particles emerging from the source must therefore be negatively charged. Quite moderate magnetic fields are found to produce large deflections. A magnetic field of 10^{-2} T (produced, for instance, by a pair of

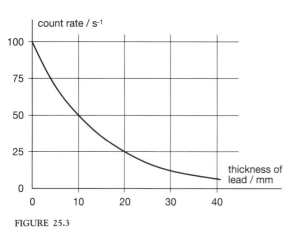

FIGURE 25.3

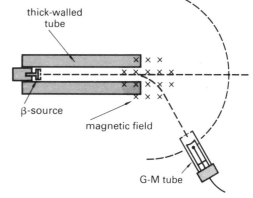

FIGURE 25.4

magnadur magnets) is quite sufficient for this demonstration.

A similar experiment with γ-rays shows no deflection, of course, and with α-particles the experiment can not be done unless the distance from source to detector is less than 5 cm or so. At this distance a *very* strong magnetic field, over 1 T, is needed to show any deflection. In a vacuum weaker fields can be used and we can confirm that α-particles are positively charged.

EXAMPLE

Calculate the wavelength of γ-rays of energy 1.3 MeV. Take $c = 3.0 \times 10^8 \, \text{m s}^{-1}$, $e = 1.6 \times 10^{-19} \, \text{C}$ and $h = 6.6 \times 10^{-34} \, \text{J s}$.

$$1.3 \, \text{MeV} = 1.3 \times 10^6 \times 1.6 \times 10^{-19} \, \text{J}$$
$$= 2.1 \times 10^{-13} \, \text{J}$$

As $E = hf$ (see Section 26.1) and $c = f\lambda$

then $E = hc/\lambda$ and $\lambda = hc/E$

$$\therefore \quad \lambda = \frac{(6.6 \times 10^{-34} \, \text{J s})(3.0 \times 10^8 \, \text{m s}^{-1})}{2.1 \times 10^{-13} \, \text{J}}$$

$$= 9.4 \times 10^{-13} \, \text{m} \quad \text{or} \quad 94 \, \text{pm}$$

25.2 Background radiation

Even without any special radioactive source in the neighbourhood some **background radiation** is recorded by a GM tube. This is caused partly by:

- naturally occurring radioactive sources in rocks, soil and buildings;

- the radioactive gas radon which is a decay product of these sources; and

- cosmic radiation entering the Earth's atmosphere from outer space.

The first and last sources in the above list produce γ-rays; radon is an α-source.

The **background count** rate of a GM tube usually amounts to between 20 and 50 per minute, though with lead shielding it can be reduced to about 10 per minute. Its value is very dependent

on the local geological structure of the Earth.

When measuring the background count the *random* nature of radioactivity becomes very obvious. Figure 25.5 shows typical fluctuations in the background count measured over 1 minute time intervals. In general we find that the higher the total count in each interval chosen, the lower the fluctuations, when expressed as a percentage of the count. If the count is 100, the variation to be expected is ±10 (i.e. 10%) but if the count is 10 000, the variation to be expected is ±100 (i.e. only 1%).

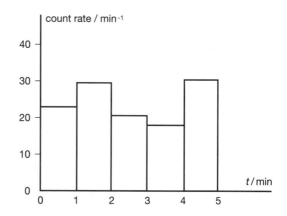

FIGURE 25.5

Radiation and the body

The existence of background radiation means that each of us is constantly being exposed to 'radioactivity'. To measure the cumulative effect of this radiation it is not the count rate in a detector which is important but the total ionising effect occurring *in* the body. Thus breathing a radioactive gas such as radon is much more dangerous than sitting next to a rock which gives off α- and β-particles (your clothes would prevent the α-particles from reaching your body); though the situation is not as simple as this as the rock may produce radon, an α-source, as one of its radioactive decay products.

We measure the effective background dose in **sievert** (Sv), and the map on page 343 shows the geographical variation of the cumulative background dose in a year over England, Wales and Scotland. To get some idea of the size of a millisievert, an average chest X-ray produces ionisation in the body equivalent to 2 mSv while a dose of 100 mSv, which you might get as a result of being too close to a nuclear explosion, could result in temporary sterility in a man. So 1 mSv from the environment in a year is a very low cumulative radiation dose.

The food we eat also contains traces of radioactive materials and these contribute to the overall dose. You receive perhaps 15% from food and 10%, on average, from medical X-rays during your life – the remaining 75% coming from the list at the beginning of this section. The contribution of fallout from nuclear tests and accidents such as that at Chernobyl will not amount to more than 1% or 2% in the United Kingdom.

Safety precautions

Radioactive sources, such as those in schools and colleges, must be treated with the greatest possible care – particularly because, unlike an X-ray tube, they are never 'switched off', and it is all too easy to forget their presence. By storing α- and β-sources in lead boxes their radiations are prevented from causing damage and, by handling with special tools and by screening β-sources with a layer of perspex during experiments, any danger is minimised.

The chief danger in handling radioactive materials comes from γ-rays. Even a layer of lead 10 mm thick is only sufficient to absorb about 50% of the γ-radiation incident on it, and so it is difficult and expensive to provide adequate screening. All γ-sources must therefore be as weak as possible, and must be kept at a sufficient distance from people in the laboratory to reduce the radiation dose to well below the acceptable limit.

For those who work with X-rays and γ-rays in hospitals and for all who work in the nuclear industry their exposure to ionising radiations is monitored by a **film badge** which consists of a piece of photographic film in a special holder. It depends on the different absorbing properties of the materials shown on the film badge in figure 25.6.

This badge is worn during working hours and is tested and replaced periodically. It measures the worker's exposure to each kind of radiation and consequently enables the effective cumulative dose to be calculated in mSv.

EXAMPLE

(a) Describe how you would demonstrate that the intensity of γ-rays from a radioactive source obeys an inverse square law.
(b) If the count rate in a detector was 6210 counts min^{-1} when it was 40 cm from the source, at what distance will it fall to less than 100 counts min^{-1}? Take the background count to be 30 counts min^{-1}.

(a) Set up the sources and GM tube and take readings of the counts in 100 s, C, for various values of distance x. Repeat three times at each x and find the average C.

Suppose the background count is C_0 and that the effective distance from the source to the detector is $x + x_0$. For the inverse square law to be valid

$$C - C_0 \propto \frac{1}{(x + x_0)^2}$$

As x_0 is not known (the exact position of the radioactive material inside its protective case is

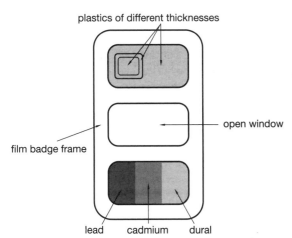

plastics of different thicknesses

film badge frame

open window

lead   cadmium   dural

FIGURE 25.6

not known, nor is the position inside the GM tube at which the γ-photon is detected) this can be rearranged to

$$x + x_0 \propto \frac{1}{\sqrt{(C - C_0)}}$$

and a graph of x against $(C - C_0)^{-1/2}$ plotted. If it is linear then the γ-rays obey an inverse square law.

(b) Suppose it is moved to a distance r before the count falls to 100 counts min^{-1} and assuming that in this case x_0 can be taken as zero:

$$\frac{r^2}{(40\,\text{cm})^2} = \frac{6210 - 30}{100 - 30}$$

$$\Rightarrow \qquad r = 380\,\text{cm}$$

25.3 Ionising radiations

An atom or molecule is said to be ionised when it has an electron removed from its structure. This requires energy which it could gain from a collision with another atom, with an electron, with a photon or with the products of radioactive decay. α-, β- and γ-radiations are all therefore ionising radiations. The α-particle, β-particle or γ-photon loses energy and the ionised atom or molecule becomes positively charged. The electron which is removed usually becomes attached to another atom or molecule which then becomes negatively charged.

You can demonstrate this **ionisation** by placing a radioactive source near the cap of a charged leaf electroscope. The deflection immediately starts to decrease whether the electroscope is positively or negatively charged. This shows the existence of +ve or −ve ions in the air and these must have been produced by the radiation from the source.

To produce a pair of +ve and −ve ions in air needs about 30 eV ($\approx 5 \times 10^{-18}$ J) or in the inert gas argon exactly 15.7 eV. As the products of radioactive decay have energies nearly a million times these figures you can see that a single α- or

β-particle or γ-photon can cause lots and lots of ionisation before it has used up all its energy. For example a single 4.5 MeV α-particle could cause $4.5 \times 10^6 \div 30 = 150\,000$ ionisations, and thus 150 000 ion pairs, in air.

The density of ionisation

Any α- or β-particle or any γ-photon of energy 4.5 MeV will produce the 150 000 ion pairs in air calculated above. How far it travels depends, however, on how many ionising collisions it makes along each mm of its path.

◆ **α-particles** produce very dense ionisation; about 3000 ion pairs per mm in air. This explains why they are so readily absorbed. It also warns us that an α-particle will cause very dense ionisation in skin or in the lining of the lung and it is this which makes handling an α-source or breathing in radon gas so dangerous, as these dense patches of ionised (damaged) molecules can become the site of cancerous tissue.

Figure 25.7 is a cloud chamber photograph showing α-particle tracks. They are almost all the same length which is evidence for the α-particles all having the same energy and the straight, densely ionised tracks indicate that the α-particles (positively charged helium nuclei) are not deflected in their collisions with the air molecules in the chamber.

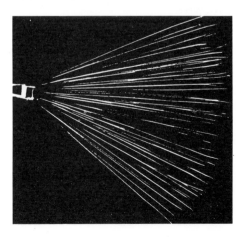

FIGURE 25.7

◆ **β-particles** produce fewer ion pairs per mm than α-particles, nearer 3 than 3000 and thus travel about 1000 times further before losing all their energy. Figure 25.8 shows the track of a fast β-particle. Towards the end of its track (not shown) a β-particle (an electron) will be deflected in its collisions with air molecules and the path will become increasingly random.

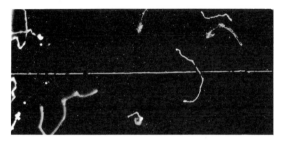

FIGURE 25.8

◆ The ionising ability of **γ-photons** depends very much on their energy but the number of ions per mm which they produce in air is negligible and their paths in cloud chambers are very difficult to detect. In solid materials and in flesh they ionise more densely.

The cloud chamber

A simple diffusion **cloud chamber** such as that shown in figure 25.9 can be readily set up to show α-tracks and, with care, β-tracks. The random nature of radioactive decay is very obvious when viewing α-tracks.

The floor of the chamber consists of a metal plate, which is cooled by a layer of 'dry ice' (solid carbon dioxide) packed beneath it. The ethanol condenses on the plate, and there is thus a steady diffusion of vapour from the top to the bottom of the chamber. A short distance above the metal plate the vapour passes through a narrow region in which it becomes supercooled. In this condition it will condense on any ions that happen to be present. A *string of droplets* therefore forms along any track in which ionisation occurs, and in this way the paths of any ionising particles are made visible.

By rubbing the top with a duster an electric field can be produced to sweep the ions out of the chamber so that new tracks can be seen. Bubble chambers are modern developments of the cloud chamber. They can be very large and produce long tracks from high-energy charged particles.

The GM tube

The GM tube and scaler (the **geiger counter**), like all radiation detectors, depends on ionisation. A typical design is shown in figure 25.10. The anode consists of a fine rod, which runs along the axis of the cylindrical cathode. A large electric field is produced near the surface of the rod by a p.d. of about 400 V between cathode and anode. In this region any free electrons are sufficiently accelerated to cause further ionisation and an 'avalanche' of electrons. Thus any ionisation of the low-pressure argon gas in the tube (even a single ion-electron pair) can be sufficient to trigger off an appreciable pulse of current which produces a voltage pulse in the external circuit. This is amplified and can be registered on a scaler – a counting device – or made to produce an audible click. The maximum

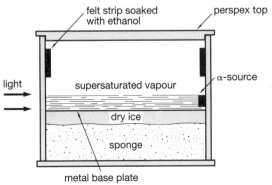

FIGURE 25.9

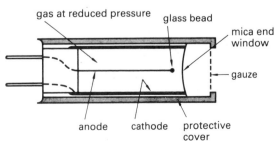

FIGURE 25.10

useful count rate is about $5000\,\text{s}^{-1}$.

The mica end window allows α-particles to be detected. γ-rays approaching from any direction will penetrate the tube, though only about 1% of the γ-photons passing through the tube will produce ionisation and be detected.

EXAMPLE

An α-source is placed in an ionisation chamber as shown in the diagram. The smallest current which can be detected in the circuit is $1\,\text{pA}$ $(1 \times 10^{-12}\,\text{A})$. If each α-particle from the source produces 1.2×10^5 ion pairs in the chamber, what is the weakest α-source which could be detected in this manner?

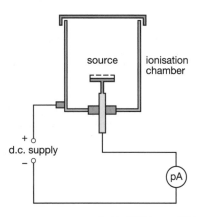

A current of $1 \times 10^{-12}\,\text{A}$ means a rate of flow of charge of $1 \times 10^{-12}\,\text{C}\,\text{s}^{-1}$. This charge is carried by ions each of which has a charge of $\pm 1.6 \times 10^{-19}\,\text{C}$.

The rate of flow of ions

$$= \frac{1 \times 10^{-12}\,\text{C}\,\text{s}^{-1}}{1.6 \times 10^{-19}\,\text{C}}$$

$$= 6.3 \times 10^6\,\text{s}^{-1}$$

There are 12×10^4 ion pairs per α-particle so the rate at which α-particles are emitted, n, is given by

$$n = \frac{6.3 \times 10^6\,\text{s}^{-1}}{12 \times 10^4} = 52\,\text{s}^{-1}$$

Thus the source would need to produce just over 50 α-particles per second.

Other detectors

The tracks of all three kinds of radiation may also be studied in photographic films. The emulsion reduces the range of α-particles to about 0.2 mm, and even β-particle tracks are only about 1 mm in length. A microscope must therefore be used to study them. Special nuclear emulsions are used for this purpose which are thicker than normal and have a higher density of silver bromide grains. The method has the advantage of automatically making a permanent record of the events studied. It is used in film badges – figure 25.6 – and in Tastrak, a film for detecting radon. The use of Tastrak by school students has produced a detailed survey of radon densities throughout the United Kingdom.

For detecting γ-rays a scintillation detector and photomultiplier – a **scintillation counter** – is often used. This device uses a crystal of sodium iodide which produces a flash of visible light when a γ-photon produces ionisation in it. These visible photons are then arranged to produce photoelectrons which are multiplied before producing a voltage pulse which can be counted in the normal manner.

25.4 The nucleus

The material of the **nucleus** is very, very dense (about $10^{16}\,\text{kg}\,\text{m}^{-3}$) and occupies a tiny volume at the centre of an atom. It contains protons and neutrons, the positively charged protons being held against their electrical repulsion by nuclear forces. Almost all (>99.9%) of the mass of an atom is in the nucleus – the electrons filling the rest of the 'space'.

Scattering experiments

The **nuclear model** of the atom described above was first devised by Rutherford after learning of the results of experiments in which a beam of α-particles was aimed at thin metal foil and the angles at which the α-particles were scattered were measured.

Figure 25.11 shows a modern version of the experiment. A detailed analysis of the number of α-particles at various values of θ confirm the nuclear model and also enable the charge on the nucleus of the metal atom to be found.

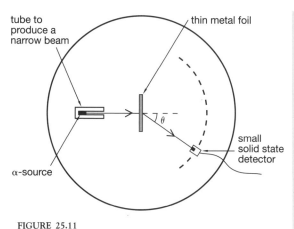

tube to
produce a
narrow beam

thin metal foil

α-source

small
solid state
detector

θ

FIGURE 25.11

Figure 25.12, a stroboscopic photograph taken as a steel sphere is rolled up a specially shaped hill, illustrates the path which an α-particle would take close to a nucleus. The varying distances between the positions of the steel sphere show how its kinetic energy changes along its path.

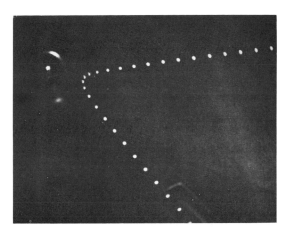

FIGURE 25.12

EXAMPLE

An α-particle emerges from a radioactive source with energy of 4.2 MeV and travels directly towards the nucleus of a gold atom. If the charges on the α-particle and gold nucleus are 2e and 79e respectively, where $e = 1.6 \times 10^{-19}$ C, calculate how close the α-particle gets to the nucleus before its motion is reversed. State any

assumptions you make.
Take $\epsilon_0 = 8.85 \times 10^{-12} \, \text{F m}^{-1}$.

$$4.2 \, \text{MeV} = 4.2 \times 10^6 \times 1.6 \times 10^{-19} \, \text{J}$$
$$= 6.7 \times 10^{-13} \, \text{J}$$

This is the kinetic energy of the α-particle before it reaches the gold nucleus. If the least distance between the α-particle and the nucleus is r_0, then at r_0 the α-particle has zero k.e. and all the energy is converted momentarily to electric potential energy. This assumes that the nucleus is fixed, i.e. that it does not recoil.

$$\text{e.p.e.} = \frac{1}{4\pi\epsilon_0} \frac{Q_1 Q_2}{r}$$

$$6.7 \times 10^{-13} \, \text{J} =$$
$$\frac{(2 \times 1.6 \times 10^{-19} \, \text{C})(79 \times 1.6 \times 10^{-19} \, \text{C})}{4\pi(8.85 \times 10^{-12} \, \text{F m}^{-1})r_0}$$

$$\Rightarrow \quad r_0 = 5.4 \times 10^{-14} \, \text{m}$$

This distance of closest approach is, presumably, a measure of the radius of the gold nucleus. The diameter of a nucleus is usually taken to be about 10^{-14} m.

Atomic number and mass number

The charge on the nucleus is **quantised**, it is a multiple of the electronic charge $e = 1.6 \times 10^{-19}$ C and is carried by the protons in the nucleus. The number of these photons we call the atomic number Z and different elements have different atomic (or proton) numbers, e.g. hydrogen Z = 1, helium Z = 2, carbon Z = 6, gold Z = 79 etc.

Although α-scattering experiments tell us that all the positive charge is in a tiny nucleus it was the work of a young British physicist H.G.J. Moseley (1887–1915) studying X-ray spectra which first established the order of the elements in the periodic table and hence the correct values of Z for the elements. The chemical properties of an element are determined by its electrons, especially by those in the outermost orbital. A neutral atom has Z electrons, i.e. the positive charge on the nucleus is Ze and the negative electron charge is $-Ze$.

The masses of atoms make it look at first as if mass is also quantised; it looks as if the protons and neutrons which make up the nucleus each have the same mass. This is not so; the mass of the proton and of the neutron are

$m_p = 1.007\,276\,u$ and

$m_n = 1.008\,665\,u$

where $1\,u$ is called the **atomic mass unit** and is equal to $1.660\,566 \times 10^{-27}\,kg$. It is a convenient unit of mass at the nuclear level. As the mass of an electron is only $m_e = 0.000\,549\,u$ you can see that an atom with Z protons and N neutrons will have a mass of approximately $(Z + N)\,u$, and $Z + N$ gives the mass number A of the atom:

$$Z + N = A$$

A is sometimes called the the nucleon number and N the neutron number.

Nuclides

Atoms which contain the same number of protons but different numbers of neutrons are called **isotopes**. Because not all atoms of, for example, carbon are identical – there are carbon atoms with 6 and with 7 neutrons, i.e. with $A = 12$ or 13 – we need a word to describe an atom with a particular nuclear structure. The word we use is **nuclide**, and we use the convention

$$\text{mass number} \quad {}^{}_{}X \quad \text{or} \quad {}^{A}_{Z}X$$
$$\text{atomic number}$$

TABLE 25.1 Data on selected nuclides

Particle or element	Atomic number Z	Mass number A	Symbol	Atomic mass/u	% Abundance
proton	1	1	${}^{1}_{1}p$	1.0073	
neutron	0	1	${}^{1}_{0}n$	1.0087	
electron	-1	0	${}^{0}_{-1}e$	0.0006	
hydrogen	1	1	${}^{1}_{1}H$	1.0078	99.99
		2	${}^{2}_{1}D$	2.0141	0.01
helium	2	1	${}^{3}_{2}He$	3.0160	<0.01
		2	${}^{4}_{2}He$	4.0026	≈100
carbon	6	12	${}^{12}_{6}C$	exactly 12	98.9
		13	${}^{13}_{6}C$	13.0034	1.1
		14	${}^{14}_{6}C$	14.00	trace
nitrogen	7	14	${}^{14}_{7}N$	14.0031	99.6
		15	${}^{15}_{7}N$	15.0001	0.4
calcium	20	40	${}^{40}_{20}Ca$	39.9626	97.0
		42	${}^{42}_{20}Ca$	41.9586	0.6
		43	${}^{43}_{20}Ca$	42.9588	0.1
		44	${}^{44}_{20}Ca$	43.9555	2.1
		46	${}^{46}_{20}Ca$	45.9537	0.003
		48	${}^{48}_{20}Ca$	47.9525	0.2
radium	88	223*	${}^{223}_{88}Ra$	223.0186	
		224*	${}^{224}_{88}Ra$	224.0202	
		226*	${}^{236}_{88}Ra$	226.0254	
		228*	${}^{238}_{88}Ra$	228.0312	
uranium	92	234*	${}^{234}_{92}U$	234.0409	trace
		235*	${}^{235}_{92}U$	235.0439	0.7
		236*	${}^{236}_{92}U$	238.0508	99.3

to represent a nuclide of element X. Thus $^{12}_{6}C$ and $^{13}_{6}C$ are two different nuclides of carbon – they are isotopes.

The convention is easy to use, $^{23}_{11}Na$ is a nuclide of sodium; it has 11 protons and $23 - 11 = 12$ neutrons. As Z is fixed for any one element we often simply refer to the mass number, e.g. for uranium which has $Z = 92$, uranium-235 and uranium-238 refer to nuclides with $235 - 92 = 143$ neutrons and $238 - 92 = 146$ neutrons.

Table 25.1 gives some data on selected nuclides. You will need to make use of data books for further information.

Once you know A for a nuclide you can calculate its approximate mass. The mass of $^{14}_{7}N$ is $14\,u = 14 \times 1.66 \times 10^{-27}\,kg = 23.2 \times 10^{-24}\,g$ and a mole of $^{14}_{7}N$ atoms, 6.02×10^{23} atoms, has a mass

$$(23.2 \times 10^{-24}\,g)(6.0 \times 10^{23}) = 14.0\,g$$

This is not a fluke, the value of the atomic mass unit u is chosen to make the mass of 1 mole of carbon-12 atoms equal to exactly 12 grams. For any nuclide of mass number A, a mass of A grams of the nuclide contains 6.02×10^{23} atoms, i.e. it is a mole of the nuclide. Fractions of a mole can thus be expressed as the mass in grams divided by A grams.

More precise values of atomic masses are given in the table as a multiple of u, together with the percentage abundances of isotopes. An asterisk indicates an unstable nuclide for which a percentage abundance cannot be quoted except in the case of very long-lived radioactive isotopes such as those of uranium.

25.5 Nuclear reactions

Nuclear reactions are represented by equations which note the composition of the nuclei involved.

For **naturally occurring** radioactivity these may be grouped into α-decay or β-decay.

α-decay

Taking radium 226 as an example:

$$^{226}_{88}Ra \rightarrow\, ^{222}_{86}Rn + ^{4}_{2}He$$

because an α-particle contains two protons and two neutrons it is identical with a helium nucleus.

Radon, Rn-222, is itself radioactive and decays by α-emission. Because radon is a gas it may not remain trapped in the radium (a metal). For this reason radium sources are sealed. The decay process, however, does not stop there. A *chain* of radioactive decay occurs and consequently a radium source is in practice a mixture of nine nuclides including the parent radium.

In general, for α-decay

$$^{A}_{Z}X \rightarrow\, ^{A-4}_{Z-2}Y + ^{4}_{2}He$$

The experimental evidence that α-particles are $^{4}_{2}He$ nuclei came initially from experiments where the α-particles, after gaining two electrons, were shown to form helium gas, its identity being convincingly demonstrated by its emission line spectrum.

Confirmation is shown in the cloud chamber photograph of figure 25.13. Here the cloud chamber is filled with helium instead of air and the incoming particle is an α-particle. The angle between the tracks of α-particles and helium nuclei after the collision is 90°, convincing evidence that the particles have the same mass.

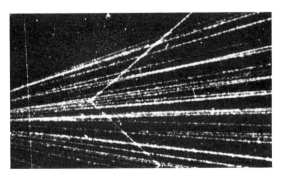

FIGURE 25.13

β-decay

Taking lead-214 as an example:

$$^{214}_{82}\text{Pb} \rightarrow ^{214}_{83}\text{Bi} + ^{0}_{-1}\text{e}$$

or, carbon-14 to nitrogen-14

$$^{14}_{6}\text{C} \rightarrow ^{14}_{7}\text{N} + ^{0}_{-1}\text{e}$$

which is the reaction on which carbon dating is based (see page 356).

The β-particles from naturally occurring radioactive materials are always negatively charged – we call them β⁻ particles, but we now know that β⁺ particles also exist. These have the same specific charge as the electron but the charge is positive (we talk of beta minus and beta plus). For both, the fact that they have a range of energies is explained by the emission of another particle – an **antineutrino** for β⁻-decay (see figure 25.14) and a **neutrino** for β⁺-decay – the total available energy being shared between the two emitted particles. The total energy of the particles is constant for a given radioactive decay process.

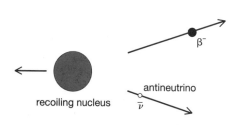

FIGURE 25.14

The antineutrino, $\bar{\nu}$, has no charge and only a very, very tiny mass which is not known precisely as it interacts so weakly with matter. The $\bar{\nu}$ is sometimes omitted, as above, from nuclear equations.

In β⁻-decay a nuclear neutron splits into a proton and an electron (and an antineutrino). The electron is emitted as the β⁻-particle. The number of nucleons and the mass number is unchanged but the nucleus is now that of a new element with one more proton and one fewer neutron. The atomic number Z goes up by 1.

Figure 25.15 shows a plot of neutron number N against atomic number Z; each dot on this represents a known stable nucleus. For most elements there are several dots above each value of Z, representing the different possible stable isotopes of that element. Unstable isotopes are not shown; they would lie mostly either above or below the fuzzy line formed by the plot. When they decay by radioactive emission they do so in such a way as to bring the resulting nucleus *nearer to* the region of stability close to the line. Thus, nuclei above the line are those that have a surplus of neutrons; these will tend to decay by β⁻-emission, which increases the proton number by one at the expense of the neutron number. Nuclei below the line have a surplus of protons, and tend to decay by β⁺-emission, which has the opposite effect.

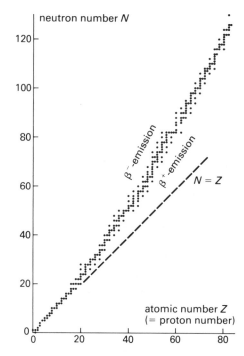

FIGURE 25.15

You need not learn any of these nuclear equations; it is enough to be able to deduce the nuclear composition of the decay product given the mode of decay.

The α-, β-, and γ-radiations from naturally occurring radioactive decay processes carry a lot of energy (of the order of MeV per disintegration).

This energy comes from the nucleus, which is *unstable*; we call it **nuclear energy**, and there should be an indication on the right-hand side of each of the above equations that there is excess energy involved. If you look up the atomic *masses* of the nuclides involved in any one transformation, you will find that there is a loss of a mass Δm. For example:

$$^{226}_{88}\text{Ra} \rightarrow {}^{222}_{86}\text{Ru} + {}^{4}_{2}\text{He} + 4.9\,\text{MeV}$$

with a mass loss of 0.0052 u, or

$$^{214}_{82}\text{Pb} \rightarrow {}^{214}_{83}\text{Bi} + {}^{0}_{-1}\text{e} + 1.04\,\text{MeV}$$

with a mass loss of 0.0011 u.

The link between the excess energy and the mass loss is the mass–energy equation $\Delta E = c^2 \Delta m$, where $c = 3.00 \times 10^8\,\text{m s}^{-1}$, the speed of light. This is developed more fully in Section 25.7 where the calculations for the α-decay of radium-222 are given in full (page 362).

Artificial transformations

When a nuclear particle (an alpha particle, a proton, other positive ions or a neutron) is fired at a nucleus there is a chance that it will be absorbed. The resulting nucleus is usually unstable and subsequently decays. The first such artificial transmutation was studied in 1919: it was

$$^{4}_{2}\text{He} + {}^{14}_{7}\text{N} \rightarrow {}^{17}_{8}\text{O} + {}^{1}_{1}\text{H}$$

Another way of writing this sort of reaction is

$$^{14}\text{N}\,(\alpha, \text{p})\,^{17}\text{O}$$

where the bracketed part indicates an absorbed alpha-particle and an emitted proton. The mass of the products on the right-hand side is here greater than the mass of those on the left and the mass difference is 0.00129 u or 2.14×10^{-30} kg, the equation $\Delta E = c^2 \Delta m$ represents an energy of 1.93×10^{-13} J or 1.20 MeV. This reaction, therefore, requires α-particles with a minimum kinetic energy of 1.20 MeV in order to proceed. Notice that nuclear reactions involve MeV of energy but chemical reactions involve only a few eV; a million times less. This is why mass differences in chemical reactions are never detectable.

EXAMPLE

Radioactive phosphorus-30 can be produced by the α-bombardment of aluminium, a neutron being emitted. The phosphorus-30 subsequently decays to silicon-30.

(a) Deduce (i) the mass number of the aluminium and (ii) the mode of decay of the phosphorus (P). Take the atomic numbers of aluminium (Al) and silicon (Si) to be 13 and 14 respectively.

(b) The energy available from the decay of the phosphorus-30 is 1.6 MeV. Explain why the energy of the emitted particle will be less than this.

(a) (i) The reaction is of the form

$$^{4}_{2}\text{He} + {}^{x}_{13}\text{Al} \rightarrow {}^{30}_{y}\text{P} + {}^{1}_{0}\text{n}$$

nucleon numbers: $4 + x = 30 + 1$

$\therefore$ A for aluminium is $x = 27$

proton numbers: $2 + 13 = y + 0$

$\therefore$ Z for phosphorus is $y = 15$

(ii) The reaction is of the form

$$^{30}_{15}\text{P} \rightarrow {}^{30}_{14}\text{Si} + {}^{0}_{1}\text{X}$$

X must be a β^+-particle, a positron.

(b) When a positron is emitted so also is a neutrino, ν. The energy is shared between them so the positron energy will be less than 1.6 MeV.

Also the Si nucleus will recoil a little and this will absorb a very small percentage of this energy.

The maximum energy of bombarding particles from natural radioactive decay processes is about 10 MeV. By using particle accelerators protons and other positive ions can be given much more energy. By 1940 ions from a cyclotron reached energies of 32 MeV and by 1950, 900 MeV ions were available. The particle accelerators at CERN, Geneva, were producing 300 keV protons by the mid-1970s and have now broken through to TeV energies (TeV = 10^{12} eV). The range of nuclear reactions induced by such energetic particles is vast. In all of them, however, the number of nucleons is conserved and the mass–energy conservation rule holds. Charge is also conserved. Figure 25.16 shows a photograph showing the results of collisions between the high-energy nuclear particles in a bubble chamber. It is from a study of such photographs that our knowledge of **fundamental particles** has developed.

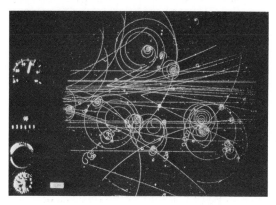

FIGURE 25.16

Carbon dating

One source of high-energy particles is the **cosmic rays** which continually hit the Earth. The action of these cosmic rays in the Earth's atmosphere causes a steady production of the radioactive isotope carbon-14 from nitrogen. Because of this a very small proportion of the carbon in the atmosphere is in this radioactive form. There is a continual interchange of carbon (in the form of carbon dioxide) between the atmosphere and all living creatures, whose cells therefore contain the same small proportion of carbon-14, about 1 part in 10^{12}.

However, when the creature dies, the interchange of carbon with the atmosphere ceases, and the carbon-14 content of the remains starts to decrease. By measuring the activity of the carbon-14 in a specimen (such as wood from a building or parchment from a manuscript) we can find the total amount of carbon-14 still remaining in it; and so we can find its age. The half-life of carbon-14 is 5730 years, which is ideal for archaeological purposes. The technique has enabled ages from 500 years up to about 10 000 years to be found, in some

cases with uncertainties of less than 100 years and has helped in the study of objects such as the Turin Shroud (figure 25.17) which was believed to date from the time of Christ but is now believed to be medieval in origin.

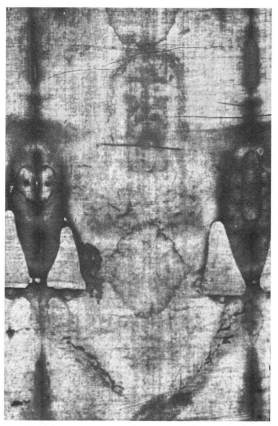

FIGURE 25.17

Neutron-induced reactions

By using beams of neutrons (symbol ^1_0n) many new kinds of nuclear reaction can be produced. A *charged* particle entering a nucleus needs to have sufficient energy to overcome the intense electrical repulsion. Protons and α-particles must therefore have large energies to do this. But the neutron, being *uncharged*, can pass readily through a nucleus, whatever its energy; and it may in the process be absorbed by it.

One such reaction provides one of the most convenient methods of detecting neutrons;

$$\ce{^{10}_{5}B + ^{1}_{0}n -> ^{7}_{3}Li + ^{4}_{2}He}$$

or $$\ce{^{10}B(n, \alpha)^{7}Li}$$

The boron-10 nucleus absorbs neutrons very readily, and then disintegrates into a lithium nucleus and an α-particle, both of which are heavily ionising particles. The ionisation produced by neutrons themselves in air is negligible so that an ordinary GM tube scarcely responds to them at all. But by filling the tube with boron trifluoride gas a high proportion of the incident neutrons can be detected.

Cobalt-60, a radioactive isotope widely used in radiotherapy, is produced when natural cobalt absorbs a neutron:

$$\ce{^{59}_{27}Co + ^{1}_{0}n -> ^{60}_{27}Co^{*}}$$

the asterisk showing the nuclide is unstable. The cobalt-60 is a γ-emitter with a half-life of just over five years:

$$\ce{^{60}_{27}Co^{*} -> ^{60}_{27}Co + \gamma}$$

Figure 25.18 shows a **gamma knife** which concentrates the effect of 200 individually harmless γ-emitters at a single point to irradiate some cancerous tissue in the body (here, a brain tumour).

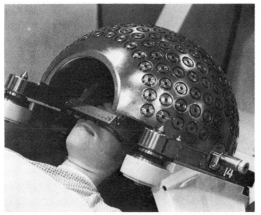

FIGURE 25.18

The neutron was first discovered only in 1932 though its existence had then been long suspected. Nowadays nuclear reactors produce a large flux of neutrons which can be used to produce many of the radioactive isotopes used for research and in medicine. Some of these applications are taken up in the next section.

25.6 Radioactive decay

In the previous section various uses for radioisotopes were mentioned, together with their half-lives. Radioactive nuclides have a half-life because each atom undergoes a spontaneous decay. We cannot predict when a *particular* radioactive nucleus will decay: all we can say is that there is a fixed chance that it will decay in a certain time. Although the behaviour of a particular nucleus is unpredictable, we *can* make predictions when we are dealing with large numbers of the same kind of nucleus (just as we cannot tell which students will be ill on a particular day, but we can make a reasonable guess at the total number who will be away ill). The **half-life** of a radioactive nuclide is the expected time for half of the original number of undecayed nuclei to decay. For example, the half-life of iodine-131 is 8.1 days, so if we started with 1 000 000 of these nuclei, there would (probably) be 500 000 left after 8.1 days. But the process continues: after another 8.1 days, there would again be half left, but this time half of 500 000, so there would be 250 000 left after a total of 16.2 days. This shows that the number decreases **exponentially** with time, since the numbers, after equal times, decrease in a constant ratio.

EXAMPLE

Iron-59 is a radioisotope used in medical diagnosis. Each one of its atoms has a 50% chance of decaying by α-emission in 46 days. Estimate what fraction of the atoms decay within the first 100 days.

Suppose there are N atoms to start with.

After 46 days there will be $N/2$ left
After 92 days there will be $N/4$ left
After 138 days there will be $N/8$ left

The sketch graph shows that after 100 days about 220 atoms remain, i.e. 780 or 78% have decayed.

357

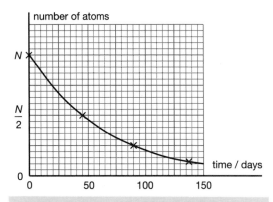

It is often possible to make such estimates by sketching a graph or interpolating data rather than by attempting complex calculations. Notice that the initial value of N does not matter. Whatever it is, 78% of the atoms would remain after 100 days.

Activity

The rate at which a radioisotope emits particles is called its **activity**, A. Thus, when there are N decaying atoms

$$A = -\frac{dN}{dt}$$

where dN/dt simply means that the rate of change of N and the minus sign tells us that N is getting smaller.

The unit of activity is the **becquerel** (Bq) and is a name for one (random) disintegration per second. The activities of sources used during your course are never greater than 5×10^5 Bq and are often as small as 5×10^3 Bq. However, in medicine, e.g. for deep radiotherapy, cobalt-60 γ-sources of over 10^{12} Bq are used. (You may have come across sources with activities marked in μCi, which is short for microcurie, an old unit for activity. $1\,\mu$Ci $= 3.7 \times 10^4$ Bq.)

Everyday substances are often weakly radioactive. EC law defines as radioactive any substance with an activity per unit mass of more than $400\,\text{Bq}\,\text{kg}^{-1}$. Coffee powder, for example, is radioactive with an activity of about $1500\,\text{Bq}\,\text{kg}^{-1}$.

The decay equation

As already explained the activity of any source containing a radioactive nuclide depends upon the number of atoms of that nuclide present

i.e.

$$-\frac{dN}{dt} \propto N$$

so

$$\frac{dN}{dt} = -\lambda N$$

or

$$A = -\lambda N$$

where λ is called the **decay constant** for that nuclide, e.g. for radon-220, $\lambda = 0.014\,\text{s}^{-1}$. The unit of λ will be per second. The use of calculus allows us to solve this equation to get N in terms of t,

$$N = N_0 e^{-\lambda t}$$

and it is this equation which describes the curve in the above example. N_0 is the number of radioactive atoms present at $t = 0$. This is an exponential decay curve and N falls by half every $t_{\frac{1}{2}}$, the half-life of the nuclide. As $N = \frac{1}{2}N_0$ when $t = t_{\frac{1}{2}}$, then

$$\frac{N_0}{2} = N_0 e^{-\lambda}$$

$$\therefore \qquad e^{-\lambda t_{\frac{1}{2}}} = \frac{1}{2}$$

or

$$e^{\lambda t_{\frac{1}{2}}} = 2$$

$$\Rightarrow \qquad \lambda t_{\frac{1}{2}} \ln e = \ln 2$$

which as

$$\ln e = 1$$

$$\Rightarrow \qquad \lambda t_{\frac{1}{2}} = \ln 2$$

so that if you know $t_{\frac{1}{2}}$ you can calculate λ, and vice versa.

Measuring half-life

The half-lives of known radioactive substances cover an enormous range – from less than 10^{-15} seconds to more than 10^{10} years. The direct measurements of long half-lives is unthinkable. The half-life of radium-226 is 1620 years; and even this is trifling compared with that of uranium-238 (4.5×10^9 years). The amount of radium-226 in a given sample would have decreased by only 5% during the time since this element was discovered by the Curies; so this cannot be made the basis for a measurement of its half-life.

However, if we can estimate the number of atoms N in a sample, we can derive the half-life from a measurement of its activity because, as

$$A = \lambda N \text{ numerically}$$

then

$$\lambda = \frac{A}{N}$$

and we can then work out $t_{\frac{1}{2}}$.

EXAMPLE

A speck of dust contains $1.00\,\mu g$ of plutonium-239 (an α-emitting substance). This is found to have an activity of $2300\,s^{-1}$. Calculate the half-life of plutonium. Take the Avogadro constant $= 6.02 \times 10^{23}\,mol^{-1}$.

One mole of plutonium atoms has a mass of $239\,g$. Therefore $1.00\,\mu g$ (i.e. $1.00 \times 10^{-9}\,kg$) of plutonium contains a number N of atoms given by

$$N = \frac{1.00 \times 10^{-9}\,kg \times 6.02 \times 10^{23}\,mol^{-1}}{0.239\,kg\,mol^{-1}}$$

$$= 2.52 \times 10^{15}\text{ atoms}$$

Hence $\lambda = \dfrac{A}{N} = \dfrac{2300\,s^{-1}}{2.52 \times 10^{15}}$

$$= 9.13 \times 10^{-13}\,s^{-1}$$

$$\Rightarrow \quad t_{\frac{1}{2}} = \frac{0.693}{\lambda} = \frac{0.693}{9.13 \times 10^{-13}\,s^{-1}}$$

$$= 7.59 \times 10^{11}\,s$$

and in years,

$$t_{\frac{1}{2}} = \frac{7.59 \times 10^{11}\,s}{(3600 \times 24 \times 365)\,s\,yr^{-1}}$$

$$= 24\,000\text{ years}$$

The *direct* measurement of half-life is possible only for short-lived radioactive isotopes. Many of these are too dangerous for us to experiment with, because being short-lived, they have high activities. However, radon-220, formed as one of the decay products of thorium-232, can be handled safely in a school or college laboratory. Figure 25.19 shows the arrangement. Firstly the picoammeter should be switched on and set to zero; this eliminates the background count which may be greater than usual because of

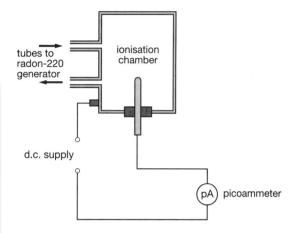

FIGURE 25.19

contamination of the chamber after previous experiments.

The mass of radon-220 transferred from the generator to the ionisation chamber is probably about $10^{-20}\,kg$, quite undetectable by chemical means. However, such a sample could contain 5×10^4 atoms, and these will decay initially at a rate of about 700 per second. Each α-particle produces about 1.5×10^5 ion pairs, so that the total charge passing through the ionisation chamber per second is about

$$(1.5 \times 10^5)(700\,s^{-1})(1.6 \times 10^{-19}\,C)$$

$$= 1.7 \times 10^{-11}\,C\,s^{-1}$$

$$= 1.7 \times 10^{-11}\,A$$

This **ionisation current** I is proportional to the activity A of the sample, which in turn is proportional to the number of atoms of radon-220 remaining in it. We therefore expect to find I decaying exponentially in the same manner as N. Thus

$$I = I_0 e^{-\lambda t}$$

where I_0 is the initial current (where $t = 0$). The best way to test an exponential decay is to plot the equivalent logarithmic graph

$$\ln I = -\lambda t + \ln I_0$$

If this is a straight line, the decay is shown to be exponential: the gradient is $-\lambda$, so λ can be calculated.

Figure 25.20 shows a graph of data taken from such an experiment with radon-220.

$$\text{gradient} = \frac{0.2 - 2.8}{200\,\text{s} - 0}$$

$$= -1.3 \times 10^{-2}\,\text{s}^{-1}$$

$$\Rightarrow \quad \lambda = +1.3 \times 10^{-2}\,\text{s}^{-1}$$

Hence
$$t_{\frac{1}{2}} = \frac{\ln 2}{\lambda} = 53\,\text{s}$$

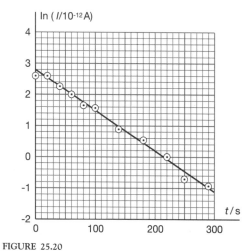

FIGURE 25.20

Radioactive tracers

The amount of radon-220 used in the above experiment would be regarded, by any other standards, as undetectably small. However, the number of atoms in the sample is large enough to provide a measurable ionisation current. Radioactive samples used in this way are called **radioactive tracers**. They have become an indispensable tool of research in many fields. By using a GM tube it is possible to follow a radioactive isotope through a physical or chemical process, and its presence can be detected by the radiations it emits; but only a minute quantity of the isotope need be used, as long as it is sufficient to ensure that the count rate is significantly greater than the background count. Tracers are used in medical diagnosis and to follow the flow of mud in rivers or water leaking from underground pipes. They can also be used to measure wear in bearings and pistons.

A **gamma camera** can be used to detect the position of iodine-131, a radioactive tracer, in the patient's kidneys during a kidney scan. In selecting a suitable nuclide for use as a tracer you need to consider the mode of decay and the half-life, as well as any non-radioactive toxic qualities it may possess. A tracer used inside the body, for example, should have a half-life of not more than a few days so that its ionising effect falls off rapidly after it has served its purpose. (Some of it may be excreted from the body, thus lowering its 'effective' half-life.) Obviously a γ-source is needed as otherwise no radiation would be detectable outside the body.

Other uses for radionuclides

- ♦ **α-sources** are used in luminous watches and also in smoke detectors. In the former the α-particles interact with zinc sulphide to produce visible photons. In the latter the α-particles ionise the air across a gap in an electrical circuit. If the current falls, e.g. because smoke particles prevent the movement of the ions, the alarm is set off. Again it is obvious that α-sources with long half-lives are needed.

- ♦ **β-sources** can be used to monitor the thickness of materials, e.g. paper or polythene, being produced as a continuous strip. They can also detect when a tin or box is not filled to the correct level. Such sources should have a very long half-life; otherwise a change in the activity of the source will have to be allowed for in the day-to-day setting of the feedback systems involved.

- ♦ **γ-sources** are used to detect cracks in cast metal components. They are used, just like the X-rays used in studying cracks in human bones, together with suitable photographic film. γ-rays are used to kill bacteria in foods for patients who need a sterile diet. The same process can also increase the shelf-life of fresh foods, e.g. mushrooms. Another biological application is in pest control where male pupae of tsetse flies for example are γ-irradiated. These develop into sterile flies which, if repeatedly released in large numbers, can reduce or eliminate the local fly population.

25.7 Nuclear energy

The mass–energy equation

FIGURE 25.21

One of the results of Einstein's (figure 25.21) **special theory of relativity** is that energy and mass are equivalent. The theory predicts that a body gaining energy ΔE thereby increases in mass by an amount Δm given by

$$\Delta E = c^2 \Delta m$$

where c is the speed of light. The quantities of energy handled in everyday life are far too small to involve detectable changes of mass. Thus 1 kg of water absorbs 4.2×10^5 J of energy to raise its temperature from 0°C to 100°C. The theory of relativity therefore predicts that its mass will increase accordingly by an amount Δm, given by

$$\Delta m = \frac{4.2 \times 10^5 \, J}{(3.0 \times 10^8 \, m\,s^{-1})^2}$$

$$\approx 5 \times 10^{-12} \, kg$$

In chemical reactions also the changes of energy are of this order of magnitude, involving at the

most a few eV of energy per atom; and the associated changes of mass are again quite undetectable.

However, in nuclear processes the quantities of energy released are of the order of a few MeV per nucleus, and are sufficient to produce significant changes in the masses of the particles concerned. The mass equivalent Δm of 1.00 MeV of energy may be calculated as follows:

$$1.00 \, MeV = 1.60 \times 10^{-13} \, J$$

Hence
$$\Delta m = \frac{\Delta E}{c^2}$$

$$= \frac{1.60 \times 10^{-13} \, J}{(3.00 \times 10^8 \, m\,s^{-1})^2}$$

$$= 1.78 \times 10^{-30} \, kg$$

thus 1.00 MeV of energy (of any form) is equivalent to 1.78×10^{-30} kg of mass. (This is nearly *twice* the mass of an electron.) A 1 MeV γ-ray carries this mass away from the nucleus that emits it. In the same way a 1 MeV β-particle has a total mass *three* times the rest mass of a stationary electron (its rest mass + the mass of its kinetic energy). So the emission of a β-particle not only carries away from a nucleus the rest mass of an electron, but also the mass equivalent of its kinetic energy.

Pair production

The minimum energy that might produce an electron–positron pair is $2m_e c^2$, where m_e is the mass of an electron (or positron). This is equal to

FIGURE 25.22

1.02 MeV. Any γ-photon above this energy is capable of causing **pair production**, and at very high energies pair production accounts for most of the absorption of γ-rays by matter.

In figure 25.22 a group of particles has been produced, at left, by cosmic rays. Since the cloud chamber is in a magnetic field the negative electrons curve in one direction and the positrons in the other.

Mass defect

As we have seen in the previous section the processes of radioactive decay always involve appreciable reductions in the total rest mass of the nuclei and particles concerned. The changes are still small compared with the mass of a nucleon, so the atomic masses, expressed in atomic mass units, of all atoms remain close to whole numbers.

Consider for instance the α-decay of radium-226 discussed on page 355. The masses of one atom of the decay product radon-222 and of helium-4 are given by

$$m_{Rn} = 222.0176\,u$$

$$m_{He} = 4.0026\,u$$

so that $\qquad m_{Rn} + m_{He} = 226.0202\,u$
but the mass of the original radium-226 was

$$m_{Ra} = 226.0254\,u$$

which is more than the mass of the products by $0.0052\,u$, where $1\,u = 1.66 \times 10^{-27}\,kg$.

This is the **mass defect** for this nuclear reaction.

$$\Delta m = 0.0052 \times 1.66 \times 10^{-27}\,kg$$

so that the surplus energy

$\Delta E = c^2 \Delta m$

$\qquad = (3.00 \times 10^8\,m\,s^{-1})^2 (0.0052 \times 1.66 \times 10^{-27}\,kg)$

$\qquad = 7.77 \times 10^{-13}\,J$

As $1\,MeV = 1.60 \times 10^{-13}\,J$ this energy surplus can be expressed as $4.85\,MeV$. Not all this becomes kinetic energy; a small amount is carried away by a γ-photon shortly after the α-particle is expelled.

A calculation of this kind can be applied to any of the nuclear reactions mentioned in the previous

section. In general the *greater* the mass defect of a spontaneous reaction, the *shorter* the half-life of the reaction. When an artificial reaction is triggered by a bombarding particle, the mass equivalent of the incoming kinetic energy $(\Delta m = k.e./c^2)$ needs to be considered in calculating the net mass defect. It can also be applied to calculate the energy which would be needed to pull apart the particles within a nucleus against the nuclear forces which hold them together.

EXAMPLE

Calculate the total mass of 8 protons and 8 neutrons. If the mass of an oxygen-16 nucleus is $15.9949\,u$ calculate the mass defect per nucleon for oxygen-16.

$$m_p = 1.007\,28\,u, \quad 8m_p = 8.058\,24\,u$$

$$m_n = 1.008\,76\,u, \quad 8m_n = 8.069\,36\,u$$

$$8m_p + 8m_n = 16.127\,60\,u$$

giving a total mass defect of $0.13268\,u$ and a mass defect per nucleon of $0.00829\,u$.

Binding energy

If we calculate the mass defect per nucleon for all nuclides we get a graph such as that shown in figure 25.23.

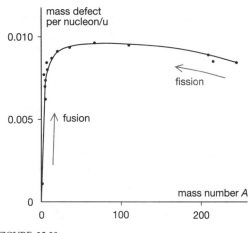

FIGURE 25.23

There are therefore two possible ways in which the nuclear energy of matter can be tapped. Either we seek to break down the more massive nuclei (A = 230 to 250) into smaller fragments (i.e. nearer to the minimum at A = 60); or we endeavour to fuse together the lighter nuclei into more massive units. Both these processes are possible; they are called fission and fusion respectively.

Uranium fission

When uranium is bombarded with neutrons, it is found that some of the nuclei absorb neutrons and immediately break up into two fragments, a process known as **nuclear fission**. The two fragments are nuclei of elements of medium atomic mass, nearer therefore to the minimum energy condition for nucleons. The energy released in the fission of one uranium nucleus is about 200 MeV. In addition to the two main fragments several neutrons are emitted at the same time. Figure 25.24 shows a neutron-rich nucleus undergoing fission. The small blobs are neutrons.

$$^{235}_{92}U + {}^{1}_{0}n \rightarrow \begin{matrix} \text{fission} \\ \text{fragments} \end{matrix} + \begin{matrix} \text{2 or 3} \\ \text{neutrons} \end{matrix} + \text{energy}$$

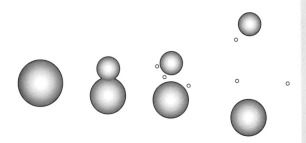

FIGURE 25.24

This provides the possibility of starting a **chain reaction** in a piece of uranium. The fission of one nucleus can produce neutrons to cause the disintegration of several more nuclei; these in turn can trigger the fission of an even larger number, and so on. In a sufficiently large lump of uranium this can happen spontaneously. In this way a substantial fraction of the energy locked up in the uranium nuclei can be released.

EXAMPLE

One possible nuclear reaction which occurs in the fission of uranium-235 is

$$^{235}_{92}U + {}^{1}_{0}n \rightarrow {}^{138}_{55}Cs + {}^{96}_{37}Rb + 2{}^{1}_{0}n$$

Use the following information to show that the energy released by one such reaction is 2.7×10^{-11} J. Hence calculate the energy available from the fission of 1.00 kg of uranium-235.

Avogadro constant = 6.02×10^{23} mol^{-1}
speed of light *in vacuo* = 3.00×10^{8} m s^{-1}
unified atomic
mass constant (u) = 1.66×10^{-27} kg
mass of neutron = 1.009 u
masses of
relevant nuclei: U-235 = 235.044 u
Cs-138 = 137.920 u
Rb-96 = 95.932 u

mass loss in each fission = $m_U - m_{Cs} - m_{Rb} - m_n$

= 0.183 u

= $0.183 \times 1.66 \times 10^{-27}$ kg = 3.04×10^{-23} kg

∴ energy released = $c^2 \Delta m$
= $(3.00 \times 10^8$ m s$^{-1})^2 (3.04 \times 10^{-28}$ kg)
= 2.7×10^{-11} J

1.00 kg of U-235 contains $1000 \div 235$ mol of uranium

i.e. $\dfrac{1000}{235} \times 6.02 \times 10^{23}$ atoms = 2.6×10^{24} atoms

∴ energy available = $2.6 \times 10^{24} \times 2.7 \times 10^{-11}$ J

= 7.0×10^{13} J

In practice there is a further 0.8×10^{-13} J per fission which comes from the β⁻-decay of Cs and Rb. They have too many neutrons and will spontaneously undergo beta-minus decay.

In the atomic bomb (properly called a nuclear bomb) this energy is converted in an uncontrolled way – with devastating results. But in a **nuclear reactor** the rate of reaction is controlled so that the internal energy produced from the nuclear energy can be carried away steadily by circulating fluids, and used to drive the turbines of a power station. The energy in the above example is enough to run a 200 MW power station for 3.5×10^5 s $\approx$ 4 days!

A diagrammatic representation of the core of nuclear fission reaction is shown in figure 25.25.

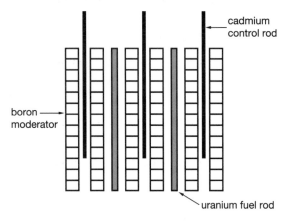

FIGURE 25.25

The fuel rods, containing uranium-235, emit energetic neutrons. These are moving too quickly to be captured by other uranium nuclei and so are slowed down by the **moderator** which is usually made of an element of low Z, e.g. boron. To ensure that the rate at which the fission takes place can be controlled, cadmium **control rods** which absorb slow neutrons are arranged in such a way that they can be raised (to speed up the reaction) or lowered (to slow down the reaction). In the event of some overheating they fall, thus shutting down the reactor.

A serious practical problem in the operation of nuclear reactors is the safe disposal of the large quantities of radioactive waste formed: some of this will retain its activity for thousands of years, during which time it must not be allowed to contaminate the environment.

Hydrogen fusion

In this type of reaction light nuclei are fused together into larger units. By using high-energy beams of protons, deuterons, etc., many such reactions have been discovered. In each case the condition for the reaction to proceed is that the two reacting particles should approach one another at sufficient speed and therefore with sufficient energy to overcome the electrical repulsion between them. To achieve this on a large scale it is necessary only to raise the temperature of the ingredients to the point at which some of the particles have the required speed. For this reason these are known as **thermonuclear reactions**; unfortunately temperatures of 10^8 K are needed to achieve them. The Joint European Taurus is one experimental facility pursuing such temperatures in gas plasmas.

The only places in which such reactions normally proceed are the interiors of the Sun and stars. A variety of thermonuclear reactions take place there whose net effect is the conversion of hydrogen into helium.

In the Sun the following cycle of reactions occurs:

$$_1^1\text{H} + {_1^1}\text{H} \rightarrow {_1^2}\text{H} + {_1^0}\text{e} + \nu$$

$$_1^2\text{H} + {_1^1}\text{H} \rightarrow {_2^3}\text{He}$$

$$_2^3\text{He} + {_2^3}\text{He} \rightarrow {_2^4}\text{He} + 2({_1^1}\text{H})$$

At each stage energy is released; a total of 27 MeV for each $_2^4$He helium nucleus formed.

More than 0.8% of the rest mass of the original particles is converted to internal energy in the fusion process. The Sun converts about 4×10^6 tonnes of matter per second in this way; and there seems no reason why it should not continue in this profligate style for the next 10^{10} years. If and when we learn how to produce controlled nuclear fusion, a source of cheap and abundant energy will be available for the human race.

Exercises on each section of this chapter may be found in the companion textbook, **Practice in Physics**.

SUMMARY

At the end of this chapter you should be able to:

◆ remember the nature of naturally occurring alpha, beta and gamma radiations together with their relative charges and masses.

◆ remember the relative ionising properties of the radiations and their typical penetrating abilities.

◆ understand what is meant by the half-value thickness of an absorbing material.

◆ understand that nuclear reactions involve energy changes of the order of MeV and remember that $1\,\text{MeV} = 1.6 \times 10^{-13}\,\text{J}$.

◆ explain that the cumulative effect of background radiation varies from place to place.

◆ describe how to use a diffusion cloud chamber and a GM tube and scaler in the study of ionising radiations.

◆ explain the experimental evidence for the nuclear atom.

◆ use the notation $^{A}_{Z}\text{X}$ for the atomic (proton) number Z and the mass (nucleon) number A when describing nuclides.

◆ use the atomic mass unit u and remember that it is equal to one twelfth of the mass of the nuclide $^{12}_{6}\text{C}$.

◆ draw up nuclear equations of the form:

$$^{4}_{2}\text{He} + ^{14}_{7}\text{N} \rightarrow ^{17}_{8}\text{O} + ^{1}_{1}\text{H} \qquad \text{or} \qquad ^{14}\text{N}(\alpha, \text{p})^{17}\text{O}$$

in which both charge and mass are conserved.

◆ explain beta minus and beta plus decay in terms of a surplus or deficit of neutrons.

◆ use the mass–energy conservation equation $\Delta E = c^{2}\Delta m$ to calculate the energy available from a nuclear reaction.

◆ understand that radioactive decay is a random process.

◆ explain that $dN/dt = -\lambda N$ is a differential equation the solution of which is an exponential decay curve.

◆ remember that the activity of a radioactive source is equal to the number of disintegrations per second and is measured in becquerel.

◆ use radioactive decay curves to find the half-life and the decay constant of radioactive nuclides.

◆ remember the relationship $\lambda t_{\frac{1}{2}} = \ln 2$.

◆ use the equation describing radioactive decay

$$N = N_{0}e^{-\lambda t}$$

◆ explain how to measure the half-life of long-lived radioactive nuclides.

◆ remember how radioactive nuclides can be used, for example as tracers or in smoke detectors.

◆ explain the terms binding energy and mass defect of a nucleus.

◆ understand that both nuclear fission and nuclear fusion increase the mass defect per nucleon of the nuclei involved in the reaction.

26 Photons and Electrons

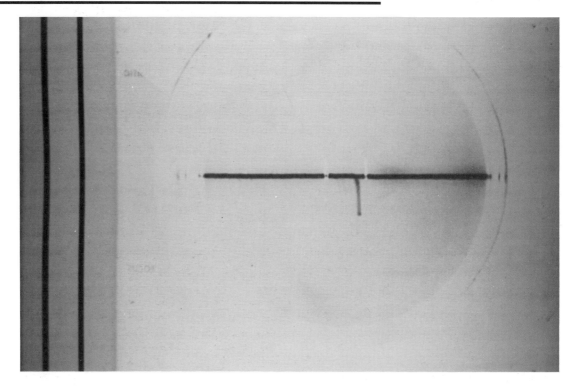

Electromagnetic radiation does not come in a continuous stream: it comes in lumps. There was a time when we could buy milk in any amount we wanted: the milkman poured it out into a can. Nowadays he delivers it in bottles or cartons of a certain size: we can have a pint, or a quart, but not 1.37 pints or 0.68 of a pint. Light is like that. When we see a sheet of paper lit by a lamp, it is not being bathed by energy smeared out all over it: it is being bombarded by very very many small bits of energy in amounts of different size, arriving randomly. It is rather like the 'steady' pressure which we experience from the air around us: in fact it only appears steady because the number of tiny, randomly-arriving particles which strike us each second is so very great. The photograph shows a voltage spike which corresponds to the arrival of a single 'bit' of light energy. These bits are called photons. We refer to the experiment again in Section 26.2.

26.1 Photons

You know that if you want to read something more clearly you move nearer to a lamp. The intensity I of the light is greater when you are closer to the lamp. It varies with distance r from a lamp of power P according to the equation

$$I = \frac{P}{4\pi r^2}$$

as we saw for mechanical waves in Chapter 21. This applies strictly only to a point source, but many sources (e.g. a street lamp if we are far enough away from it) approximate to point sources.

Quantisation

The energy of electromagnetic radiation falling on a surface does not arrive in a continuous stream, although it seems as if it does. The energy is emitted, travels and is absorbed in packets called quanta (plural of **quantum**). The size of a quantum is proportional to the frequency of the radiation. If E is the energy of one quantum,

$$E = hf$$

where f is the frequency and h is the **Planck constant**. The value of h is 6.63×10^{-34} J s, and a simple calculation will show you why it is hard to tell that light is quantised. A typical wavelength for yellow light is about 500 nm, so using $c = f\lambda$, we have

$$f = \frac{c}{\lambda} = \frac{3.0 \times 10^{8} \text{m s}^{-1}}{500 \times 10^{-9} \text{m}} = 6 \times 10^{14} \text{Hz}$$

and $E = hf$ gives $E = 4.0 \times 10^{-19}$ J which is a very small amount (about 2.5 eV). However, radiation of higher frequency, such as X-radiation and γ-radiation, has quanta which are much larger: γ-radiation of frequency 10^{19} Hz would have quanta whose energy was of the order of MeV. This is why γ-radiation and X-radiation can be detected as *individual* quanta by Geiger–Müller tubes: each quantum carries enough energy to ionise the gas in the tube. This radiation is much 'lumpier' and we can tell easily that the radiation is travelling in quanta. We use the word **photon** for a quantum. It sounds like the names for other *particles* (protons, neutrons, etc.) and so reminds us that the energy has particle-like properties.

It is important to realise that the energy of a photon cannot be split up: a photon cannot give up some of its energy to a particle and continue on with the rest of it: if it gives up any of its energy it must give up all of it.

EXAMPLE

Two of the wavelengths of the light produced by the mercury vapour in fluorescent tubes are 546 nm (green) and 405 nm (violet). What is the energy carried by photons of these wavelengths? Planck constant $h = 6.63 \times 10^{-34}$ J s; speed of light $c = 3.00 \times 10^{8}$ m s^{-1}.

$$E = hf \text{ and } c = f\lambda, \text{ so } E = \frac{hc}{\lambda}$$

for the green light,

$$E = \frac{(6.63 \times 10^{-34} \text{J s})(3.00 \times 10^{8} \text{m s}^{-1})}{546 \times 10^{-9} \text{m}}$$

$$= 3.44 \times 10^{-19} \text{J}$$

for the violet light,

$$E = \frac{(6.63 \times 10^{-34} \text{J s})(3.00 \times 10^{8} \text{m s}^{-1})}{405 \times 10^{-9} \text{m}}$$

$$= 4.91 \times 10^{-19} \text{J}$$

EXAMPLE

A low-pressure sodium street lamp has a power of 300 W. Assuming that it radiates as a point source, calculate how many photons per second enter the eye pupil (of 4 mm diameter) of a person standing 30 m from the lamp. The wavelength of the sodium light is 589 nm: assume that the lamp is 30% efficient, i.e. that 30% of the energy converted by the lamp is emitted as light. $h = 6.63 \times 10^{-34}$ J s, $c = 3.00 \times 10^{8}$ m s^{-1}.

Rate of emission of energy as light
$= 0.30(300 \text{W})$
$= 90 \text{W} = 90 \text{J s}^{-1}$.

$$\text{Energy of one photon} = \frac{hc}{\lambda}$$

$$= \frac{6.63 \times 10^{-34} \text{J s})(3.00 \times 10^{8} \text{m s}^{-1})}{589 \times 10^{-9} \text{m}}$$

$$= 3.38 \times 10^{-19} \text{J}$$

so number of photons emitted per second

$$= \frac{90 \text{J s}^{-1}}{3.38 \times 10^{-19} \text{J}}$$

$$= 2.67 \times 10^{20} \text{s}^{-1}.$$

Imagine a spherical surface drawn round the lamp with a radius of 30 m. At this distance the photons are passing through a spherical area A given by $A = 4\pi r^2 = 4\pi(30 \text{m})^2 =$

$1.13 \times 10^4 \, \text{m}^2$. The person's eye pupil has an area of $\pi(2.0 \times 10^{-3} \, \text{m})^2 = 1.26 \times 10^{-5} \, \text{m}^2$, so the number of photons which enter the eye per second is

$$\left(\frac{1.26 \times 10^{-5} \, \text{m}^2}{1.13 \times 10^4 \, \text{m}^2}\right)(2.67 \times 10^{20} \, \text{s}^{-1}) = 3.0 \times 10^{11} \, \text{s}^{-1}.$$

Notice the very large number of photons ($>10^{20}$) emitted each second by this street lamp: it is not surprising that we think that the energy is transmitted continuously.

The momentum of a photon

There are other experiments which support the particle-like nature of light. Einstein's mass–energy relationship ($E = mc^2$) can be used to derive an expression for the momentum p of a photon:

If the energy can be written as either mc^2 or hf, then

$$mc^2 = hf$$

But $f = \dfrac{c}{\lambda}$ therefore $mc^2 = h\dfrac{c}{\lambda} \Rightarrow mc = \dfrac{h}{\lambda}$

and the momentum p of a photon is given by

$$p = mc$$

so

$$p = \frac{h}{\lambda}.$$

A collision between a photon and a free electron (i.e. one not bound to an atom) can be perfectly described by thinking of a photon as a particle with this momentum p and energy hf. In this type of

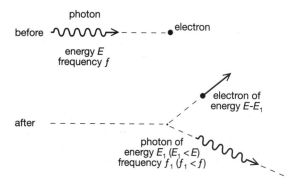

FIGURE 26.1

collision (which is called **Compton scattering**), the result (as shown in figure 26.1) is that the electron moves off with some kinetic energy, and another photon is emitted with less energy than the original photon. Momentum and energy are conserved. It really does seem to be a collision between two *particles*.

26.2 The photoelectric effect

You have seen that an ordinary lamp radiates a very large number of photons each second: so many that it is hard to tell that the energy is quantised. It is, however, possible to use a lamp of low power, place a shield round it so that the light passes through a small hole, and then use filters to absorb all but about one-millionth of the radiation passing through the hole. When this very weak light is examined it is found that individual photons can be detected. The photograph at the start of this chapter showed the result: the photons strike a photomultiplier tube which amplifies the effect until a pulse of p.d. is large enough to be recorded on an oscilloscope screen. What you are seeing in that photograph is the result of a single photon arriving independently of any others.

You would not be able to do that kind of experiment in your laboratory, but you could use one of the experiments which led scientists to the conclusion that light must be quantised. If you use a freshly cleaned zinc plate as the cap of a negatively charged electroscope, as shown in figure 26.2, you will find that the electroscope is discharged immediately if ultra-violet radiation is shone on it. The electroscope is not discharged if some other combinations of properties are used: e.g. if you tried the experiment with a brass cap, negatively charged, it would not be discharged if green light were shone on it.

How can we make sense of this **photoelectric effect**? With a simple 'classical' wave theory, which assumes that energy is delivered continuously, we would expect the electromagnetic wave to deliver energy to the charged particles in the metal, making them oscillate, and giving them enough energy to be able to escape from the metal. So we

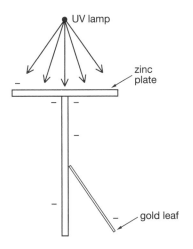

UV lamp

zinc plate

gold leaf

FIGURE 26.2

might expect there to be electrons emitted. But there are several observations which the classical wave theory cannot explain but which the quantum theory can. These are listed below.

- ◆ Observation: if the intensity of the radiation is very low (e.g. about $10^{-10} \, \text{W m}^{-2}$) the rate of production of electrons is low but they are emitted immediately after the radiation is switched on.
  Explanation:
  (i) *Classical* – none, since energy would be delivered to all the electrons equally and it would take days for any one electron to gain enough energy to be able to escape.
  (ii) *Quantum* – one photon delivers all its energy to a single electron, so emission of electrons can start as soon as the first photon arrives.

- ◆ Observation: for a particular frequency, and metal, there is a definite maximum k.e. for the electrons.
  Explanation:
  (i) *Classical* – none.
  (ii) *Quantum* – the electron cannot have more k.e. than the energy of one photon, minus the energy needed to remove the electron from the metal.

- ◆ Observation: below a particular frequency (for a particular metal) no electrons are emitted.

Explanation:
(i) *Classical* – none, since there is nothing 'special' about any particular frequency.
(ii) *Quantum* – the photon energy depends on the frequency, so if the frequency is not high enough to be able to supply enough energy to pull an electron away from the surface, no electrons can be emitted.

There seems no doubt that we need the quantum theory to explain the photoelectric effect.

The photoelectric equation

When a photon arrives on a metal surface and emits an electron we can write

$$hf = \phi + \tfrac{1}{2}mv^2_{\text{max}}$$

where hf is the energy of the photon and ϕ is the work function of the metal. The **work function** is the *minimum* energy needed to remove an electron from the metal, i.e. *from the surface* of the metal. It will require more energy to remove an electron which is just below the surface. So some of the energy of the photon is used to remove the electron, and the rest is available to the electron as kinetic energy. The *maximum* k.e. occurs when the *minimum* energy ϕ is used to remove the electron from the atom: the emitted electrons will have kinetic energies ranging from this maximum down to zero, depending on where they have come from.

Testing the equation

The equation can be tested with a photoelectric cell which consists of an evacuated glass bulb containing a layer of potassium as its cathode and a wire ring as its anode, as shown in figure 26.3(a). When light falls on the potassium surface, electrons are emitted and can travel across the bulb to the wire ring. The photoelectric cell is placed in

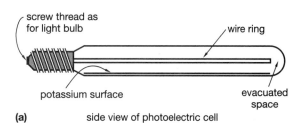

screw thread as for light bulb

wire ring

potassium surface

evacuated space

(a)   side view of photoelectric cell

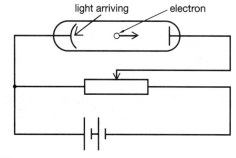

(b)

FIGURE 26.3

the circuit shown in figure 26.3(b), so that the potential divider supplies a p.d. which makes the wire ring negative relative to the potassium surface. Light of a particular frequency is then shone on to the potassium surface. One way of doing this is to use a mercury vapour lamp, which emits light of just four frequencies of visible radiation, and four narrow-band filters, each of which lets light through of just one of the four frequencies. The four frequencies correspond to photon energies of 2.15 eV (yellow–orange), 2.28 eV (green), 2.85 eV (violet) and 3.07 eV (deep violet).

It is difficult to measure the speed and therefore the k.e. of the emitted electrons (but not impossible, using different apparatus), but we can avoid having to do so. Imagine balls, of different kinetic energies, being rolled along a plank of wood, the far end of which can be tilted upwards, as shown in figure 26.4. As the plank is tilted, the balls begin to stop and roll back; this happens when the g.p.e. they gain as they roll up the plank is equal to the k.e. they had when they started.

shown. As this p.d. is increased, more and more of the electrons fail to reach the wire ring: at a certain p.d., all the electrons are stopped and the current falls to zero. Then the k.e. of the fastest electrons has all been converted to electric potential energy. The p.d. at which this happens is called the **stopping potential** V_S, and the electric p.e. of the fastest electrons will then be eV_S. So we can rewrite the equation as

$$hf = \phi + eV_S \quad \text{or} \quad V_S = \frac{h}{e}f - \frac{\phi}{e}$$

If we take measurements of V_S for the four different frequencies f we can plot a graph of V_S against f and expect to find a straight line of slope h/e and an intercept of $-\phi/e$ on the y-axis, as shown in figure 26.5. The intercept on the x-axis is the minimum frequency f_0 at which electrons will be emitted from the metal being tested: this is called the **threshold frequency**. Then $V_S = 0$ and so $hf_0 = \phi$. The value of f_0 will, of course, be different for different metals, but all graphs will have the same slope h/e, whichever metal is used. Some graphs for different named metals have been shown: notice that for zinc, the minimum frequency is in the ultra-violet part of the spectrum.

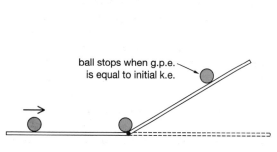

FIGURE 26.4

Eventually, as the plank is tilted more, all the balls are stopped. This is like what happens in the photoelectric cell which has a p.d. applied as

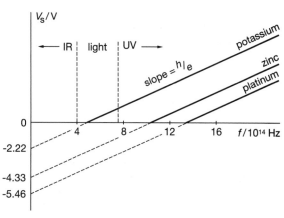

FIGURE 26.5

EXAMPLE

Light of wavelength 436 nm is shone on to a caesium surface and it is found that a p.d. of 0.71 V is needed to stop the current flowing. If $h = 6.63 \times 10^{-34}$ J s, $e = 1.60 \times 10^{-19}$ C and $c = 3.00 \times 10^8$ m s^{-1}, find the work function of caesium (in eV), the threshhold frequency, and the maximum wavelength which can cause the emission of electrons.

The frequency f of the radiation is given by

$$f = \frac{c}{\lambda} = \frac{3.00 \times 10^8 \, \text{m s}^{-1}}{436 \times 10^{-9} \, \text{m}} = 6.88 \times 10^{14} \, \text{Hz}$$

The work function $\phi = hf - eV_S$
$$= (6.63 \times 10^{-34} \, \text{J s})(6.88 \times 10^{14} \, \text{Hz})$$
$$- (1.60 \times 10^{-19} \, \text{C})(0.71 \, \text{V})$$

$$= 3.43 \times 10^{-19} \, \text{J} = \frac{3.43 \times 10^{-19} \, \text{J}}{1.60 \times 10^{-19} \, \text{J eV}^{-1}} = 2.14 \, \text{eV}$$

$hf_0 = \phi$ so the threshhold frequency

$$f_0 = \frac{\phi}{h} = \frac{3.43 \times 10^{-19} \, \text{J}}{6.63 \times 10^{-34} \, \text{J s}}$$

$$= 5.17 \times 10^{14} \, \text{Hz}$$

This is the minimum frequency: the maximum wavelength λ_{max} is given by

$$\lambda_{\text{max}} = \frac{c}{f_0} = \frac{3.00 \times 10^8 \, \text{m s}^{-1}}{5.17 \times 10^{14} \, \text{Hz}}$$

$$= 5.80 \times 10^{-7} \, \text{m} = 580 \, \text{nm}$$

Light of wavelength 580 nm is an orange–yellow colour, so photons of red light or infra-red radiation will not have enough energy to liberate electrons from caesium.

26.3 Energy levels

Figure 26.6 shows a continuous spectrum of the visible radiation produced by a filament lamp, and the line spectrum produced by hydrogen gas in an electrical discharge tube. You can see that the hydrogen gas produces just a few wavelengths in the visible part of the spectrum. Why is this, and what decides what the wavelengths will be? You

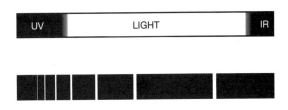

FIGURE 26.6

know that the frequency of electromagnetic radiation tells us how much energy the photons have. So let us look at the red line in the hydrogen spectrum. We can work out the energy of the photons of red light, given that the wavelength is 656.3 nm:

$$E = hf = \frac{hc}{\lambda} = \frac{(6.625 \times 10^{-34} \, \text{J s})(2.998 \times 10^8 \, \text{m s}^{-1})}{656.3 \times 10^{-9} \, \text{m}}$$

$$= 3.026 \times 10^{-19} \, \text{J} = 1.889 \, \text{eV}$$

Why do these photons have precisely this energy? It is possible for atoms to be **excited**, i.e. to have their energy increased from what they have in their normal or ground state, because one of the electrons has moved up from one orbital to an orbital where it has more energy. This **excitation** can happen in a number of ways: violent collisions between the atoms, or bombardment by fast-moving electrons, or irradiation by photons which have just the right amount of energy. (If a lot of energy is given to an electron, it may leave the atom altogether: this is called **ionisation**.)

There are only a few sharply defined states of energy in which an atom can exist: i.e. it can have energies E_1, E_2, E_3 etc., but no others. So an atom can absorb energy only in very precisely defined amounts, just as someone climbing a ladder can stop in only a few places (when standing on a rung) and have just a few particular values of gravitational p.e. If the energy of the atom changes, e.g. from a state with energy E_2 to a state with energy E_3, it will absorb an amount of energy $E_3 - E_2$. In this excited state it is unstable, and the electron will soon drop back to its original level, and the energy which it is losing is emitted in the form of electromagnetic radiation. Since the energy is precisely defined, so is the frequency of the radiation.

The zero for potential energy

If we want to talk about an electron having a certain amount of electric potential energy we have to choose to say where the electric potential energy is zero. In this situation it is usual to say that it is zero when it has been pulled completely away from the atom, i.e. when the atom has been ionised. Since the potential energy of the electron increases as it moves further away from the centre of the atom, it must have less than zero anywhere else, i.e. at all other places its potential energy is negative. (Just the same sort of thing would happen if we chose to say that the zero for gravitational potential energy was the ground floor of a building: then a mass of 5 kg taken down 3 m into the basement would have about −150 J of g.p.e.)

EXAMPLE

Some of the energy levels for the mercury atom are −2.67 eV, −3.70 eV, −5.51 eV and (the ground state) −10.40 eV. Describe what might happen if cool mercury vapour is bombarded with (a) electrons whose k.e. is 3.00 eV, (b) electrons whose k.e. is 5.00 eV, (c) light of wavelength 253 nm, (d) light of wavelength 300 nm.

The diagram shows a sketch of the energy levels. If the mercury vapour is 'cool', nearly all the atoms will be in the ground state. Assume that they all are.

(a) The electrons cannot give any of their energy to the atom, since the atom needs at least 4.89 eV to raise it from the ground state to the next highest energy level. So if the moving electrons strike the electrons in the atoms, they make elastic collisions and move on, keeping all their k.e. of 3.00 eV.

(b) The electrons can give 4.89 eV to the mercury atom; if they make a collision in which this happens, they will come away with a k.e. of 0.11 eV. Each atom which has gained this energy will soon return to the ground state, emitting a photon of energy 4.89 eV.

(c) Light of wavelength 253 nm: using $c = f\lambda$, we find $f = 1.186 \times 10^{15}$ Hz, and using $W = hf$ we find that $W = 7.83 \times 10^{-19}$ J $= 4.89$ eV. So the light has exactly the right wavelength to be absorbed by the atoms. The photons which meet electrons may therefore be absorbed by the atoms. As in (b), each atom which has gained this energy will soon return to the ground state, emitting a photon of energy 4.89 eV.

(d) Light of wavelength 200 nm has photons of energy 6.19 eV. This is too much energy to raise the atom to the next energy level, and too little to raise it to the energy level above that. A photon must transfer all or none of its energy, so these photons are not absorbed.

0 ——————————

-2.67 eV ——————————
-3.70 eV ——————————

-5.51 eV ——————————

-10.40 eV ——————————— ← assume all atoms in ground state

Energy levels for the hydrogen atom

Figure 26.7 shows some of the energy levels for the hydrogen atom marked out according to this idea: in the ground state (n = 1) the electron has −13.598 eV of electric potential energy. You can see now how the photon of red light comes to be emitted from excited hydrogen: 1.889 eV is exactly the difference between the energies of the levels for which n = 2 and n = 3. The arrows on the energy level diagram show the possible energy changes. There is a series of changes which end at

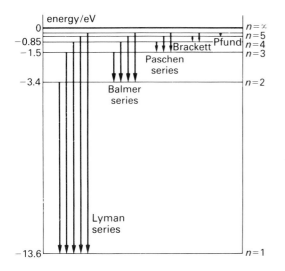

FIGURE 26.7

the level for which n = 1: these energy changes are all relatively large, and produce photons in the ultra-violet range. The series which ends at the level for which n = 2 includes the energy difference 1.889 eV which we have already identified with the red line in the hydrogen spectrum: the other members of this series all have larger energy differences, and hence higher frequencies and shorter wavelengths. At least three give photons of visible radiation. Others produce photons in the ultra-violet part of the spectrum. There are other series all of which are in the infrared part of the spectrum.

Hydrogen is the simplest atom, so its line spectrum is relatively simple too. But the same principles apply to all atoms. Examining the line spectra of excited gases gives us a vast amount of information about the structure of the atom. And since the wavelengths can be measured very precisely (often to 6 significant figures) the information about the energy levels is also very precise.

There is, however, other evidence that these energy levels exist. Many different types of experiment have been done in which electrons have been accelerated through known p.d.s (so that the electrons have had known amounts of kinetic energy). The electrons have then struck atoms and excited them, and measurements of the electrons' kinetic energy after the collisions tell us how much energy was absorbed by the atom, and

therefore what at least some of the energy differences are for that atom.

EXAMPLE

The hydrogen atom contains the following energy levels (among others): $-3.400\,\text{eV}$, $-1.511\,\text{eV}$, $-0.849\,\text{eV}$, $-0.544\,\text{eV}$ and $-0.378\,\text{eV}$.

(a) How many different lines in the emission spectrum would be produced by energy changes between these levels?
(b) Which energy change would have the shortest wavelength, what is that wavelength, and in what part of the spectrum does it lie?
($c = 2.998 \times 10^8\,\text{m s}^{-1}$, $h = 6.625 \times 10^{-34}\,\text{J s}$, $1\,\text{eV} = 1.602 \times 10^{-19}\,\text{J}$.)

(a) An electron can fall from $-0.378\,\text{eV}$ to 4 other levels, from $-0.544\,\text{eV}$ to 3 other levels, from $-0.849\,\text{eV}$ to 2 other levels, and from $-1.511\,\text{eV}$ to one other level: i.e. there are 10 possible changes.
(b) The shortest wavelength corresponds to the highest frequency and therefore to the largest energy change, which is
$(-0.378\,\text{eV}) - (-3.400\,\text{eV}) = 3.022\,\text{eV}$.

$$E = hf \text{ and } c = f\lambda, \text{ so } \lambda = \frac{hc}{E}$$

$$= \frac{(2.998 \times 10^8\,\text{m s}^{-1})(6.625 \times 10^{-34}\,\text{J s})}{(3.022\,\text{eV})(1.602 \times 10^{-19}\,\text{J eV}^{-1})}$$

$$= 4.10 \times 10^{-7}\,\text{m} = 410\,\text{nm}.$$

This is violet light, just within the boundary of the visible spectrum.

26.4 X-rays

X-ray tubes

Figure 26.8 shows a photograph of an X-ray tube and a diagram of its essential parts. It consists of an evacuated glass envelope with a heated tungsten filament which emits electrons. A p.d. of between 20 kV and 1 MV is used to accelerate the electrons along the tube so that they strike an anode. The

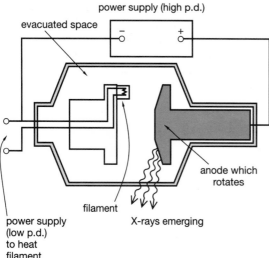

FIGURE 26.8

X-ray line spectra

When the electrons strike the anode several different things may happen. Most (perhaps 99%) of the electrons slow down gently when they strike the target because they do not come very close to the nuclei of the atoms in the target. It is these low-energy interactions which convert most of the electrons' kinetic energy into internal energy. But the only effect of this that the target becomes hot, and this process does not produce X-radiation.

A more important effect is that the electron may penetrate the atoms of the target material and knock an electron out of one of the electronic shells. The anode is usually made of a metal with a high atomic number so that there are many electrons per unit volume, which makes it more likely that one of the accelerated electrons will meet an electron in the anode. Because the electrons have been accelerated through a large p.d., they have enough energy to knock electrons out of even the inner shells of the atom, as shown in figure 26.9. These shells are labelled K, L, M, etc., K being the innermost. So a gap then exists in one of the inner shells, and one of the electrons in an outer shell is pulled in to fill the gap. When this happens, it loses energy, which is given out as a photon, just as excited gases give out photons when electrons fall back to their original positions.

movement of charge means that there is an electric current in the tube. The size of the current depends on the rate of production of electrons emitted by the filament: it is typically between 10 mA and 100 mA. Most of the kinetic energy of the electrons is converted into internal energy in the anode. The anode is therefore made of tungsten, which has a high melting-point, but in addition the anode is rotated at about 3000 r.p.m. so that it is not always the same spot on the anode that is being struck by electrons. The central core of the anode is made of copper so that the internal energy can be conducted away to the oil which surrounds the tube. In tubes where the p.d. is higher, and the amount of energy generated therefore greater, oil is pumped through the copper core to cool it. The whole tube is surrounded by a lead-lined steel casing to protect users. The X-rays emerge through a small window.

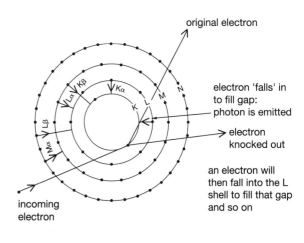

Not to scale

FIGURE 26.9

374

There will then be a gap where that electron came from, so another electron will be pulled in to fill *that* gap, and so on. Each movement inwards of an electron causes the emission of a photon. But the sizes of the energy changes between the innermost levels are much greater, so these photons have much more energy. They have enough to give them a frequency of between 10^{18} Hz and 10^{20} Hz, which corresponds to a wavelength of 10^{-10} m to 10^{-12} m.

A **line spectrum** is produced, since there are just a few possible energy changes, so just a few particular frequencies, as shown in figure 26.10.

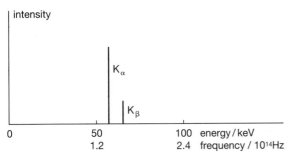

FIGURE 26.10

Here we use the word *spectrum* differently from the way in which it is used with light, since we cannot 'see' the X-radiation: here we use the word spectrum to describe a graph which shows how the intensity of radiation varies with either wavelength or frequency. The peaks of intensity are still called 'lines', to correspond with the lines of particular wavelength or frequency in a spectrum of light. Most of these X-ray photons are produced when electrons move down to the K shell, and the lines are therefore called K lines: the two commonest, caused by electrons moving down from the L and M shells, are called the K_α and K_β lines, as shown in figure 26.10. There are also L and M lines, as shown in the figure.

X-ray continuous spectrum

If an electron does not make a direct hit on one of the electrons it may still come so close to the nucleus that a large force is exerted on it and it has a large acceleration, either because it slows down or because it changes direction. When charged particles accelerate they emit electromagnetic radiation. Here this radiation is called **bremsstrahlung** (German for 'braking radiation'). The braking may be rapid or gentle: the more rapid it is, the more energetic the photon emitted, but of course the energy of the photon cannot be greater than the energy of the electron. So if the p.d. in the tube is 50 kV, the electrons will have a kinetic energy of 50 keV, and the photons may have any energy up to, but not greater than, 50 keV. One electron may pass near several nuclei, and may therefore cause the emission of several photons, but the total energy of these photons cannot be greater than the k.e. of the electrons. This effect gives a continuous spectrum of X-radiation, as shown in figure 26.11, which has a maximum frequency which depends only on the p.d. in the tube.

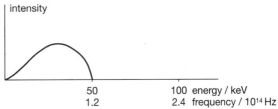

FIGURE 26.11

Complete X-ray spectra

The X-radiation from a particular target material therefore consists of a line spectrum superimposed on a continuous spectrum. The *line spectrum* is characteristic of the target material: the extent of the *continuous spectrum* depends on the p.d. used in the X-ray tube. Figure 26.12 shows the complete spectrum for (a) a molybdenum target at 50 kV, (b) a tungsten target at 50 kV, and (c) a tungsten target at 100 kV. There are several points to notice about these spectra.

◆ The sharp cut-off at a particular frequency for all the spectra.

◆ The maximum frequency is proportional to the X-ray tube p.d.

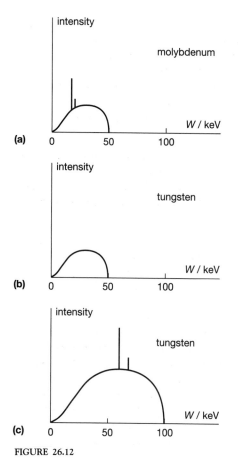

(a)

(b)

(c)

FIGURE 26.12

- The positions of the K-lines for molybdenum do not depend on the X-ray tube p.d., since they are characteristic of the *element*.

- The K-lines for molybdenum occur at lower energy than those for tungsten: this is because molybdenum's atomic number is 42 (tungsten's is 74) so the charge on the nucleus is less, and so the corresponding energy changes are smaller.

- The K-lines for tungsten do not appear when the tube p.d. is 50 kV: the electrons then do not have enough energy to remove electrons from the K shells in tungsten.

EXAMPLE

An X-ray tube uses a p.d. of 100 kV and a current of 70 mA, which is switched on for 0.20 s.

(a) How much energy is delivered to the anode?
(b) The beam strikes an area which is about 2 mm square. Assume that the volume heated, if the anode does not rotate, is 1.0 cm^3. What temperature rise is produced? (Density of tungsten $= 19400\,\mathrm{kg\,m^{-3}}$, s.h.c. $= 142\,\mathrm{J\,kg^{-1}K^{-1}}$.)
(c) What is the maximum energy of the photons emitted and what is their frequency? Deduce the corresponding minimum wavelength? ($h = 6.6 \times 10^{-34}\,\mathrm{J\,s}$, $c = 3.0 \times 10^8\,\mathrm{m\,s^{-1}}$, $1\,\mathrm{eV} = 1.6 \times 10^{-19}\,\mathrm{J}$.)

(a) Energy delivered $=$ power $\times$ time

$= VIt = (100 \times 10^3\,\mathrm{V})(70 \times 10^{-3}\,\mathrm{A})(0.20\,\mathrm{s})$

$= 1400\,\mathrm{J}$

(b) Mass of tungsten $= \rho V$
$= (19400\,\mathrm{kg\,m^{-3}})(1.0 \times 10^{-6}\,\mathrm{m^3}) = 0.0194\,\mathrm{kg}$

$$\text{Temperature rise} = \frac{\Delta Q}{cm}$$

$$= \frac{1400\,\mathrm{J}}{(142\,\mathrm{J\,kg^{-1}K^{-1}})(0.0194\,\mathrm{kg})} = 508\,\mathrm{K}$$

(c) Maximum photon energy

$=$ k.e. of electron
$= 100\,\mathrm{keV}$
$= (100 \times 10^3\,\mathrm{V})(1.6 \times 10^{-19}\,\mathrm{J\,eV^{-1}})$
$= 1.6 \times 10^{-14}\,\mathrm{J}$

Maximum frequency

$$f = \frac{E}{h} = \frac{1.6 \times 10^{-14}\,\mathrm{J}}{6.6 \times 10^{-34}\,\mathrm{J\,s}} = 2.4 \times 10^{19}\,\mathrm{Hz}$$

Minimum wavelength

$$\lambda = \frac{c}{f} = \frac{3.0 \times 10^8\,\mathrm{m\,s^{-1}}}{2.4 \times 10^{19}\,\mathrm{Hz}} = 1.3 \times 10^{-11}\,\mathrm{m}$$

In part (b) the volume heated by the electrons might in practice be less than 1.0 cm^3: the figure given in the question allows for some conduction by the tungsten, which is about half as good a conductor as copper. You can see that if the anode were stationary there would be unacceptably large temperature rise.

Note in part (c) that a maximum frequency corresponds to a minimum wavelength: the photons cannot have a wavelength of less than 1.2×10^{-11} m. Graphs of X-ray intensity (like those in figure 26.13) are sometimes drawn with wavelength on the x-axis. You should try drawing the I against λ graph which corresponds to figure 26.13(c): it will have a sharp cut-off showing minimum wavelength, and tail off indefinitely towards the higher wavelengths.

Using X-rays

♦ *Medical diagnosis*. X-rays penetrate matter: this is the principle of the X-ray process which most of us have undergone in a hospital. X-rays are passed through the patient, and a shadow cast on photographic film on the far side. Different kinds of tissue absorb radiation to different extents, and bone absorbs particularly well, so the film records shadows behind the patient which indicate the nature of the interior of the body. In this way bone fractures can be examined and tumours discovered. A more sophisticated technique, called **computed tomography** (CT), gives a picture of the interior *in one plane* within the patient's body: figure 26.13 shows a CT scan of a human head. You can see a cross-section of the eyeballs at the top of the photograph.

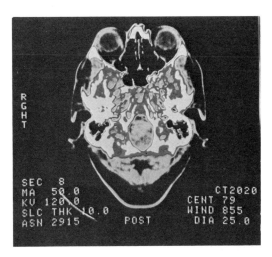

FIGURE 26.13

♦ *Medical therapy*. All cells in the human body can be damaged by X-radiation, so our exposure to it must be limited. But equally X-rays can be used to deliver energy to destroy malignant cancer cells, especially as these are particularly susceptible to damage by X-radiation. For these purposes much higher p.d.s (of up to 40 MV) are used.

♦ *Industrial uses*. X-rays can be used to examine structures for defects in manufactured products, such as cylinder blocks, pistons and connecting rods in engines. (γ-rays can also be used.)

♦ *Crystallography*. The spacings between layers of atoms in solids are about the same size as the wavelengths of X-rays, so X-rays which pass through, or are scattered from the surface of, crystalline materials show diffraction and interference effects. If the K-lines (of known wavelength from a particular target) are used, information can be obtained about the spacing of the layers in crystals, and hence about the size of the atoms and the structure of the crystals. The ability of crystalline structures to diffract waves whose wavelength is of the order of 10^{-10} m will be mentioned again later.

26.5 Waves and particles

What is light?

In this chapter we have seen how electromagnetic radiation can be thought of as having particle properties. The radiation comes in packets of a certain size. But earlier in the book we saw that the radiation also had wave properties: electromagnetic radiation diffracts, superposes to give interference patterns, and does all the things that waves do. So is the radiation particles or waves? Figure 26.14 shows an experiment in which it seems to be both: radiation from a mercury vapour lamp passes through a diffraction grating and the light of different colours emits electrons (of different k.e.) from a potassium surface in a photoelectric cell. The direction in which the light travels is decided by its wavelength; the energy of the emitted

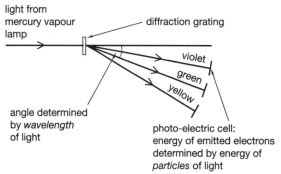

light from
mercury vapour
lamp

diffraction grating

violet

green

yellow

angle determined
by *wavelength*
of light

photo-electric cell:
energy of emitted electrons
determined by energy of
particles of light

FIGURE 26.14

electrons is decided by the energy of the 'particles' which we call photons. How can light be both wave and particle? It does not seem to make sense. The mistake lies in trying to think of light as if it was something like other kinds of wave we know (like a water wave) or other kinds of particle (like a tennis ball). Light is not like anything else.

To make it clearer that light does not behave like anything else we know, think about this situation. Very dim light travels towards a double slit, on the far side of which is some photographic film, as shown in figure 26.15. As expected, the typical two-slit interference pattern is recorded on the film. But the light is so dim that there is never more than one photon at a time in the space between the slits and the film. So how can superposition take place? How can the photons superpose when they are by themselves? If a photon passes through one slit, how does it 'know' that there is another slit nearby which decides where the bright and dark places are in the interference pattern? What decides that the photon will go to *one* of the places where light is recorded, rather than another? These are all impossible questions to answer if we think of the

wave-nature of light is being like the wave-nature of water waves, and the particle-nature of light as being like the particle-nature of tennis balls. Light (and the rest of the electromagnetic spectrum) is like nothing else that we know about.

Electrons have wave properties

Earlier in the chapter we saw that the wavelength λ of a photon is given by

$$\lambda = \frac{h}{p}$$

where h is the Planck constant and p is its momentum. In 1924 Louis de Broglie speculated that if light, which had once been thought of as a wave, could be thought of as a particle, then also a particle might be thought of as a wave, with its wavelength given by the same equation $= h/p$. The example which follows will show what sort of wavelengths might be expected.

EXAMPLE

An electron is accelerated through p.d.s. of (a) 5000 V, (b) 50 V. What is its wavelength? Assume $\lambda = h/p$ for the electron. (Electronic charge $= 1.6 \times 10^{-19}$ C, mass of electron $= 9.1 \times 10^{-31}$ kg, Planck constant $= 6.6 \times 10^{-34}$ J s.)

(a) For the accelerated electron $eV = \frac{1}{2}mv^2 \Rightarrow$
$$2meV = (mv)^2$$

So $p = mv = \sqrt{(2meV)}$
$= \sqrt{(2 \times 9.1 \times 10^{-31}\,\text{kg})(1.6 \times 10^{-19}\,\text{C})}$
$\qquad\qquad\qquad\qquad\qquad (5000\,\text{V})$
$= 3.82 \times 10^{-23}\,\text{N s}$

$$\lambda = \frac{h}{p} = \frac{6.6 \times 10^{-34}\,\text{J s}}{3.82 \times 10^{-23}\,\text{N s}} = 1.7 \times 10^{-11}\,\text{m}$$

(b) If the p.d. is 50 V, i.e. one hundredth as much, p is only one-tenth as much, so the wavelength is 10 times greater, i.e. 1.7×10^{-10} m.

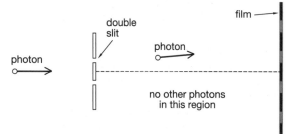

film

double
slit

photon

photon

no other photons
in this region

FIGURE 26.15

So if the equation $\lambda = h/p$ holds for electrons, the wavelength associated with electrons accelerated through p.d.s of this size would be similar to that of X-radiation. Crystalline structures (with the spacing between atomic layers of the order of 10^{-10} m) could be used to test whether electrons had wave properties, and to measure the associated wavelength. A year after de Broglie's suggestion, Davisson and Germer, using 54 eV electrons striking a nickel crystal, found that electrons were indeed diffracted in such a way as to support the relationship $\lambda = h/p$.

An electron diffraction experiment

You can observe electron diffraction in your own school or college laboratory. A beam of electrons, accelerated through a p.d. of a few kV, is arranged to hit a thin film of graphite (i.e. carbon) in an evacuated tube, as shown in figure 26.16(a). Many of the electrons arrive near the centre of the screen but others arrive at two or three particular distances from the centre of the screen, as shown by the rings in the photograph of the end of the tube in figure 26.16(b). The only way in which we can explain this observation is to think that the electrons must have wave properties, and that they are diffracting and superposing as they pass through

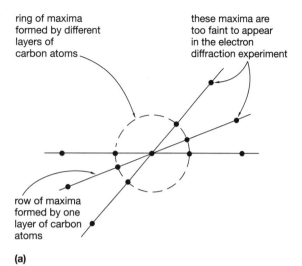

(a)

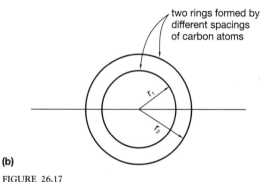

(b)

FIGURE 26.17

the graphite. The graphite has a crystalline structure and the gaps between the rows of carbon atoms must be acting as slits in a diffraction grating. With a beam of light (e.g. from a laser) passing through an optical diffraction grating you would expect a row of 'spot' maxima in a line, as in figure 26.17(a). In the electron diffraction experiment only the first-order maxima, nearest the centre, are visible but the graphite film contains many layers, with different orientations, of carbon atoms and so is more like many diffraction gratings laid on top of each other so that you get the spots out along many lines, and not just one. The many spots together make up a ring, as shown in figure 26.17(a).

However, unlike a diffraction grating, where there is only one spacing of the slits, there is more than one spacing of carbon atoms in the graphite, and each of these spacings causes another ring: this

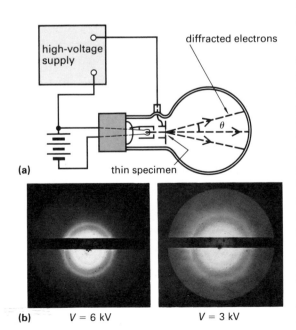

FIGURE 26.16

is why there is more than one ring.

For a particular measured accelerating p.d. V we can measure the radii r of the two prominent rings. The graphite film is approximately at the centre of the spherical surface of the screen of the tube: let us call this distance from the graphite film to the screen R. Then $\theta = r/R$. For a diffraction grating we also know that the angle θ at which there is a maximum is given by $\sin\theta = \lambda/s$, where s is the spacing of the slits, so making the approximation $\sin\theta \approx \theta$ (the angles are $<15°$) we have

$$\frac{r}{R} = \frac{\lambda}{s}$$

So we would expect to find $r \propto 1/s$. The two largest spacings s between rows of carbon atoms in graphite are 2.13×10^{-10} m and 1.23×10^{-10} m, i.e. a ratio of $2.13/1.23 = 1.73$. When you do the experiment you will find that the ratio of the radii of the rings *is* approximately 1.73. You could repeat the experiment using different accelerating p.d.s. The radii alter but the ratio remains the same.

When V changes the radii change because the wavelength changes. We have

$$\lambda = \frac{h}{p} = \frac{h}{mv} = \frac{h}{\sqrt{(2meV)}}$$

so since $r \propto \lambda$, $r \propto \dfrac{1}{\sqrt{V}}$

The table shows some sample results (the *diameter* d of the rings was measured to achieve greater precision):

V/kV	3.0	4.0	5.0
d/mm	34	28	26

You would ideally like more readings, with smaller p.d.s. to get a greater range of values, but the intensity of the rings on the screen is too low with lower p.d.s. However, if you plot a graph of r against $1/\sqrt{V}$ you will probably agree that the points lie on a best-fit straight line which passes through the origin, thus showing that $\lambda \propto 1/p$.

What are electrons?

The section began with the question 'What is light?' You are probably now wondering 'What are

electrons?' The question should really be 'What are particles?' since protons, neutrons and other 'particles' also can be shown to diffract. You used to think of such things as particles. They certainly seem to behave like particles most of the time. Even in the electron diffraction experiment which has just been described, the electrons were accelerated like particles in the tube: we could use $F = ma$ to calculate their acceleration. All we can say is that it is a mistake to want to try to imagine a particle (like an electron) as if it was like anything else that we know about. It seems that when we are dealing with the world of very small objects there are additional laws which govern their behaviour. When we are dealing with the forces on particles like electrons, and their energy changes, we can treat them as particles, but when we want to know where they are, we have to use wave ideas, just as we do with photons. The phrase **wave–particle duality** is used to describe the two-sided nature of electromagnetic radiation, and the two-sided nature of particles.

Similarities and differences

Figure 26.18 shows, side by side, photographs of diffraction patterns taken by passing (a) X-rays and

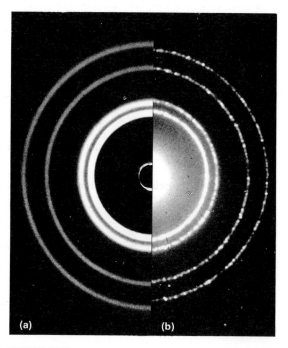

(a)　　　(b)

FIGURE 26.18

(b) electrons through the *same* very thin aluminium foil. It was arranged that the X-rays and the electrons would have the same wavelength, and you can see that the X-rays and the electrons diffract identically. But do not think that X-rays and electrons are the same thing. X-rays are electromagnetic waves: they all travel through a vacuum with the same high speed. They are very penetrating. Electrons are constituents of atoms: they have different speeds in empty space, and even fast-moving electrons (e.g. β-particles) travel through only a few mm of aluminium.

Exercises on each section of this chapter may be found in the companion textbook, ***Practice in Physics***.

SUMMARY

At the end of this chapter you should be able to:

♦ use the equation $I = P/4\pi r^2$ to calculate intensity of radiation.

♦ understand that the energy E of electromagnetic radiation is quantised according to the equation $E = hf$.

♦ understand how quantisation explains the photoelectric effect.

♦ use the photoelectric equation $hf = \phi + \frac{1}{2}mv_{max}^2$.

♦ understand that the maximum k.e. $\frac{1}{2}mv_{max}^2$ may be equated to the electrical potential energy eV_S, where V_S is the stopping potential.

♦ understand that an atom can exist in a few sharply-defined states of energy because there are only certain 'positions' where its electrons may be.

♦ understand that the energy of an atom may be changed by bombardment by electrons or photons, or by the emission of photons.

♦ calculate the frequency of radiation emitted when an electron moves from one energy level to another.

♦ understand the principle of the production of X-radiation.

♦ understand why, for a particular accelerating p.d., there is a maximum frequency (and minimum wavelength) for the X-radiation.

♦ understand that we cannot describe light in terms of other phenomena (i.e. it is not 'like' water waves or bullets).

♦ understand that just as light has particle properties, so electrons (and other particles) have wave properties, and diffraction experiments are evidence for this.

♦ use the equation $\lambda = h/p$ to calculate the wavelength of a particle.

27 Electronics

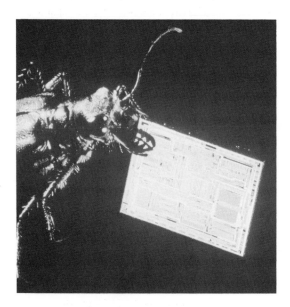

Richard Dawkins, in his book **The Selfish Gene**, suggests that human beings are little more than carriers for DNA and that it is the self-replicating DNA molecule which is the winner in the fight for survival. In the concluding chapter he throws out the idea that at some future time silicon chips might take over the lead in the evolutionary race. Looking at the photograph, could it be that integrated circuits will provide the social 'beings' of the future and that beetles, along with other animals, will be seen as filling an unimportant niche or even, like the dodo, failing to survive at all?

This may seem in the realms of fantasy, but the development of integrated circuits since the late 1960s has been almost as unbelievable, especially when you realise that the first transistor, made by Bardeen, Brattain and Shockley in 1947, was at first believed to be commercially unproducible because of the incredibly high degree of chemical purity required in its manufacture.

27.1 Analogue and digital systems

Electronics depends on an understanding of the physics you have met elsewhere in this book, particularly basic circuit theory. But it is the way in which we use **electronic systems** to design useful devices that makes electronics so important in today's world. Electrical and electronic sensors can detect and measure a vast range of physical quantities such as light intensity or humidity. Electronic devices can also control machines and robots and can process information as they do in hand-held calculators and personal computers.

However, the jump from understanding the physics of a light-dependent resistor (LDR) to designing a system for monitoring and storing data about sunlight is a huge one and we are *not* going to pursue the detailed physics (it is called solid state physics) of electronic devices in this chapter.

Figure 27.1 (notice the 40 μm scale) shows the detailed structure of the surface of a typical **integrated circuit** (i.c.), a surface constructed from layers of semiconducting material which can be deposited to form resistors, diodes, capacitors and

FIGURE 27.1

transistors. Such i.c.s are the tools of modern electronics and you will learn to *use* systems such as operational amplifiers and NAND gates without asking how they are designed, i.e. without being concerned with how you would make one for yourself out of semiconducting materials.

Transducers

FIGURE 27.2

The quantities which we want to measure in physics – pressure, temperature, magnetic flux density etc. – all vary continuously. The output (it is most useful as a variation in voltage) of a **sensor** used to detect one of these quantities we describe as an **analogue signal**. Three sensors or transducers are illustrated in figure 27.2. The microphone (left) detects pressure waves in air; its output is an alternating voltage with peaks of a few hundred μV. The bead thermistor (centre) has a resistance which varies with temperature. Its variation, a few tens of ohms per kelvin, can readily be converted to a variation of voltage of a few mV K^{-1}. The Hall probe (right) measures the magnetic field, typically a few hundred millitesla, perpendicular to its surface; the voltage between its terminals might be a few mV. These voltage variations do not jump about, they vary smoothly as the physical condition they are sensing varies. We will need to learn how to amplify their outputs before they can drive an analogue meter or a chart recorder.

If we want to store or to process information such as the output of these transducers, it is often best to convert it to digital form, i.e. to convert an analogue voltage into a series of 1s and 0s representing a binary number. A 4-bit **analogue to digital converter** (ADC) has one input line and four output lines: these output lines may be at either 5 V (which represents the binary digit 1) or 0 V (which represents 0). For the maximum input voltage for which it is designed all four output lines are at 5 V, i.e. the highest binary number the output can show is 1111, which is 15 in decimal notation (you should check that you understand why) and so to use this 4-bit ADC the input voltage is sampled and put into one of 16 states, the lowest state might, for example, cover input voltages in the range 0–1 mV and the output would be 0000, the range 5 mV to 6 mV would be 0101 etc., the maximum range being obviously 15–16 mV – see page 410.

This sampling will need to be done many times at regular intervals in order to follow the varying voltage produced by a typical transducer. Just how often of course depends on how quickly the pressure, temperature, magnetic field, etc. is varying. The binary numbers can then be displayed or stored and can be moved around by electronic devices, e.g. computers, which deal only in one of two voltage states, high (1) or low (0).

Block diagrams

Figure 27.3 shows, in the crudest way, how we represent an electronic system by using block diagrams. These show how information flows through the circuits with the words inside the boxes describing the *function* of each block; what it does to the signals fed to it.

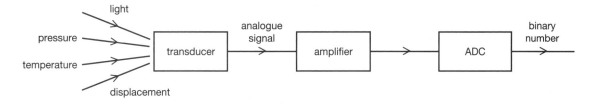

FIGURE 27.3

Figure 27.3 tells us that we need to learn about transducers and amplifiers – these are part of analogue or linear electronics. The decoders and display units which form the numbers you read from your calculator are part of the **digital** electronics; we will not develop these very far in the rest of the chapter. They form the basis of courses in computer science or specialist electronics and do not form a part of most physics courses.

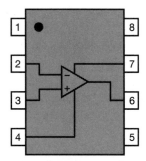

(a)

Electronic jargon

Like many branches of technology electronics has developed words and phrases which have a clear meaning to the expert but are mysterious when you first meet them. We will try in this chapter to avoid using all but the most common ones and concentrate on using the language of physics. We will, however, use the word voltage when we mean potential difference. Thus, for example,

input voltage (V_{in})
= p.d. between input terminal and $0\,V$

voltage at $P\,(V_p)$
= p.d. between a point P and $0\,V$

etc. You will quickly get used to this.

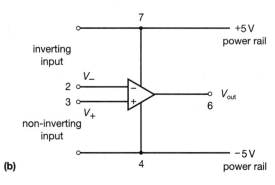

(b)

FIGURE 27.4

27.2 Inverting amplifiers

The operational amplifier

The operational amplifier or **op-amp** is the basic component for processing analogue signals.

A typical op-amp, the TL081 (figure 27.4), is a complex circuit of transistors, resistors and diodes enclosed in a black plastic box which has eight terminals or pins. Two, numbers 4 and 7, are for connecting to power supply rails which can range from $\pm 3\,V$ to $\pm 18\,V$ and must be steady. We shall use $\pm 5\,V$ throughout this chapter. There are two input terminals: number 3, called the non-inverting, and number 2, called the inverting input. There is one output terminal, number 6. We will mention the use of the other terminals as the need arises. The triangular shape is the circuit symbol for an amplifier.

Open loop gain

Op-amps amplify the potential differences between V_+ and V_-, the two inputs. This amplification can be represented by the equation

$$V_{out} = A(V_+ - V_-)$$

where A is called the open-loop gain of the op-amp: it has no units. (It is sometimes written A_{OL}.) For a TL081, A might be 200 000, a huge gain. If the inverting input is earthed, V_- is zero, and then $V_{out} = AV_+$; if the non-inverting input V_+ is earthed, i.e. $V_+ = 0\,V$, then $V_{out} = -AV_-$, which is why we call the inputs by the name given. Do not confuse the minus in V_- with the value of the voltage there; V_- could be $-2.0\,\mu V$, which with V_+ earthed would give

$$V_{out} = -(200\,000)(-2.0 \times 10^{-6}\,V)$$

$$= +0.4\,V$$

One further point before we get onto useful circuits involving op-amps – the pin connections number 4 and 7 are power connections *to the inside of the op-amp*. These connections are automatically made when you connect the op-amp to its power supply and they are usually omitted from circuit diagrams.

But their presence means that there can be currents, i.e. a flow of charge and of energy, into and out of an op-amp along connections which are not visible on your circuit diagrams.

Negative feedback

An amplifier with a gain of 200 000 sounds like an electronic engineer's dream. But it is 'too much of a good thing'. The gain turns out to vary with frequency (see page 391), to depend slightly on the size of the (very small) input voltages, and to limit the signals to be amplified to tiny voltages as the output cannot be greater than the voltage on the supply rails, $\pm 5\,V$. Nor is it stable; the behaviour of the amplifier varies as the properties of its components change very slightly, e.g. with temperature. A more useful amplifier is achieved by the use of **negative feedback** which reduces the gain dramatically. Figure 27.5 shows an **inverting amplifier** with negative feedback. The non-inverting input is at $0\,V$, earthed, and a *feedback resistor* R_f is connected from the output to the inverting input. An *input resistor* R_{in} is also connected as shown.

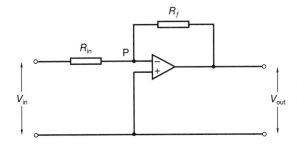

FIGURE 27.5

The theory of this arrangement depends on the very high open-loop gain of a TL081 (or other op-amp, e.g. a 741), and the fact that the resistance of the op-amp is itself very high, typically $10^{12}\,\Omega$ for a TL081 (and $>10^6\,\Omega$ for a 741), i.e. there can be only tiny currents into the non-inverting input; 50 pA is the maximum current into either the inverting or the non-inverting input of a TL081.

Suppose V_{out} was $-5\,V$, its minimum value:

then $\qquad V_+ - V_- = -5\,V \div 200\,000$

$$= -2.5 \times 10^{-5}\,V$$

$$\text{or } -25\,\mu V$$

As the non-inverting input is earthed, i.e. $V_+ = 0\,V$, the voltage at P in figure 27.5 is

$$V_P = V_- = +25\,\mu V$$

which is the *maximum* value it can take. (A similar argument holds for positive values of V_{out}.)

Compared with $V_{in} \approx 0.1\,V$ or even $0.01\,V$ this $25\,\mu V$ is so tiny that P is called a **virtual earth** (literally we say it is virtually at earth potential). In practice we treat the op-amp as if $V_+ = V_-$. *How* this is achieved by adding a feedback resistor depends on the internal design of the op-amp, and we are not going to attempt to discuss that. We shall state two rules for dealing with op-amp circuits: the first one we have just explained so:

♦ **Rule 1** states that with feedback: $V_+ = V_-$.

Of course, if the output voltage is less than $\pm 5\,V$, the approximation of Rule 1 holds even more strongly. Rule 2 is much easier to understand:

♦ **Rule 2** states that any current in R_{in} all goes on into R_f or vice versa.

Suppose the current in R_{in} is $20\,\mu A$ from left to right. At most 50 pA enters the inverting input of the op-amp so

$$20 \times 10^{-6}\,A - 50 \times 10^{-12}\,A \approx 20 \times 10^{-6}\,A$$

goes on to R_f.

We can now calculate the gain, it is called the closed-loop gain G or A_{CL} for this inverting feedback amplifier.

As $V_P = 0$ (rule 1)

the p.d. across $R_{in} = V_{in}$

$$\therefore \text{ the current in } R_{in} = I = \frac{V_{in}}{R_{in}}$$

All this current goes on to R_f (rule 2)

i.e. the current in R_f is V_{in}/R_{in}

$\therefore$ the p.d. across $R_f = IR_f = \dfrac{V_{in}}{R_{in}}R_f$

But $V_P = 0$

$\therefore V_{out} = 0 - \dfrac{V_{in}}{R_{in}}R_f$

$\Rightarrow \qquad\qquad \dfrac{V_{out}}{V_{in}} = -\dfrac{R_f}{R_{in}}$

This very simple result for the closed-loop gain does not contain the open-loop gain A. If $R_f = 47\,k\Omega$ and $R_{in} = 4.7\,k\Omega$ the gain is -10. The minus sign explains why this is called an inverting amplifier: if V_{in} goes up from 0.10 V to 0.11 V, the output goes from -1.0 V to -1.1 V, i.e. it goes down.

EXAMPLE

In the diagram P is a virtual earth, a point where the potential is held at zero but where there is no current to earth. Calculate V_B in each of the following cases.

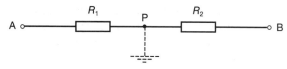

(a) $V_A = +0.3$ V,
  $R_1 = 2.2\,k\Omega, R_2 = 4.7\,k\Omega,$
(b) $V_A = -0.1$ V, $R_1 = 47\,k\Omega, R_2 = 47\,k\Omega,$
(c) $V_A = 0.2$ V, $R_1 = \frac{1}{3}R_2.$

(a) $V_{AP} = +0.3$ V

$\therefore I_{AP} = \dfrac{0.3\,V}{2.2\times10^3\,\Omega} = I_{PB}$

All this current continues to B, no charge leaks to earth at P as there is no circuit around which it can flow. This means that
  $I_{PB} = 0.14\,mA$
$\therefore \qquad V_{PB} = (0.14\times10^{-3}\,A)(4.7\times10^3\,\Omega)$

  $= 0.66$ V

and, as $V_P = 0$, $V_B = -0.66$ V

(b) $V_{AP} = -0.1$ V

  $I_{AP} = I_{BP}$ and $R_1 = R_2$

$\therefore \quad V_{BP} = -0.1$ V

but $V_P = 0$, $\therefore V_B = 0.1$ V

(c) $R_2 = 3R_1$
  $\therefore V_{PB} = 3V_{AP}$

$\Rightarrow \qquad V_{PB} = 3\times0.2\,V = 0.6$ V

but $V_P = 0$, $\therefore V_B = -0.6$ V

You can see that each of these calculations in the previous example is nothing more than a potential divider calculation where the 'middle' is held at 0 V.

You can represent the results in the above example diagrammatically as shown in figure 27.6 and this is what happens in the amplifier of figure 27.5 with R_{in} deciding the 'slope' on the diagram, the input current, and R_f deciding the size of the final potential.

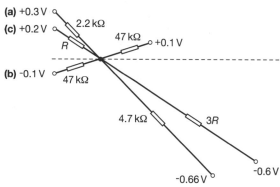

(a) +0.3 V
(c) +0.2 V
2.2 kΩ
47 kΩ +0.1 V
R
(b) -0.1 V
47 kΩ
4.7 kΩ
3R
-0.66 V
-0.6 V

FIGURE 27.6

EXAMPLE

A sinusoidal signal of peak voltage 0.10 V and frequency 500 Hz is fed to the input of an inverting amplifier with a gain of 5. Draw, on the same graph axes, the input and output voltages over two cycles.

  $f = 500\,Hz,$

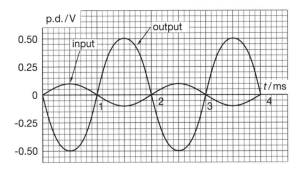

therefore

$$T = \frac{1}{f} = 0.0020\,s$$

The output must be *minus* 5 times the input voltage at every instant.

Transfer characteristics

Figure 27.7 shows how you could study the amplifying properties of a negative feedback inverting amplifier. If the gain, R_f/R_{in}, is set at about 10 you will need small input voltages and these are best produced with the potential divider shown in figure 27.7 using a fixed resistor of value $10R$ and a rheostat of value R. Remember that V_{out} is negative and so you must connect the output voltmeter accordingly. To get negative input voltages the $-5\,V$ supply rail can replace the $+5\,V$ one shown.

A typical set of results (called a **transfer characteristic**) is shown in figure 27.8 for the case where the gain is -10, the gradient of the centre part of the graph. The op-amp **saturates** when V_{out} is just under $\pm5\,V$, i.e. at a voltage just below that of the supply rails $\pm5\,V$. This occurs when V_{in} is just under $\pm0.5\,V$.

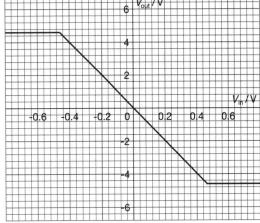

FIGURE 27.8

You can study the properties of this amplifier for a.c. signals using the same circuit but replacing the $+5\,V$ connection with one from a signal generator, the other being earthed, and replacing the input and output voltmeters with a double beam oscilloscope.

For a single frequency, e.g. $1\,kHz$, the gain – which can be measured simply as (length of output trace)/(length of input trace) with the time-base

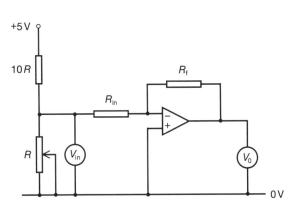

FIGURE 27.7

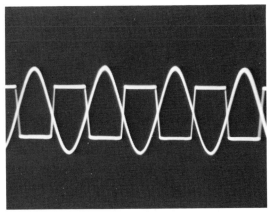

FIGURE 27.9

387

switched off – is constant providing the input voltage does not exceed a certain value.

At higher values of V_{in} the output signal becomes clipped. It has a flat top. At very high values of V_{in}, the output signal is a square wave. A typical result is shown in figure 27.9. You should be able to explain what is happening by arguing from the transfer characteristic of figure 27.8.

EXAMPLE

The diagram shows a photodiode connected to the inverting input of a negative feedback amplifier. When illuminated the photodiode conducts and the current is proportional to the intensity of illumination. Explain why this arrangement acts as a lightmeter.

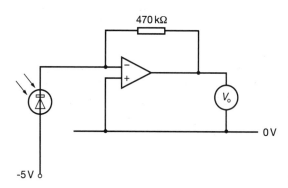

When the photodiode conducts there is a current from the inverting input to the $-5\,V$ rail. This current cannot come from the op-amp. It comes from the $470\,k\Omega$ feedback resistor. There is therefore a p.d. across the feedback resistor, e.g. if the current is $1\,\mu A$ the p.d. is $0.47\,V$. As $V_+ - V_-$ must be zero and V_+ is zero then $V_- = 0$, so the potential at the output of the op-amp is $+0.47\,V$. This is registered by the output voltmeter.

As the illumination changes so does the current and hence so does the voltmeter reading. In this case the voltmeter reading varies by about $0.5\,V$ per μA in the photodiode.

This use for an op-amp is sometimes called a **current-to-voltage converter.**

EXAMPLE

In the circuit shown R is a $1\,M\Omega$ resistor and C a $2.2\,\mu F$ capacitor. Describe what happens in the first second after the switch S is closed.

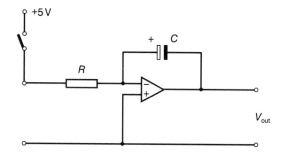

The inverting input remains effectively at $0\,V$ as the open loop gain is very high (rule 1), so the current in R is

$$I_{in} = \frac{5\,V}{1 \times 10^6\,\Omega} = 5\,\mu A$$

All this current goes to charge C; none enters the op-amp which has a very high input resistance (rule 2).

In $0.1\,s$ a charge of $0.5\,\mu C$ flows onto the capacitor, and so the p.d. across it rises to

$$V = \frac{Q}{C} = \frac{0.5\,\mu C}{2.2\,\mu F}$$

$$= 0.23\,V$$

so V_{out} is then equal to $-0.23\,V$.

The same charge flows to C during each $0.1\,s$ interval and so during the first second V_{out} falls from 0 to $-2.3\,V$.

Such an arrangement is called an **integrator**. This is because a steady V_{in} gives rise to a steadily falling V_{out}. You should try to draw graphs of V_{in} against t and V_{out} against t in the above case.

In fact

$$V_{out} = -\frac{1}{RC} \int V_{in}\,dt$$

This, and other circuits involving capacitors, enable op-amps, the basic building block of analogue electronics, to solve differential equations.

The summing amplifier

Figure 27.10 shows how to add two voltages V_1 and V_2. The current in both R_1 and R_2 continues to R_f (rule 2). Here P is a virtual earth (rule 1).

$$I_1 = V_1/R_1 \text{ and } I_2 = V_2/R_2$$

so

$$I_1 + I_2 = \frac{V_1}{R_1} + \frac{V_2}{R_2}$$

and

$$I_f = -\frac{V_{out}}{R_f}$$

∴

$$\frac{V_1}{R_1} + \frac{V_2}{R_2} = -\frac{V_{out}}{R_f}$$

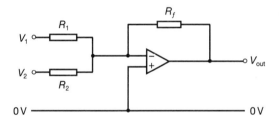

FIGURE 27.10

Simply to add V_1 and V_2 we make $R_1 = R_2 = R_f$ giving $V_1 + V_2 = -V_{out}$. If a positive result is required we must invert the output by using a second inverting amplifier with a gain of -1. To subtract, $V_1 - V_2$, all we need to do is to invert V_2 *before* adding.

As many inputs as you like can be used: in a **mixer** up to 20 voltages from different microphones may be connected to the same op-amp. Such circuits can also be used as digital-to-analogue converters (DAC) where the values of R_1, R_2, R_3, etc., are in the ratio 1, 2, 4, etc. For example, if $R_1 = 200\,\text{k}\Omega$, $R_2 = 400\,\text{k}\Omega$ and $R_3 = 800\,\text{k}\Omega$, when $R_f = 16\,\text{k}\Omega$, then for $V_1 = 5\,\text{V}$, $V_2 = 0\,\text{V}$ and $V_3 = 5\,\text{V}$ (the binary number 101 if we treat $+5\,\text{V}$ as 1 and $0\,\text{V}$ as 0 – see page 412)

$$V_{out} = 16\,\text{k}\Omega \left(\frac{5\,\text{V}}{200\,\text{k}\Omega} + \frac{0\,\text{V}}{400\,\text{k}\Omega} + \frac{5\,\text{V}}{800\,\text{k}\Omega} \right)$$

$$= 0.4\,\text{V} + 0.1\,\text{V} = 0.5\,\text{V}$$

which could represent the number 5. Try drawing the circuit diagram for this DAC and check some other examples, e.g. that 011 ($V_1 = 0\,\text{V}$, $V_2 = 5\,\text{V}$, $V_3 = 5\,\text{V}$) gives 0.3 V etc.

27.3 Non-inverting amplifiers

The theory of the inverting amplifier is straightforward, but as the input resistor might be as little as $1\,\text{k}\Omega$, a current will be drawn from an input voltage source. This current might be as much as $1\,\text{mA}$, which represents quite a large drain from it and to provide it the voltage levels within the source may change. Figure 27.11 shows the basic circuit for a **non-inverting amplifier** which always drains very little current from the source of input voltage, because the input voltage source is connected directly to the non-inverting input of the op-amp which has a very high resistance.

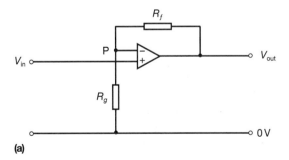

(a)

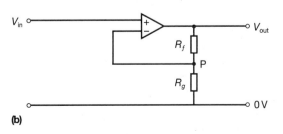

(b)

FIGURE 27.11

Two arrangements are given: the op-amp is drawn 'upside down' in (b). You should be sure you can see that these two circuits are electrically identical (if you had (a) on the bench with flexible wire connections you could pull and twist the wires to give (b) without disconnecting anything). In both R_f is the feedback resistor and R_g is the resistor to earth or ground. Notice that it is still negative feedback which is used, i.e. the feedback is to the inverting input on the op-amp. If any part of the output voltage were to find its way to the non-inverting input (positive feedback) there would be a danger of the amplifier becoming unstable. This is a potential danger with non-inverting amplifiers but not a serious one.

Using our two rules from page 385

$$V_{in} = V_+ = V_- \qquad \text{(rule 1)}$$

and
$$V_- = V_P$$

$$\text{Current in } R_g = \text{current in } R_f \qquad \text{(rule 2)}$$

so
$$V_P = \frac{R_g}{R_f + R_g} V_{out}$$

∴
$$V_{in} = \frac{R_g}{R_f + R_g} V_{out}$$

which gives

$$\frac{V_{out}}{V_{in}} = 1 + \frac{R_f}{R_g}$$

Notice that this expression does not contain the open loop gain A. The transfer characteristic for a non-inverting amplifier is similar in shape to that of figure 27.8 but the section where the amplifier does not saturate has a positive slope.

EXAMPLE

A particular thermocouple gives an e.m.f. of $42\,\mu V$ for a temperature difference of one kelvin. Draw a circuit to show how the e.m.f. produced by the thermocouple could be displayed on a voltmeter with a full-scale deflection of $1.0\,V$ if the thermocouple was being used over a temperature range of $0°C$ to $100°C$. Explain how you choose the values of circuit components.

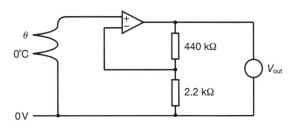

Connect the thermocouple as shown.

For $\theta = 100°C$ the thermocouple e.m.f.

$$= 100\,K \times 42 \times 10^{-6}\,V\,K^{-1}$$

$$= 4.2 \times 10^{-3}\,V$$

so a gain of 200 would give a maximum output voltage of $0.84\,V$, just right for the voltmeter. The resistance values on the circuit would produce a gain of

$$1 + \frac{440\,k\Omega}{2.2\,k\Omega} = 201$$

$2.2\,k\Omega$ is a preferred value resistor but $440\,\Omega$ is not; the nearest preferred value is $470\,\Omega$ which gives a gain of 215 and a maximum V_{out} of just over $0.9\,V$, i.e. still below the f.s.d. of the voltmeter.

Voltage followers

Suppose in the circuits of figure 27.11, $R_f = 0$ and $R_g = \infty$ (infinitely large, a resistance value which is achieved by not making a connection); the circuit would become that of figure 27.12 and the gain would be exactly 1. Such a circuit is called a **unity**

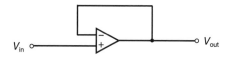

FIGURE 27.12

390

gain **voltage follower** or simply a follower. What can the use of such a circuit possibly be? $V_{out} = V_{in}$ at all times, so what does a follower achieve? We must remember the connections to the op-amp from the $\pm 5\,V$ power rails which are not shown in our circuit diagrams. These connections mean that for a TL081 the output current may be as high as $10\,mA$ but the input current is only $50\,pA$. There can therefore be a marked **power gain**. For $V_{in} = V_{out} = 0.5\,V$, for example, you can see that the

input power $= I_{in}V_{in}$

$$= 50 \times 10^{-12}\,A \times 0.5\,V$$

$$= 2.5 \times 10^{-11}\,W$$

and the output power

$$= 10 \times 10^{-3}\,A \times 0.5\,V$$

$$= 5 \times 10^{-3}\,W$$

a power gain of 2×10^8! Of course, the output power is only $5\,mW$, which may be adequate for a chart recorder, but we may need to use the follower to activate a switch to connect power to a lamp or heater (see page 394).

The coulombmeter

Because the input to a non-inverting amplifier is connected directly to V_+ where the resistance is $10^{12}\,\Omega$, there is only a very tiny current drain from the voltage source, about $50\,pA$ at most. As we have seen, this is the key to the use of the follower as an isolating device and it also explains how we can use it as a coulombmeter for measuring charge. Figure 27.13 shows a coulombmeter or electronic electrometer.

The switch S is first pressed to discharge the

capacitor. Then $V_{in} = 0$ and $V_{out} = 0$ also, so the meter reads zero. The charged object is then touched to V_{in}, the input terminal. Suppose its charge Q raises the p.d. across the capacitor to V: then $Q = CV$, so measuring V can give us Q.

For the $1.0\,\mu F$ capacitor shown a charge of $2000\,nC$ (nanocoulombs) will (using $Q = CV$) produce $V_{in} = V_{out} = 2.0\,V$, which will appear on the display as $2000\,mV$ and can be read as $2000\,nC$. The use of a coulombmeter such as that shown in the photo of figure 27.14 is described on page 196.

FIGURE 27.14

Frequency response

Let us look more closely at the negative feedback in figure 27.11 – it is easier to see this in (b). The fraction β of the output voltage V_{out} which is fed back to V_- is given by

$$\beta V_{out} = \frac{R_g}{R_f + R_g} V_{out}$$

$$= V_-$$

and, here,

$$V_{in} = V_+$$

Substituting these values of V_- and V_+ in the definition, $V_{out} = A(V_+ - V_-)$, of open-loop gain for an op-amp we have

$$V_{out} = A(V_{in} - \beta V_{out})$$

$$\Rightarrow \qquad G = \frac{V_{out}}{V_{in}} = \frac{A}{1 + \beta A}$$

which, as before, we call the closed-loop gain.

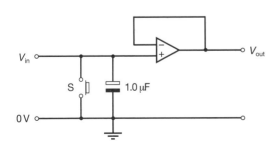

FIGURE 27.13

(This equation relates the closed-loop gain G or A_{CL} with the open-loop gain A or A_{OL}.)

If $\beta A \gg 1$ then $G \approx A/A\beta = 1/\beta$. Rule 1 works only when βA is much greater than one. As A = 200 000 this is usually no problem unless $1/\beta$ becomes greater than 2000; then βA becomes less than 100.

However, the open-loop gain of an op-amp depends on the frequency of the incoming signal. This is because of capacitive effects inside the op-amp which increase as f increases. Figure 27.15 shows (a) how A_{OL} varies with f and (b) how A_{CL} varies with f for two values of β. The use of $k(\times 10^3)$ and $M(\times 10^6)$ simplifies the axes but note that they are both logarithmic.

In (b) the closed-loop gain is independent of frequency until $A_{CL}\beta$ approaches 1 when the curve breaks to follow the open-loop dashed line. Thus the greater A_{CL} or $1/\beta$ the more the gain drops off at high frequencies and hence the greater the danger of distortion. The **bandwidth** of the amplifier is the range of frequencies over which it has a constant gain. For example, in (b) it is about 3 kHz for the upper curve and 30 kHz for the lower.

EXAMPLE

The diagram shows an amplifying system which has input frequencies in the range 20 Hz to 20 kHz and an input signal of 0.3 μV r.m.s. Calculate the overall gain of the system and explain why it suits its purpose well.

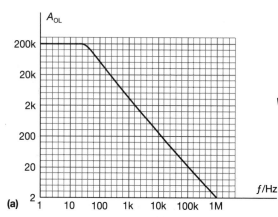

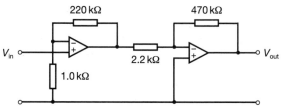

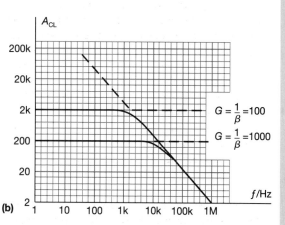

(a)

(b)

FIGURE 27.15

The non-inverting amplifier offers a very high resistance to the signal source and consequently drains very little current from it. It acts as a power amplifier with a voltage gain of

$$G_1 = 1 + \frac{220\,k\Omega}{1.0\,k\Omega} = 221$$

The inverting amplifier has a gain of

$$G_2 = \frac{470\,k\Omega}{2.2\,k\Omega} = 214$$

and ensures that the system is stable. The overall gain is $G_1 \times G_2 \approx 50\,000$ so the output will be about 15 mV. The bandwidth of the non-inverting amplifier is greater than the required 20 kHz as $\beta A = (1/221) \times 200\,000 \approx 900$, so the signal will not be distorted.

27.4 Switching devices

Comparators

An op-amp *without* feedback can be used to tell you which of two voltages is the larger – it acts as a **voltage comparator**. The basic equation

$$V_{out} = A(V_+ - V_-)$$

with $A = 200\,000$ makes it obvious that if $V_+ > V_-$ then V_{out} will be large and positive while if $V_+ < V$, V_{out} will be large and negative. The op-amp saturates when there is a p.d. of more than about $20\,\mu V$ between the input terminals. (Remember that we are using supply rails at $\pm 5\,V$ and that the op-amp saturates a little below these voltages.) You will have great difficulty in getting a voltmeter connected between the output and the earth rail to register anything but $+5\,V$ or $-5\,V$ even with the circuit of figure 27.16 as the resistors are very unlikely to be identical to within the tolerance needed to make $V_+ = V_-$ to 1 part in 10^5.

In practice even with four identical resistors the output voltage may not be zero as it is not possible to make op-amps with perfectly balanced inputs. Two other terminals (pins 1 and 5 in figure 27.1), not so far mentioned, can be used to give $V_{out} = 0$ when $V_+ = V_-$. Some op-amps, mounted on circuit boards, have a potentiometer connected between these pins to make this easy to do. It is called a **null adjustment**.

A balanced op-amp can be used with a bridge circuit similar to that of figure 27.16 as a **differential amplifier**. For example one of the

resistors could be in the form of a strain gauge in a pressure cell. As the pressure difference across the cell changes the resistance of the strain gauge changes. A strain of 0.001% might produce a change of resistance large enough to alter $V_{XY} = V_+ - V_-$ by $5\,\mu V$ and this, amplified $200\,000$ times, is $1.0\,V$ which is readily fed via a voltage follower to a chart recorder or data logging device.

Usually, however, comparators saturate. The output will then have one of two states, and can be thought of as a digital output: either ON or OFF. Figure 27.17 illustrates how a comparator used with an LDR and an LED might help you to locate a keyhole, for example after dark.

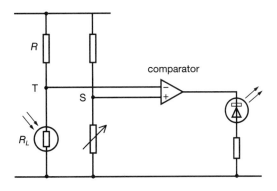

FIGURE 27.17

You can see how it works. When it is daylight the resistance of the LDR R_L is low and so the potential at T is low. As the light level falls R_L rises and V_T rises. (Notice how useful the shorthand of R_L and V_T is – always try to label your circuit diagrams.) When V_T exceeds the potential V_S at the non-inverting input the op-amp suddenly switches from saturation at $+5\,V$ to saturation at $-5\,V$ and the LED, which in this example would be located in or near the keyhole, comes on. The switching level is set by the position of the rheostat, i.e. by the reference potential V_S. (LEDs can be damaged if there is too large a current in them and a resistor – it is about $500\,\Omega$ – is connected in series with them to limit the current to about $10\,mA$.)

The bridge circuit for the pressure cell and the key locating system both use potential dividers for the inputs to the op-amp. Any pair of components

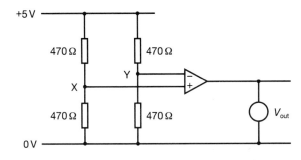

FIGURE 27.16

the resistance of which varies because of a change in temperature, light level, pressure, humidity, etc. can be used, together with a comparator, as an electronic switch.

Drivers

The switching action of a comparator is an example of the use of a **driver**, a device which can sense when an input voltage reaches a critical level and then switch on a large current without draining (much) current from the input voltage source. Another device which can be used as a driver is the **transistor**. Figure 27.18 shows an n-p-n (silicon) transistor switch.

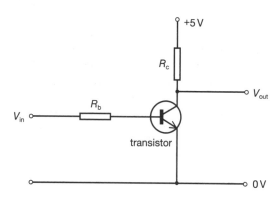

FIGURE 27.18

R_b is called the *base* resistor and R_c the *collector* resistor. Both are typically a few kΩ. The other connection to the transistor is called the *emitter*. You will find that V_{out} varies with V_{in} as follows:

V_{in}/V	0	0.5	1.0	1.5	etc.	5.0
V_{out}/V	5.0	5.0	0.2	0.2	etc.	0.2

i.e. the output of transistor circuit switches between V_{in} = 0.5 V and 1.0 V. The switching action is not as sharp as that in a comparator circuit but the currents available at V_{out} are larger.

We shall not discuss transistors any further – they are widely used as amplifiers, for example – except to repeat that the first transistor was made in 1948. There are now thousands of different types produced for specialist purposes. Transistors form the vital component in integrated circuits (i.c.) of which the op-amp is just one example, and an i.c. driver such as the NE555 can itself be used as a switch. It, like the n-p-n transistor, needs

+5 V and 0 V rails and has a very clearly defined (i.e. sharp) switching action.

Figure 27.19 shows the general circuit symbol for a driver together with an ideal driver characteristic. It is often called a NOT gate, that is, a device for which the output is low (OFF) when the input is high (ON) and vice versa. The connections to the power rails are often omitted (as they are with op-amps).

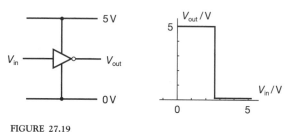

FIGURE 27.19

The relay

A more direct approach to the problem of switching is to use an **electromagnetic relay**. The relay (figure 27.20) consists of a coil of many turns wound on an iron core. When the core is magnetised by a small current (perhaps 10 mA) in the coil, it attracts a hinged iron plate, which carries the switch arm. Two circuit symbols for a relay are shown alongside it.

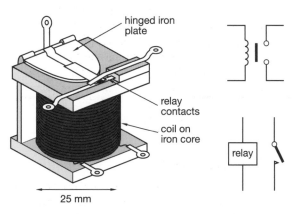

25 mm

FIGURE 27.20

A relay can thus be driven by the output of a semiconductor driver and such switches are often used together when a mains-operated device such as an external floodlight or a fan needs to be switched on or off. A safety diode (not shown in figure 27.20) is usually connected across the input of a relay in order to protect the driver from the large induced e.m.f.s which arise when the current in the relay coil is switched off.

EXAMPLE

Describe the working of the circuit shown below. Suggest a use for this circuit.

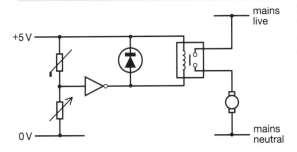

When the temperature rises the resistance of the thermistor falls and so the p.d. across it falls as well. This raises the input voltage to the driver which switches from a high to a low output at a temperature determined by the setting of the rheostat in the potential divider. The driver energises the relay, attracting the contacts together and hence switches on the motor. One possible use for the circuit would be to operate a fan to help to keep a room cool in hot weather, or to blow cold air over part of a machine which is overheating. When the temperature falls the driver switches to a high output and the relay is de-energised, thus switching the motor off. When the current in the relay coil is switched off, the large induced e.m.f. continues to send a current in the same direction, but the diode short circuits the coil, and thus protects the semiconductor driver.

Coupling circuits

In this chapter so far a number of ways of providing inputs to electronic systems have been described and so have some of the devices which are

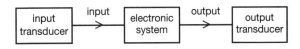

FIGURE 27.21

operated by the systems. In block diagram form what we have been describing, in its most general form, looks like figure 27.21.

Maximum power transfer

On page 115 a source of e.m.f. with internal resistance r is shown to transfer the maximum power to a load of resistance R when $r = R$. This is of especial significance when alternating signals from microphones or into loudspeakers are being considered (we should then speak of impedances Z rather than resistances). A signal generator will have two outputs; one for connection to low R (or Z) devices, e.g. a loudspeaker, and a second for connection to high R (or Z) devices, especially an oscilloscope. In general $Z_{out} = Z_{in}$ for maximum power transfer – figure 27.22.

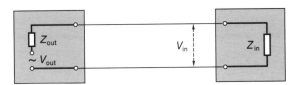

FIGURE 27.22

If we simply want $V_{in} = V_{out}$ the condition is different. As

$$V_{in} = \frac{Z_{in}}{Z_{in} + Z_{out}} V_{out}$$

the best arrangement now is for $Z_{in} \gg Z_{out}$ and this is the usual case in electronic circuits where, typically, $Z_{in} > 10^4 Z_{out}$. You might wonder why, given this equation, there is any need for a separate high impedance output on a signal generator. It is provided because the output signal is then very precisely shaped, an operation which is difficult to achieve at low impedances.

EXAMPLE

A microphone with a resistance of $33\,k\Omega$ is connected to an oscilloscope and the peak-to-peak voltage registered is $22\,mV$. What will be the new peak-to-peak voltage if a $47\,k\Omega$ resistor is connected across the input terminals of the oscilloscope? State any assumptions you make.

Assuming that the oscilloscope has an infinite input resistance, the microphone is developing $22\,mV$ peak-to-peak. With the $47\,k\Omega$ connected there will be a current in the circuit and the $22\,mV$ will be split between the two resistors in the ratio of their resistances.

New peak-to-peak voltage

$$= \frac{47\,k\Omega}{47\,k\Omega + 33\,k\Omega} \times 22\,mV$$

$$= 13\,mV$$

assuming inductive and capacitative effects are negligible.

Filters

When an a.c. signal is superimposed on a d.c. voltage the a.c. component can be decoupled from the d.c. by using a **blocking capacitor**. For example, the input of an oscilloscope contains such a capacitor and an a.c./d.c. switch which enables

you to use direct coupling in which the entire signal is displayed (d.c.) or capacitor coupling in which only the varying part of the signal is displayed (a.c.) – see figure 9.16. This blocking capacitor is an extreme example of a **frequency filter**. Figure 27.23 shows a cross-over filter circuit commonly used at the output of an audio amplifier, i.e. at the input to the speaker unit, to enable the speakers designed for high frequencies (tweeters) and low frequencies (woofers) to be used.

You will appreciate the principle if you have studied Section 19.3. For a sound which contains a range of frequencies those with low f, a few hundred Hz, will produce higher voltages across the capacitor but those with high f, a few thousand Hz, will produce higher voltages across the inductor.

Feedback

The negative feedback used in operational amplifier circuits involves reducing the input by adding to it a part of the output which is inverted. In a.c. terms the feedback signal is in antiphase with the original input signal. If the feedback signal is instead in phase with the original input signal we say there is **positive feedback**.

Negative feedback is the key to **control systems** of all kinds. Riding a bicycle along a straight road involves you in steering a little to the left if you find yourself going right and so on. You are compensating, often unwittingly, for small steering errors. Robots are controlled, and babies feed themselves, by a series of correcting motions, i.e. by the use of negative feedback.

Positive feedback, by contrast, is usually a nuisance. The howl you get when the microphone and speaker of an intercom system are placed near one another is the result of the following sequence of events. Any signal reaching the microphone is amplified and emitted as sound by the speaker. If these sound waves now reach the speaker in phase with the original signal a positive feedback loop can be established. As energy is added each time by the amplifier, the sound can build up to a howl if this added energy makes the output greater than it was without the positive feedback on each round trip. We sometimes describe this by saying that the loop (energy) gain is greater than one.

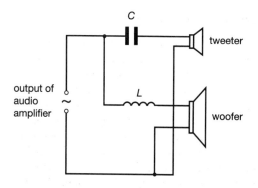

FIGURE 27.23

27.5 Logic gates

Logic gates are components which deal only in signals which are high ($\approx 5\,V$ = ON) or low ($0\,V$ = OFF). We describe these as **logic level 1** and **logic level 0**, or simply 1 and 0. They can be produced using relays, transistors or integrated circuits. We shall assume you are using i.c. logic gates, two types of which are commonly available: **TTL**, which operate extremely quickly, and **CMOS** which use very little power but which are slightly slower. Whichever you use in learning digital electronics matters little, but you will need to get used to the fact that an unconnected input to a TTL gate automatically becomes a high (1) while that to a CMOS gate behaves erratically and so CMOS circuits incorporate resistors which ensure that unused inputs are definitely 1 (or 0).

Each gate has connections to the $+5\,V$ and $0\,V$ power rails but these are not usually shown in circuit diagrams incorporating logic gates.

The names for the gates are obvious. The symbols are very odd; perhaps the dee-shape for AND and an oar-shape for OR were originally in mind. Two of the gates have a little circle before the output. You can see that the circle indicates inversion; it turns AND into NAND and OR into NOR. Also notice that if *both* inputs to a NAND or a NOR gate are high (1) the output of each is low (0) so that used in this way both NAND and NOR gates act as inverters. These facts can be represented by diagrams combining gates as in figure 27.26. The last row shows the symbol for a non-inverting driver, an i.c. amplifier with a gain of $+1$. You may also have come across symbols all of which are square boxes but we will not use them in this book.

NOT

input	output
0	1
1	0

FIGURE 27.24

The behaviour of a single input NOT gate is described by a **truth table** as in figure 27.24. A NOT gate is sometimes called an **inverter**. Other logic gates have more than one input but all have only one output. The symbols for four gates and their truth tables are summarised in figure 27.25.

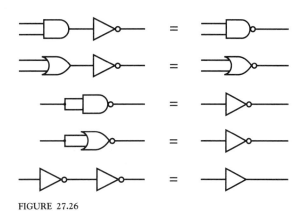

FIGURE 27.26

Experimental work

It is more economical to supply simple logic gates in groups of four or six, *integrated* into a simple silicon chip measuring perhaps 1 mm square. This chip is enclosed by a piece of plastic fitted with pins (often 14 of them) so that connections can be made to the terminals of the gates. One single pair of pins provides connection from the power supply to all the logic gates on the chip. That leaves 12 pins for the gate terminals, so you might get six inverters (two pins each) or 4 two-input NAND gates (three pins each).

inputs		AND	OR	NAND	NOR
A	B	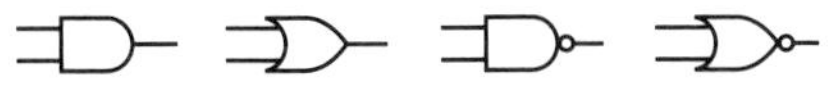			
0	0	0	0	1	1
0	1	0	1	1	0
1	0	0	1	1	0
1	1	1	1	0	0

FIGURE 27.25

Figure 27.27 shows a diagram of a quad NAND gate together with a photograph of it mounted on a board so that connections may be made to it.

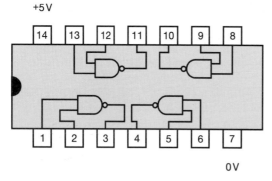

+5 V

0 V

FIGURE 27.27

The photograph shows the +5 V and 0 V rails. One of the NAND gates has its inputs connected – one high and one low. To test the state of the output you need only connect the output via an LED and safety resistor to the 0 V rail – they are usually represented on circuit diagrams by the symbol for an indicator lamp ⊗. Any individual logic gate can be similarly investigated but it is the properties of linked logic gates which are of greater significance. You should set up circuits using LEDs or oscilloscopes to test some of the applications suggested in the rest of this chapter.

Timing diagrams

Figure 27.28 shows two streams of pulses: the input to A is a sequence of random pulses from a GM tube which have been amplified and squared; the input to B can be thought of as a control signal. If

B is high for 10 s, then low for 10 s etc., you can see that the output pulse stream gives the number of particles detected in the GM tube in 10 s. These could be counted and displayed before the next 10-s interval is sampled. Of course, the control signal could be on for 100 s and off for 10 s etc. Timing diagrams like these usually omit the axes which are voltage (up) and time (along) respectively. The logic gate you need for this function is a simple two-input AND gate.

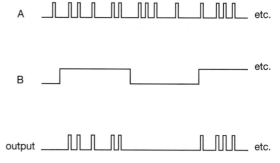

FIGURE 27.28

Information which is digitally encoded is similarly represented by a stream of pulses on a timing diagram. Figure 27.29(a) shows a sequence of 1 s and 0 s such as that which may be picked up by a laser barcode reader at a supermarket output.

(a) 1 0 0 1 1 0 1 0 1 0 0 0 0 0 1 1 0 0 1 0 0

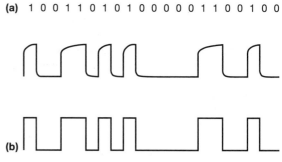

(b)

FIGURE 27.29

A digital signal like this may not initially be a perfect set of square pulses but one of the general characteristics of **digital information processing** is that a signal can be clipped at top and bottom and then amplified to give the ideal shape, that shown

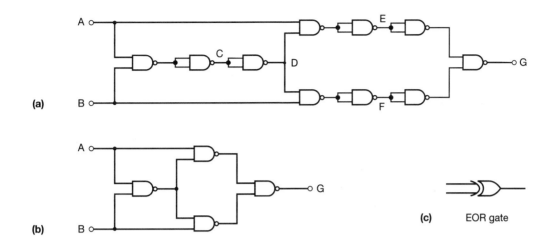

(a)

(b)

(c) EOR gate

FIGURE 27.30

in (b). In this way any small imperfection in the received signal or any degradation of the original signal caused, perhaps, by dispersion in an optical fibre or picked-up noise can be removed. An analogue signal cannot be 'cleaned' in this manner. See also page 407.

Combining logic gates

It can be very hard to design useful systems which involve more than two or three logic gates, and it is a skill you need not develop. What you do need to be able to do is to analyse a given system and recognise its usefulness. If you are given a set of gates with only the inputs and the final output labelled it is vital to add letters to intermediate points before trying to work out what is happening.

EXAMPLE

Draw up a truth table for the system shown in the diagram.

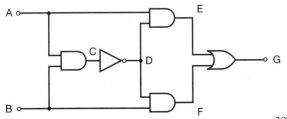

A	B	C	D	E	F	G
0	0	0	1	0	0	0
0	1	0	1	0	1	1
1	0	0	1	1	0	1
1	1	1	0	0	0	0

The output is high when either, but not both, of the inputs are high.

The system in the above example used three AND gates, one NOT gate and one OR gate. As you buy chips with at least four gates of one kind on each, you can see that there will be a lot of redundant gates when you come to set up this system. In practice any logic gate can be made up from a number of NAND gates – you should immediately be able to see how a NOT gate can be made from one NAND gate and how an AND gate can be made from two NAND gates. An OR gate takes three NAND gates and a NOR gate four NAND gates. So it looks as if the system in the example would need $1 + (3 \times 2) + 3 = 10$ NAND gates. The circuit would look like figure 27.30(a) above.

Where there is a pair of NOT gates (NAND gates with the two inputs connected) in series they do not perform any useful function and the pair can be omitted. So our system can actually be made from only four NAND gates – figure 27.30(b) – from a single quad NAND chip in fact. It is a useful logic gate and is called an exclusive OR gate or EOR gate. It has the special symbol shown in (c) and is available commercially as a single gate.

The EOR gate helps us to design logic systems which can add binary numbers. You can see that the truth table (A B G in the example) has output 0 when A + B = 0, and output 1 when A or B = 1. When A and B are both 1, i.e. A + B = 10 (binary) the output of the EOR gate is 0. If A and B are also fed to an AND gate its output will be 0 except when A + B = 10 (binary). In this way the EOR provides the *sum* and the AND gate the *carry* for a simple addition. Such a circuit is called a **half-adder** – it uses six NAND gates. Building up from here we can design full-adders and ultimately all the arithmetic procedures which go on in your calculator where streams of binary numbers are processed using the many thousands of NAND gates on the chip at the heart of the calculator.

Using logic gates

We will give only one further example of the use of logic gates here. In this case gates are used as part of a sensing system; you should be able to see how the inputs to the op-amp circuits of Section 27.3 can make use of logic gates to provide warnings of smoke *and* high temperature, of cold *or* dark, etc.

Figure 27.31 shows an elementary red/green light warning system for telling you when a person who has sat on a car seat has not fastened his or her seat belt. You should be able to deduce that the light changes from red to green again only when the seat belt is clicked into place. An inverted EOR gate, incidentally, is sometimes called a **parity gate**, i.e. a gate which has a high output only when both inputs are the same.

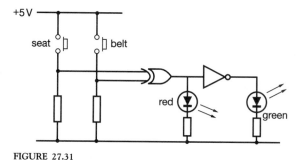

FIGURE 27.31

Exercises on each section of this chapter may be found in the companion textbook, *Practice in Physics*.

SUMMARY

At the end of this chapter you should be able to:

◆ explain the difference between an analogue and a digital system.

◆ draw block diagrams for simple electronic systems.

◆ remember that the open-loop gain of an op-amp is very high and that the resistance of the op-amp itself is extremely high.

◆ draw circuit diagrams for inverting and for non-inverting op-amp circuits using negative feedback.

◆ explain what is meant by negative and positive feedback in electronic systems.

◆ understand that for an op-amp with negative feedback V_+ remains virtually equal to V_- and that there is no current into the inverting or the non-inverting inputs to the op-amp.

◆ use the equations for closed-loop gain in negative feedback amplifiers:

inverting amplifier

$$\frac{V_{out}}{V_{in}} = -\frac{R_f}{R_{in}}$$

non-inverting amplifier

$$\frac{V_{out}}{V_{in}} = 1 + \frac{R_f}{R_g}$$

◆ explain how an inverting amplifier can be used as a summing amplifier.

◆ explain how a non-inverting amplifier can be used as a voltage follower.

◆ remember that the open-loop gain of a non-inverting amplifier depends on frequency but that the closed-loop gain is independent of frequency up to a maximum which is controlled by the amount of feedback.

◆ understand the use of an op-amp without feedback as a voltage comparator.

◆ remember the symbols for logic gates and their associated truth tables.

28 Communications

If a person from Shakespeare's time were able to visit the world of the late twentieth century, what would he find most remarkable? Almost certainly it is our ability to transfer information instantly and accurately to and from anywhere on the Earth's surface. Telephones, television systems and faxes have made a greater difference to our lives than did trains, cars or aeroplanes. Underpinning these telecommunications ('tele' means 'at a distance') is electrical and electronic technology but the concepts required for a study of communications are separate from the ideas dealt with in the previous chapter.

The position of telecommunications in the present information technology revolution is crucial. In the future we are likely to see high definition TV, videophones, an increasing use of the cordless telephone, and the decentralisation of offices and businesses as more and more office work is done from the worker's home. Most of the latest advances depend upon the use of the digital encoding of information and hence the principles of analogue-to-digital conversion, and vice versa, are a central feature of this chapter.

28.1 Analogue modulation

Information can be transmitted, either over insulated conducting wires or cables called transmission lines, using moving charges, or through the atmosphere and space using electromagnetic waves. A radio-frequency **carrier signal** is used in both cases (see the table on page 306). But a sinusoidal signal does not contain any **information**. To transfer information we have to modify the shape of the signal; some form of **modulation** is necessary. The techniques used in the past were confined to analogue methods. Nowadays pulse methods are increasingly being developed which use a digital or pulse train as the carrier signal.

Noise

Any signal will pick up unwanted electrical signals called **noise** as it propagates and the ratio of the received signal power to the noise power is ultimately the main factor by which the performance of a transmission system is judged.

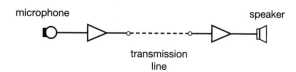

microphone   speaker

transmission line

FIGURE 28.1

Figure 28.1 shows the bare bones of a one-way telephone link (without modulation). Noise in this system may be the result of many effects including:

- thermal noise resulting from the random motion of electrons and atoms in transducers and amplifiers.

- fluctuation noise resulting from natural or man-made disturbance, e.g. electric storms, car ignition systems.

- crosstalk, where signals are transmitted from one line to a neighbouring line.

The **signal-to-noise** ratio, S/N, is usually measured in decibels (dB) and is calculated as

$$\frac{S}{N} = 10\lg\left(\frac{\text{signal power}}{\text{noise power}}\right) \text{dB}$$

For the public switched telephone network (PSTN) the minimum acceptable value of S/N is about 40 dB while for television transmission systems it is 50 dB. Whenever the signal travels in air or in space ionospheric and cosmic effects produce further unwanted noise. **Attenuation** of a signal in a cable and the inverse square law for waves in space further reduce the signal power and make the effects of **white noise** – the general name for the sorts of noise we have discussed so far – proportionally greater.

EXAMPLE

A baby alarm radio telephone system requires a signal-to-noise ratio of 20 dB. Explain what this means.

A signal-to-noise ratio of 20 dB means that

$$10\lg\left(\frac{\text{signal power}}{\text{noise power}}\right) \text{dB} = 20 \text{dB}$$

$$\lg\left(\frac{\text{signal power}}{\text{noise power}}\right) = 2$$

$$\therefore \qquad \text{signal power} = 100 \times \text{noise power}$$

As $P \propto V^2$, the r.m.s. signal voltage must be equal to 10 times the r.m.s. noise voltage at the speaker – the output transducer.

Signal losses along transmission lines are also expressed in **decibels**. For an initial power P_0 which falls to a power P the expression

$$10\lg\left(\frac{P}{P_0}\right) \text{dB}$$

gives the loss, e.g. if $P/P_0 = 10^{-3}$ then the power loss is -30 dB, i.e. the power is reduced by 30 dB. Using decibels we can add losses and gains. Thus if a signal is reduced by 30 dB and then amplified by 20 dB the net reduction is simply 10 dB.

Amplitude modulation (AM)

Medium-wave and long-wave broadcasting in Great Britain encodes the information to be transmitted on to a high frequency **carrier signal** by varying the amplitude of the carrier. Figure 28.2 shows a block diagram of an AM radio transmitter together with the waveforms of the signal at each of the positions numbered 1, 2 and 3. These are

1 – the carrier wave (it will be of much higher frequency relative to the audio signal than that shown).

2 – the analogue audio signal from the microphone (after amplification).

3 – the modulated signal (this is the shape of the transmitted signal).

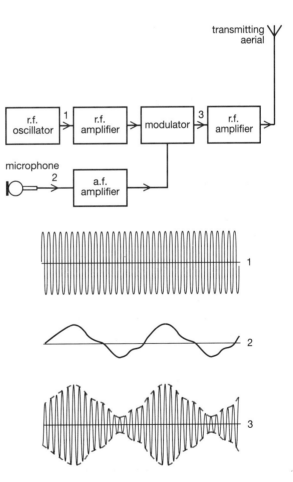

FIGURE 28.2

In a line transmission system the same principle would be used before introducing a speech signal. The final stage would then be the introduction of the modulated r.f. signal into a cable rather than into space.

At the receiving end the signal is selected by a tuning circuit which is a resonant LC loop circuit – see figure 19.21 on page 271. The selected frequency is given by $f = 1/2\pi\sqrt{LC}$. It is then demodulated – the original audio signal is said to be **detected** – by a diode suitable for use at high frequencies. Figure 28.3 shows the block diagram and the waveforms at 4 and 5. (The waveform at 3 will be an attenuated version of the transmitted signal but, of course, the aerial picks up lots of other unwanted signals.) The waveforms are

4 – the demodulated signal.

5 – the audio signal fed to the loudspeaker (after amplification).

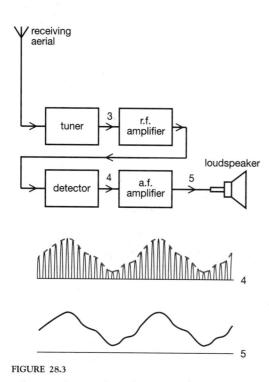

FIGURE 28.3

No noise is shown in any of the waveforms 1–5. The **depth of modulation** is a measure of the ratio (peak a.f. signal)/(peak r.f. signal). In figure 28.2 it is about 0.4. If it became greater than 0.5 the

modulated signal would not carry all the information from the audio signal (can you see why?) and the modulated signal would be distorted.

Bandwidth

When an r.f. carrier wave of frequency f_c is modulated by a single a.f. tone of frequency f_m the resulting AM wave contains three frequency components. They are of frequencies $f_c - f_m$, f_c and $f_c + f_m$. If the a.f. signal covers a range of frequencies, e.g. the 300 Hz to 3400 Hz range used for telephones, the AM wave will contain numerous frequencies ranging:

from $\quad f_c - 3400\,\text{Hz}$ to $f_c - 300\,\text{Hz}$;
and from $\quad f_c + \ 300\ \text{Hz}$ to $f_c + 3400\ \text{Hz}$,

as well as the carrier frequency f_c. You do not need to understand how this arises.

These **sidebands** for an AM signal are usually represented as a **frequency spectrum** as shown in figure 28.4. The carrier frequency shown here is equal to that used for what is now called Radio 4, namely 200 kHz. The amplitude of every part of the sidebands is shown to be the same as we are only describing the frequency ranges.

FIGURE 28.4

The range of frequencies to be transmitted is called the **signal bandwidth**. In figure 28.4 it is from 196.6 kHz, i.e. $(200 - 3.4)$ kHz to 203.4 kHz, so the signal bandwidth is 203.4 kHz – 196.6 kHz = 6.8 kHz or twice the maximum frequency of the a.f. signal. With basic amplitude modulation it is always true that the signal bandwidth = $2f_{max}$. For a.f. signals covering a wider range of frequencies the bandwidth increases, e.g. to transmit music faithfully a range from 50 Hz to 15 kHz is required and hence the signal bandwidth is 30 kHz.

The word bandwidth is also used to mean the complete range of frequencies that can be transmitted by a given communications channel, e.g. a VHF band radio which operates in the range 88 MHz to 108 MHz. The **channel bandwidth** in this case is simply 20 MHz and its significance is that, as we can use different carrier wave frequencies for different stations, the number of stations that can be 'fitted in' depends on both the signal bandwidth and the channel bandwidth. In practice any reduction in the former or increase in the latter is valuable.

It is not possible to reduce $2f_{max}$ but it is possible to transmit only one of the sidebands and to filter out, and hence suppress, the other sideband and, if necessary, the carrier frequency itself.

The process of fitting many AM signals into one channel is called **frequency division multiplexing.** It essentially involves transmitting a number of a.f. signals by using spaced r.f. carriers. Over 10 000 telephone calls can be carried by a single cable using multiplexing while 25 000 can be sent via satellite links using microwaves.

Frequency modulation (FM)

With frequency modulation the amplitude of the carrier wave remains constant and the information is encoded by altering its frequency. Figure 28.5 shows the result for a single modulating pulse which first goes positive and then negative.

EXAMPLE

In the MF band the allowed range for radio broadcasts is 526 kHz to 1606 kHz. A carrier wave of 1053 kHz is amplitude modulated with a signal in the range 300 Hz to 3400 Hz. (a) What are the channel bandwidth and the signal bandwidth in this case? (b) If a bandwidth of 1.4 times the signal bandwidth must be allowed for transmission, how many a.f. signals with this range could be transmitted in the MF band? How could this number be increased?

(a) Channel bandwidth

$$= 1606\,kHz - 526\,kHz$$

$$= 1080\,kHz$$

Signal bandwidth

$$= 2f_{max} = 2 \times 3.4\,kHz$$

$$= 6.8\,kHz$$

(b) Number of a.f. signals possible

$$= \frac{1080\,kHz}{1.4 \times 6.8\,kHz}$$

$$= 110$$

By transmitting only one sideband, which contains all the required information, the number of possible signals could be doubled to 220.

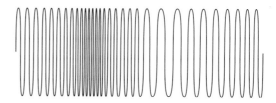

frequency
increased

frequency
decreased

FIGURE 28.5

VHF broadcasting uses this method of modulation and is increasingly being used in the U.K. where channels in the frequency range 30 MHz to 300 MHz are used. You can see that any noise will tend to alter the amplitude rather than the frequency of the modulated carrier signal and for this reason the signal-to-noise ratio for FM signals may be allowed to be higher than for AM signals of the same signal power. Typically, for the same operating conditions, FM is 20 dB better than AM in this respect.

Frequency modulated signals do, however, have a much larger signal bandwidth. In theory, infinitely large: in practice, the bandwidth is taken to be 180 kHz for sound broadcasting of a.f. frequencies in the range 50 Hz to 15 kHz. Hence VHF needs to be used for FM signals as it provides the required very large channel bandwidth of about 270 MHz. Many signals, including stereo signals, can be broadcast simultaneously without the danger of overlapping sidebands. A disadvantage of VHF broadcasts is that the VHF signals have a relatively small range. They do not diffract much to follow the curvature of the Earth because of their relatively small wavelength and consequently there is a need for a large number of transmitters. One or two low frequency (large wavelength) AM transmitters, on the other hand, cover the whole of the United Kingdom and can be picked up reasonably well in mid-France. Also see page 307.

28.2 Pulse modulation

Pulse modulation uses a digital carrier signal consisting of a train of pulses which can be modulated in a number of different ways to carry

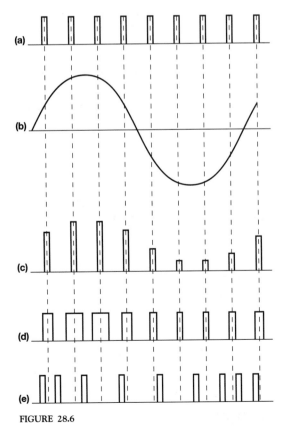

FIGURE 28.6

the required information. Each pulse samples the analogue signal from the transducer and its amplitude, width or position is then altered to produce the modulated signal. In figure 28.6:

(a) represents the unmodulated pulse carrier waveform. The horizontal axis is time and the gaps between pulses may be of the order of $100\,\mu\text{s}$.

(b) is the modulating signal which will generally be of a much lower frequency than the pulse rate in the carrier. (The modulating signal is shown here as a simple sine wave for simplicity.)

(c) the amplitude of the pulses is varied to match the amplitude of the modulation signal. This is called **pulse amplitude modulation** or PAM.

(d) the duration of the pulses is varied to match the amplitude of the modulating signal. For positive values the pulse is widened and for negative values it is narrowed. This is called **pulse duration modulation** or PDM.

(e) the position of the pulses is varied to match the amplitude of the modulating signal. For positive values the pulse is advanced and for negative values it is retarded. This is called **pulse position modulation** or PPM.

In order to reproduce the original signal after demodulation it is taken as a rule of thumb that the sampling frequency must be at least twice the maximum frequency in the original analogue signal. Thus, for a telephone circuit transmitting only one sideband of width 4.5 kHz, the sampling frequency must be at least 9.0 kHz, i.e. just under 10 000 pulses per second.

Of the three types of pulse modulation described above, PPM gives a better signal-to-noise performance than PAM or PDM as noise has a smaller disturbing effect on the time position of a pulse than on its height or its width. This advantage is, however, at the expense of a larger required bandwidth.

Pulse code modulation

The principle of pulse code modulation or PCM can be described by a block diagram such as that shown in figure 28.7. The signals at 1, 2 and 3

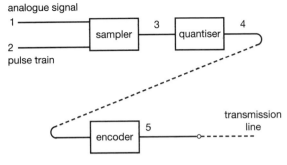

analogue signal

pulse train

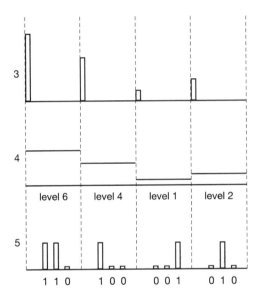

FIGURE 28.7

There may be 128 or 256 discrete levels (2^7 or 2^8) and you can see that the more levels there are, the more faithfully the analogue signal is read for a given sampling frequency. When 128 levels are used the largest binary number for transmission is 1111111 or 127. This seven-bit number and others like it is transmitted and a decoder and demodulating system is required at the receiving end.

The encoding in figure 28.7 shows only seven possible levels. The numbers given are 110, 100, 001 and 010 representing levels 6, 4, 1 and 2 respectively. It is assumed that the first bit (not shown) of the four bit code is used to indicate the start of a new number. An alternative to using, as here, a positive pulse for 1 and no pulse for 0 is to use positive and negative pulses for 1 and 0 respectively. An additional 1 bit is always needed to mark for the receiving equipment the start of the transmission of a new number; this gives a series of 4-bit numbers (or 8-bit **bytes** in the case of 127 levels).

When only a few levels are used there is a large possible error between the level chosen to represent a given pulse height and its actual value – it is usual to quantise at the nearest level but in some systems the lower level is always chosen. This error leads to what is called **quantisation noise** but it is not high with 256 levels and high sampling frequencies. As always the penalty for high signal-to-noise ratios is a large signal bandwidth.

Regeneration

correspond to those at (a), (b) and (c) in figure 28.6; (c) is thus a set of amplitude modulated pulses. The other two are:

4 – the sampled pulses rounded off to the nearest whole number voltage level.
5 – the values of the quantised voltage levels encoded into binary form.

The key to PCM then is that the height of the pulses are matched with a set of chosen voltage levels. This process is called **quantisation**. The chosen levels, numbered 0, 1, 2, 3, etc. are represented by binary numbers so that only a train of 1s and 0s, highs and lows, is transmitted.

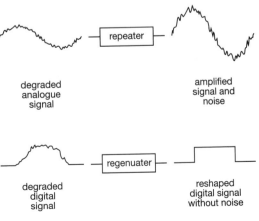

FIGURE 28.8

Signals on transmission lines become attenuated and pick up noise. It is therefore necessary to boost the signal amplitude using **repeater** circuits placed at intervals along the line. Figure 28.8 illustrates the difference between the results of boosting an analogue and a digital signal. Our ability to **regenerate**, i.e. to amplify and to reshape a digital signal, means that there is no decrease in the signal-to-noise ratio along the transmission line, unless the noise power rises to 50% of the signal power! By contrast any amplification of an analogue signal results in the noise being amplified by the same factor as the modulated information signal. This is the single most significant reason for the choice of PCM and of **digital signalling** in communications networks.

EXAMPLE

A VHF signal with a power of 6.0 W is transmitted over a coaxial cable system in which the attenuation losses are 30 dB km^{-1}. The signal picks up noise amounting to 2.0×10^{-8} W betweeen booster stations. If the minimum acceptable signal-to-noise ratio at the booster is 40 dB, calculate their maximum separation.

If the minimum signal power at the regenerator is S, then

$$10 \lg \left(\frac{S}{2.0 \times 10^{-8}\,\text{W}} \right) \text{dB} = 40\,\text{dB}$$

so $\qquad \lg \left(\frac{S}{2.0 \times 10^{-8}\,\text{W}} \right) = 4$

$\therefore \qquad \dfrac{S}{2.0 \times 10^{-8}\,\text{W}} = 10^4$

$\Rightarrow \qquad\qquad S = 2.0 \times 10^{-4}\,\text{W}$

In a distance d, the attentuation losses would be $(30\,\text{dB km}^{-1})\,d$,

$$\therefore 10 \lg \left(\frac{6.0\,\text{W}}{2.0 \times 10^{-4}\,\text{W}} \right) \text{dB} = \left(30 \frac{\text{dB}}{\text{km}} \right) d$$

$$10(4.447)\,\text{dB} = \left(30 \frac{\text{dB}}{\text{km}} \right) d$$

$\Rightarrow \qquad\qquad d = 1.5\,\text{km}$

As all **computer data** are in the form of digital pulses, 1s and 0s, the electronic technology for handling digital signals is highly developed and the subject of much research. Thus pulse code modulation is becoming the standard method for the encoding of analogue information for transfer over cable and (see below) fibre communication systems.

Bit rates

Suppose in the example given in figure 28.6 the analogue signal was sampled every 125 µs, i.e. the sampling frequency is 8000 Hz, which is the standard frequency used for the telephone network. As there are 4 bits needed for each encoded level then the transmission bit rate = 4×8000 bits s^{-1}. A bit s^{-1} is sometimes called a **baud**. This number, here 32 kbaud, is in general found as follows:

$$\text{bit rate} = \left(\begin{array}{c} \text{number of bits} \\ \text{per sample} \end{array} \right) \left(\begin{array}{c} \text{sampling} \\ \text{frequency} \end{array} \right)$$

A 625-line television system requires a bit rate of over 100 Mbits^{-1} and hence a much greater sampling rate though various technical 'tricks' can reduce this somewhat. For ordinary telephone signals a bit rate of 64 kbits^{-1} is adequate but a modern transmission line or satellite link can carry several Mbit s^{-1} with integrity.

Consider what happens when 8-bit (one byte) pulse trains are fed into the line at the rate of 1.0 Mbits^{-1} but the sampling frequency is only 8.0 kHz. Each byte will occupy only 8 µs so there is a 'gap' of 117 µs between bytes – see figure 28.9. In principle we could place many more bytes into this gap and this is indeed done using a system called **time division multiplexing** or TDM. You can see that up to 15 separate channels could be sent along one line providing there is a method of sorting them at the input and output. The usual

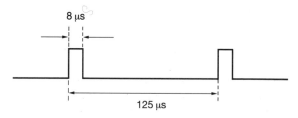

FIGURE 28.9

arrangement is to use a multiplexer which uses **logic gates** to steer successive channel signals to the correct place in the pulse sequence. Nowadays the smallest commercial TDM systems use 30 channels and the maximum number rises as time goes on.

28.3 Fibre links

The use of **optical fibres** in commercial communication systems began only in the mid-1970s. Their advantages are numerous. For instance optical fibres:

◆ offer an extremely high bandwidth for transmission,

◆ are small in diameter and light in weight,

◆ produce very low signal attenuation,

◆ do not suffer from fluctuation noise or crosstalk.

Bandwidth

The wavelengths used in optical communications are in fact in the infra-red part of the electromagnetic spectrum. Light emitting diodes (**LEDs**) and injection laser diodes (which we will refer to simply as **lasers**) emitting at $0.85\,\mu m$, $1.3\,\mu m$ and $1.5\,\mu m$ are common. The channel bandwidths available using these are of the order of $10^{10}\,Hz$ or $10^4\,MHz$ which enables voice, data and television signals all to be transmitted, using time division multiplexing techniques, over the same fibre. As the agreed signal bandwidth for television signals in the U.K., including the sound element, is $8\,MHz$ wide, you can see that an optical fibre could carry hundreds of TV channels.

Dispersion

Some facts about optical fibres are given in section 24.3 where the physics of total internal reflection is used to explain why optical fibres can act as light 'pipes' or optical **waveguides**. The dispersion associated with **multimode** fibres is also discussed there, in the example on page 332. Therefore optical fibres for long range transmission (more

than ten kilometres or so) are monomode in structure (see page 331). When LED sources are used there is a wavelength spread of a micrometre or so. As this propagates along the monomode fibre different wavelengths travel at different speeds and what starts out as a square pulse suffers 'colour' or **material dispersion** and spreads a little, the danger being that you may not be able to resolve two adjacent pulses at the receiver. Laser pulses do not suffer in this manner as the wavelength spread is less than a nanometre. A further advantage of lasers for long-distance transmission is that they have a highly directional output with a power of tens of milliwatts as opposed to LEDs which have a fraction of a milliwatt of output power which is non-directional.

Attenuation

Figure 28.10 shows the attenuation loss in optical fibres made from very pure glass. Losses of less than $1\,dB\,km^{-1}$ are possible, making 10 kilometres of this glass as clear as 10 millimetres of typical window glass! The losses increase on the low wavelength side by a process known as **scattering**, the degree of which depends on $1/\lambda^4$, where λ is the wavelength. On the high wavelength side the glass absorbs the infra-red waves by molecular resonance, the degree of which increases rapidly with λ. Because of the low losses found in modern monomode fibres the distance between repeater stations is far less than that for coaxial cables, typically tens of kilometres as opposed to only one or two.

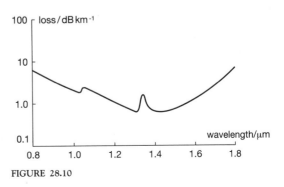

FIGURE 28.10

The attenuation loss in a monomode optical fibre is $0.5\,dB\,km^{-1}$. The cable is laid with joins every kilometre and with regenerators every 30 km. At each join there is a power loss of $0.2\,dB$ and the connections to the input and output of the regenerator each involve a further loss of $1.0\,dB$.

(a) Calculate the total loss on each 30-km section of the fibre link.

(c) Suggest a value for the voltage gain required at each regenerator.

(a) Over 30 km the attenuation loss
$$= 30\,km\ 0.5\,dB\,km^{-2} = 15\,dB$$
For 30 joins the power loss
$$= 30 \times 0.2\,dB = 6\,dB$$
For regenerator connections the power loss
$$= 2\,dB$$
$\therefore$ The total power loss per 30-km section
$$= (15 + 6 + 2)dB$$
$$= 23\,dB$$

Losses and gains expressed in decibels can simply be added taking due care to count gains as positive and losses as negative. This is why quoting them in decibels is so useful.

(b) A gain of more than $23\,dB$ is needed at each regenerator. A power gain of about $30\,dB$ would more than suffice. As $V^2 \propto P$, the voltage gain required is only half this value, $(\lg P \propto 2\lg V)$ i.e. the amplifiers in the regenerator need to have a net voltage gain of $15\,dB$.

Security

Immunity to interference noise is clearly a great advantage and the lack of crosstalk also means that fibre communications are very **secure**. It is not possible to tap the information; a vital property for military applications and for the transfer of sensitive data between computers in, for example, banking systems.

28.4 ADC and DAC

The quantisation process described in PCM is really a process which turns an analogue voltage – the amplitude of the PAM pulse – into a digital number. **Analogue to digital conversion (ADC)** is involved. Equally at the receiving end of a communications system the process needs to be reversed and a **digital to analogue conversion (DAC)** performed. There are three levels at which you might wish to approach an understanding of ADC and DAC. You may simply appreciate the need to do both and assume that the necessary technology exists; you may study the underlying principles which enable both tasks to be achieved but delve no further into real systems; or you may become an electronics/communications/computer expert and design the systems yourself. We shall follow the middle route in this section and will assume that you have studied Chapter 27 – *Electronics*. If you have not studied electronics beyond GCSE you should still be able to understand the ADC called an integration converter which is based on a ramp generator opposite – and this is enough for you to see, in principle, how PCM is achieved and how digital voltmeters and digital instruments of all kinds work.

Flash converters

The circuit shown in figure 28.11, converts a voltage input at P into a two-bit binary number at

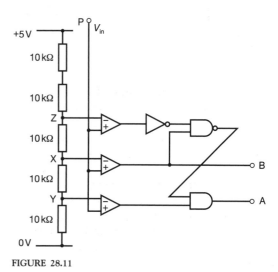

FIGURE 28.11

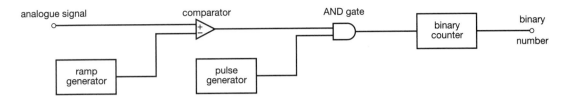

FIGURE 28.13

BA. The inverting inputs to the three comparators are held at $V_X = 1.0\,V$, $V_Y = 2.0\,V$ and $V_Z = 3.0\,V$ by the resistance chain. When an input voltage, V_{in}, of between $0\,V$ and $1.0\,V$ is applied, each of the comparators will be saturated (see page 387) with an output of $-5\,V$ and both B and A will register 0. As V_{in} rises to between $1.0\,V$ and $2.0\,V$ the output of the X-comparator switches to $+5\,V$ and A switches to 1. As V_{in} continues to rise you can see that the states of A and B are as shown in Figure 28.12. The analogue input voltage has thus been **quantised** at one of four levels. This circuit is called a flash converter because it responds very quickly, in less than $50\,\mu s$, to changes in V_{in}.

Integration converters

Another approach to designing an ADC again uses a comparator together with what is called a **ramp generator**. Let us look at it first in terms of a semi-block diagram, i.e. consider the functions of each part (see figure 28.13). The ramp generator produces a voltage which rises steadily with time; its design is that of the integrator explained in the example on page 388.

The comparator's output is held high by the input analogue voltage until the ramp voltage reaches the same value, at which instant it

level	V_{in}/V	B	A
0	0 - 1.0	0	0
1	1.0 - 2.0	0	1
2	2.0 - 3.0	1	0
3	3.0 - 5.0	1	1

FIGURE 28.12

The obvious drawback to this ADC is that it only resolves V into one of four levels and hence the quantisation errors are very high. This can be improved by increasing the number of comparators. To produce an 8 bit output would need 127 comparators (!) but then the quantisation noise would be lower than 1%. This is just what is needed in a high frequency PCM system such as that used for encoding analogue signals for television (PAL) transmission, though to achieve quantisation in this manner is technologically complex.

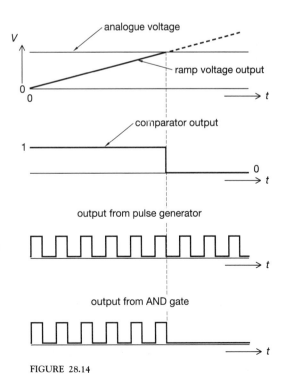

FIGURE 28.14

switches rapidly to low. During this time interval pulses from the pulse generator have been passing through the AND gate and have been counted. As soon as the comparator switches, the pulse train from the AND gate ceases and the number then held in the binary counter is the binary equivalent of the analogue voltage.

These steps are shown diagrammatically in figure 28.14. The counter has been registering each pulse and so has gone through 000, 001, 010, 011, 100, 101, 110 and stops there as only six pulses have passed through the AND gate. The resolution of such an ADC device depends only on the frequency of the pulse generator. Of course a great deal has been left unsaid. The values of the resistors and capacitors in the ramp generator have to be very precise and stable, with the latter not being easy to achieve. And the whole system has to be switched on and off at 8000 Hz, the sampling rate needed for many PCM communications systems.

EXAMPLE

An ADC based on a ramp generator produces an 8-bit digital output 8000 times per second. It is designed to deal with a maximum input voltage of 200 mV. Calculate (a) the rate of rise of voltage in the ramp generator and (b) the bit rate of the pulse generator.

(a) The rise time of the ramp voltage

$$= \frac{1}{8000\,\text{Hz}}$$

$$= 125\,\mu\text{s}$$

The ramp voltage must rise at

$$= \frac{200 \times 10^{-3}\,\text{V}}{125 \times 10^{-6}\,\text{s}}$$

$$= 1600\,\text{V}\,\text{s}^{-1}$$

to reach the maximum of 200 mV.

(b) The pulse generator will need to produce $2^8 (= 256)$ pulses in order to generate an 8-bit output.

The pulse produce is producing

$$\frac{256\,\text{bit}}{125 \times 10^{-6}\,\text{s}} = 2.048\,\text{Mbit}\,\text{s}^{-1}$$

DACs

The process of digital to analogue conversion was outlined on page 389. Figure 27.10 on that page shows a summing amplifier using an **operational amplifier** with negative feedback. A generalised circuit for a 4-bit DAC is shown in figure 28.15 where MSB and LSB refer to the most and the least significant bit respectively. Suppose a 4-bit binary number is represented by 1s which are each always at a voltage V or by 0s which are always zero voltage. The voltage at X is always zero, so for the decimal number 15, i.e. binary 1111, the current in the feedback resistors is given by (see page 385 for an explanation of the flow of charge in such circuits):

$$I_f = \frac{V}{R} + \frac{V}{2R} + \frac{V}{4R} + \frac{V}{8R}$$

$$= \frac{V}{8R}(8 + 4 + 2 + 1)$$

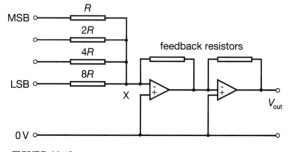

FIGURE 28.15

If one of the binary digits is zero, there is no voltage at that input so the current is correspondingly smaller; you can see that the currents are:

I_f for 1111 = 15V/8R = 15I
I_f for 1110 = 14V/8R = 14I
I_f for 1101 = 13V/8R = 13I

I_f for 0010 = 2V/8R = 2I
I_f for 0001 = V/8R = I
I_f for 0000 = zero = 0

The current in the feedback resistors is proportional to V_{out} so V_{out} is a quantised analogue

voltage representing the initial 4-bit binary
number. You should think of V_{out} as a *ladder of
voltages* rising in equal steps from zero. To operate
on 8-bit binary numbers you need a group of eight
parallel resistors ranging in value from R to $128R$.
The latter will be very large (several MΩ) and such
resistors are not easy to make with low tolerances
so the steps on the ladder may not be exactly
equal, giving rise to quantisation errors.

ADCs

Many analogue to digital converters, including
those used for PCM at low bit rates, e.g. for
telephone systems, and those used for **digital
voltmeters**, incorporate a ramp generator which
rises in discrete steps. Such **counter-ramp
converters** use a DAC to achieve this – figure
28.16 illustrates the basic idea for an 8-bit
converter.

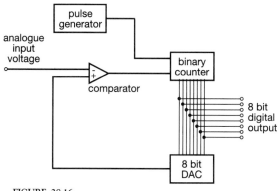

FIGURE 28.16

Further components are needed to hold the
digital output if it is being displayed on a seven
segment liquid crystal display. By introducing a
small offset current in the comparator the
quantisation in a PCM system using this ADC can
be made so that the level chosen at each sampling
point is the nearest level rather than the nearest
lower level. This reduces the quantisation noise in
such a system. We will not pursue these ideas
further, but they are central to specialist courses in
electronics or computer studies.

Data transmission

Information encoded using PCM and the data
processed by computers all consist of trains of
pulses with high bit rates. The commonest
computer code is the **ASCII** (American Standard
Code for Information Transfer) **code** which has a
seven-bit binary coding for all letters, decimal
digits, punctuation marks etc. What does a binary
number *look like* as it moves along a fibre or
transmission line?

Figure 28.17 gives one example for each: (b)
shows an amplitude modulated (AM) light pulse
and (c) a phase shift keyed (PSK) carrier signal,
both representing the binary 6-bit number 101100
shown in (a). The frequency associated with the
light pulse is of the order of 10^{14} Hz and so no sine
wave is shown as the maximum bit rates for data
transmission are not going to be more than a few
Mbit s^{-1}. For the transmission line the data pulses
are used to change the **phase** of the r.f. carrier
signal. This can be done in a number of ways; here
a phase change of 180° occurs when there is a
change from 0 to 1 or from 1 to 0.

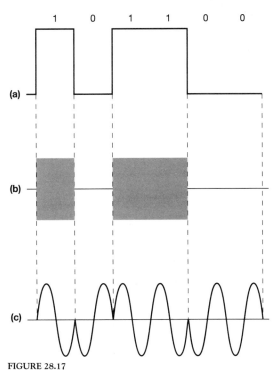

FIGURE 28.17

Exercises on each section of this chapter may be found in the companion textbook, *Practice in Physics*.

SUMMARY

At the end of this chapter you should be able to:

◆ understand the difference between analogue and pulse modulation.

◆ list the types of noise affecting communications links.

◆ know how to define signal-to-noise ratio and be able to perform calculations involving decibels.

◆ draw a block diagram of an a.f. radio transmitter and receiver.

◆ draw frequency spectra to explain sidebands.

◆ know that for AM signals, the signal bandwidth is twice the maximum frequency of the a.f. signal.

◆ be able to explain the nature of amplitude modulation and frequency modulation.

◆ understand how pulse code modulation is achieved.

◆ appreciate that signals attenuate and pick up noise as they propagate.

◆ know that bit rate is calculated as

$$\text{number of bits per sample} \times \text{sampling frequency}$$

◆ explain the nature of frequency division and time division multiplexing.

◆ understand the advantage of transmitting digital rather than analogue signals.

◆ list and explain the advantages of using optical fibres for information transfer.

◆ appreciate the need for analogue to digital and digital to analogue conversion.

◆ describe an ADC based on a ramp generator.

◆ describe a DAC using a summing amplifier.

Index

A page number followed by the letter e refers to an Example